W9-CRB-357

This book is the first to give a comprehensive view on the polaron and bipolaron theory of high-temperature superconductivity, one of the most significant discoveries in physics in the past decade.

With the discovery of high-temperature superconductors, research into polarons and bipolarons has attracted much attention. It appears that carriers in some high-temperature superconductors are strongly correlated, both in the normal and in the superconducting states. In the strong-coupling limit the Fermi-liquid ground state may be destroyed by the formation of small polarons and bipolarons. The experimental and theoretical study of such particles is a central issue in current research into high-temperature superconductivity.

Polarons and bipolarons have been observed previously in magnetic semi-conductors and transition metal oxides. Thorough investigation of these non-superconducting materials has contributed greatly to the basic understanding of the physical properties of both polarons and bipolarons. This book contains a series of authoritative articles on the most advanced research on polarons and bipolarons in high-temperature superconductors and related materials.

This book will be of great interest to researchers in condensed matter physics, and especially to those working in the field of superconductivity.

POLARONS AND BIPOLARONS IN HIGH-T_c SUPERCONDUCTORS AND RELATED MATERIALS

POLARONS AND BIPOLARONS IN HIGH-T_c SUPERCONDUCTORS AND RELATED MATERIALS

Edited by

E. K. H. SALJE, A. S. ALEXANDROV AND W. Y. LIANG

CAMBRIDGE
UNIVERSITY PRESS

Published by the Press Syndicate of the University of Cambridge
The Pitt Building, Trumpington Street, Cambridge CB2 1RP
40 West 20th Street, New York, NY 10011-4211, USA
10 Stamford Road, Oakleigh, Melbourne 3166, Australia

First published 1995

Printed in Great Britain at the University Press, Cambridge

A catalogue record for this book is available from the British Library

Library of Congress cataloguing in publication data

Polarons and bipolarons in high-T_c superconductors and related
materials / edited by E. K. H. Salje, A. S. Alexandrov and W. Y. Liang.
p.   cm.
Includes index.
ISBN 0 521 48175 9
1. High temperature superconductors.   2. Polarons.   I. Salje,
Ekhard K. H.   II. Alexandrov, A. S.   III. Liang, W. Y.
QC611.98.H54P65   1995
537.6′231–dc20   94-49684 CIP

ISBN 0 521 48175 9 hardback

Contents

Preface

With the advent of high-temperature superconductors, research on polarons and bipolarons has gained renewed attention. It appears that carriers in some high-temperature superconductors are strongly correlated both in the normal and in the superconducting state. In the strong-coupling limit the Fermi-liquid ground state may be destroyed by the formation of small polarons and bipolarons. Experimental and theoretical analysis of such particles is a central issue for current research in superconductivity.

This book contains a series of authoritative articles on the most advanced research on polarons and bipolarons. They were invited for presentation during a workshop on 'Polarons and Bipolarons in High-T_c Superconductors and Related Materials', which was held at the Interdisciplinary Research Centre (IRC) in Superconductivity, Cambridge, UK, between 7th and 9th April 1994. Over 50 participants from ten countries took part in this workshop, representing the major research centres currently working in this field. The workshop was held in honour of Sir Nevill Mott in recognition of his important contributions to the physics of polarons.

The editors are grateful to all contributors for their overwhelmingly positive response to the idea of publishing this book. We are indebted to Ken Diffey and Margaret Hilton who ensured the smooth running of the workshop, to all of the referees and to William Beere for their help editing the manuscripts. We also thank Simon Capelin of Cambridge University Press for his encouragement and help throughout the publication of this book.

E. K. H. Salje
A. S. Alexandrov
W. Y. Liang

ix

1

A polaron theory of high-temperature superconductors

N. F. MOTT

Interdisciplinary Research Centre in Superconductivity, University of Cambridge, Madingley Road, Cambridge, CB3 0HE, UK

Abstract

The present status of polarons and bipolarons in physics is discussed in connection with the theory of high-temperature superconductors.

The concept of the polaron, introduced into physics by Landau in 1933 [1], has recently gained renewed importance on account of its application to the high-temperature superconductors. We give here an outline of how this has come about.

The polaron with which we deal is in the usual notation 'small'. While polarons can exist in non-polar material (as for instance in liquid or solid rare gases in which holes can be self-trapped), the high-temperature superconductors are materials whose static dielectric constant ε_0 is much greater than that ε at high frequency. For these we use the theory given by Fröhlich [2], Holstein [3] and others [4]. In such a theory we introduce a distance r_p, beyond which the medium is fully polarised so that the potential energy of the self-trapped electron is

$$V_p = -\frac{e^2}{k_p r_p}, \qquad r < r_p, \tag{1}$$

$$V_p = -\frac{e^2}{k_p r}, \qquad r > r_p, \tag{2}$$

where $1/k_p = 1/\varepsilon - 1/\varepsilon_0$.

A self-consistent calculation by Fröhlich and Allcock [5] gives for r_p

$$r_p = \frac{5\hbar^2 k_p}{me^2}, \tag{3}$$

where m is the effective mass of the carrier before polaron formation. This calculation is not correct if the value of r_p obtained is smaller than the distance between ions in the solids, as may be the case if $m \gg m_e$. For this case Bogomolov *et al.* [6] give

$$r_p = \frac{1}{2}\left(\frac{\pi}{6N}\right)^{1/3},$$ (4)

where N is the number of cells per unit volume. From these formulae we find that the energy gained per polaron formation is

$$W_p \simeq \frac{e^2}{2k_p r_p}.$$ (5)

At temperatures above $\Theta_{\text{Debye}}/2$ the polaron moves by a hopping process, the number of hops per unit time being (in the adiabatic approximation)

$$\omega \exp\left(-\frac{W_H}{k_B T}\right),$$ (6)

where ω is a phonon frequency and

$$W_H = \tfrac{1}{2} W_p.$$ (7)

At temperatures below $\Theta_{\text{Debye}}/2$ it behaves like a heavy particle with effective mass

$$m_p \simeq \frac{\hbar}{2\omega R^2} \exp\left(\frac{W_p}{\hbar\omega}\right),$$ (8)

where R is the hopping distance.

It is important to realise that, because of relationship (5), $W_p/\hbar\omega$ is not necessarily very large; effective masses of order $5m_e$ are possible.

Polarons were first envisaged as existing in conduction or valence bands of semiconductors. Although two polarons of the same sign will of course repel each other at large distances, attraction sets in when the polarised regions overlap, and bound pairs (bosons) may form, especially in two-dimensional systems, for which any attraction must lead to a bound state.

For application to high-temperature superconductors, it is essential to consider a degenerate gas of polarons or bipolarons. As far as we know, such a situation was first considered by Mott [7] and Mott and Davis [4] and evidence given there shows that $SrTiO_3$ and other similar materials can be described in this way. In an ionic crystal, owing to competition between each carrier for ions with which to form a polaron, a concentration of polarons greater than about

10% of the available sites is unlikely to occur. However, this does not lead to a drop in the effective mass for high concentration [7].

Successful introduction of polarons into the theory of superconductors dates from the work of Alexandrov and Ranninger [8] before the discovery of the new materials.[1] Alexandrov and Ranninger consider the BCS formula for the energy gap

$$\Delta \simeq 2\hbar\omega \exp\left(-\frac{1}{\lambda}\right) \tag{9}$$

with $\lambda = VN(0)$, V being the electron–phonon interaction. This originated in the demonstration by Fröhlich that, in a metal, phonons could lead to an electron–phonon interaction, and the experimental discovery of an isotope effect. λ is about 0.3 for metallic superconductors. Many authors considered the behaviour of materials with strong interaction, so that λ increases. T_c is given by

$$k_B T_c \simeq \hbar\omega \exp\left(-\frac{1+\lambda}{\lambda}\right) \tag{10}$$

and remains fairly low (< 40 K) for reasonable values of ω and λ. The new point is that $\lambda \simeq 1$ is the condition for polaron formation, and it is shown that, for a value of λ in the neighbourhood of unity, there is a fairly sharp transition to a situation in which all the carriers form polarons, so that the gap extends across the whole Fermi distribution. As pointed out by Alexandrov [10] the basic phenomenon that allows the high T_c is polaronic narrowing of the band followed by increase in density of states $N(0)$, which eliminates the small exponential factor in the BCS formula (9), Fig. 1. We thus have to consider the properties of a gas of charged pairs, of small dimensions (the observed correlation being small) and small overlap (unlike the Cooper pairs of the BCS theory). The pairs, being bosons, obey Bose–Einstein statistics, and must be treated according to an analysis similar to that for liquid ^{4}He. If interaction between the pairs is neglected, then the critical temperature is given by

$$k_B T_c = \frac{3.3\hbar^2 n^{2/3}}{m_b} \tag{11}$$

where m_b is the boson mass, $m_b = (m_1 m_2 m_3)^{1/3}$ (Prelovsek, Rice and Zhang [11]) if masses are different in the three crystallographic directions. In spite of presumed large effects of interaction, (11) gives a fair approximation.

The discovery of the new superconductors by Bednorz and Müller [12] in

---

[1] Chakraverty [9] discussed bipolarons in connection with superconductivity, but considered that bipolarons would be immobile and therefore inactive as superconductors.

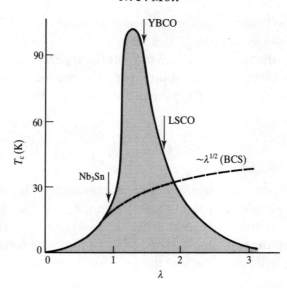

Fig. 1. The maximum in the critical temperature of a polaronic superconductor at intermediate values of the coupling constant (Alexandrov, A. S. 1988 *Phys.Rev.***B 38**, 925).

1986 led to renewed interest in this model, sometimes called the Schafroth model because of the early proposal [13] before the BCS theory. Most of the experimental work has been in the copper oxides, for which we believe that the carriers are somewhat more complicated than a simple bipolaron. Other materials showing high T_c are the cubic bismuth oxides and carbon–metal compounds. We believe that these materials may be described by a bipolaron model, but the experimental evidence is far less complete than for the copper oxide materials, and the remainder of this article will be confined to them.

We take first the substance $La_{2-x}Sr_xCuO_4$. For $x=0$ this is an antiferromagnetic insulator, highly anisotropic with Néel temperatures $T_N \simeq 300$ K. On adding Sr, T_N drops very rapidly to zero at about $x=0.01$, suggesting that each Sr atom adds a rather large defect. Then, as shown in Fig. 2, there is a non-metallic range that shows hopping conduction leading to a transition to a superconducting state. T_c then rises to a maximum after which it decreases to zero.

It is sometimes stated that the behaviour of a doped 'Mott' insulator – that is an antiferromagnetic material in which the moments have value $S=\frac{1}{2}$ is an unsolved problem; but we do not consider this to be the case for antiferromagnetic materials in general. In materials such as EuO doped with excess Eu, or $Gd_{3-x}V_xS_4$ where V stands for a vacancy, the carriers in the conduction band each form a spin polaron – sometimes called a ferron; that is a group of

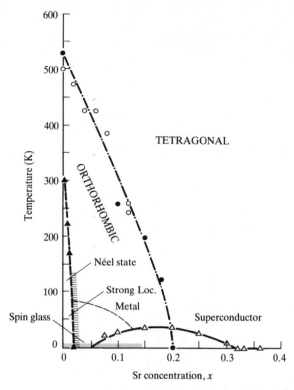

Fig. 2. The phase diagram of $La_{2-x}Sr_xCuO_4$.

moments oriented antiparallel to that of the carrier, as illustrated in Fig. 3 (Von Molnar *et al.* [14]). This entity, introduced into physics by de Gennes [15], has a mass increasing exponentially with the number of moments that it contains. We shall assume that this remains so for a simple Mott insulator such as La_2CuO_4, if holes in the oxygen 2p band are introduced by doping with strontium (these are thought to be hybridised with Cu $3d^9$).

Kamimura and co-workers [16] and Wood and Cooke [17] have independently introduced spin polaron models of high-T_c materials. These models, however, differ from ours, particularly in that they do not admit bosons above T_c, and will not be discussed further here.

We think, then, that our polarons must be of complex hybrid type. In the centre is a spin polaron, containing perhaps six moments, but in these highly dielectric materials, they will be surrounded by a polarised region. Both will contribute to the mass enhancement (Mott [18]).

Each strontium atom contributes one of these, which is localised by disorder in the hopping region, where they are already paired into bosons (Doniach and

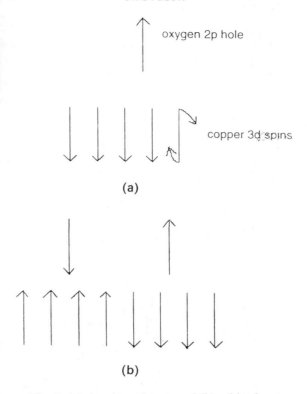

Fig. 3. (a) A spin polaron and (b) a bipolaron.

Inui [19], Mott [20]). There are therefore twice as many sites as carriers, so the spin-glass region is a compensated semiconductor, localised in the Anderson sense by disorder. The transition to metallic superconducting behaviour is thus of Anderson (rather than Mott–Hubbard) type, for which there is independent evidence from the specific heat (Mott [21]).

One can obtain independent evidence of this hybrid structure from the observed isotope effect, that is the shift of T_c upon adding a heavy isotope of oxygen. In the polaron model this must occur through a change of m_p in equation (8), which is only possible if the dielectric sheath occurs around the carriers. We think that the saturation of T_c shown in Fig. 2 is a result of overcrowding, of which the first effect should be there being no room for the carriers. So we expect the isotope effect to drop for overdoping. This is just what is observed, according to the results of Franck [22] and others illustrated in Fig. 4.

The model described here predicts that the current above T_c is carried by a non-degenerate gas of bipolarons each with charge $q = 2e$. Consequently the

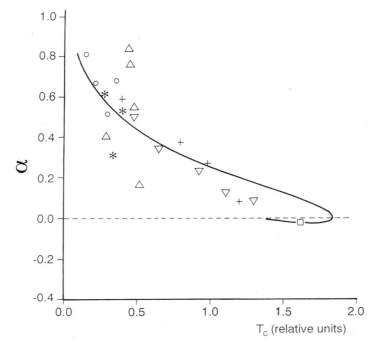

Fig. 4. The isotope effect: ($\bigcirc$) $La_{2-x}Ca_xCuO_4$, ($\triangle$) $La_{2-x}Sr_xCuO_4$, ($*$) $La_{2-x}Ba_xCuO_4$ (M. K. Crawford *et al.*, *Science* **250**, 1309 (1990)); ($\triangledown$) $Y_{1-x}Pr_xBa_2Cu_3O_{6.92}$ (J. P. Franck *et al.*, *Physica* C **185–189**, 1379 (1991)); ($+$) $YBa_2Cu_{4-x}Ni_xO_8$ (H. J. Bornemann *et al.*, *Physica* C **185–189**, 1359 (1991)); ($\square$) $Bi_{1.6}Pb_{0.4}Ca_2Sr_2Cu_3O_{10}$ (H. J. Bornemann *et al.*, *Physica* C **182**, 132 (1991). The theoretical curve is after [23].

contribution made by the carriers to the thermal conductivity K will be, by the Wiedermann–Franz law.

$$K = L_B \sigma T \tag{12}$$

with the bosonic Lorentz number L_B being a quarter of the value for fermions. Unfortunately the contribution from the phonons is larger than that from the carriers, and can be estimated by the value for non-conducting specimens. The available evidence, also for $T < T_c$, is discussed by Alexandrov and Mott [24] and strongly favours $q = 2e$.

The correctness of a theory must be judged by the number of phenomena that it can explain. Alexandrov *et al.* [25] have discussed the infra-red absorption of YBCO. A peak in the absorption is observed, which is interpreted as exitation of the bipolarons. The frequency and width of the peak are the same above and below T_c, but above T_c the intensity drops with increasing T. This was interpreted by Dewing and Salje [26] (see also Mott [28]) as being

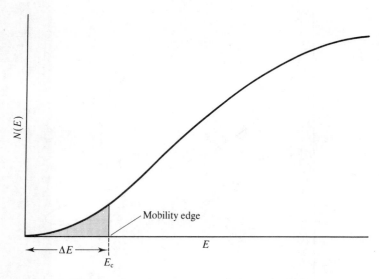

Fig. 5. The density of states for partly localised bipolarons.

due to the presence of an excited triplet state. This is shown also by a change in the slope of the resistivity curve observed by Bucher *et al.* [27], and explained by Mott [28]) The broadening of the NMR lines $1/T_1$ is also caused by the triplets, an assumption that fits the temperature-dependence of $1/T_1$ (Alexandrov [29]).

With the exception of $YBa_2Cu_3O_7$, all the copper oxide superconductors are disordered; thus in $La_{2-x}Sr_xCuO_4$ the strontium alone and the La vacancies are on random sites. Just as in the well-known theory of doped silicon, this random potential produces Anderson localisation, this time of the bosons. Mott [21] has pointed out that, with a random potential of the kind used by Anderson, charged bosons will behave very like fermions, their mutual repulsion preventing more than one (or another small number) being in any localised state. States will therefore be occupied up to a limited value E_c, which behaves like a Fermi energy. Only bosons with energies above this limit can take part in conduction. The density of states at the insulator–metal transition could thus be as in Fig. 5, E_c being the pseudo-Fermi energy and also the mobility edge.

Alexandrov, Bratkovskii and Mott [30] deduce from this model that the number of free bosons above T_c will be proportional to T, and hence explain the behaviour of the Hall coefficient, which drops linearly with increasing T. The ρ–T curve is explained by a time of relaxation proportional to T^2 caused by boson–boson collisions, and with the number of mobile bosons proportional to T one finds ρ proportional to T. A residual resistance is found if there are scatterers (like zinc atoms) in the copper oxide planes.

The extensive measurements of specific heat by Loram and co-workers [31]

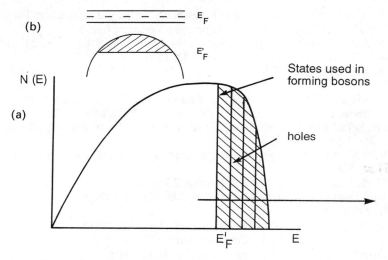

Fig. 6. The density of states, showing photoemission.

show entropy at the transition considerably less than $nk_B \ln 2$, the value of a non-interacting Bose gas of density n; this in our view is because a high proportion of the carriers is still localised at T_c.

Angle-resolved photoemission enables a Fermi energy to be mapped, the result being similar to the band calculations. The conclusion, which is sometimes drawn, that the carriers are fermions, is, we believe, false. Our explanation is illustrated in Fig. 6(a), which shows the density of states in doped semiconductors as it would be if the holes did not form bosons. Electrons are excited across the gap and a Fermi surface at E'_F observed, but in fact the holes in the shaded area drop down and form bosons, so that there are no states in the area. At E'_F then, no excitation is possible and the behaviour measured has no direct relation to the current. A similar experiment could be carried out for Si : B, where a pseudo-Fermi energy could be observed, the holes dropping into localised states produced by the B ions, Fig. 6(b).

References

[1] L. D. Landau, 1933 *Phys. Z. Sowjetunion* **3**, 664.
[2] H. Fröhlich, 1952 *Adv. Phys.* **3**, 325.
[3] T. Holstein, 1959 *Ann. Phys.* **8**, 325.
[4] See for instance Mott, N. F. and Davis, E. A. 1979 *Electronic Processes in Non-crystalline Materials*, 2nd ed. Oxford University Press, Chap.3.
[5] G. R. Allcock, 1956 *Adv. Phys.* **5**, 412.
[6] V. N. Bogomolov, E. A. Kudinov and Yu. A. Firsov, 1968 *Sov. Phys. Solid State* **9**, 2502.

 [7] N. F. Mott, 1973 in *Cooperative Phenomena* dedicated to Herbert Frohlich, ed.
       Haken, H. and Wagner M., Springer-Verlag, Chap.1.
 [8] A. S. Alexandrov and J. Ranninger, 1981 *Phys. Rev.* B **23**, 1796; **24**, 1164.
 [9] B. K. Chakraverty, 1981, *J. Physique* **42**, 1351.
[10] A. S. Alexandrov, 1983 *Zh. Fiz. Khim.* **57**, 273 (1983 *Russ. J. Phys. Chem.* **57**,
       167).
[11] P. Prelovsek, T. M. Rice and F. C. Zhang, 1987 *J. Phys.* C **20**, L289.
[12] J. G. Bednorz and K. A. Müller, 1986 *Z. Phys.*B **64**, 189.
[13] M. R. Schafroth, 1955 *Phys. Rev.* **100**, 463; J. M. Blatt and S. T. Butler, *Phys.
       Rev.* 1955 **100**, 476.
[14] S. Von Molnar, A. Briggs, J. Floquet and G. Remenyi, 1983 *Phys. Rev. Lett.*
       51 **57**, 706.
[15] P. G. de Gennes, 1962 *J. Phys. Radium* **23**, 630.
[16] H. Kamimura, M. Eto, S. Matsuno and H. Ushio, 1992 *Comments Condens.
       Mat. Phys.* **15**, 303.
[17] R. F. Wood and J. F. Cooke, 1992 *Phys. Rev.* B **45**, 5585.
[18] N. F. Mott, 1993 *J. Phys.: Condens. Matter* **5**, 3487.
[19] S. Doniach and M. Inui, 1990 *Phys. Rev.* B **41**, 6668.
[20] N. F. Mott, 1994 *Phil. Mag. Lett.* **69**, 155.
[21] N. F. Mott, 1990 *Phil. Mag. Lett.* **62**, 37.
[22] J. P. Franck *et al.* 1989 *Physica* C **162–164**, 51.
[23] A. S. Alexandrov, 1992 *Phys. Rev.* B **46**, 14932.
[24] A. S. Alexandrov and N. F. Mott, 1993 *Phys. Rev. Lett.* **71**, 1075.
[25] A. S. Alexandrov, A. M. Bratkovsky, N. F. Mott and E. K. H. Salje, 1993
       Physica C **215**, 359.
[26] H. L. Dewing and E. K. H. Salje, 1992 *Supercond. Sci. Technol.* **5**, 50.
[27] B. Bucher, P. Steiner, J. Karpinski, E. Kaldis and P. Wachter, 1993 *Phys. Rev.
       Lett.* **70**, 2012.
[28] N. F. Mott, 1993 *Phil. Mag. Lett.* **68**, 245.
[29] A. S. Alexandrov, 1992 *J. Low Temp. Phys.* **87**, 721.
[30] A. S. Alexandrov, A. M. Bratkovskii and N. F. Mott, 1994 *Phys. Rev. Lett.* **72**,
       1734.
[31] J. W. Loram, K. A. Mirza, J. R. Cooper and W. Y. Liang, 1993 *Phys. Rev.
       Lett.* **71**, 1740.

2

On the possibility of non-BCS superconductivity

G. M. ELIASHBERG

*Laboratoire des Champs Magnetique Intenses, MPI/CNRS Grenoble; Permanent address:
L. D. Landau Institute for Theoretical Physics, Chernogolovka, 142 432 Russia*

Abstract

There exists a well-established empirical trend, namely that the best superconductors are among the bad conductors, in which electrons are essentially
localized on atomic orbitals. Just this electronic structure is requisite for the
metal–insulator transition predicted by N. Mott. I show in this paper that the
coexistence of these two remarkable phenomena within the same set of
materials is not accidental.

1 Introduction

It was understood long ago that there exists some relationship between
superconductivity and Bose–Einstein condensation. According to Schafroth *et
al.* to get the Bose particles the electrons should be bound somehow into quasi-
molecules (pairs) [1], and 'the only obstacle' to achieving an explanation of
superconductivity was the nature of these quasi-molecules. The original belief
was that they should have an atomic size to maintain Bose statistics when their
concentration is high (of the order of one per unit cell), and the problem of how
to overcome Coulomb repulsion seemed insurmountable. However, this had
nothing to do with local pairs in the case of the superconductors known in the
middle of the 1950s. Being good metals, these superconductors above $T_c =$
1–10 K display almost-free-electron behaviour with the Fermi energy not less
than 5 eV, and it was clear that the superconducting transition here concerns
only a very thin shell around the Fermi surface. BCS theory [2] was addressed
precisely to these superconductors. It was shown that, at $T = 0$, the normal state
is unstable with respect to formation of Cooper pairs [3], when in the vicinity of
the Fermi surface there exists an arbitrary weak attraction between electrons.
The source of attraction was known already to be the electron–phonon
interaction, which may be strong enough to compete with Coulomb repulsion,

11

but only for the initial and final scattering states close to the Fermi level. On the base of Migdal's analysis [4] it was shown that the BCS approach provides accuracy limited not by interaction strength, but only by the condition of adiabaticity, which is well fulfilled in ordinary metals [5]. As a result, quantitative agreement with experiment was achieved also for the relatively high-T_c superconductors like Pb and Nb and even for A-15 binary systems such as Nb_3Sn.

The success of BCS theory during the years after 1957 was so impressive that any criticism was almost suppressed. Nevertheless the idea of superconductivity with local pairing survived. Moreover, due to the Cooper effect it was understood that the overlap between pairs is not an obstacle for Bose condensation, so that the problem became rather quantitative. After superconductivity had been found in compounds like $(Ba, Pb)BiO_3$ and Chevrel phases this idea acquired a second life and it is now being considered among other proposals concerning the nature of high-T_c superconductivity (see [6, 7] for example). New materials differ strongly from the simple metals to which BCS theory was originally addressed. They are so-called narrow band conductors, and it is usually believed that the Fermi energy here is much smaller than that in simple metals, and that the effective radius of pairs, or the coherence length, as estimated within BCS phenomenology, is often less than 100 Å (even 20 Å, in high-T_c cuprates). Therefore it seems that the situation has indeed changed in such a way as to favour the local-pairing mechanism.

However, the narrow band conductors differ from simple metals not only by their smaller Fermi energy and, probably, by stronger electron–phonon interaction, and the path from Cooper pairs to quasi-molecules is not a straight line marked by values of the coherence length. In 1949 Mott began to publish his famous analysis of the Coulomb correlations in the narrow band conductors [8], and the culmination of this analysis was the prediction of the metal–insulator transition and splitting of the conduction band [9]. The physics discovered by Mott looks quite transparent. The conduction electrons in these materials are essentially confined on the atomic orbitals, which, due to their mutual overlap, constitute a conducting network. When there is one electron per orbital (the ideal metallic composition!), then the formation of Bloch states is inevitably followed by short-ranged charge fluctuations, in particular, of two electrons with opposite spins on the same orbital (on-site fluctuations). These fluctuations cost some energy, and therefore short-ranged correlations tend to suppress the formation of Bloch states and thus give rise to further localization of the electrons on atomic orbitals. According to Mott, one can expect complete localization already when some characteristic Coulomb energy (Hubbard U, for example) is of the order of the band width, defined through the overlap between orbitals. If this happens then only activated conductivity is possible, which means that a gap has arisen in the one-electron density of states around the Fermi level.

These arguments seem convincing, but it is important to emphasize that, within the framework of conventional band theory, there are no appropriate terms with which to discuss what is going on with the electron spectrum on the way to Mott localization. The present author has recently stressed that there exists a counterpoise to Mott localization, namely the Pauli principle, which is manifested in the sum rule derived by Luttinger and Ward (known also as the Luttinger theorem) [13,14]. It has been shown that, within the scope of the Luttinger theorem, it is possible to maintain the band description valid for the correlated electron system and thus account for the peculiar duality of the electron states, when the electrons involved in band dynamics are to some extent localized on the atomic orbitals [10–12].

In the framework of this approach, the problem announced in the title is discussed, and for convenience in Section 2 the necessary formalism is presented. The main goal is to show the relation between superconductivity and the Mott transition. As has been mentioned, the effect of Coulomb correlations is the suppression of short-ranged charge fluctuations, which are intrinsic for one-electron current carriers. When the correlated system approaches the Mott transition, the one-electron density of states decreases within some vicinity of the Fermi level, which is a consequence of increasing localization of electrons on the atomic (or on more complicated) orbitals. However, the states remaining in this vicinity, as will be shown, become even more itinerant, which is reflected in sharper dispersion of the band spectrum, as defined below. There exist different possible ways to eliminate these states completely and thus to realize the Mott transition. The insulating state with a broad energy gap may arise as a result of an antiferromagnetic transition or relatively small lattice distortion. Such examples are well known, and some of them will be discussed in this paper. However, without any change of translational symmetry the only possible way to eliminate the residual one-electron states from the Fermi level is to violate the Luttinger–Ward sum rule. This is just what happens in the superconducting state (Section 3): Bose condensation (the macroscopic phase coherence) is incompatible with the Pauli principle, whose manifestation is the Luttinger theorem. The supercurrent is not followed by any local low-energy charge fluctuations, and therefore it is quite natural that the superconducting state is one of the possible final stages of Mott localization. W. von der Linden and the present author [15] have obtained an instructive illustration of this statement by analysing the numerical data for a one-dimensional Hubbard cluster (Section 3). From this point of view the evolution of the BCS physical picture together with an enhancement of the correlation effects will be discussed (Section 4).

2 The band spectrum for correlated electrons

The text-book band theory is valid only within the self-consistent-field approximation. To account for the correlation effects, one should express the spectral properties in terms of Green functions, which are defined here as follows:

$$\hat{G}_R(\varepsilon) = i \int_0^{+\infty} \exp(i\varepsilon t)\,\hat{A}(t)\,dt$$

$$\hat{G}_A(\varepsilon) = i \int_{-\infty}^{0} \exp(i\varepsilon t)\,\hat{A}(t)\,dt \tag{1}$$

$$\hat{A}(t) \equiv A(t;x,x') = \langle \psi(x,t)\psi^+(x',0) + \psi^+(x',0)\psi(x,t)\rangle$$

where ψ and ψ^+ are the creation and the annihilation operators in the Heisenberg representation with the evolution operator $\exp[-i(H-\mu N)t]$; x, x' include spatial coordinates r, r' and spin variables σ, σ'; $\langle \ldots \rangle$ is the expectation value for the grand canonical distribution. The chemical potential μ as a function of temperature T for a given number of particles N in a volume V is defined by the equation

$$\mu = \int [\exp(\varepsilon/T)+1]^{-1}\mathrm{Tr}[\hat{\rho}(\varepsilon)]\,d\varepsilon, \tag{2}$$

where the spectral density operator

$$\hat{\rho}(\varepsilon) = (1/2\pi i)[G_R(\varepsilon) - G_A(\varepsilon)] \tag{3}$$

satisfies the condition of completeness, which follows immediately from Eq. (1):

$$\int \hat{\rho}(\varepsilon)\,d\varepsilon = \delta(\mathbf{r}-\mathbf{r}')\delta_{\sigma\sigma'}. \tag{4}$$

The operator $\hat{\rho}(\varepsilon)$ contains all the information concerning the one-electron spectral properties. For non-interacting particles in the crystal lattice

$$\hat{G}^{-1}(\varepsilon) = \hat{h} - \mu - \varepsilon \tag{5}$$

so that $\hat{\rho}(\varepsilon)$ and the one-particle Hamiltonian $\hat{h}$ have the same set of Bloch eigenstates ϕ_{nk}, n and k being the band index and the wave number, correspondingly. The eigenvalues of $\rho(\varepsilon)$ are in this case

$$\rho_{nk}(\varepsilon) = \delta(\varepsilon - \varepsilon_{nk}). \tag{6}$$

The same remains valid within the Hartree–Fock approximation. However, the correlation effects destroy this simple band picture, according to which the electrons are completely involved in the band states, which is manifested in Eq. (6). In the presence of the correlations it is impossible to express the spectral density (as is done in Eq. (6)) in terms of a somehow renormalized one-particle spectrum. Moreover, it is impossible to diagonalize the operator $\hat{\rho}(\varepsilon)$ at different values of ε simultaneously. However, there exists an important class of correlated electron systems for which it is possible to maintain an essential part of the band phenomenology. These are the electron systems governed by the Luttinger theorem [10–12]. In the ground state, $T = 0$, they obey the following property:

$$\hat{G}_R^{-1}(0) = \hat{G}_A^{-1}(0) \equiv \hat{\xi}. \tag{7}$$

Comparing Eq. (7) with Eq. (5) we see that Hermitian operator $\hat{\xi}$ is equal to $\hat{h} - \mu$ in the absence of correlations. Therefore it is some generalized one-electron 'Hamiltonian', and it is a unique one, because the Luttinger theorem is formulated in terms of its eigenvalues:

$$N = \sum_{nk} n(\xi_{nk}), \tag{8}$$

where $n(x) = 1$ for $x < 0$, $n(x) = 0$ for $x > 0$. The band index n contains spin, Kramer's index and so on, depending on the magnetic structure of the Bloch eigenstates.

To derive the sum rule Eq. (8) Luttinger and Ward obtained the remarkable expression for the thermodynamic potential $\Omega(\mu, T)$ that we reproduce here in terms of the Green functions of Eq. (1):

$$\Omega(\mu, T) = \left[T \int n(\varepsilon) \, \text{Tr} \{ \ln [\hat{G}_R^{-1}(\varepsilon)] + \hat{\Sigma}_R(\varepsilon) \hat{G}_R(\varepsilon) \} - \text{c.c.} \right] \frac{1}{2\pi i} + \Omega' \tag{9}$$

where $\hat{\Sigma}$ is the self-energy operator, defined by the equation

$$\hat{G}^{-1}(\varepsilon) = \hat{h} - \mu - \varepsilon - \hat{\Sigma}(\varepsilon), \tag{10}$$

$n(\varepsilon)$ is the Fermi function and Ω' contains the whole set of irreducible diagrams. The first variation of Ω with respect to Σ is equal to zero, and differentiating Ω with respect to μ one can immediately get Eq. (2). The derivation of Eq. (8) is not so straightforward, but careful examination of the procedure used by Luttinger and Ward shows that only those properties of G-functions that are related to their ε-dependence are important. These are the general analytical properties, finiteness of $\Sigma(\varepsilon)$ at $\varepsilon = 0$ and Eq. (7). For the ground state, $T = 0$, we come to the expression

$$N = (1/2\pi i)\{\mathrm{Tr}\,[\ln \hat{G}_R^{-1}(0)] - \mathrm{c.c.}\} \qquad (8')$$

which is the same as Eq. (8). The derivation remains valid when the operator $\hat{G}$ has a non-trivial spin structure, so that the sum rule Eq. (8) is fulfilled also for the ground state with magnetic order.

We see, therefore, that, for systems governed by the Luttinger theorem, it is possible to define the band spectrum ξ_{nk}. In terms of this spectrum we can, as usual, distinguish between metals and insulators. We deal with the metal case, when some ξ_{nk} change sign within a Brilouin zone, $\xi_{nk} = 0$ being a Fermi surface. Using the well-known representation

$$\hat{G}_R(\varepsilon) = \int \frac{\hat{\rho}(E)}{E - \varepsilon - i\delta}\,\mathrm{d}E \qquad (11)$$

we can get a simple relation between the spectral density at the Fermi level and the spectrum ξ_{nk}:

$$\rho_{nk}(0) = \delta(\xi_{nk}). \qquad (12)$$

In contrast to Eq. (6), this relation is exact for a correlated system only at the Fermi level. Away from the Fermi level, at non-zero ε the operator $\hat{\rho}(\varepsilon)$ becomes non-diagonal, and at a given k only

$$\rho_k(\varepsilon) = \mathrm{Tr}_n \hat{\rho}_k(\varepsilon) \qquad (13)$$

has invariant meaning while the contribution of each given band is not a well-defined quantity.

It is necessary to clarify the nature of the density of states of which we speak. It follows from the definition of G-functions, Eq. (1), that $\hat{\rho}(\varepsilon)$ characterizes the spectral density of the electron which enters the crystal from outside (affinity levels), or leaves the crystal (ionization levels). In the case of a metal the lowest affinity level and the highest ionization level coincide at the Fermi level. Thus, to probe this spectral density one should use photoemission or tunnelling measurements, for example.

In the Fermi-liquid theory, proposed by Landau [16], we deal with the density of states of the quasiparticles (Fermi excitations), which differs from the one defined by Eq. (10). This difference, which is a very important characteristic of the correlated Fermi system, is, as a rule, forgotten (together with the spectrum ξ_{nk}), which gives rise to a lot of misunderstandings. To derive the Fermi-liquid theory one should suppose that the Green function $G_p(\varepsilon)$, which is diagonal in p-representation for the uniform liquid, has a pole in the

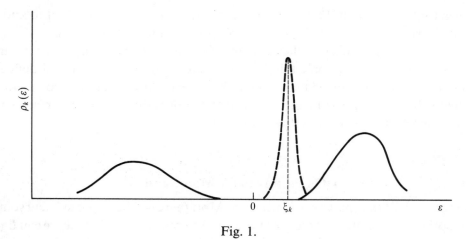

Fig. 1.

vicinity of the Fermi level. Neglecting damping of excitations, one can write the following expression for the spectral density $\rho_p(\varepsilon)$:

$$\rho_p(\varepsilon) = Z\delta(\tilde{\xi}_p - \varepsilon) + \rho'_p(\varepsilon) \tag{14}$$

where $\rho'_p(\varepsilon)$ is the so-called incoherent part, which goes to zero together with ε. The excitation spectrum $\tilde{\xi}_p$ is simply connected with the spectrum defined above, Eq. (7):

$$\tilde{\xi}_p = Z\xi_p \tag{15}$$

It has been shown (see, [17] for example), that, by means of the appropriate renormalization, the quantity Z, the residue of the pole of the G-function, can be completely excluded from the Fermi-liquid theory. This means that it is impossible to get any information on Z by measuring those properties of the Fermi liquid that are within the scope of this theory. The phenomena related to particle extraction from (or injection into) the liquid are outside its scope.

The spectral distribution Eq. (14) is shown schematically in Fig. 1. From the condition of completeness Eq. (4) it follows (provided that the wave functions, the plane waves for a liquid, are normalized in the volume V) that:

$$\int_{-\infty}^{+\infty} \rho_p(\varepsilon)\,d\varepsilon \equiv 1 = Z + \int_{-\infty}^{+\infty} \rho'_p(\varepsilon)\,d\varepsilon \tag{16}$$

The spectral density is positive at any values of ε, and therefore Z is less than 1 in the case of a liquid: as a result of the correlation effects the part $1 - Z$ of the total spectral weight is pushed out from the Fermi-level vicinity. We see that, even for liquid helium-3, the unique Fermi liquid, it will be important to know how large this part is and how it depends, for example, on pressure.

This is especially true for the electrons in solids. By introducing the spectrum ξ_{nk} into the band phenomenology we acquire an improved and well-defined language, in terms of which we can characterize the variety of correlation effects. In what follows I restrict myself to the problem of what happens with the ξ-spectrum when the system approaches the Mott transition.

3 On the way to the Mott transition

In the spirit of the analysis performed by Mott there will be discussed here some hypothetical evolution of the ground state, different stages of which we can find among the manifold of electron systems. Nobody before has made any attempt to get information concerning this spectrum. W. von der Linden was the first to calculate the ξ-spectrum for the one-dimensional Hubbard cluster [14]. The data even for this simplest model have appeared very instructive; part of them will be presented below. First consider some general trends.

Starting from the simple metallic case, we define the ξ-spectrum, which will be close to the usual band spectrum, calculated within the Hartree–Fock approximation. Some of the bands ξ_{nk} will cross the Fermi level, and the total density of states (per unit volume) on the Fermi surface according to Eq. (12) will be

$$\rho_F = 2 \oint \frac{1}{v} \, dS_F \qquad (17)$$

where $v = d\xi_{nk}/dk_\perp$. The density of states in a wide energy region will be almost exhausted by the spectrum ξ_{nk}, so that $Z \approx 1$. When the system evolves in the direction of the Mott transition, the correlation induced localization becomes more pronounced, and it is very natural to expect that the density of states at the Fermi level, Eq. (17), will decrease. This means that some averaged Fermi velocity increases, and the dispersion of the ξ-spectrum becomes sharper. This is contrary to usual expectations based on the Fermi-excitation picture. Regarding the latter, we can expect, of course, an increase in the incoherent part of the spectral density $\rho_k(\varepsilon)$, and as soon as Fermi excitations survive we can write down an expression like Eq. (15). However, for the crystal case $\rho'_k(\varepsilon)$ includes, according to Eq. (13), contributions from different bands, and how the coefficient Z will evolve depends on the specific properties of a given system, such as the electronic structure of the ions, the nature of the chemical bonds and so on. Moreover, we cannot be sure that Z will always be less than 1, which is definitely the case only for a single-band model. We mention that Z depends on k on the Fermi surface, especially when more than one band crosses the Fermi level. Nevertheless it seems that typically some averaged value of Z decreases together with the enhancement of correlations, and that even so the Fermi velocity of the excitations $\tilde{v} = Zv$ becomes smaller (the quasiparticles become heavier). This effect is well known and is usually interpreted as a narrowing of the band. However, we understand now that this is a misunderstanding and that the band structure should be formulated in terms of the ξ-spectrum, Eq. (7), and the density of states around the Fermi level for 'real' electrons will, as a rule, become smaller with increasing localization.

Close to the Mott transition the dispersion of ξ_{nk} around the Fermi surface should be sharp, but as soon as a ξ-spectrum exists the density of states ρ_F remains non-zero. It may happen that this strongly correlated ground state will be unstable with respect to some change of the symmetry (antiferromagnetic order, lattice distortion), and that the new ground state after rearrangement will have a larger unit cell. The 'Hamiltonian' $\hat{\xi}$, Eq. (7), acquires the symmetry of the new ground state, so it is natural that some of the bands ξ_{nk} will now be completely below, and others completely above, the Fermi level. This situation is still within the scope of the Luttinger theorem, and in this sense the gap in the new spectrum will be the usual insulating gap, although such an insulator is a strongly correlated one (a Mott insulator).

Suppose that a system already close to the Mott transition is stable with respect to any change of translational symmetry. The only possibility to eliminate the residual one-electron density of states around the Fermi level is to violate the Luttinger theorem, Eq. (7) and Eq. (8). This will happen when the system undergoes a superconducting transition. It seems very natural that the Luttiger theorem does not work in the superconducting state: the Pauli principle, a fundamental condition of this theorem, is incompatible with Bose condensation, the phenomenon closely related to superconductivity. We can, of course, establish this fact by revising the Luttinger–Ward derivation. It was shown long ago [18,19] that Feinman perturbation expansion remains valid for the superconducting state in terms of the matrix Green function, which includes the anomalous functions F and F^+ introduced by Gor'kov [20]. The expression for $\Omega(\mu, T)$, Eq. (9), remains almost the same in terms of matrix G-functions: the only difference is that Tr should be substituted by $(\frac{1}{2})$Tr (the additional matrix index is included). Such an expression for Ω was used in the particular case of the electron–phonon model [21]. The insurmountable obstacle, which prevents derivation of Eq. (8), is that both combinations $\varepsilon + \mu$ and $\varepsilon - \mu$ enter into the Green functions in the superconducting state. The nature of this obstacle is clear: this is 'broken gauge symmetry', a new physical

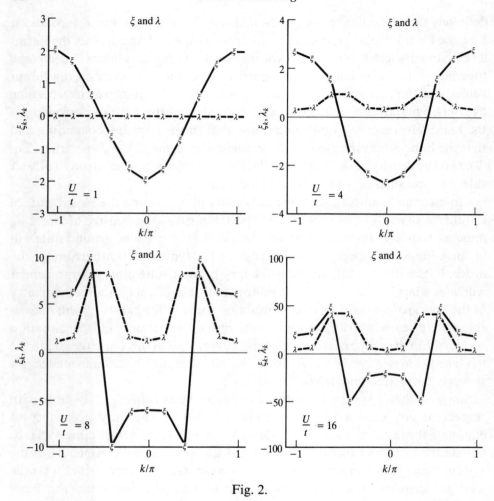

Fig. 2.

quantity exists in the superconducting state, namely the chemical potential of the electron pairs, which is zero at equilibrium.

It is known that the band splitting occurs at a half-filling in the one-dimensional Hubbard model [22], and it is interesting what happens with the ξ-spectrum in this case. Shown in Fig. 2 are numerical data for a cluster (ten lattice sites on the ring) [15] at several values of $(U/t) = 1, 4, 8$ and 16 and for $N = 10$ (a half-filled band). The solid lines are for ξ_k and the dashed lines for $\lambda_k = (\mathrm{d}\Sigma_k(\varepsilon)/\mathrm{d}\varepsilon)$ at $\varepsilon = 0$ $(Z = (1 + \lambda)^{-1}$, all lines are guides for the eyes). For $(U/t) = 1$ the effect of interaction is almost not seen: $\xi_k \approx 2t\cos(ka)$ and $\lambda \approx 0$. The gap-like behaviour is seen at $(U/t) = 4$, and for larger U we have a dramatic transformation of the dispersion minimum. Whether this is a trend towards

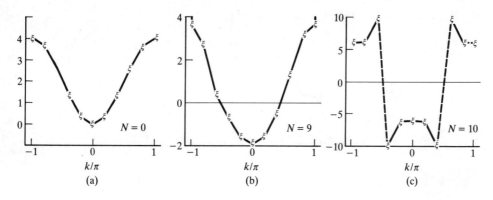

Fig. 3.

divergence of the ξ-spectrum at the Fermi boundary $k_F = (\pi/2)$, or there exists a finite jump in ξ_k for the macroscopic limit, it would be very interesting to check by analysing the exact Leib and Wu solution, or, at least, calculating for a larger cluster. The only question to be answered, is that of whether $G^{-1}(0, k_F) = 0$, or whether it has a finite discontinuity. In any case, the established trend is very important, because it allows us to understand to some extent how we should consider the cluster spectral properties for a system that is a superconductor in the macroscopic limit. To clarify this point, let us look at $G_R^{-1}(\varepsilon, k)$ at $\varepsilon = 0$ for a BCS single-band superconductor [20]:

$$G_R^{-1}(0, k) = v(k - k_F) + \frac{\Delta^2}{v(k - k_F) + i\delta}$$

We see that this is a complex quantity, and Eq. (5), which is requisite for the Luttinger theorem, is not fulfilled. The real part of this quantity, which should represent the ξ-spectrum, has the same divergence at $k = k_F$, the trend to which we have seen in the above example. This means that, on calculating the ξ-spectrum for a sufficiently big superconducting cluster, we will find this type of anomalous behaviour.

We shall continue this discussion in the concluding section and here return to some other data for the one-dimensional Hubbard model. In Fig. 3 is shown how the ξ-spectrum evolves depending on filling at $U = 8t$. The band width increases considerably at $N = 9$ with respect to the empty band, so that the density of states for real particles around the Fermi level decreases, as expected (Fermi-liquid excitations do not exist in this model). The most remarkable effect here is a very strong change in behaviour of the ξ-spectrum from $N = 9$, panel (b), to the half-filled band, $N = 10$, panel (c). The corresponding data for

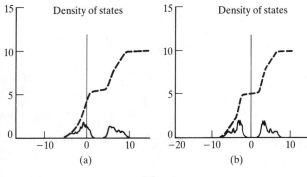

Fig. 4.

$\rho(\varepsilon)$, which is the sum over k of partial densities $\rho_k(\varepsilon)$, are shown in Fig. 4. The picture looks very like that for a usual doped insulator, as was stressed by Meinders *et al.* [23], who have calculated $\rho(\varepsilon)$ (but not ξ_k) for the same cluster. However, Fig. 3 undoubtedly shows that the behaviour of the ξ-spectrum for the one-dimensional Hubbard model should be continuous, so that calculating for a much larger cluster to get, say, $N/N_a = 0.98$, we find intermediate behaviour: ξ_k should cross the Fermi level (there is no gap in this model at any filling except for $N/N_a = 1$), as is shown by the dashed line in Fig. 3, panel (c); the Fermi level will be set within the gap-like region where the density of states will be small, but non-zero. This is in fact a very fragile metallic state. For example, the effect of finite temperature will be different from that in a usual metal. Suppose that we have such a state for U/t essentially larger than 1. The exchange interaction J is in this case of the order of t^2/U, it is much less even than t, and, because the band is almost half-filled, the spin entropy will achieve the value $\ln 2$ already at $T = T^* \geq J$. At higher T we will certainly have no evidence of metallic properties. Let us forget for a moment that we are dealing with a one-dimensional model and suppose that such a state is realized in some real system. At $T \ll T^*$ it will be a metal with a very large linear term in the specific heat. The velocity v of the ξ-spectrum is large in this state (see Fig. 3(c), dashed line), and therefore the coefficient Z should be very small: the density of states of the excitations should be large, while the density is small for 'real' electrons that enter the crystal from outside. We have here a situation somewhat reminiscent of heavy fermions.

It is outside the scope of the present paper to discuss further this interesting problem, but this example shows that, in the framework of the proposed phenomenology, it is possible to characterize a wide spectrum of strongly correlated states. The one-dimensional Hubbard model is very instructive, but it is a marginal case. We can use this model as a branch point, starting from which it is possible to choose different ways for its complication. We have seen that, being half-filled, this model is either a 'marginal antiferromagnet' (ξ_k has a finite jump at k_F) or a 'marginal superconductor' (ξ_k diverges at k_F). Away from but close to half-filling, this model represents a highly correlated metal. It will be very interesting, for example, to consider a three-dimensional crystal, which consists of weakly connected one-dimensional Hubbard chains: for a strongly correlated metallic state the low density of states, defined by the ξ-spectrum, will essentially influence electron transfer between chains.

4 From BCS to Mott superconductors

The above discussion highlights the limitations of a single-band model. We have seen that, in strongly correlated states the band width of the ξ-spectrum becomes large. This means that, for any real many-band spectrum, the conduction band will inevitably cross some of the upper and lower bands: the electron, being already essentially localized, uses any possibility to move. In the superconducting state the Luttinger theorem does not work and, rigorously speaking, here there is no ξ-spectrum at all. However, this is the only spectrum in terms of which it is possible to get the band description of a strongly correlated electron system. Therefore a consistent theory of the superconducting ground state should not deal with such terms as 'band filling', 'Fermi surface' and so on.

This conclusion does not mean, of course, that it is necessary to refuse completely to use the band phenomenology in further development of the theory of superconductivity. BCS theory works well within its appropriate scope of metals. Here it is quite natural to speak about 'the normal state of the superconductor': above T_c we have a normal state that we can precisely extrapolate to $T = 0$. Moreover, for phonon-induced superconductivity we have a model for this normal ground state: it is almost the same metal, but with an infinitely heavy lattice. In terms of the band spectrum calculated for this normal metal we can, within the adiabatic approximation [4], express all the properties of the superconductor using the BCS approach. Coulomb interaction appears in this theory only in the form of some repulsive pseudo-potential. As was shown recently by the present author [11,12], it is still possible to maintain the essential part of BCS theory even when Coulomb correlations are moderately strong, but in this case instead of a Coulomb pseudo-potential we deal with some renormalized scattering amplitude, concerning which it is impossible *a priori* to draw a conclusion as to whether it is repulsive or attractive. Therefore for these moderately correlated superconductors we can expect BCS-like properties, but the 'mechanism of superconductivity' remains an open problem, which should be solved in every special case. Probably new materials, among which are the high-T_c superconductors, are of this type [11,12], but it seems that already there it is necessary to distinguish different problems: the problem of a superconducting ground state and the problem of T_c. Even if it turns out to be possible to express the superconducting state in terms of some hypothetical normal state, the extent to which this state differs from that above T_c will remain unclear.

Nevertheless, let us consider the problem of high T_c from another point of view. Above I have made an attempt to explain that superconductivity is incompatible with the Luttinger theorem and, thus, with the one-electron metallic spectrum. From this point of view the best conditions for superconductivity exist among those materials that have a pronounced trend towards Mott localization. Some of them should have a superconducting ground state, with no one-electron ξ-spectrum at all. We can call these materials 'Mott superconductors'. As has been discussed, one should try to find them among those materials that, being strongly correlated, are stable with respect to any doubling of the unit cell, e.g., antiferromagnetic transitions, Peierls distortions etc. Probably, this is why it is necessary to dope La_2CuO_4 or $BaBiO_3$, but after such a doping the materials become farther from the Mott transition, which in itself is not the best way to get higher T_c.

Mott superconductors will have unusual properties above T_c as is the case for the Mott antiferromagnet La_2CuO_4 above T_N. Some of the unusual features we have already in high-T_c cuprates, but here they are not so strongly pronounced. The theory of T_c for a Mott superconductor becomes a special problem. The superconducting state is formed in this case due to appropriate phase coherence between localized electron states, and at finite temperatures this coherence is gradually destroyed. Actually, it is very like what we have in a Mott antiferromagnet, but here we have another structure of local states, and another phase relation between them.

It seems that the most important problem in the pursuit of higher T_c is to understand the structures of these local states, which are appropriate for the superconducting phase correlations.

Acknowledgements

I would like to express my gratitude to W. von der Linden for permission to show the numerical data for the one-dimensional Hubbard model. This paper was completed during my stay at the High Magnetic Field Laboratory, MPI/ CNRS Grenoble, and I am grateful to P. Wyder and G. Martinez for their kind hospitality.

References

[1] R. M. Schafroth, S. T. Butler and J. M. Blatt, *Helv. Phys. Acta* **30**, 93 (1957).

[2] J. Bardeen, L. N. Cooper and J. R. Schrieffer, *Phys. Rev.* **108**, 1175 (1957).

[3] L. N. Cooper, *Phys. Rev.* **104**, 1189 (1956).

[4] A. B. Migdal, *Sov. Phys. JETP* **7**, 996 (1958).

[5] G. M. Eliashberg, *Sov. Phys. JETP* **12**, 1000 (1960).

[6] R. Micnas, J. Ranninger and S. Robaszkiewicz, *Rev. Mod. Phys.* **62**, 113 (1990).

[7] A. S. Alexandrov and N. F. Mott, *Supercond. Sci. Technol.* **6**, 215 (1993).

[8] N. F. Mott, *Proc. Phys. Soc. London* **62**, 416 (1949).

[9] N. F. Mott, *Phil. Mag.* **6**, 287 (1961).

[10] G. M. Eliashberg, in *Electronic Properties of High*-T_c *Superconductors* eds. H. Kuzmany, M. Mehring and J. Fink, Springer Series in Solid State Science, vol. 113, p. 385 (Springer, Berlin 1993).

[11] G. M. Eliashberg, *Physica* A **200**, 95 (1993).

[12] G. M. Eliashberg, *J. Supercond.* **7**, 525 (1994).

[13] J. M. Luttinger and J. C. Ward, *Phys. Rev.* **118**, 1417 (1960).

[14] J. M. Luttinger, *Phys. Rev.* **119**, 1153 (1960).

[15] G. M. Eliashberg and W. von der Linden, *Sov. Phys. JETP Lett.* **59**, 441 (1994).

[16] L. D. Landau, *Sov. Phys. JETP* **3**, 920 (1957).

[17] A. A. Abrikosov, L. P. Gor'kov and I. E. Dzyaloshinskii, *Quantum Field Theoretical Methods in Statistical Physics* (Pergamon, New York and Oxford, 1965).

[18] Y. Nambu, *Phys. Rev.* **177**, 648 (1960).

[19] G. M. Eliashberg, *Sov. Phys. JETP* **11**, 696 (1960).

[20] L. P. Gor'kov, *Sov. Phys. JETP* **7**, 505 (1958).

[21] G. M. Eliashberg, *Sov. Phys. JETP* **16**, 780 (1963).

[22] E. H. Lieb and F. Y. Wu, *Phys. Rev. Lett.* **25**, 1445 (1968).

[23] M. B. J. Meinders, H. Eskes and G. A. Sawatzky, *Phys. Rev.* B **48**, 4302 (1993).

3

A bipolaron Bose liquid in high-T_c superconductors

A. S. ALEXANDROV

Interdisciplinary Research Centre in Superconductivity, University of Cambridge, Madingley Road, Cambridge, CB3 0HE, UK

Abstract

The BSC theory is extended to strong electron–phonon coupling for $\lambda > 1$. In this limit carriers are charged 2e bosons (singlet and triplet inter-site bipolarons). The Anderson localisation of the bosons resulting from disorder is also considered. Several non Fermi-liquid features of copper-based high-T_c oxides, in particular the spin gap in NMR and neutron scattering, the temperature-dependent Hall effect, linear resistivity and divergent $H_{c2}(T)$ are explained.

1 The strong-coupling extension of the BCS theory

The electron–phonon coupling constant λ in the BCS theory is the ratio of the characteristic interaction energy $V = 2E_p$ of carriers with a bosonic field, for instance of phonons, which is responsible for the coupling to their kinetic energy E_F, $\lambda \simeq V/(2E_F)$. At the point $\lambda \simeq 1$ the characteristic potential energy due to the local lattice deformation exceeds the kinetic energy. This is a condition of small-polaron formation which has been known for a long time as a solution for a single electron on a lattice coupled with lattice vibrations. So long as $\lambda > 1$ the kinetic energy remains smaller than the interaction energy and a self-consistent treatment of a many-electron system strongly coupled with phonons is possible with the '$1/\lambda$' expansion technique [1]. This possibility results from the fact, which has been known for a long time, that there is an exact solution for a single electron in the strong-coupling limit $\lambda \to \infty$. Following Lang and Firsov (1962) one can apply the canonical transformation $\exp(S_1)$ to diagonalise the single-electron Fröhlich Hamiltonian (under the 'Fröhlich Hamiltonian' we assume that any electron–phonon interaction occurs with its matrix element depending on the phonon momentum). The diagonalisation is exact if the bare hopping integral $T(m) \sim E_F$ is zero (or $\lambda = \infty$).

Using $1/\lambda$ as a small parameter for a many-electron system coupled with phonons

$$H = \sum_{k,s} E(k)c^\dagger_{k,s}c_{k,s} + (2N)^{-1/2} \sum_{k,q,s} \gamma(q)\omega(q)c^\dagger_{k+q,s}c_{k,s}(d_q + d^\dagger_{-q}) + H_{ph} + V_c \quad (1)$$

Alexandrov and Ranninger [2] derived the bipolaronic Hamiltonian, describing the repulsion and tunnelling of small bipolarons, which are hard-core charged 2e bosons on a lattice:

$$H_b = \sum_{m,m'} (-t_{m,m'}b^\dagger_m b_{m'} + v^{(2)}_{m,m'}n_m(n_{m'}-1) + 4v_{m,m'}n_m n_{m'}) + H_{ph}, \quad (2)$$

where $H_{ph} = \Sigma_q \omega(q)d^\dagger_q d_q$ is the phonon energy, $n_m = b^\dagger_m b_m$ is the bipolaronic density operator and $b_m = c_{m,\uparrow}c_{m,\downarrow}$. The interaction $v_{m,m'}$ does not depend on the kinetic energy and includes the direct Coulomb repulsion (V_c) and the attraction via phonons between two small polarons on different cells m, $t_{m,m'}$ and $v^{(2)}_{m,m'}$ are the bipolaron transfer integral and the bipolaron repulsion due to the virtual polaron exchange, correspondingly, both of second order in the electronic kinetic energy $E(k)$:

$$t_{m,m'} = 2i \int_0^\infty d\tau \langle \hat{\sigma}_{m,m'}(\tau)\hat{\sigma}_{m,m'}(0)\rangle_{ph} \exp(-i\Delta\tau), \quad (3)$$

$$v^{(2)}_{m,m'} = 2i \int_0^\infty d\tau \langle \hat{\sigma}_{m,m'}(\tau)\hat{\sigma}_{m',m}(0)\rangle_{ph} \exp(-i\Delta\tau), \quad (4)$$

where

$$\hat{\sigma}_{m,m'}(\tau) = T_{m,m'}e^{iH_{ph}\tau} \exp\left(\frac{1}{\sqrt{2N}}\sum_q \gamma(q)d^\dagger_q[e^{-iq\cdot m} - e^{-iq\cdot m'}] - \text{H.c.}\right)e^{-H_{ph}\tau}, \quad (5)$$

where $\langle \ldots \rangle_{ph}$ denotes averaging with respect to the phonon density matrix and $T_{m,m'} = (1/N)\Sigma_k E(k)\exp[ik\cdot(m-m')]$ is a bare electron transfer integral between sites m and m'.

The transformation of the Fröhlich Hamiltonian, Eq. (1), into a bipolaronic one, Eq. (2), is performed with two canonical transformations, $\exp S_1$ and $\exp S_2$:

$$S_1 = (2N)^{-1/2} \sum_{q,m,s} \gamma(q)c^\dagger_{m,s}c_{m,s}(d^\dagger_q e^{-iq\cdot m} - \text{H.c.}), \quad (6)$$

$$(S_2)_{f,p} = \sum_{m,m',s} \frac{\langle f|\hat{\sigma}_{m,m'}(0)c^\dagger_{m,s}c_{m',s}|p\rangle}{E_f - E_p}, \quad (7)$$

where $E_{f,p}$ and $|f\rangle$, $|p\rangle$ are the energy levels and eigenstates of the original Fröhlich Hamiltonian, Eq. (1) without the kinetic energy term.

The first canonical transformation eliminates the Fröhlich interaction and the second one eliminates one-electron hopping. The bipolaronic Hamiltonian, Eq. (2) describes low-energy excitations of the strongly-coupled ($\lambda > 1$) many-electron–phonon system at low enough temperature $T \ll \Delta = 2E_p - V_c$, where E_p is the familiar polaron shift of the electron band:

$$E_p = \frac{1}{2N} \sum_q \gamma(q)^2 \omega(q), \tag{8}$$

with $\omega(q)$ the phonon frequency and N the total number of sites (cells). There are additional internal quantum numbers (spin and orbital momentum), which should be included in the definition of the bipolaron creation and annihilation operators in case of *intersite* bipolarons.

2 A basic model for copper-based high-T_c oxides

The solution of the two-particle Schrödinger equation for the CuO_2 plane with two different types of short-range correlations (repulsion on copper and attraction on oxygen) yields the radius of the bound state as less then two lattice constants in the relevant region of correlations [3]. It is also important that the solution of the two-particle problem with the electron–phonon interaction and dielectric constants ε_∞ and ε_0 characteristic for copper oxides shows that the bipolaron is bound more easily in two dimensions than in three, and that the mean value of the pair radius is a few ångström units ([4–6]). Taking this into account and also the microscopic consideration discussed above, Alexandrov and Mott [7] proposed a simple model for the copper–oxygen plane, which is a key structural element of all copper-based high-T_c compounds.

Our assumption is that *all electrons* are bound in small singlet or triplet *intersite* bipolarons stabilised by the lattice and spin distortion. Because the undoped plane has a half-filled Cu $3d^9$ band there is no space for bipolarons to move if they are intersite ones. Their Brillouin zone is half of the original electron one and completely filled with hard-core bosons. *Hole* pairs, which appear with doping, have enough space to move and they are responsible for the low-energy charge excitations of the CuO_2 plane. Above T_c a material such as YBCO contains a non-degenerate gas of these hole bipolarons in singlet or in triplet states with a slightly lower mass due to the lower binding energy, Fig. 1.

The main part of the electron–electron correlation energy and the electron–phonon interaction is included in the binding energy of bipolarons and in their band-width renormalisation. The rest, including the boson–boson repulsion, is

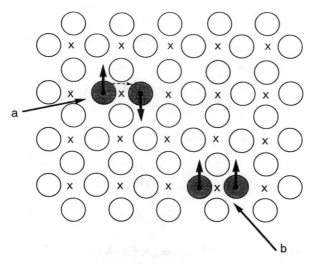

Fig. 1. Inter-site singlet (a) and triplet (b) bipolarons on oxygen sites (circles) surrounded by copper (crosses).

considered as the perturbation for extended hole states resulting in the canonical Boltzmann kinetics or in Bogolubov excitations in the superconducting state. In the normal state corrections to the single-particle spectrum due to interaction are small if the RPA parameter r_s for the Coulomb forces is not very large. To describe the kinetics and the metal–insulator transition one should also take into account the Anderson localisation of bipolarons. The exchange interaction between singlet electron bipolarons could be responsible for the antiferromagnetic ground state of undoped materials.

The low-energy band structure consists of two bosonic bands (singlet and triplet), separated by the singlet–triplet exchange energy J, estimated to be of the order of a few dozen milli-electron-volts, Fig. 2. The half-bandwidth $w = z t_{m,m+a}$ is of the same order. The bipolaron binding energy is assumed to be large, $\Delta \gg T$ and therefore single polarons are irrelevant in the temperature region under consideration. This band structure is applied for hole pairs. The singlet–triplet exchange energy for electron pairs might be different from that of hole pairs. In general one should not distinguish between 'copper' and 'oxygen' carriers because, due to hybridisation of the d and p orbitals, the bipolarons are somewhere between copper and oxygen sites.

We argue that many features of spin and charge excitations of metal oxides can be described within our simple model, Fig. 2.

The population of singlet and triplet bands is controlled by the chemical potential $\mu = T \ln y$, where y is determined from the thermal equilibrium of singlet and triplet bipolarons if $T > T_c$

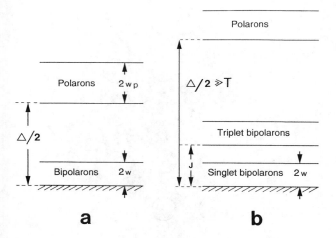

Fig. 2. On-site (a) and intersite (b) bipolaronic energy bands.

$$\ln\left(\frac{1-y\mathrm{e}^{-2w/T}}{1-y}\right)+3\ln\left(\frac{1-y\mathrm{e}^{(-2w-J)/T}}{1-y\mathrm{e}^{-J/T}}\right)=\frac{2nw}{T},\qquad(9)$$

and $y=1$ if $T<T_c$. Here n is the total number of pairs per cell and the density of states is assumed to be energy-independent within bands.

3 NMR and neutron scattering

Taking the usual contact hyperfine coupling of nuclei with electron spins and performing transformation to polarons and bipolarons, one obtains the effective interaction of triplet bipolarons with nuclear spins [8]

$$H_{\mathrm{int}}\simeq\sum_{\boldsymbol{m},l,l'}b^{\dagger}_{\boldsymbol{m},l}b_{\boldsymbol{m},l'}+\mathrm{H.c.},\qquad(10)$$

where only on-cell terms $\boldsymbol{m}=\boldsymbol{m}'$ are kept and $l=0,\pm1$. The NMR width due to spin-flip scattering of triplet bipolarons on nuclei is obtained with Fermi's golden rule:

$$\frac{1}{T_1}\simeq\int\mathrm{d}\xi\,N_t^2(\xi)f(\xi)(1+f(\xi))\qquad(11)$$

with $f(\xi)=\{\exp[(\xi-\mu)/T]-1\}^{-1}$ and $N_t(\xi)$ the DOS in the triplet band. With $N_t\simeq1/(2w)$ one obtains

$$\frac{1}{T_1}\simeq\frac{AT\sinh(w/T)}{w^2\{\cosh[(w+J)/T-\ln y]-\cosh(w/T)\}},\qquad(12)$$

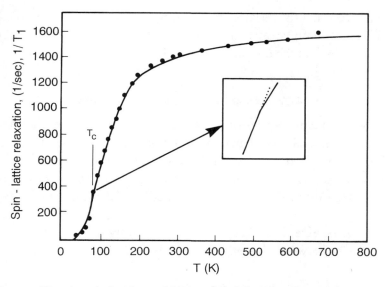

Fig. 3. The non-Korringa behaviour of $1/T_1$ and the absence of the Hebel–Slichter peak of the Cu NMR rate of $YBa_2Cu_4O_8$ (T. Machi, I. Tomeno, T. Miyatake, and S. Tanaka, *Physica* C **173**, 32 (1991)), the theoretical curve is after [8].

where A is a temperature-independent hyperfine coupling constant of the same order as that in simple metals.

With Eq. (12) one can understand the main features of the nuclear spin relaxation rate in copper-based oxides: the absence of the Hebel–Slichter coherent peak below and the temperature-dependent Korringa ratio $1/(TT_1)$ above T_c and the large value of $1/T_1$ due to the small bandwidth w, and one can fit the experimental data with reasonable values of the parameters w and J, Fig. 3. The fitting parameters w and J vary with the chemical formula. In particular, in the optimally doped $YBa_2Cu_3O_{6.95}$, the singlet–triplet exchange energy is very small. The Knight shift, which measures the spin susceptibility of carriers, drops well above T_c in many copper oxides. This confirms our interpretation of the NMR and magnetic susceptibility data as due to triplet bipolarons with temperature-dependent density.

The value of the singlet–triplet bipolaron exchange energy $J \simeq 10$–20 meV, determined from the fit to the experimental NMR width, is close to the 'spin' gap observed above and below T_c in the underdoped YBCO with unpolarised [9] and polarised [10] neutron scattering, which reflects their both having the same origin. The disappearance of magnetic scattering for energy transfer above approximately 50 meV [9] is due to the narrowness of the bipolaron bands, as is shown in Fig. 2. For energy transfer above $2w + J$ there are no final states available. The characteristic q-width of the magnetic susceptibility, determined from neutron scattering, is comparable with the reciprocal lattice vector. This fact demonstrates the local character of magnetic excitations of the CuO_2 plane, which is the characteristic feature of small bipolarons.

4 Anderson localisation of bipolarons

To describe kinetic properties of bipolarons in copper oxides one should take into account their localisation in a random potential. Because of low dimensionality (2D rather then 3D), any random potential leads to localisation no matter how weak it is. Coulomb repulsion limits the number of bosons in each localised state, so that the distribution function will show a mobility edge E_c [11].

The density of extended (free) bosons is given by (for simplicity we do not distinguish here between singlets and triplets):

$$n_b(T) = \frac{T}{2w} \ln \left(\frac{1 - ye^{-2w/T}}{1-y} \right). \tag{13}$$

To calculate the density of localised bosons $n_L(T)$ one should take into account the repulsion between them. One cannot ignore the fact that the localisation length ξ generally varies with energy and diverges at the mobility edge. One would expect that the number of *hard-core* bosons in a localised state near the mobility edge diverges in a similar way to that in which the localisation length does. However, in our case of *charged* bosons their number in a single potential well is determined by competition between their long-range Coulomb repulsion $\simeq 4e^2/\xi$ and the binding energy $E_c - \varepsilon$. If the localisation length diverges with the critical exponent $v < 1$: ($\xi \sim (E_c - \varepsilon)^{-v}$), then one can apply a 'single well–single particle' approximation, assuming that one can place only one boson in each potential well. Within this approximation localised *charged* bosons obey the Fermi–Dirac statistics:

$$n_L(T) = \int_{-\infty}^{E_c} \frac{N_L(\varepsilon)\,d\varepsilon}{\exp(\varepsilon - \mu/T) + 1}, \tag{14}$$

where $N_L(\varepsilon)$ is the density of localised states. Near the mobility edge it remains constant at $N_L(\varepsilon) \simeq n_L/\gamma$ with γ of the order of a binding energy in a single random potential well and n_L the total number of localised states per unit cell. The integral in Eq. (14) gives

$$n_L(T) = n_L - n_L \sum_{k=1}^{\infty} (-1)^{k-1} \frac{y^k}{1 + \gamma k/T}. \tag{15}$$

The second term in Eq. (15), which is the number of empty localised states, turns out to be linear as a function of temperature over a wide temperature range $T < 2w$, γ because the chemical potential is locked in this temperature region near the mobility edge, which one can choose equal to zero, $\mu \simeq E_c = 0$. This follows from conservation of the total number of bosons per cell, $n = n_b(T) + n_L(T)$, which gives for the chemical potential

$$\frac{T}{2w} \ln\left(\frac{1}{1-y}\right) - \frac{n_L T}{\gamma} \ln(1 + y^{-1}) = n - n_L. \tag{16}$$

If $T \ll (\gamma, 2w)$, then the solution of this equation is $y \simeq 1$ with the exception of a very narrow region of concentration $n - n_L \ll T/(2w)$, where y decreases. The density n_b depends on y logarithmically; therefore its temperature dependence remains practically linear up to $T \simeq \gamma$:

$$n_b(T) = n - n_L + n_L b T, \tag{17}$$

with temperature-independent $b = \ln(1 + y^{-1})/\gamma$.

In a more general case the localised bipolaron statistics is different from both Fermi and Bose–Einstein ones. This 'para-statistics' is important for some kinetic and thermodynamic properties like the thermoelectric power and specific heat.

5 The Hall effect, resistivity and spin gap: the connection with high-T_c

Solving the Boltzmann equation with a weak magnetic field for extended bosons, scattered by acoustical phonons, by each other and by unscreened random potential, one obtains the canonical expressions for the Hall ratio R_H, resistivity ρ and $\cot \Theta_H = \rho/R_H$ [12]:

$$R_H = \frac{v_0 <\tau^2>}{2 e n_b(T) <\tau>^2}, \tag{18}$$

$$\rho = \frac{v_0 m^{**}}{4 e^2 n_b(T) <\tau>}, \tag{19}$$

$$\cot \Theta_H = \frac{<\tau>}{\omega_c <\tau^2>}, \tag{20}$$

where $m^{**} = \pi/(wa^2)$ is the in-plane boson mass with a the in-plane lattice constant, $\omega = 2eH/m^{**}$, τ the transport relaxation time, v_0 the volume of an elementary cell ($0.167\,\mathrm{nm}^3$ for $\mathrm{YBa_2Cu_3O_7}$), and $<\ldots>$ means an average with respect to energy and the derivative of the Bose–Einstein distribution function.

The transport relaxation rate due to 2D boson–phonon scattering has been shown to be energy-independent and linear in temperature [8]:

$$\frac{1}{\tau_{\text{b–ac}}} = m^{**}C_{\text{ac}}T, \tag{21}$$

where the constant C_{ac} is proportional to the deformation potential.

In the case of Bose–Einstein statistics and $T \ll 2w$ *umklapp* scattering can be neglected, so scattering between extended bosons does not contribute to the resistivity. However, the inelastic scattering of an extended boson by localised bosons makes a contribution because the momentum is not conserved in two-particle collisions in the presence of the impurity potential and the carriers are heavy. In the 'single well–single particle approximation' the role of the Pauli exclusion principle is played by the dynamical repulsion between bosons. That is why the boson–boson relaxation rate has the same temperature-dependence as the fermion–fermion scattering. The relaxation rate is proportional to temperature squared because only localised bosons within the energy shell of the order of T near the mobility edge contribute to the scattering and because the number of final states is proportional to temperature:

$$\frac{1}{\tau_{\text{b–b}}} = \frac{e^2 b n_{\text{L}}}{m^{**}\alpha} T^2, \tag{22}$$

with α as a constant.

Unoccupied impurity wells of density $b n_{\text{L}} T$ also contribute to the scattering giving rise to the energy-independent elastic relaxation rate, which is linear in temperature:

$$\frac{1}{\tau_{\text{b–im}}} = m^{**}C_{\text{im}}n_{\text{L}}T, \tag{23}$$

with C_{im} a constant.

Substitution of Eqs. (21–23) into Eqs. (18)–(20) yields

$$R_{\text{H}} = \frac{v_0}{2e(n - n_{\text{L}} + b n_{\text{L}} T)}, \tag{24}$$

$$\rho = \frac{(m^{**})^2 C v_0}{4e^2} \frac{T + \sigma_{\text{b}} n_{\text{L}} T^2}{n - n_{\text{L}} + b n_{\text{L}} T}, \tag{25}$$

with $C = C_{\text{ac}} + n_{\text{L}} C_{\text{im}}$ and $\sigma_{\text{b}} = e^2 b/(m^{**}\alpha C)$ and

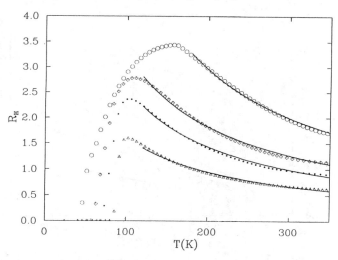

Fig. 4. The Hall coefficient (in units $10^{-9} \, C^{-1} \, m^3$) for $YBa_2Cu_3O_{7-\delta}$ [13] compared with the theory Eq. (24) (solid lines) for different δ: 0.05 ($\triangle$); 0.19 ($\bullet$); 0.23 ($\diamond$) and 0.39 ($\bigcirc$). The parameters of the model are presented in Table 1.

$$\cot \Theta_{\mathrm{H}} = \frac{(m^{**})^2 C}{2eH} \, (T + \sigma_b n_{\mathrm{L}} T^2). \qquad (26)$$

These formulae contain important information about the number of bosons, localised states and the relative strength of different scattering channels.

There are two fitting parameters, n and n_{L}, if no significant variation of $(m^{**})^2 C$, b and σ_b is expected with doping. Because of the chains the number of in-plane carriers is not fixed by the chemical formula, at least in '1–2–3' YBCO.

There are in-plane kinetic data for single homogeneous crystals of $YBa_2Cu_3O_{7-\delta}$ in a wide range of doping and temperatures [13,14]. The theoretical fit with parameters n and n_{L} is shown in Figs. 4–6. These parameters as well as other 'constants' are presented in Table 1. The number of extended bosons increases with doping and the scattering cross-sections and their relative contribution depend slightly on the doping. Underdoped samples ($\delta > 0.2$) are practically 'compensated' in the sense that the total number of bosons is very close to the number of localised states. This should be the case if every additional oxygen ion gives a single localised state. Boson–phonon scattering is mainly responsible for the linear temperature-dependence of ρ at low temperature while boson–boson scattering and the temperature-dependent concentration $n_b(T)$ are responsible for the linear ρ at higher temperatures. *The residual resistivity is taken to be zero.* The slope of $\cot \Theta_{\mathrm{H}}$ at low temperature increases with the impurity concentration n_{L} due to elastic scattering by the random unoccupied potential wells.

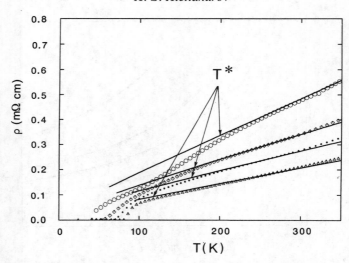

Fig. 5. Resistivity for $YBa_2Cu_3O_{7-\delta}$ [13] compared with the theory, Eq. (25). For parameters see Table 1.

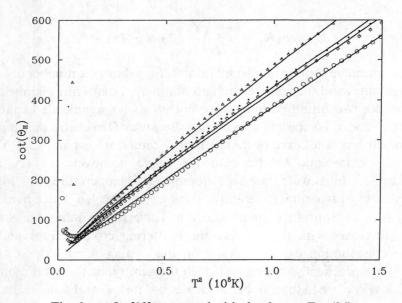

Fig. 6. $\cot \Theta_H$ [13] compared with the theory, Eq. (26).

Table 1. *Microscopic parameters of the model determined from experimental data [13].*

δ	$n - n_L$	$bn_L\ (10^{-3}\ \mathrm{K^{-1}})$	$(m^{**})^2 C\ (10^{-18}\ \mathrm{kg\ s^{-1}\ K^{-1}})$	$\sigma_b(10^{-2}\ \mathrm{K^{-1}})$
0.05	0.103	2.22	0.81	1.0
0.19	0.041	1.61	0.69	1.1
0.23	0.035	1.25	0.62	1.2
0.28	0.023	1.11	0.46	1.6
0.39	0.007	0.85	0.46	1.6

It should be mentioned that the dissociation of a boson on two single non-degenerate polarons at a sufficiently high temperature has no influence on the temperature dependence of R_H and ρ if the effective mass of a boson is close to the effective mass of two single polarons. This is also the case for triplet bipolarons. However, if triplets are lighter than singlets then the slope of resistivity (proportional to $(m^{**})^2 C$) diminishes with temperature because singlets are thermally excited into triplets. The characteristic small deviation from linearity (see Fig. 5) should appear at temperature T^*, where a 'spin' gap in NMR and in neutron scattering appears. The change in slope of resistivity at $T = T^*$ was measured by Bucher *et al.* [15] and explained by Mott [16]. Ito *et al.* [14] observed a correlation in the slope of resistivity with the temperature-dependent Korringa ratio in $YBa_2Cu_3O_{7-\delta}$. These observations support our explanation of the 'spin' gap as a singlet–triplet bipolaron exchange energy.

There is a direct link between the proposed normal state kinetic theory and high-T_c superconductivity. If T_c is the temperature of the Bose–Einstein condensation of charged bosons then the carrier density conservation at $\mu = 0$ yields

$$\int \frac{N(\varepsilon)\mathrm{d}\varepsilon}{\exp(\varepsilon/T_c) - 1} = n - n_L + \frac{n_L T_c \ln 2}{\gamma}. \tag{27}$$

With three-dimensional corrections to the bipolaron band dispersion taken into account [8]

$$E_k = \frac{k_\parallel^2}{2m^{**}} + 2t_\perp [1 - \cos(k_\perp d)] \tag{28}$$

the density of the extended states $N(\varepsilon)$ is given by

$$N(\varepsilon) = \frac{1}{2w\pi} \arccos\left(1 - \frac{\varepsilon}{2t_\perp}\right) \tag{29}$$

for $0 < \varepsilon < 4t_\perp$ and $N(\varepsilon) = 1/(2w)$ for $4t_\perp < \varepsilon < 2w$ with $t_\perp \ll w$ the interplane hopping and d the interplane distance.

Substitution of $N(\varepsilon)$ in Eq. (27) with the assumption $T_c \ll 2w$ yields

$$\frac{T_c}{2w\pi} \int_0^{4t_\perp/T_c} dx \frac{\arccos[1 - xT_c/(2t_\perp)]}{\exp x - 1} - \frac{T_c}{2w} \ln\left[1 - \exp\left(-\frac{4t_\perp}{T_c}\right)\right]$$

$$= n - n_L + \frac{T_c n_L \ln 2}{\gamma}. \tag{30}$$

This equation is simplified if $T_c \gg t_\perp$:

$$\frac{T_c}{2w} \ln\left(\frac{T_c \exp(1 - 2n_L w \ln 2/\gamma)}{2t_\perp}\right) = n - n_L. \tag{31}$$

Depending on the compensation, T_c changes from

$$T_c = \frac{2w(n - n_L)}{L} \tag{32}$$

for

$$n - n_L \gg \frac{t_\perp \exp(2n_L w \ln 2/\gamma - 1)}{w} \tag{33}$$

with

$$L = \ln\left(\frac{(n - n_L)w \exp(1 - 2n_L w \ln 2/\gamma)}{t_\perp}\right) \tag{34}$$

to

$$T_c \simeq 2t_\perp \exp(2n_L w \ln 2/\gamma - 1) \tag{35}$$

if $n = n_L$. With a reasonable value of the bipolaron half-bandwidth $w \simeq 450$ K in Eq. (32) or with the value of $2t_\perp \simeq 45$ K in Eq. (35) one can explain the high value of $T_c \simeq 90$ K (with bn_L from Table 1 the *exponent* in Eq. (35) is of order unity).

6 The upper critical field

An ideal charged Bose-gas in a magnetic field cannot be condensed because of the one-dimensional character of particle motion within the lowest Landau level. However, the interacting charged Bose-gas is condensed in a field lower than a certain critical value H^* because the interaction with impurities [17], or between bosons [18], broadens the Landau levels and thereby eliminates the one-dimensional singularity of the density of states. The critical field of Bose–Einstein condensation has an unusual positive curvature near T_{c0}, $H^*(T) \sim (T_{c0} - T)^{3/2}$ and diverges at $T \to 0$, where $T_{c0} \simeq 3.31 n^{3/2}/m$ is the critical temperature of Bose–Einstein condensation of an ideal gas in zero field.

It was shown recently [19] that localisation drastically changes the low-temperature behaviour of the critical field. H^* saturates with the temperature lowering at some value of the impurity concentration and at higher concentrations the re-entry effect to the normal state occurs.

H^* is determined as the field in which the first non-zero solution of the linearised stationary Ginzburg–Pitaevskii equation [20] for the macroscopic condensate wave function $\psi_0(r) = \langle N|\hat{\psi}(r, \tau)|N+1 \rangle$, ($N \to \infty$, $N/V = n = $ constant) appears:

$$\left(-\frac{1}{2m^{**}} (\nabla - 2ieA(r))^2 + U_{imp}(r) \right) \psi_0(r) = \mu \psi_0(r), \tag{36}$$

where $2e$ is the charge of a boson, $A(r)$, $U_{imp}(r)$ and μ are the vector, random and chemical potentials, respectively.

The definition of H^*, Eq. (36) is identical to that of the upper critical field H_{c2} of BCS superconductors of the second kind. Therefore H^* determines the upper critical field of a bipolaronic or any 'bosonic' superconductor.

In general the energy spectrum of the Hamiltonian Eq. (36) contains discrete levels (localised states) and a continuous part (delocalised states). The density of delocalised states $\tilde{N}(\varepsilon, H^*) \sim \text{Im} \Sigma(\varepsilon)$ and the lowest delocalised energy E_c (the mobility edge: $\tilde{N}(E_c, H^*) = 0$) can be found with the random phase ('ladder') approximation for the one-particle self-energy:

$$\Sigma(\varepsilon) = \frac{4\pi^2 n_{im} f^2}{m} \int \frac{N(\varepsilon', H^*) \, d\varepsilon'}{\varepsilon - \varepsilon' - \Sigma(\varepsilon)}, \tag{37}$$

where n_{im} is the impurity concentration, f is the scattering amplitude in zero field and

$$N(\varepsilon, H) = \frac{\sqrt{2}(m^{**})^{3/2} \omega}{4\pi^2} \text{Re} \sum_{N=0}^{\infty} \frac{1}{[\varepsilon - \omega(N+1/2)]^{1/2}} \tag{38}$$

is the density of states for a non-interacting system with $\omega = 2eH^*/m^{**}$.

The solution of Eq. (37) yields

$$\tilde{N}_0(\varepsilon, H^*) = \frac{\sqrt{6}(m^{**})^{3/2} \omega}{8\pi^2 \sqrt{\Gamma_0}} \left\{ \left[\frac{\tilde{\varepsilon}^3}{27} + \frac{1}{2} + \left(\frac{\tilde{\varepsilon}^3}{27} + \frac{1}{4} \right)^{1/2} \right]^{1/3} - \left[\frac{\tilde{\varepsilon}^3}{27} + \frac{1}{2} - \left(\frac{\tilde{\varepsilon}^3}{27} + \frac{1}{4} \right)^{1/2} \right]^{1/3} \right\}. \tag{39}$$

and

$$E_c = \frac{\omega}{2} - \frac{3\Gamma_0}{2^{2/3}}$$ (40)

with $\Gamma_0 = (n_{im}8\pi f^2 eH^*)^{2/3}/(2m^{**})$ and $\tilde{\varepsilon} = (\varepsilon - \omega/2)/\Gamma_0$.

Eq. (39) describes the energy-dependence of the density of states of the lowest Landau level $(N = 0)$ near the mobility edge. Since the square root singularity of the density of states of upper levels is integrated out (see below) one can neglect their quantisation using the zero field density of states for $\varepsilon > \omega$:

$$N(\varepsilon) \simeq \frac{(m^{**})^{3/2}\sqrt{\varepsilon}}{\sqrt{2}\pi^2}.$$ (41)

The first non-trivial *delocalised* solution of Eq. (36) appears at $\mu = E_c$. Thus the critical curve $H^*(T)$ is determined from conservation of the number of particles n under the condition that the chemical potential coincides with the mobility edge:

$$\int_{E_c}^{\infty} \frac{\tilde{N}_0(\varepsilon, H^*)\,d\varepsilon}{\exp[(\varepsilon - E_c)/T] - 1} = n\left[1 - \left(\frac{T}{T_{c0}}\right)^{3/2} - \frac{n_L(T)}{n}\right],$$ (42)

where $n_L(T)$ is the number of localised bosons.

The left-hand side of Eq. (42) is the number of bosons on the lowest Landau level, while the second term on the right-hand side is the number of bosons on all upper Landau levels, calculated with the classical density of states, Eq. (41).

Substitution of Eq. (39) into Eq. (42) yields the final expression for the critical field of Bose–Einstein condensation:

$$H^*(T) = H_d\left(\frac{T_{c0}}{T}\right)^{3/2}\left[1 - \left(\frac{T}{T_{c0}}\right)^{3/2} - \frac{Tn_L}{\gamma n}\beta\left(\frac{T}{\gamma}\right)\right]^{3/2},$$ (43)

with

$$\beta(x) = \sum_{k=1}^{\infty} \frac{(-1)^k}{x+k}$$ (44)

and temperature-independent $H_d = \phi_0/(2\pi\xi_0^2)$. The 'coherence' length ξ_0 is determined by both the mean free path $l = (4\pi n_{im}f^2)^{-1}$ and the inter-particle distance:

$$\xi_0 \simeq 0.8(l/n)^{1/4}.$$ (45)

$\phi_0 = \pi/e$ is the flux quantum.

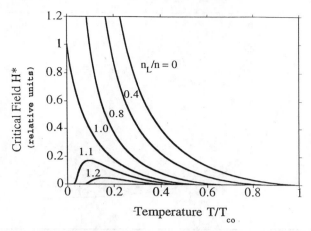

Fig. 7. The temperature-dependence of the critical magnetic field of Bose–Einstein condensation (in units of $H_d(T_{c0}\ln 2/\gamma)^{3/2}$) for the different relative number of localized states n_L/n and $\gamma/T_{c0}=0.2$.

The localisation does not change the positive '3/2' curvature of the critical magnetic field near T_c[17]. I believe that this curvature is a universal feature of a charged Bose gas, which depends neither on a particular scattering mechanism nor on the approximations made. The number of bosons at the lowest Landau level is proportional to the density of states near the mobility edge $\tilde{N}_0$ $\sim H/\sqrt{\Gamma(H)}$, where the 'width' of the Landau level is also proportional to the same density of states $\Gamma(H)\sim H/\sqrt{\Gamma(H)}$. Hence $\Gamma(H)\sim H^{2/3}$ and the number of condensed bosons is proportional to $H^{2/3}$. On the other hand, this number should in the vicinity of T_c be proportional to T_c-T (the total number minus the number of thermally excited bosons). That gives the '3/2' law for H^*.

At low temperature $T\ll\gamma$ the temperature-dependence of H^* turns out to be drastically different for different impurity concentration, Fig. 7.

An upward curvature of H_{c2} near T_c has been observed in practically all superconducting oxides, including cubic ones. The transition in a magnetic field for a wide temperature range starting from the millikelvin level up to T_c has recently been reported by Mackenzie *et al.* [21]. Resistively determined H_{c2} values from $T/T_c=0.0025$ to $T/T_c=1$ in a $T_c=20$ K single crystal of $Tl_2Ba_2CuO_6$ follow a temperature-dependence that is in good qualitative agreement with the type of curve shown in Fig. 7 for $n_L/n\simeq1$, Fig. 8.

Osofsky *et al.* [22] also observed a divergent upward temperature-dependence of the upper critical field $H_{c2}(T)$ for thin BSCO films, which was five times that expected for a conventional superconductor at the lowest temperature. To describe the data they applied the formula Eq. (43).

A charged Bose liquid of bipolarons turns out to be a simple but far-

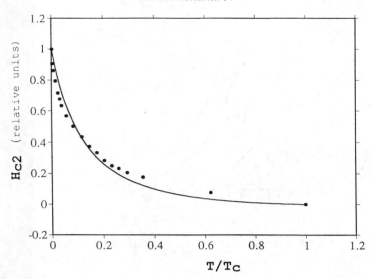

Fig. 8. The upper critical field of $Tl_2Ba_2CuO_6$ [21] compared with the theoretical curve Eq. (43) for $n_L = n$ and $\gamma/T_{c0} = 0.2$.

reaching model of $H_{c2}(T)$ of both Bi-Sr-Cu-O and Tl-Ba-Cu-O high-T_c superconductors.

7 Conclusion

In recent years several experimental results have been shown to be consistent with the existence of small (bi)polarons in high-T_c superconductors: anomalous near- and mid-infrared optical absorption [23–26], temperature-dependence of the vibration energy of atoms engaged in polaron formation [27] and EXAFS results on the radial distribution function of apical oxygen ions [28] interpreted as evidence for inter-site pairs [29,30]. Some other features of high-T_c copper oxides like the anomalous isotope effect [31], the λ-like heat capacity [32], the power-law London penetration depth [33] and the 'T_c–n/m^*' plot [34] are also characteristic of the bipolaron Bose liquid.

In this paper I have shown that extension of the BCS theory to the strong-coupling limit and Anderson localisation of bipolarons give a quantitative explanation of many features of low-frequency spin and charge kinetics of high-T_c superconductors both in the normal and in the superconducting states.

Acknowledgements

I am grateful to Sir Nevill Mott for his support. His ideas about the localisation of charged bosons laid the groundwork for elaborating the normal state kinetics of copper-based oxides in our joint publications. The author thanks A. Bratkovsky, J. Cooper, R. Haydock, M. Kaveh, D. Khmelnitskii, W. Liang, G. Lonzarich, J. Loram, A. Mackenzie, E. Salje and J. Wheatley for stimulating discussions and appreciates the financial support of the Leverhulme Trust.

References

[1] A. S. Alexandrov, *Phys. Rev.* B **46**, 2838 (1992).
[2] A. S. Alexandrov and J. Ranninger, *Phys. Rev.* B **23**, 1796–801 (1981); **24**, 1164–9 (1981).
[3] A. S. Alexandrov and P. E. Kornilovich, *Z. Phys.* B **91**, 47 (1993).
[4] D. Emin, *Phys. Rev. Lett.* **62**, 1544 (1989); D. Emin and S. Hillery, *Phys. Rev.* B **39**, 6575 (1989).
[5] G. Verbist, F. M. Peeters and J. T. Devreese, *Phys. Rev.* B **43**, 2712 (1991).
[6] F. Bassani, M. Geddo, G. Iadonisi and D. Ninno, *Phys. Rev.* B **43**, 5296 (1991).
[7] A. S. Alexandrov and N. F. Mott, Invited paper at the MOS Conference, Eugene, Oregon (July (1993)); *Supercond. Sci. Technol.* **6**, 215 (1993).
[8] A. S. Alexandrov, *J. Low Temp. Phys.* **87**, 721–9 (1992); *Physica* C **182**, 327–32 (1991).
[9] J. Rossat-Mignod, L. P. Regnault, P. Bourges, C. Vettier, P. Burlet and J. Y. Henry, *Physica Scripta* **45**, 74–80 (1992).
[10] H. A. Mook, M. Yethiraj, G. Aeppli, T. E. Mason and T. Armstrong, *Phys. Rev. Lett.* **70**, 3490 (1993).
[11] N. F. Mott, *Physica* C **205**, 191 (1993).
[12] A. S. Alexandrov, A. M. Bratkovsky and N. F. Mott, *Phys. Rev. Lett.* **72**, 1734 (1994).
[13] A. Carrington, D. J. C. Walker, A. P. Mackenzie and J. R. Cooper, Preprint IRC (1992).
[14] T. Ito, K. Takenaka and S. Uchida, *Phys. Rev. Lett.* **70**, 3995 (1993).
[15] B. Bucher, P. Steiner, J. Karpinski, E. Kaldis and P. Wachter, *Phys. Rev. Lett.* **70**, 2012 (1993).
[16] N. F. Mott, *Phil. Mag. Lett.* **68**, 245 (1993).
[17] A. S. Alexandrov, Doctoral thesis, Moscow Engineering Physics Institute, Moscow (1984); A. S. Alexandrov, J. Ranninger and S. Robaszkiewicz, *Phys. Rev.* B **33**, 4526 (1986).
[18] A. S. Alexandrov, D. A. Samarchenko and S. V. Traven, *Zh. Eksp. Teor. Fiz.* **93**, 1007 (1987) (*Sov. Phys. JETP* **66**, 567 (1987)).
[19] A. S. Alexandrov, *Phys. Rev.* B **48**, 10571 (1993).
[20] E. M. Lifshitz and L. P. Pitaevskii, *Statistical Physics. Part 2*, Pergamon Press (1980), p. 117.
[21] A. P. Mackenzie, S. R. Julian, G. G. Lonzarich, A. Carrington, S. D. Hughes, R. S. Liu and D. C. Sinclair, *Phys. Rev. Lett.* **71**, 1238 (1993).
[22] M. S. Osofsky *et al.*, *Phys. Rev. Lett.* **71**, 2315 (1993)
[23] D. Mihailovic, C. M. Foster, K. Voss and A. J. Heeger, *Phys. Rev.* B **42**, 7989 (1990).
[24] G. A. Thomas, D. H. Rapkine, S. L. Cooper, S.-W. Cheong, A. S. Cooper, L. F. Schneemeyer and J. V. Waszczak, *Phys. Rev.* B **45**, 2474 (1992).
[25] X. X. Bi and P. C. Eklund, *Phys. Rev. Lett.* **70**, 2625 (1993).
[26] A. S. Alexandrov, A. M. Bratkovsky, N. F. Mott and E. K. H. Salje, *Physica* C **215**, 359 (1993).

[27] H. A. Mook, M. Mostoller, J. A. Harvey, N. W. Hill, B. C. Chakoumakos and B. S. Sales, *Phys. Rev. Lett.* **65**, 2712 (1990).

[28] J. Mustre de Leon, S. D. Conradson, I. Batistic and A. R. Bishop, *Phys. Rev. Lett.* **65**, 1675 (1990).

[29] J. Mustre de Leon, I. Batistic, A. R. Bishop, S. D. Conradson and S. A. Trugman, *Phys. Rev. Lett.* **68**, 3236 (1992).

[30] J. Ranninger and U. Thibblin, *Phys. Rev.* B **45**, 773 (1992).

[31] A. S. Alexandrov, *Phys. Rev.* B **46**, 14 932 (1992).

[32] A. S. Alexandrov and J. Ranninger, *Solid State Commun.* **81**, 403–6 (1992).

[33] A. S. Alexandrov and S. V. Traven, *Pis'ma Zh. Eksp. Teor. Fiz.* **48**, 426 (1988) (*JETP Lett.* **48**, 468 (1988)).

[34] Y. J. Uemura *et al., Nature* **353**, 605 (1991).

4

Spin polarons in high-T_c superconductors

R. F. WOOD

Solid State Division, Oak Ridge National Laboratory, Oak Ridge, TN 37831-6032, USA

Abstract

The spin-polaron concept is introduced in analogy to ionic and electronic polarons and the assumptions underlying the author's approach to spin-polaron-mediated high-T_c superconductivity are discussed. Elementary considerations about the spin-polaron formation energy are reviewed and the possible origin of the pairing mechanism illustrated schematically. The electronic structure of the CuO_2 planes is treated from the standpoint of antiferromagnetic band calculations that lead directly to the picture of holes predominantly on the oxygen sublattice in a Mott–Hubbard/charge transfer insulator. Assuming the holes to be described in a Bloch representation but with the effective mass renormalized by spin-polaron formation, equations for the superconducting gap, Δ, and transition temperature, T_c, are developed and the symmetry of Δ discussed. After further simplifications, T_c is calculated as a function of the carrier concentration, x. It is shown that the calculated behavior of $T_c(x)$ follows the experimental results closely and leads to a natural explanation of the effects of under- and over-doping. The paper concludes with a few remarks about the evidence for the carriers being fermions (polarons) or bosons (bipolarons).

1 Introduction

A carrier (electron or hole) moving through an ionic lattice will induce displacements of the ions and, under certain conditions, the carrier plus ionic displacements may form a good quasi-particle, i.e., the ionic polaron [1]. Similarly, electronic polarons may form when the carriers induce polarization of localized or quasi-localized electronic distributions. In an analagous manner, a spin polaron is a spin $\frac{1}{2}$ carrier moving in a magnetic medium accompanied by deviations of localized ionic spins. An ionic bipolaron consists of two

45

polarons that move through the lattice bound together on the same or nearby lattice sites. The energy associated with ionic displacements of the bound pair is enough to overcome the repulsive electrostatic energy. The possibility that ionic bipolarons can lead to superconductivity has been discussed at length in the literature [2].

The question of whether or not pairing of spin polarons might also lead to superconductivity arises and is of obvious relevance to the high-T_c superconductors because of the long-range antiferromagnetic (AF) ordering of Cu spins in the CuO_2 planes of the parent materials. While the introduction of holes into the CuO_2 planes by doping, nonstoichiometry, etc. destroys the long-range AF ordering, short-range order persists and may serve as a basis for superconducting pairing. The assumption that a strong component of k-space pairing is involved leads to an approach like that developed by the author and his co-workers [3]. On the other hand, the assumption that the spin polarons (and accompanying lattice distortions) are tightly bound in real space to form bosons leads to the approach studied extensively by Mott and collaborators [4] at Cambridge. In this paper, the first approach is considered and some of the work done on it by the author and his collaborators is discussed.

Of course, numerous papers [5] in which magnetic interactions in one way or another form the basis for high-T_c superconductivity have now appeared in the literature. Some of these invoke a 'spin' or 'magnetic' polaron approach while others specifically reject it. Confusion often arises because of the considerable variance in the way in which these terms are used. Crucial to the concept of spin polarons, as it is used here, is that one or more spin flips should be involved so that a small ferromagnetic region may accompany the carrier. In contrast, many of the papers [6] in the literature start from some form of a two-dimensional Hubbard model, which at half-filling generates an antiferromagnetic ground state. A carrier (electron or hole) is then added and the spin density distribution recalculated. The carrier plus the change in spin density is referred to as either a spin or a magnetic polaron. Since this approach does not allow for actual spin flips, it does not generate the type of spin polaron to be discussed here. Perhaps it would be better to call these polarons 'ferrons' as suggested by Nagaev (see below).

The following assumptions form the basis for the present discussion: (1) antiferromagnetic band calculations for the undoped materials yield a generalized Mott–Hubbard (M–H) or charge-transfer insulator; (2) doping (e.g., by Sr in $La_{2-x}Sr_xCuO_4$) produces holes at the top of the M–H valence band; (3) the spins of the holes polarize the localized spins on neighboring ions to form spin polarons, thereby destroying the long-range AF ordering of the undoped crystal; (4) to a first approximation the spin polaron states are given by AF band calculations, but with the polaron effective mass; and (5) pairing occurs because spin deviations of the individual polarons interact antiferro-magnetically to lower the energy. Pairing occurs when holes are at restricted positions with respect to one another, consistent with the short-range order.

The first two assumptions seem well established by experiment, while the last three are controversial. Although the long-range AF order is indeed destroyed by doping, and spin-polaron formation will cause this, other mechanisms may operate and it is not at all clear how best to formulate the problem theoretically. Likewise, it is not evident that a band description is appropriate if the spin polarons do form or that an AF band calculation is an appropriate starting point for a heavily doped material. Finally, although a potential spin-polaron-mediated pairing mechanism can be identified, rigorous theoretical treatment is exceedingly difficult. In view of these limitations, the approach here will often be intuitive and schematic.

The plan of the paper is as follows. In the next section, the spin-polaron concept is reviewed and a simplified calculation of the formation energy is given. Spin-polaron pairing in one- and two-dimensions is discussed schematically. In Section 3 parameterized AF band calculations for the CuO planes are illustrated, dispersion curves are given, and Fermi surfaces sketched. In Section 4, equations for the superconducting gap, Δ, and the transition temperature, T_c, are set up and solved. The various symmetries of Δ that follow from the treatment are illustrated. Section 5 is devoted to discussions of some of the results and in the final section a summary and some concluding observations on the polaron/bipolaron question are made.

2 Spin polarons

A brief overview of the development of the spin-polaron concept is of interest. In 1951, Zener [7] discussed a 'double exchange' mechanism in which an effective exchange interaction between two localized spins in a transition metal oxide is mediated by an intervening O ion. It occurs because, in addition to configurations in which the O 2p shell is closed to give O^{2-}, it is also necessary to include configurations in which an electron from an O^{2-} ion is placed in a degenerate d orbital on the transition metal site. A few years later (1955), Anderson and Hasegawa [8] extended the work of Zener and demonstrated that double exchange is fundamentally different from the ordinary $S_i \cdot S_f$ Heisenberg-type exchange.

The double-exchange mechanism was first applied to certain types of antiferromagnetic crystals by de Gennes [9] (1960), who recognized that it led to polarization of the localized spins of the lattice by the spin of the mobile carrier. In 1971, Nagaev [10] studied mobile ferromagnetic clusters induced by carriers in semiconducting antiferromagnets and introduced the terms 'magnetic polarons' and 'ferrons.' In the same year, Mott and Davis [11] provided a clear and concise discussion of the spin-polaron concept in different types of magnetic semiconductors. Brinkman and Rice [12] studied the electronic band states in doped magnetic insulators but they rejected the idea of spin-polaron formation. Nagaev in 1974 and again in 1992 has provided review articles [13] on his approach to spin polarons. Extensive references to other work in this area can be found in the papers cited above.

Evidence for spin polarons is not difficult to find experimentally. For example, Mott [14] cites the work of Shapira *et al.* [15] who measured the temperature-dependence of the resistivity, ρ, of EuTe as a function of magnetic field. With no magnetic field on the sample, ρ was very high at low T but decreased rapidly with increasing T. With a field of 84 kOe on the sample, $\rho(4\,\mathrm{K})$ dropped more than two orders of magnitude. The interpretation is that, with a strong magnetic field applied, spin polarons with their high effective mass cannot form since the localized Eu spins are ferromagnetically aligned. Torrance *et al.* [16] suggested that the metal–insulator transition in Eu-rich EuO is due to formation of bound magnetic polarons. They argued that, above T_c the *ferromagnetic* ordering temperature of the Eu spins, an electron associated with an O vacancy can polarize neighboring spins, thereby increasing its binding energy to the vacancy. Below T_c, when the Eu spins are ferromagnetically aligned spin polarons cannot form and the energy of the electron is raised to the point at which it is no longer bound to the vacancy and the material becomes metallic-like.

The formation of spin polarons is readily understood from Fig. 1a, which shows a 1D array of AF-aligned, localized spins. These spins are assumed to occupy nondegenerate spatial orbitals. Let one additional ↑-spin electron be added to this array at site 1. When this electron is at a ↓-spin site it can occupy the $3\mathrm{d}(x^2-y^2)$ orbital to form a d^{10} configuration, but it cannot hop to the neighboring ↑-spin at site 2 unless it goes into a highly excited state. Note that the on-site Coulomb repulsion term is the same for each site. Since it is energetically unfavorable to go into the excited state, there is a strong tendency for the electron to be localized at ↓-spin site 1. However, if the energy to flip an ↑-spin to a ↓-spin orientation (dashed below) is small, then the electron may gain more than enough delocalization energy (reduced kinetic energy) to compensate for the spin-flip energy. Consequently, within some small region the spins of the ions may become F aligned with one another and AF aligned with the spin of the added electron to form the spin polaron. The spin flip at site 2 allows the electron to spread over sites 1–3 in Fig. 1a. Two points should be

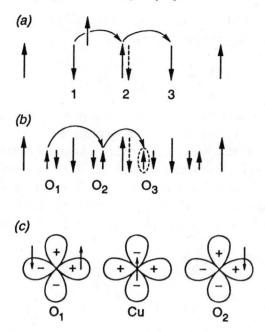

Fig. 1. A schematic illustration of spin polaron formation (a) in an AF array of local spins and (b) in a linear CuO array. If a Cu spin is reversed, as indicated by the dashed arrows, then an added carrier can decrease its kinetic energy by becoming more delocalized. The dashed ellipse in (b) indicates the creation of a hole by removing an electron on an oxygen site. In panel (c), the ↑-spin electron in the $2p\sigma$ orbital on O$_1$ can be shared with the Cu and O(2) sites if the Cu spin is reversed from ↑ to ↓.

emphasized here. First, the interaction that produces the spin alignment is a kinetic energy effect and is not directly related to any particular quantum mechanical exchange integral. Second, with reference to Fig. 1a, complete initial AF alignment of the localized spins is unnecessary for formation of spin polarons since the tendency for alignment of the spins of the carrier and of the ions will persist even when the localized spins are disordered.

2.1 Formation energy

Following ionic polaron theory, let us consider a 3D analogue of Fig. 1a and express the kinetic energy associated with confinement of the carrier to a region of radius R as $\pi^2/(2m^*R^2)$, with m^* the band effective mass. Within this region there are $(4\pi/3)(R/a^*)^3$ spins, where a^* is the nearest neighbor distance in the magnetic unit cell. Half of these spins must be reversed at a cost in energy of J_N per reversed spin. The gain in delocalization energy, $\mathscr{E}_d$, is zI where z is the coordination number and I is the hopping integral in a simple band calculation. The total energy of the spin polaron is

$$\mathscr{E}(R) = \frac{\pi^2}{2m^*R^2} + \frac{1}{2}\left(\frac{4\pi}{3}\right)\left(\frac{R}{a^*}\right)^3 J_N - zI. \qquad (1a)$$

The last term on the right-hand side does not depend on R because only a single electron is involved. At any one site the delocalization energy is weighted by the probability that the electron is there, but then a spatial integration over R must be made, by which zI is recovered. The radius, R_p, of the polaron is found by minimizing $\mathscr{E}(R)$. If the polaron is to form, $\mathscr{E}(R_p)$ must be negative. A rough estimate of the effective mass of the spin polaron, m_p^*, can be obtained from an equation given by Mott and Davis [11,14], i.e., $m_p^* = m^* \exp(\gamma R_p/a^*)$ in which $\gamma \simeq 1$.

The 2D analog of Eq. (1a) is

$$\mathscr{E}(R) = \frac{\pi b^2}{2m^*R^2} + \frac{1}{2}\pi\left(\frac{R}{a^*}\right)^2 J_N - \mathscr{E}_d \qquad (1b)$$

in which $\mathscr{E}_d$ is the above-mentioned delocalization energy (corresponding to zI in Eq. (1a)) and b is a numerical factor due to confinement of the carrier in two dimensions. Considerations of space preclude a discussion of $\mathscr{E}_d$ here, but estimates indicate that it is about 1–2 eV, large enough to cause the spin polarons to form.

To apply these ideas to holes in the CuO_2 planes we again consider a 1D array, as in Fig. 1b. This shows the same array as that in Fig. 1a, but with doubly occupied O^{2-} ions (indicated by the short arrows) inserted between the Cu sites. At the O_3 site, the ↑-spin electron has been removed to form a hole. For the hole at O_3 to move to O_2, an ↑-spin electron at O_2 must move to O_3. If no spin flip were to occur, it would be necessary for the electron to go into a highly excited state on the intervening Cu site already occupied by an ↑-spin $3d(x^2-y^2)$ electron. However, if a spin flip were allowed, thus forming a spin polaron, then the hole may spread over three sites, as indicated. In Fig. 1c the same effect is represented in terms of atomic orbitals. If the Cu spin is reversed, the hole on O_2 can be shared with O_1. The difficulty with the real case of the CuO_2 planes is that overlapping of the O 2p orbitals on different sites greatly complicates the situation.

2.2 Pairing of spin polarons

The basic idea behind the spin-polaron pairing mechanism of [3] is illustrated in Fig. 2, which shows schematic representations of the wave functions of two

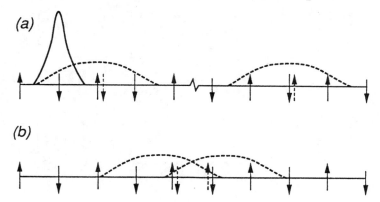

Fig. 2. On the left-hand side of panel (a), wave functions of an added ↑-spin electron before (solid curve) and after (dashed curve) spin-polaron formation are illustrated. The second dashed curve to the right shows a ↓-spin polaron. In panel (b) the spin polarons are allowed to interact. Pairing will occur if the energy associated with the recovery of the AF alignment is sufficient to overcome any increase in Coulomb repulsion energy *relative to the unpaired configuration.*

electrons. The solid curve in Fig. 2a shows the wave function for an electron before the spin reversal, while the dashed curves hold after the spin reversals indicated by the dashed arrows. It can be understood from the form of the wave functions how the kinetic energy of an added electron is reduced by reversing one of the localized spins. The reversal of more spins would reduce the kinetic energy further but the increase in spin-reversal energy rapidly limits the process. Let us assume that each polaron has only one reversed spin associated with it. Then two polarons may experience an attractive interaction if their associated spin reversals re-establish an AF alignment. How this might come about is ilustrated in Fig. 2b, where the polarons are arranged so that their accompanying spin reversals are adjacent to one another. For this to happen, the lowering of the energy produced by the indicated AF alignment (dashed arrows) must be greater than any increase in kinetic and Coulombic energy associated with the alignment.

We turn now to a consideration of the corresponding effects in the CuO$_2$ plane and consider the schematic diagrams of Fig. 3. It was shown in [3] that a hole (dashed ellipse) on an O site will always tend to flip a spin, as indicated in the first array. The flipped spin leads to a ferromagnetic 'defect,' denoted by F, accompanying each hole. The energy $E(F)$ of this defect corresponds to the second term on the right-hand side of Eq. (1), while $E(F_h)$ corresponds roughly to the other terms. Hence, the 1D polaron energy is given here by

$$W_p(1D) = E(F_h) + E(F).$$
(2)

(a)

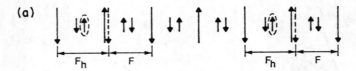

(b)

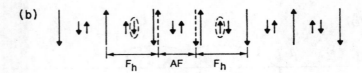

(c)

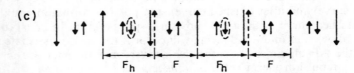

Fig. 3. A schematic diagram of 1D pairing configurations. Panel (a) shows two well-separated holes and the induced Cu spin deviations, (b) two holes forming a singlet at 2nn separation, (c) triplet pairing at 2nn, F (AF) stands for ferromagnetic (antiferromagnetic) coupling with no hole on the bridging O^{2-} ion, F_h stands for F-coupling with a hole on the O^{2-} ion.

In the array in Fig. 3b it is seen that, when the two holes are at a second nearest neighbor (2nn) distance with respect to one another in a singlet configuration, the ferromagnetic defects accompanying the individual holes are eliminated or 'healed' and the AF pairing of the Cu spins regained. Neglecting any change in Coulomb repulsion energy due to pairing, the 1D hole-pairing energy $V(1D)$ is

$$V(1D) = 2[E(F_h) + E(F)] - 2E(F_h) = 2E(F). \tag{3}$$

$E(F)$ is related to the spin-wave exchange parameter J by $E(F) = J/2$ so that

$$V(1D) = J. \tag{4}$$

For a 2nn triplet alignment of the hole spins, the array in Fig. 3c, there is no repair and hence no pairing energy.

Similar considerations [3] hold in two dimensions. In this case, the energy for an isolated polaron is given by

$$W_p(2D) = E(F_h) + 3E(F) \tag{5}$$

and the energy of the interacting pair by $2[E(F_h) + 2E(F)]$, but the pairing energy is again found to be $V(2D) = 2E(F) = J$.

2.3 *Formal theories of spin polarons*

Detailed consideration of formal theories of spin polarons is outside the scope of this paper, but the analogy with small ionic polarons, suggested by some of the above results, warrants a few remarks.

Following work of Sewell [17] on ionic polarons, the Hamiltonian for a system consisting of an extra carrier added to an AF insulator composed of localized spins is

$$\mathcal{H} = H_c + H_{mag} + H_{int} \tag{6}$$

in an obvious notation. As with the phonon case, a canonical transformation can be carried out that eliminates the first term of H_{int} to first order and transforms H_c to the Hamiltonian of a *localized* spin polaron. The transformed Hamiltonian can be written as

$$\mathcal{H} = \varepsilon \sum_{\mu} c_{\mu}^{\dagger} c_{\mu} + H_{mag} + \sum_{v \neq \mu} c_{\mu}^{\dagger} c_{v} \Phi_{\mu v}(a, a^{\dagger}), \tag{7}$$

where ε, $c^{\dagger}$ and c now refer to the spin polarons instead of the bare carriers. $\Phi_{\mu v}(a, a^{\dagger})$ is a complicated function in which are embedded a number of effects including the renormalization of the magnetic excitations. From it can be extracted terms that give a renormalized transfer integral, $\tilde{I}_{\mu v}$, which involves a thermal average of the expectation values of products of spin-excitation operators, i.e., a spin–spin correlation function. This correlation function will presumably depend not only on temperature but also on the large fluctuations inherent to a quasi-2D system and on the presence of the spin polarons, which tend to destroy the AF ordering. Thus the effect of $\tilde{I}_{\mu v}$ is to increase the effective mass of the spin polaron over the band effective mass (determined by $I_{\mu v}$) and make it temperature-dependent.

A second approach to a formal theory, developed extensively by Nagaev [10,13], involves a transformation to a 'spin polaron' representation. It is assumed that, when an electron is located on a magnetic ion, the intra-atomic exchange energy is much greater than the width of the conduction band and both of these are much greater than the exchange energy between the spins of the magnetic ions. As a consequence, it is useful to transform to a representation in which the intra-atomic exchange part of the Hamiltonian is diagonalized to generate a set of energies and eigenfunctions. A basis set for the expansion of the wave function of the entire Hamiltonian is then constructed from products of these eigenfunctions and spin eigenfunctions of the remaining magnetic ions of the lattice. This approach, which is very complex and will not be discussed further here, may serve as a useful starting point for a theory of spin polarons in the high-T_c materials.

3 Electronic structure of the CuO$_2$ planes

Many band and cluster calculations for the copper-oxide superconductors have now been published. Most of them were carried out within the framework of conventional (i.e., non-spin-polarized) band theory, usually employing a local charge density approximation. A few spin-polarized calculations, almost invariably with a local *spin* density approximation, have also appeared. These and other calculations relating to the electronic structure of the high-T_c materials have been reviewed by Pickett [18]. Unfortunately, the 'first principles' spin-polarized calculations give quite conflicting results about the observed AF ordering within the CuO$_2$ planes. Conventional band calculations fail because they do not allow for strong on-site correlations at the Cu sites so that the bands are doubly occupied, leaving one half-filled band in which the Fermi level falls. Consequently, the CuO$_2$ planes are metallic and do not have the AF ordering that is observed. The few spin-polarized calculations that have appeared usually work with a local spin density approximation without gradient corrections and this is evidently inadequate. Of course, to do the band calculations correctly is very difficult and challenging.

Here, calculations for a CuO$_2$ plane using the type of approach introduced by Slater [19] to study antiferromagnetism in 3d metals are discussed. In such calculations, the chemical unit cell (CUC) is doubled to form a magnetic cell (MUC), which contains one magnetic ion with predominantly up spin and one with predominantly down spin; correspondingly, the chemical Brillouin zone (CBZ) is halved to form the MBZ. In the 3d metals considered by Slater, there was a direct interaction between the up and down spin sites via an effective exchange interaction. In our case, the magnetic ions interact only indirectly through the intervening oxygen ions (superexchange) and the occurrence of up and down spins on the same Cu site is made energetically unfavorable by the introduction of a Hubbard U term. The effect is to modify the density of states (DOS) that would be obtained from a conventional band calculation and, in particular, to open a gap so that a material such as stoichiometric La$_2$CuO$_4$ is changed from a metal to an insulator (or semiconductor).

The basis set for the calculations consists of the 3d$(x^2 - y^2)$ orbital on the two Cu atoms and the 2px and 2py orbitals on the four O atoms. With each of these spatial orbitals goes both an up- and a down-spin function to generate a basis of twenty spin orbitals. However, since there are no spin-mixing terms in the Hamiltonian, the α and β functions do not mix and the matrix equation will be 10×10 for each spin orientation.

The Hamiltonian underlying the band calculations is

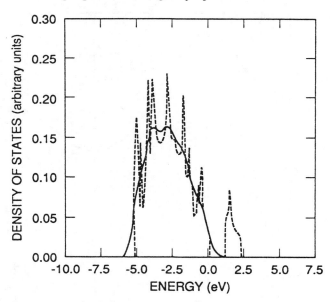

Fig. 4. The total density of states in a CuO$_2$ plane. The solid curve gives the DOS in the occupied MH valence band convoluted with a Gaussian of 0.8 eV FWHM.

$$H = \sum_{i,j,s} \varepsilon_{ig} C^\dagger_{is} C_{js} + \frac{1}{2} \sum_{\substack{i,j \\ s,s'}} U_{is,js'} C^\dagger_{is} C_{is} C^\dagger_{js'} C_{js'}. \qquad (8)$$

Diagonal terms giving the energies of the Cu 3d$(x^2 - y^2)$ and the O 2pσ and 2pπ orbitals were retained from the first sum; these are ε(d,d), $\varepsilon(\sigma,\sigma)$, and $\varepsilon(\pi,\pi)$, respectively. Four off-diagonal terms, denoted by ε(d,σ'), $\varepsilon(\sigma,\sigma')$, $\varepsilon(\pi,\pi')$ and $\varepsilon(\sigma,\pi')$, were also kept. The first of these is just the hopping integral between the d orbital and the nearest neighbor O 2pσ orbitals and results in hybridization of the B$_1$ symmetry (C_{4v} point group) orbitals at the Γ point of the BZ. Finally, there is a Hubbard term that gives the repulsive energy when an α-spin electron is located on a β-spin Cu site; since this is the only such term retained explicitly, it is referred to simply as U.

The DOS obtained in a typical calculation is shown by the dashed curve in Fig. 4. The M–H gap beginning at about 0.05 eV closes when $U = 0$ and density is shifted toward the lower energy states. Since the α- and β-spin bands are completely degenerate, the Fermi level is located by filling each set of bands with half the total number of electrons. The Fermi level then falls at the bottom of the M–H gap, making La$_2$CuO$_4$ a semiconductor, as observed; with $U = 0$ it would be a metal. Also, with $U = 0$ there is no magnetic moment on the Cu ions, whereas moments close to the value of $(0.48 \pm 0.15)\mu_B$ reported in [20] are found

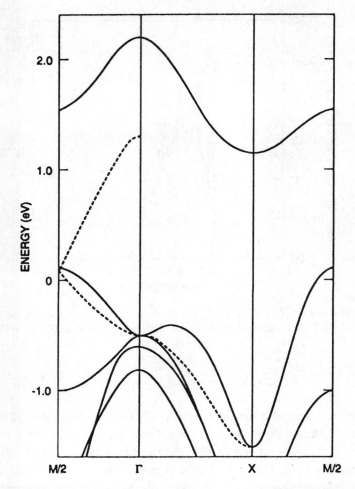

Fig. 5. Typical dispersion curves from the AF band calculations. The M–H gap at M/2 is about 1.5 eV for $U = 4$ eV. The dashed curves show the results for $U = 0$. There is no gap and the material is a metal, as found in more conventional band calculations.

for $U \simeq 3–6$ eV. The full curve in Fig. 4 shows the raw DOS data convoluted up to the Fermi level, with a Gaussian of 0.8 eV FWHM (the approximate resolution of X-ray photoemission measurements). The shape and width of the total DOS curves are in fairly satisfactory agreement with photoemission data [21] and better agreement with the width is readily obtained by small changes in the parameters, while the structure can be shifted by choice of parameters describing the various filled d bands.

Dispersion curves for a typical calculation are shown in Fig. 5. Absolute valence band maxima occur at the M/2 points with secondary maxima, about

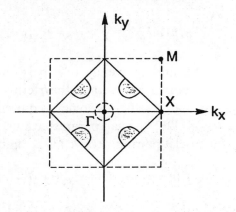

Fig. 6. The Fermi surface for doped CuO_2 planes. Approximately half-elliptical hole pockets form at the M/2 points. The dashed circular pocket about Γ comes from the O $2p\pi$ band and may or may not occur.

0.5 eV lower, along the Γ–X direction. At the M/2 maxima the square of the wave function is 0.12 $3d(x^2 - y^2)$, 0.59 O $2p\sigma$, and 0.29 O $2p\pi$. These maxima are a result of the assumed AF order. The Fermi surface to be expected from Fig. 5 is shown in Fig. 6. When $U = 0$ there is no gap and the top of the *conduction* band at Γ is mapped directly from the M point in the CBZ so that the uppermost valence band dispersion curve would monotonically increase if extended through M/2 to M. This is shown in Fig. 6 by the dashed curves.

4 Superconductivity

The reduced pair Hamiltonian [22] is given by

$$H_{\text{red}} = 2 \sum_k \varepsilon_k b_k^\dagger b_k + \sum_{kk'} V(k, k') b_{k'}^\dagger b_k, \tag{9}$$

where $V(k, k')$ is the pairing matrix element and the operators

$$b_k^\dagger \equiv c_{k\uparrow}^\dagger c_{-k\downarrow}^\dagger, \qquad b_k \equiv c_{-k\downarrow} c_{k\uparrow} \tag{10}$$

create and destroy respectively singlet Cooper pairs. The ε_k are considered to be given to a first approximation by the hole energies near the top of the valence band, renormalized by the induced spin deviations, i.e., they are the single spin polaron energies in the Bloch representation. H_{red} is commonly simplified by making a Hartree–Fock-like approximation to reduce the product of four single-particle creation and annihilation operators to a product of two. In this way, the gap parameter

$$\Delta(\boldsymbol{k}) = \sum_{\boldsymbol{k}'} V(\boldsymbol{k},\boldsymbol{k}')\langle b_{\boldsymbol{k}'}^\dagger\rangle_s = \sum_{\boldsymbol{k}'} V(\boldsymbol{k},\boldsymbol{k}')\langle c_{\boldsymbol{k}'\uparrow}^\dagger c_{-\boldsymbol{k}'\downarrow}^\dagger\rangle_s \tag{11}$$

is introduced, where $\langle\rangle_s$ indicates an average over pair states in the supercon-ducting ground state. To calculate this average, the b terms must be determined by diagonalizing H_{red}, so that a self-consistency requirement is introduced. From general considerations [23], the gap parameter must have even parity in $\boldsymbol{k}$, i.e., $\Delta(-\boldsymbol{k}) = \Delta(\boldsymbol{k})$ for singlet pairing. $V(\boldsymbol{k},\boldsymbol{k}')$ is given by

$$V(\boldsymbol{k},\boldsymbol{k}') = \int d\boldsymbol{r}_1 \, d\boldsymbol{r}_2 \, V(\boldsymbol{r}_1 - \boldsymbol{r}_2)\psi_{\boldsymbol{k}\uparrow}^*(\boldsymbol{r}_1)\psi_{-\boldsymbol{k}\downarrow}^*(\boldsymbol{r}_2)\psi_{\boldsymbol{k}'\uparrow}(\boldsymbol{r}_1)\psi_{-\boldsymbol{k}'\downarrow}(\boldsymbol{r}_2) \tag{12}$$

and we take

$$\psi_{\boldsymbol{k}}(\boldsymbol{r}) = \frac{1}{\sqrt{N}}\sum_\mu e^{i\boldsymbol{k}\cdot\boldsymbol{R}_\mu}\phi_{\boldsymbol{k}}(\boldsymbol{r} - \boldsymbol{R}_\mu). \tag{13}$$

$\boldsymbol{R}_\mu$ is a lattice vector in the magnetic lattice and $\phi_{\boldsymbol{k}}(\boldsymbol{r} - \boldsymbol{R}_\mu)$ is some linear combination of symmetrized basis functions. Substitution of Eq. (13) into $V(\boldsymbol{k},\boldsymbol{k}')$ gives an expression that can be simplified using translational invar-iance and by making the assumption that the overlap charge density of the orbitals on different sites is small compared with the site-diagonal value, i.e., $\phi_{\boldsymbol{k}}^*(\boldsymbol{r} - \boldsymbol{R}_\mu)\phi_{\boldsymbol{k}}(\boldsymbol{r} - \boldsymbol{R}_\nu) \simeq 0$ if $\boldsymbol{R}_\mu \neq \boldsymbol{R}_\nu$. Then,

$$V(\boldsymbol{k},\boldsymbol{k}') = \frac{1}{N}\sum_\mu e^{i(\boldsymbol{k}-\boldsymbol{k}')\cdot\boldsymbol{R}_\mu}$$

$$\times \int d\boldsymbol{r}_1 \, d\boldsymbol{r}_2 \, V(\boldsymbol{r}_1 - \boldsymbol{r}_2)\phi_{\boldsymbol{k}\uparrow}^*(\boldsymbol{r}_1)\phi_{-\boldsymbol{k}\downarrow}^*(\boldsymbol{r}_2 - \boldsymbol{R}_\mu)\phi_{\boldsymbol{k}'\uparrow}(\boldsymbol{r}_1)\phi_{-\boldsymbol{k}'\downarrow}(\boldsymbol{r}_2 - \boldsymbol{R}_\mu). \tag{14}$$

Results like those in Section 2 suggest that, in 2D, the singlet pairing potential is sharply peaked when the two holes are at distances a^* and $\sqrt{2}a^*$ from each other. Here we simplify by assuming that μ in Eq. (14) can run only over ions in the 1nn and 2nn unit cells of the origin. Also, $\boldsymbol{R}_\mu$ is taken to be a unit cell vector, thus neglecting the difference in phase factors between the different sites within a cell. Then, with $\boldsymbol{q} \equiv \boldsymbol{k} - \boldsymbol{k}'$,

$$V(\boldsymbol{k},\boldsymbol{k}') = \frac{V_1}{N}\sum_\mu^{1\text{nn}} e^{i\boldsymbol{q}\cdot\boldsymbol{R}_\mu} + \frac{V_2}{N}\sum_\mu^{2\text{nn}} e^{i\boldsymbol{q}\cdot\boldsymbol{R}_\mu} = V_1(\boldsymbol{k},\boldsymbol{k}') + V_2(\boldsymbol{k},\boldsymbol{k}'), \tag{15}$$

where

$$V_1(\boldsymbol{k},\boldsymbol{k}') = (2V_1/N)[\cos(q_x a^*) + \cos(q_y a^*)], \tag{16a}$$

$$V_2(\boldsymbol{k},\boldsymbol{k}') = (4V_2/N)[\cos(q_x a^*) + \cos(q_y a^*)]. \tag{16b}$$

By expanding the cosine terms, retaining only the even parity components, and projecting with functions that transform as the various irreducible representations, four pairing potentials are obtained. Specifically,

$$V_1^s(\mathbf{k}) = A[\cos(k_x a^*) + \cos(k_y a^*)] \equiv A f_1(\mathbf{k}) \qquad A_1, \qquad (17a)$$

$$V_1^d(\mathbf{k}) = B[\cos(k_x a^*) - \cos(k_y a^*)] \equiv B f_2(\mathbf{k}) \qquad B_1, \qquad (17b)$$

$$V_2^s(\mathbf{k}) = C_{xy}\cos(k_x a^*)\cos(k_y a^*) \equiv C_{xy} f_3(\mathbf{k}) \qquad A_1, \qquad (17c)$$

$$V_2^d(\mathbf{k}) = S_{xy}\sin(k_x a^*)\sin(k_y a^*) \equiv S_{xy} f_4(\mathbf{k}) \qquad B_2, \qquad (17d)$$

where A, B, C_{xy} and S_{xy} are BZ summations. V_1^s and V_2^s transform like A_1 (s-like) of the C_{4v} group, while V_1^d and V_2^d transform like B_1 [d$(x^2 - y^2)$] and B_2 [d(xy)], respectively. Hence, the even parity part of the potential of Eq. (15) can be decomposed into four terms, each of which is separable in $\mathbf{k}$ and $\mathbf{k}'$. Thus

$$V(\mathbf{k}', \mathbf{k}) = N^{-1} \sum_{n=1}^{4} \bar{V}_n f_n(\mathbf{k}') f_n(\mathbf{k}), \qquad (18)$$

with $\bar{V}_1 = \bar{V}_2 = V_1$ and $\bar{V}_3 = \bar{V}_4 = 4V_2$. The integral equation for the gap function [22] is

$$\Delta(\mathbf{k}) = \sum_{k'} \frac{V(\mathbf{k}, \mathbf{k}')\Delta(\mathbf{k}')}{2E(\mathbf{k}')} \tanh\left(\frac{\beta E(\mathbf{k}')}{2}\right), \qquad (19)$$

where $\beta = 1/(k_B T)$ and

$$E(\mathbf{k}') = [(\varepsilon(\mathbf{k}') - \mu)^2 + \Delta^2(\mathbf{k}')]^{1/2}. \qquad (20)$$

The chemical potential, μ, is usually taken to be the same as the Fermi energy, ε_F, but this may not be a good approximation here because of the proximity of ε_F to the band edge; this was found to have a significant effect only very near the zone boundary.

Inserting Eq. (18) into Eq. (19) gives

$$\Delta(\mathbf{k}) = \frac{1}{N} \sum_n \bar{V}_n f_n(\mathbf{k}) \sum_{k'} \frac{f_n(\mathbf{k}')\Delta(\mathbf{k}')}{2E(\mathbf{k}')} \tanh\left(\frac{\beta E(\mathbf{k}')}{2}\right) \qquad (21)$$

$$\equiv \sum_n f_n(\mathbf{k})\Delta_n^0. \qquad (22)$$

If $\Delta(\mathbf{k}')$ were known, then the sum over the BZ would give the numbers Δ_n^0 so that substituting Eq. (22) back into Eq. (21) and simplifying results in the set of equations

$$\sum_m \Delta_m^0 [\delta_{nm} - D_{nm}] = 0 \tag{23}$$

or

$$\det[I - D] = 0, \tag{24}$$

with

$$D_{nm}(\Delta^0) = \frac{1}{N} \sum_k \bar{V}_n \frac{f_n(\boldsymbol{k})f_m(\boldsymbol{k}) \tanh\{\frac{1}{2}\beta[(\varepsilon(\boldsymbol{k}) - \mu)^2 + (\Sigma_l f_l(\boldsymbol{k})\Delta_l^0)^2]^{1/2}\}}{2[(\varepsilon(\boldsymbol{k}) - \mu)^2 + (\Sigma_l f_l(\boldsymbol{k})\Delta_l^0)^2]^{1/2}}. \tag{25}$$

This equation could be used to determine the Δ^0 and T_c but the nonlinearities from the summations over the Δ_l^0 can make this difficult. One sees that all of the symmetry components of the gap may be coupled so that in general the gap is not of a pure symmetry type. This is an important point to keep in mind in view of the ongoing controversy in the literature about the symmetry of the gap, states in the gap, etc.

5 Numerical results

The equations for Δ and T_c are still too complicated to solve readily. Therefore, they were further simplified along lines described elsewhere [3] but will not be discussed here except to remark that a free-electron approximation was used to describe the hole pockets of Fig. 6.

With these simplifications, the summation over $\boldsymbol{k}$ in Eq. (25) can be converted into an integral over ε (using $\varepsilon = \hbar^2 k^2/(2m_p^*)$ and $\rho = 2m_p^*\pi/\hbar^2$) to get

$$D_{nm} = \sum_j \bar{V}_m \int \frac{F_{nm}(j,\varepsilon)\rho(j,\varepsilon) \tanh(\beta E(\varepsilon)/2)\, d\varepsilon}{2E(\varepsilon)}, \tag{26}$$

where j runs over the Γ and X points in the magnetic BZ and $\rho(j,\varepsilon)$ is the corresponding DOS. Since $E(\varepsilon)$ of Eq. (20) still contains the Δ_l^0, further simplifications may be required to solve the gap equation at $T = 0$ K. However, because T_c is the temperature at which the gap goes to zero, it follows that all but the largest of the Δ_n^0 have already vanished before T is increased to T_c. Thus, at T_c, Eq. (20) reduces to $E(\boldsymbol{k}') = \varepsilon(\boldsymbol{k}') - \mu$. It should be noted that, if F_{nm} and ρ in Eq. (26) were constant in a certain range of ε about ε_F, then the elementary BCS form of the gap equation would be recovered.

In Fig. 7 some of the early data [24] for $T_c(x)$ in $La_{2-x}Sr_xCuO_4$ are shown by the open circles (Tarascon et al.) and open triangles (Shafer et al.). The fall-off in $T_c(x)$ for $x \gtrsim 0.15$ was initially attributed to formation of oxygen vacancies in material grown under low ambient O_2 pressures. However, Torrance et al. [25]

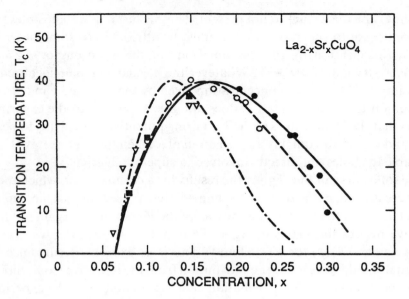

Fig. 7. $T_c(x)$ for La$_{2-x}$Sr$_x$CuO$_4$. The experimental data are from the following sources: ($\bigcirc$) Tarascon *et al.*, [24]; ($\bigtriangledown$) Shafer *et al.*, [24]; ($\blacksquare$) van Dover *et al.*, [26]; and ($\bullet$) Torrance *et al.*, [25]. The dashed curve gives the calculated results for the case of pockets at Γ and M/2 and pairing through the V_2^s potential while the dot–dashed curve is for the same potential but with no Γ pocket. The solid curve is for the case of no O 2pπ pocket at Γ and pairing in the M/2 pockets by the V_1^d potential.

established that annealing at about 100 bars of O$_2$ for long periods suppressed formation of oxygen vacancies and increased the range of accessible hole concentrations. Their data are shown in Fig. 7 by the solid circles, while the data of van Dover *et al.* [26] are given by the solid squares. These two sets, taken under similar conditions, provide $T_c(x)$ values believed (except perhaps for the solid square and circle points at $x = 0.2$ and 0.3) to be uninfluenced by O-vacancy formation.

The dashed curve in Fig. 7 gives $T_c(x)$ for a calculation in which an O 2pπ pocket at Γ begins to form shortly before O 2pσ–Cu 3d$(x^2 - y^2)$ pockets at the M/2 points (see Fig. 6) and pairing goes via the V_2^s term in the M/2 pockets. The energy off-set between the Γ and M/2 points was 0.17 eV. The band and spin-polaron effective masses were $m^* = 1.5$ and $m_p^* = 3.6m^*$ while $V_2 = 70$ meV. The value of m^* corresponds to $\rho = 0.35$ states per electron-volt for each spin direction in a magnetic unit cell. The role of the O 2pπ pocket at the Γ point in this particular calculation is to allow for the experimentally observed fact that superconductivity does not occur until $x > 0.06$.

The dot–dashed curve results from a calculation identical to the above except that the O $2p\pi$ pocket at Γ is not present. This means that the 0.06 holes that do not contribute to superconductivity must be attributed to some other origin such as localization by the random fields of the Sr atoms or *by intrinsic properties of the spin polarons*. Whatever the mechanism, 0.06 localized holes were simply added to the concentration given by the assumed band structure. The reason that this calculation does not give a good fit to the experimental data is that the form factors in the V_2^s pairing potential cause $T_c(x)$ to peak too soon and fall off too quickly as a function of x when the Γ pocket is no longer available for holes not directly involved in superconductivity.

The solid curve in Fig. 7 gives the results for a calculation in which the band structure is the same as that for the immediately preceding case, i.e., no O $2p\pi$ pocket at Γ, but the pairing goes by way of the V_1^d potential. Again holding the effective masses fixed, the strength of the potential (i.e., V_1) was varied to achieve a fit to $T_c(x)$ for $V_1 = 52$ meV. This is an expected result since the V_1 potential should have a larger contribution from the repulsive Coulombic (and kinetic energy) term. This calculation achieves a good fit to the experimental data because the smaller value of V_1 and the slower falling off of the form factor relative to that for the V_2^s potential allow a greater region of k-space to contribute to the pairing. A calculation that complements this one keeps the parameters fixed but goes back to the band structure of the first example, i.e., pockets at Γ and M/2. The results are not shown in Fig. 7 but it is easy to understand from the foregoing discussion that $T_c(x)$ peaks at a larger value of x than is observed and then falls off too slowly as x increases.

The calculations for $YBa_2Cu_3O_{7-x}$ parallel those for La_2CuO_4 but the interpretation of the results is more complicated because the distribution of holes between the CuO_3 (chain) and CuO_2 planes, especially in the well-known plateau region, is not well established and nor is the localization/mobility behavior of the holes within these units known. The details of the calculations will not be given here, but the results are quite similar to those for La_2CuO_4 as indicated in Fig. 8.

The gap at $T = 0$ K was calculated using values of the parameters that gave good fits to the $T_c(x)$ data in both materials. Not unexpectedly, it was found that the ratio $2\Delta(T=0 \text{ K})/(k_B T_c)$ gave the weak coupling value of 3.52 within the accuracy of numerical solution of the equations for Δ and T_c. Next, m_p^* was allowed to be a function of T such that it doubled on going from T_c to 0 K. This had a dramatic effect on $2\Delta(T=0 \text{ K})/(k_B T_c)$, raising it from 3.52 to 7.5 for LSCO and to 5.6 for YBCO for those values of x that gave the maximum T_c. The gap calculations indicate that the symmetry is either s-like or d-like (or some linear combination thereof) depending on whether V_2 or V_1 is larger.

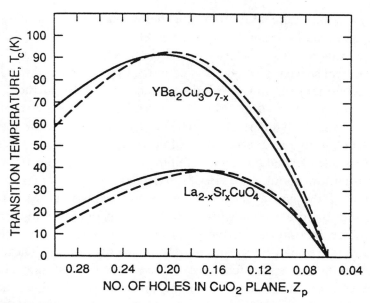

Fig. 8. The calculated $T_c(Z_p)$ for La$_{2-x}$Sr$_x$CuO$_4$ and YBa$_2$Cu$_3$O$_{7-x}$. The dashed curves are for the case of an O 2pπ pocket at Γ and O 2pσ–Cu 3d(x^2-y^2) pockets at M/2 with pairing in the latter via the V_2^s potential. The solid curves are for no Γ pocket and pairing in M/2 by the V_1^d potential.

However, even if it is s-like it may be highly anisotropic with near-zero lines in the $k_x = k_y$ directions.

6 Summary and concluding remarks

A discussion of the concept of a spin polaron and crude estimates [3] of the delocalization energy needed to form them indicated that O 2pσ–Cu 3d(x^2-y^2) polarons are likely to form in the CuO$_2$ planes with the Cu 3d^9 configuration. Schematic considerations in 1D and 2D suggested that a pairing mechanism exists because in certain configurations partial healing of the AF alignment of Cu spins can be realized. The pairing is thought to take place in *k*-space rather than real space (which would lead to spin bipolarons), partly because this would seem to lead to greater flexibility in obtaining the minimum energy configuration and partly because several types of experiments indicate the presence of a BCS-like gap. Parameterized AF band calculations were carried out in an effort to gain insight into the effects of AF order in the lattice and to study the way in which the Cu 3d(x^2-y^2), O 2pσ, and O 2pπ orbitals enter. The assumption being that even when formation of spin polarons begins to destroy the long-range AF order, the character of the electronic states will not change drastically because short-range AF order continues to exist.

The gap equation was worked out under the assumption that the pairing potential in real space is strongly peaked in the first and second neighbor cells from the reference cell, and with hole pockets in the CBZ at the Γ and M/2 points. Further simplifications, leading to a readily parameterized calculational model, were made. Numerical calculations led to good fits to the $T_c(x)$ data in $La_{2-x}Sr_xCuO_4$ using reasonable values of the parameters involved. A similar fit for YBCO is complicated by carrier trapping in the chain plane, but by forcing a fit, information about the role of carrier trapping was extracted in [3]. Calculation of the gap at $T=0$ K gave results consistent with the weak-coupling limit but did not allow us to distinguish between s and d pairing.

To conclude, let us turn to a question of particular interest, i.e., whether the charge carriers are polarons (fermions) or bipolarons (bosons). The most direct evidence that the carriers are fermions (at least in the normal state) comes from the many experiments that indicate the presence of a well-defined Fermi surface. Particularly significant are neutron scattering experiments that probe the imaginary part of the spin susceptibility $\chi(q, \omega)$. The results [27] show that there are spin-flip scattering processes occurring in $La_{1.86}Sr_{0.14}CuO_4$ for $\hbar\omega = 2$ meV at both $T = 35$ K and $T = 4.5$ K. However, the intensity at 4.5 K is greatly reduced from that at 35 K, consistent with the idea of a gap forming when the sample becomes superconducting and no gap above about 35 K. If the charge carriers were bosons, then the energy transfer, $\hbar\omega$, in the scattering process would need to be at least great enough to overcome the bipolaron binding energy even in the normal state. This is evidently not the case. The data are consistent with the traditional picture that $\chi(q, \omega)$ is associated with electron–hole pairs formed by transitions from occupied to unoccupied band states near the Fermi surface.

Acknowledgements

This research was sponsored by the Division of Materials Sciences, USA Department of Energy under contract DE-AC05-84OR21400 with Martin Marietta Energy Systems, Inc.

References

[1] Although the literature on ionic polarons is large (including many discussions in text books), to the author's knowledge there are no recent comprehensive reviews of the subject. The volume *Polarons and Excitons*, edited by C. P. Kuper and G. D. Whitfield (Oliver and Boyd, Edinburgh,

1962) remains a good source for reviews of earlier work. The discussions by Mott and Davis (N. F. Mott and E. A. Davis, *Electronic Processes in Non-Crystalline Materials* 2nd edition, Clarendon Press, Oxford, 1979 are very useful.

[2] V. L. Vinetskii, *Sov. Phys. JETP* **13**, 1023 (1961); D. M. Eagles, *Phys. Rev.* **186**, 456 (1969); S. P. Ionov *et al., Zh. Eskp. Teor. Fiz. Pis'ma* **12**, 544 (1970); S. P. Ionov *et al., Phys. Stat. Sol.* **71**, 11 (1975); *ibid.* **85**, 683 (1978); B. K. Chakraverty *et al., Phys. Rev.* B **17**, 3781 (1978); P. W. Anderson, *Phys. Rev. Lett.* **34**, 953 (1975); B. K. Chakraverty and C. Schlenker, *J. Physique* **37**, C4-353 (1976); A. Alexandrov and J. Ranninger, *Phys. Rev.* B **23**, 1796 (1981); *ibid.* B **24**, 1164 (1981); A. S. Alexandrov, *Russ. J. Phys. Chem.* **57**, 167 (1983); A. S. Alexandrov, J. Ranninger and S. Robaszkiewicz, *Phys. Rev.* B **33**, 4526 (1986).

[3] R. F. Wood, M. Mostoller and J. F. Cooke, *Bull. Am. Phys. Soc.* **33**, 260 (1988); *ibid.* **34**, 642 (1989); R. F. Wood, M. Mostoller and J. F. Cooke, *Physica* C **165**, 97 (1990); R. F. Wood and J. F. Cooke, *Phys. Rev.* B **45**, 5585 (1992).

[4] A. S. Alexandrov, A. M. Bratkovsky and N. F. Mott, *Phys. Rev. Lett.* **72**, 1734 (1994) and references therein.

[5] For early contributions see V. J. Emery, *Phys. Rev. Lett.* **58**, 2794 (1987); J. E. Hirsch, *Phys. Rev.* B **35**, 8726 (1987); P. W. Anderson, *Science* **235**, 1196 (1987); R. H. Parmenter, *Phys. Rev. Lett.* **59**, 923 (1987); G. Chen and W. A. Goddard III, *Science* **329**, 899 (1988); J. R. Schrieffer, X.-G. Wen, and S.-C. Zhang, *Phys. Rev. Lett.* **60**, 944 (1988); H. Kamimura, S. Matsuno, and R. Saito, *Solid State Commun.* **67**, 363 (1988); L. J. de Jongh, *ibid.* **65**, 963 (1988).

[6] W. P. Su, *Phys. Rev.* B **37**, 9904 (1988); H. Y. Choi and E. J. Mele, *Phys. Rev.* B **38**, 4540 (1988); W. P. Su and X. Y. Chen, *Phys. Rev.* B **38**, 8879 (1988).

[7] C. Zener, *Phys. Rev.* **82**, 403 (1951).

[8] P. W. Anderson and H. Hasegawa, *Phys. Rev.* **100**, 675 (1955).

[9] P.-G. de Gennes, *Phys. Rev.* **118**, 141 (1960).

[10] E. L. Nagaev, *Sov. Phys. Solid State* **13**, 961 (1971).

[11] N. F. Mott and E. A. Davis, *Electronic Processes in Non-Crystalline Materials* (Clarendon Press, Oxford, 1971).

[12] W. F. Brinkman and T. M. Rice, *Phys. Rev.* B **2**, 1324 (1970).

[13] E. L. Nagaev, *Phys. Stat. Sol.* B **65**, 11 (1974); *J. Magn. Magn. Mater.* **110**, 39 (1992).

[14] N. F. Mott, *Metal–Insulator Transitions* (Taylor and Francis, London, 1974).

[15] Y. Shapira *et al., Phys. Rev.* B **5**, 2647 (1972).

[16] J. B. Torrance, M. W. Shafer and T. R. McGuire, *Phys. Rev. Lett.* **29**, 1168 (1972).

[17] G. L. Sewell, In *Polarons and Excitons: Scottish Universities' Summer School – 1962*, edited by C. G. Kuper and G. D. Whitfield (Oliver and Boyd, Edinburgh, 1963); G. L. Sewell, *Phys. Rev.* **129**, 597 (1963).

[18] W. E. Pickett, *Rev. Mod. Phys.* **61**, 433 (1989).

[19] J. C. Slater, *Phys. Rev.* **82**, 538 (1951); J. C. Slater and G. F. Koster, *Phys. Rev.* **94**, 1498 (1954).

[20] D. Vaknin, S. K. Sinha, D. E. Moncton, D. C. Johnston, J. M. Newsam, C. R. Safinya and H. E. King, Jr., *Phys. Rev. Lett.* **58**, 2802 (1987).

[21] A. Fujimori, E. Takayama-Muromachi, Y. Uchida and B. Okai, *Phys. Rev.* B **35**, 8814 (1987).

[22] J. R. Schrieffer, *Superconductivity* (W. A. Benjamin, Inc., New York, 1964).
[23] G. E. Volovik and L. P. Gor'kov, *JETP Lett.* **39**, 674 (1984); *Sov. Phys. JETP*
      61, 843 (1985); E. I. Blount, *Phys. Rev.* B **32**, 2935 (1985).
[24] J. M. Tarascon, L. H. Greene, W. R. McKinnon, G. W. Hull and T. H.
      Geballe, *Science* **235**, 1373 (1987); M. W. Shafer, T. Penney and B. L.
      Olson, *Phys. Rev.* B **36**, 4047 (1987).
[25] J. B. Torrance, Y. Tokura, A. I. Nazzal, A. Bezinge, T. C. Huang and S. S. P.
      Parkin, *Phys. Rev. Lett.* **61**, 1127 (1988).
[26] R. B. van Dover, R. J. Cava, B. Batlogg and E. A. Reitman, *Phys. Rev.* B **35**,
      5337 (1987); R. M. Fleming, B. Batlogg, R. J. Cava and E. A. Reitman,
      Phys. Rev. B **35**, 7191 (1987).
[27] T. E. Mason, G. Aeppli, S. M. Hayden, A. P. Ramirez and H. A. Mook, *Phys.
      Rev. Lett.* **71**, 919 (1993).

5

The polaron scenario for high-T_c superconductors

J. RANNINGER

Centre de Recherches sur les Très Basses Températures, laboratoire associé à l'Université Joseph Fourier, C.N.R.S., BP 166, 38042 Grenoble-Cédex 9, France

Abstract

There is ample experimental evidence for localized polaronic charge carriers in high-T_c materials in the insulating phase as well as in the metallic phase at high temperatures. This would rule out *a priori* any condensation of bipolarons, since for that purpose they should be in free-particle-like states in the long-wavelength limit. Yet, provided that the localized bipolarons hybridize with a band of itinerant electrons, such a mixture of Bosons (bipolarons) and Fermion pairs (pairs of conduction electrons) can undergo an instability towards a superconducting ground state in which at high temperatures the initially localized bipolarons become superfluid upon lowering of the temperature. The experimental situation leading up to such a picture and its physical consequences are discussed.

1 Introduction

The large values of the critical temperature T_c, the small number of charge carriers together with the short coherence length, the strong dependence of T_c on n/m (n being the carrier concentration and m their mass) and the large temperature regime near T_c (with a Ginzburg temperature of $T_G \simeq 0.1$–0.01) controlled by X–Y universality strongly suggest that high-T_c superconductivity is more closely related to Bose–Einstein condensation of real-space pairs than to a BCS state of Cooper pairs. The polaronic nature of at least part of the charge carriers in these materials has been experimentally established in both the insulating and the metallic phase of these compounds. On theoretical grounds one expects small polarons to interact with each other over short distances and in a practically unretarded fashion. This makes the supposition of the existence of bipolarons – pairs of tightly bound small polarons – in these materials very natural. Given that such bipolarons exist in the metallic phase in the form of coherent band-like Bloch states, their Bose–Einstein condensation becomes conceivable, leading to what has been termed bipolaronic supercon- ductivity [1]. Following this proposition, initially stated in 1981, the feasability of bipolaronic superconductivity was subsequently thoroughly discussed in the literature. The main objection against bipolaronic superconductivity has always been the extremely small bandwidth of bipolarons, which would inevitably lead to negligibly small values for T_c. Even today this debate is not closed, and unfortunately neither on theoretical grounds nor on the basis of firm experiments can we confirm or reject the possibility of bipolaronic superconductivity. Small polarons (and this applies equally well to small bipolarons) are charge carriers that are self-trapped inside a strong and local lattice deformation that surrounds them. In the limit at which coupling between charge carriers and lattice deformation is very strong, one can envisage motion of a polaron whereby the charge carrier and the surrounding lattice deformation move together from one site to another. This leads to a strong reduction of the bare electronic band width proportional to $\exp(-\alpha^2)$ where α is the dimensionless electron–lattice coupling constant. If, on the other hand, α is of the order of unity, an abrupt but continuous changeover from small polarons to quasi-free electrons is expected. For α immediately above such a critical value, localized small polarons are still quite well-defined objects, yet their motion can in general no longer be considered to be band-like because of strong retardation between the motion of the charge carrier and that of the surrounding lattice deformation. This retardation leads to polaron hopping rates between adjacent sites that are comparable to the bare hopping rate of electrons uncoupled to the lattice. It is intuitively difficult in this case to imagine that band-like polaronic Bloch states can exist in the long-wavelength limit because of dephasing between motion of the charge carrier and that of the lattice deformation. If the charge carriers are bipolarons, then this fact is particularly detrimental because, in order for them to Bose condense, the long- wavelength excitations in the normal state should be free-particle-like, i.e. have a spectrum $\propto \hbar^2 q^2/(2m)$ for $q \to 0$, where q denotes the wave vector and m the mass of the bipolarons. Given these theoretical handicaps for the existence of bipolaronic superconductivity and the lack, up to now, of any experimental evidence of bipolaronic band states in the long-wavelength limit, we suggest here an alternative way of Bose–Einstein condensation of bipolarons. For that purpose we shall assume from the outset that bipolarons exist only as localized states, for which there is experimental evidence [2], and that, moreover, these bipolarons overlap in energy with the energies of a rather broad band of itinerant electrons into which they can decay. This picture is consistent with results obtained for intermediary electron–phonon coupling ($\alpha \simeq 1$), which is governed by sizeable fluctuations of the deformations surrounding the charge carriers, self-trapping them for part of the time and releasing them into itinerant states for the rest of the time.

After an introductory summary in Section 2 on the experimental evidence for small polarons in high-T_c superconductors we shall in Section 3 discuss the feasibility of having localized bipolarons overlap with a band of itinerant electrons.

In Section 3 we shall discuss experimental results, which possibly verify this Boson–Fermion scenario for high-T_c superconductivity. In Section 4 we discuss the circumstances under which initially localized bipolarons could become itinerant and superfluid upon lowering of the temperature. Such a picture is described by a mixture of Bosons (bipolarons) and Fermions (itinerant electrons), which model we suggested about a decade ago [3].

2 Experimental evidence for the existence of small polarons (bipolarons) in high-T_c materials

All high-T_c materials, such as cuprates, bismuthates and fullerenes, are described in their insulating parent configuration by a filled valence band separated by a charge transfer and/or Hubbard gap from an empty conduction band. Upon doping, holes are created in this valence band as soon as the doping rate exceeds a certain critical value and the materials become metallic. In the majority of cases the charge carriers are holes and hence it is the valence electrons that are responsible for metallic conduction. Doping the materials away from their undoped parent composition first of all leads to creation of localized charge carriers in small polaron states [4] and on the basis of a variety of optical and photoemission measurements we conjecture that energy levels of those polaronic states lie inside the filled valence band. For low doping, the high-T_c material hence consists of a filled valence band plus localized states of small polarons or bipolarons. On exceeding a certain critical doping rate, holes in the valence band are created, which bind additional dopant atoms. As a result, an insulator-to-metal transition takes place and the systems show high-T_c superconductivity. The fact that doping of these materials is connected above all with creation of localized polarons (or bipolarons), and only as a consequence leads to itinerant charge carriers, is a unique feature of these materials. Experimental verification of polaronic states in the metallic regime of these materials is vital to our understanding of the high-T_c phenomenon. Experimentally, polaronic states are tracked by the sizeable shifts in frequency observed in infrared absorption upon inducing carriers by photo-illumination [4]. These modes are linked to local deformation of molecular clusters, which are generally located near the dopant atoms. The large shift in frequency due to change in valency of such clusters is related to the large intramolecular deformations associated with such valence changes. The experimental verification by Raman scattering of modes in the metallic regime, with frequencies practically identical to those observed in insulating samples with photo-induced carriers, is a strong indication that polaronic states also exist in the metallic samples [5]. We hypothesize that, in the metallic regime, a dynamical charge transfer take place between the polaronic centres and the electrons in the partially emptied valence band. Such an idea is quite realistic given the characteristic features of these molecular units in the high-T_c materials that serve as centres for polaron formation. They can exist in two valence states corresponding to two distinct configurations controlled by different bond lengths, which can be close in energy.

For a number of materials these configurations are easily identifiable. In $YBa_2Cu_3O_{6+x}$ [6] they are given by square planar $Cu^{3+}-4O^{2-}$ configurations, which, by shrinking of bond lengths to two diametrically opposite oxygens go over into the chemically stable $O^{2-}(4)-Cu^+(1)-O^{2-}(4)$ dumb-bells. In $BaBiO_3$ [4,7] the Bi ion is surrounding by an octahedron of oxygens. Bi is in a Bi^{3+} state if it sits in the centre of the octahedron and it is in Bi^{5+} state if it is displaced towards one of the oxygens at the summit of a deformed octahedron. In the fullerenes, each buckey ball consists of carbons forming hexagons and pentagons. Their bond angle lies between that corresponding to the flat carbon configuration sp^2 (120°) and that corresponding to the sp^3 configuration in diamond (107°). Placing an extra electron on a buckey ball results in a C_{60}^- in which the electron is trapped inside a carbon pentagon configuration, which deforms tangentially [8]. Thus, charge fluctuations on the C_{60} framework are accompanied by local cluster deformations going either in the direction of the stable graphite or in the direction of the stable diamond configuration. For other high-T_c materials such a clear cut determination of the polaronic centres is in general as yet not available. What is known, on the contrary, is that all the cuprate materials show lattice deformations involving bond-stretching polarized principally along the c axes and frequently associated with M–O planes where M denotes the rare earth ions [5].

These dynamical fluctuations of the intramolecular deformations of those clusters favour charge transfer between a polaronic state and that of an itinerant electron from the valence band. If polarons are indeed paired up on those clusters then we can expect a charge transfer process between a Boson (bipolarons) and a pair of Fermions (a pair of itinerant electrons from the valence band). Such a process may be feasable for the cuprates and the bismuthates, but it is unlikely that it occurs on a buckey ball since C_{60}^{2-} has an energy about 1–2 eV above that of C_{60}^-.

This chemical picture can obviously only serve as a guide line in the form of ionic bonds to explain the much more realistic and complex situation of covalency in these systems. Nevertheless, this picture might be quite close to reality, as a comparison of photoemission spectra of the C_{60}^- fullerene in the gaseous phase [9] and metallic Rb_3C_{60} [10] shows. It turns out that the photoemission spectrum of the metallic state is given to roughly 80% correspondence by that of the isolated C_{60}^- [9].

3 Bipolaron–electron pair exchange

A crucial question regarding these polaronic charge carriers concerns knowledge of their location. This seems to be largely determined by the positions of the dopant atoms and their ordering. EXAFS [11], neutron powder diffraction spectroscopy [12] (measuring directly the atomic pair distribution function) and optical experiments on samples with site-selective substitution of dopants [6] provide us with valuable insight into this problem. According to these experiments, formation of polaronic states in the CuO_2 planes can be excluded on the basis that no Cu–O bond length change is observed as a function of doping or temperature. Polarons certainly form in the neighbourhood of the dopant atoms and show the characteristic features of strong local deformation of the lattice. Fully stoichiometric samples of $YBa_2Cu_4O_8$ present an intriguing exception to that. There, well-defined correlated regions of polarons seem to exist in a purely dynamical fashion [13]. For cuprate compounds, which have apex oxygens, polaron formation is seen, moreover, in shortening of the Cu(2)–O(4) (apex) distance as well as tilting of the Cu(2)–2O(2)–2O(3) complex, giving rise to an orthorhombic local symmetry change of the tetragonal phase. High-T_c superconducting samples are characterized by anomalous contributions to such local orthorhombic lattice fluctuations as determined by a combined study of pulsed and continuous neutron diffraction experiments [12]. These 'anomalous' lattice fluctuations are believed to be purely dynamical and strongly depend on temperature and doping. They are called 'anomalous' because they cannot be attributed to the obvious structural frustrations induced by the dopant atoms. In high-T_c superconducting samples they are manifest already in the normal phase above T_c. Upon entering the superconducting state by lowering the temperature, these fluctuations of local orthorhombic symmetry change and become correlated over larger distances. A similar effect has also been seen in ion channelling experiments [14].

EXAFS measurements and neutron powder diffraction [11,12] strongly indicate that these orthorhombic deformations align along certain directions and that the CuO_2 planes are composed of alternating stripes of such orthorhombic deformations and normal tetragonal regions. The alignment of these stripes, which are of locally orthorhombic symmetry, is controlled by the ordering of the dopant ions in planes parallel to the CuO_2 layers. This ordering of dopant atoms [15] seems to be a pre-requisite for the occurrence of superconductivity. The orthorhombic lattice deformations are, however, most probably the consequence rather than the origin of strong local lattice deformations around the dopant ions – which are the true centres of polaron formation.

Let us consider in more detail the case of $YBa_2Cu_3O_{6+x}$, the material in which this phenomenon has been most studied. $YBa_2Cu_3O_6$ is characterized by CuO_2 planes separated by planes of $O^{2-}(4)$–$Cu^+(1)$–$O^{2-}(4)$ molecular clusters oriented orthogonally to those CuO_2 planes where the apex $O(4)$ forms the top of a pyramid with two $O(2)$ and two $O(3)$ atoms on the base. The physics of the undoped material ($x=0$) has been successfully interpreted in terms of its being a charge transfer insulator with $Cu(2)$–$2O(2)$–$2O(3)$ bonding and antibonding bands and a $O(2)$–$O(3)$ non-bonding band. The gap occurs between the filled non-bonding and the empty antibonding $Cu(2)$–$2O(2)$–$2O(3)$ band.

Doping this material happens by introducing so-called $O(1)$ atoms between two adjacent $O^{2-}(4)$–$Cu^+(1)$–$O^{2-}(4)$ clusters and binding of the $O(1)$ occurs via a charge transfer $Cu^+(1)$–$O^0(1)$–$Cu^+(1) \rightarrow Cu^{2+}(1)$–$O^{2-}(1)$–$Cu^{2+}(1)$. This results in a sizeable change of the $Cu(1)$–$O(4)$ distance, and consequently a large shift of the local modes polarized in the c direction (from 475 to 505 cm^{-1}) and the appearance of a series of overtones. Low-doped $YBa_2Cu_3O_{6+x}$ in the insulating phase – in which the clusters involving the dopant atoms do not overlap – can hence be characterized by localized impurity type states having bipolaronic characteristics. These states have to lie inside the valence band since, upon further doping, additional $O(1)$ could otherwise not be chemically bound. This leads us to the following schematic picture of states illustrated in Fig. 1. Photoexcitation of low-doped insulating $YBa_2Cu_3O_{6+x}$ leads to creation of a hole in the valence band and simultaneously creation of an electron in the conduction band of the CuO_2 planes. The conduction band electron is subsequently transferred to an excited molecular state of $O^{2-}(4)$–$Cu^{2+}(1)$–$O^{2-}(4)$ resulting in the 475 cm^{-1} rather than 505 cm^{-1} Raman active mode. This scenario is consistent with the fact that a photoinduced MIR band is observed for photoexcited carriers in the insulating phase. It is due to the relative stability of the excited $O^{2-}(4)$–$Cu^+(1)$–$O^{2-}(4)$ units, de-exciting themselves via a cascade of one-phonon emission [16] towards $O^{2-}(4)$–$Cu^{2+}(1)$–$O^{2-}(4)$ plus a valence electron. This picture is, moreover, consistent with the fact that, quite generally in the insulating state (filled valence band) of high-T_c materials, no MIR band is observed, while in the metallic state it is.

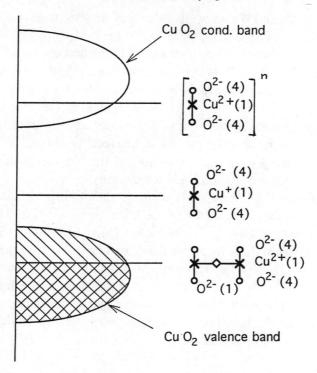

Fig. 1. The band and level scheme of $YBa_2Cu_3O_{6+x}$, $x=0$ corresponding to the filled lower band of the charge transfer insulator and $x \geq 0.4$ corresponding to this band being filled up to the level of the dopant centres (double hatched). The levels of the molecular states of the dumbbell molecular clusters in the chains of $YBa_2Cu_3O_{6+x}$ are indicated. The highest one corresponds to the non-vibrationally excited state.

Photoemission studies on Cu(1) [17] permitted measurement of the number of $Cu^+(1)$ ions converted to $Cu^{2+}(1)$ upon doping. These results together with structural analyses [18] enable one to draw up a picture of the ordering of O(1) atoms with progressive doping. If O(1) atoms were substituted randomly but such that no two adjacent O(1) states would be occupied, then the number of $Cu^{2+}(1)$ ions would increase like $2x$. This is actually observed up to $x \simeq 0.25$. Above this value the O(1) atoms rather suddenly prefer to align themselves along chains in the b direction. This is evident from the lack of any further increase of $Cu^{2+}(1)$ sites for $x \simeq [0.25, 0.4]$ and followed by an increase linear in x of $Cu^{2+}(1)$ sites for $x \simeq [0.4, 0.6]$. The reason for this ordering of dopant atoms is at present not understood. Given the fact that, in these materials, the 'impurity' binding energy is much smaller than the polaronic one, this ordering might be due to polaronic effects. It seems that it is a general phenomenon of all high-T_c superconductors and applies not only to O substitution but also to substitution of Bi and Sr as well as others in the other cuprate high-T_c superconductors.

Above $x = 0.6$, substitution of further $O(1)$ ions first continues in a random fashion (the number of $Cu^{2+}(1)$ ions increases like $2x$) and above $x \simeq 0.75$ again the dopant $O(1)$ atoms prefer to form chains and the number of $Cu^{2+}(1)$ ions suddenly ceases to increase any further. The lack of increase in number of $Cu^{2+}(1)$ atoms can easily be understood in terms of formation of chains whereby $O(1)$ ions must take the place between two already formed $O(1)$ bipolaronic centres. A detailed discussion of this dopant ordering deduced from $Cu(1)$ photoemission data has been given [19].

The only way of binding the $O(1)$ atoms in the doping regime $x = [0.25, 0.6]$ and $[0.75, 1]$ is by partial transfer of electrons from the CuO_2 planes to the chains containing the $O(1)$ ions since $Cu^{3+}(1)$ is prohibited. This will first empty out the states at the top of the valence band and eventually also the states in the bipolaronic level. When this happens, the bipolaron $\leftrightarrow$ valence electron pair exchange mechanism will lead to delocalization of the bipolarons and ultimately to their superfluidity upon lowering of the temperature. In the following we shall discuss this on the basis of a generic model involving a hybridized Boson–Fermion mixture.

4 The Boson–Fermion mixture scenario for high-T_c superconductivity

The scenario of states (see Fig. 1) described above has been deduced from the various optical, EXAFS, neutron powder diffraction and structural analyses and photoemission work. We have concluded from these experiments that, in the metallic phase, a charge transfer mechanism should be active, which involves localized bipolarons (centred near the dopant atoms) and pairs of itinerant valence electrons of the CuO_2 planes. We judiciously choose molecular units (taking into account the structural specificity of each compound) on which this charge transfer takes place. For $YBa_2Cu_3O_{6+x}$ they consist of chain sub-unit $O(1)$–$2(O(4)$–$Cu(1)$–$O(4))$ surrounded on both sides by the adjacent CuO_2 plane sub-units. We suppose, moreover, that these units are homogeneously distributed throughout the crystal and that upon doping only their concentration changes. This is not what actually occurs in real systems but will nevertheless be a guide to our understanding high-T_c superconductivity. Marking these units by site labels l we then propose the following Hamiltonian:

$$H = \sum_{\langle l \neq l' \rangle \sigma} t_{ll'} c_{l\sigma}^+ c_{l'\sigma} + \Delta \sum_l b_l^+ b_l + \frac{v}{\sqrt{\Omega}} \sum_l (b_l^+ c_{l\downarrow} c_{l\uparrow} + \text{h.c.})$$
$$- \mu \left(2 \sum_l b_l^+ b_l + \sum_{l,\sigma} c_{l\sigma}^+ c_{l\sigma} \right), \tag{1}$$

(b)

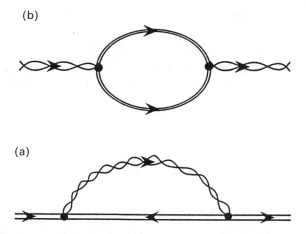

(a)

Fig. 2. The self-energy diagram for (a) Fermions and (b) Bosons.

where $c_{l\sigma}^{(+)}$ denote the annihilation (creation) operators of the itinerant electrons in the valence band of the CuO$_2$ planes and $b_l^{(+)}$ are the annihilation (creation) operators of localized bipolarons centred around the O(1) atoms. Δ is a measure of the binding energy of the bipolaronic states involving both bare impurity binding of the O^{2-}(1) state and the polaron binding caused by lattice relaxation around these O(1) ions. ν denotes the bipolaron–electron pair hybridization constant and μ the chemical potential, which is common to carriers of both subsystems because of charge conservation. μ will, however, depend on the doping rate since, as discussed in Section 2, the rate of electron transfer from the valence band to the bipolaronic level will depend sensitively on how the doping takes place. The Hamiltonian (Eq. (1)) was first introduced in 1986 [3] in order to describe a system of electrons and phonons in the intermediary coupling regime. For such systems there are strong theoretical indications that electrons find themselves on sites with large positional fluctuations of the surrounding atoms, which is suggestive of strong hybridization of itinerant electrons and localized polarons or bipolarons. Studying the Hamiltonian (Eq. (1)) within a self-consistent (fully conserving) mean field approximation – given by the Boson and Fermion self-energies depicted in Fig. 2 – enables us to follow the evolution of the electronic structure upon lowering of the temperature. As one approaches the superconducting state, it turns out that the initially localized bipolarons acquire itinerancy and that, in the spectrum of the itinerant electrons, a pseudo-gap opens up around the band of bipolaronic states [20]. On the basis of this knowledge we can then describe fairly reliably the superconducting state within the usual RPA for itinerant

Bosons and Fermions coupled via the exchange term (proportional to v in Eq. (1)).

Let us for that purpose make in Eq. (1) the change $\Delta \to \Delta + \xi_q^{B}$. Introducing $E_q \equiv \partial \xi_q^{B} + \Delta - 2\mu$ and $\varepsilon_k = \xi_k^{F} - \mu$, where $\xi_k^{F(B)} \equiv \hbar^2 k^2 / (2 m_{F(B)})$ we have the following expressions for the Fermion and Boson self-energies:

$$\Sigma_F(k, \omega) = \frac{v^2}{\Omega} \sum_q \frac{n_F(\varepsilon_{k-q}) + n_B(E_q)}{\omega - E_q + \varepsilon_{k-q} + i\eta},$$

$$\Sigma_B(q, \omega) = \frac{v^2}{\Omega} \sum_k \frac{1 - n_F(\varepsilon_k) - n_F(\varepsilon_{k-q})}{\omega - \varepsilon_{k-q} - \varepsilon_k + i\eta}, \tag{2}$$

where $m_{F(B)}$ denotes the Fermion (Boson) mass and $n_{F(B)}$ the respective distribution functions. Making in the expression for $\Sigma_F(k, \omega)$ the replacement

$$n_B(E_q) \to n_B(E_q) + \delta_{q0}(n_0 - n_B(E_q)),$$

where n_0 denotes the condensate fraction of the Bosons, we obtain in the limit $T \to 0$

$$\Sigma_F(k, \omega) \sim \frac{n_0 v^2}{\omega + \varepsilon_k + i\eta}. \tag{3}$$

Hence the dispersion for the electron quasi-particles

$$\omega_k = (\varepsilon_k^2 + n_0 v^2)^{1/2} \tag{4}$$

exhibits a BCS-like gap, which is linked in an obvious way to condensation of the Bosons. If we impose the standard requirement for Bose condensation

$$E_0 = R_B(q = 0, \omega = 0) = \frac{v^2}{2\Omega} P \sum_k \frac{1}{\varepsilon_k} \tanh\left(\frac{\beta \varepsilon_k}{2}\right) \tag{5}$$

(where $R_B(q, \omega)$ denotes the real part of $\Sigma_B(q, \omega)$), then we notice that, in the limit $T \to 0$ the integrand on the right-hand side of Eq. (5) diverges for $\varepsilon_k = 0$, unless the density of states at this point is zero – which is indeed the case in the superconducting state. This brief argument clearly shows how in the Boson–Fermion mixture Bose condensation and superconductivity are irrevocably linked together as long as the number of Bosons per site is different from values close to zero or unity. If this is not the case then we still can have superconductivity but without condensation of Bosons. Bosons then merely act as virtual states via which Cooper pairing takes place, in complete analogy with phonon-mediated BCS superconductivity. On the basis of preliminary results for T_c as a function of the total number of carriers [20] and previous studies of this

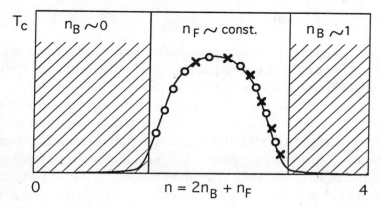

Fig. 3. The schematic behaviour of T_c as a function of n (denoting the total number of Fermions and Bosons respectively) expected for the Boson–Fermion model (Eq.(1)). The crosses and circles correspond to $T_c(n)$ for $YBa_2Cu_3O_{6+x}$ and $La_{2-x}Sr_xO_4$.

problem within a mean field approximation [3], we present a schematic plot of T_c in Fig. 3. Notice the drastic increase in T_c as soon as the density of Bosons (particles as well as holes) becomes finite.

This behaviour of T_c as a function of total carrier concentration can be regarded as a universal behaviour and qualitatively reproduces the experimental results obtained from the empirical relation between T_c and n derived on the basis of measurements of the London penetration depth [21] or the intensity of the MIR band in optical absorption [22]. The maximum of T_c corresponds to situations of roughly half filled bipolaronic levels and comparable density of Bosons and itinerant electrons. In hole-doped high-T_c superconductors the diagram in Fig. 3 has to be read from right to left when doping is increased. There is no or negligible superconductivity at low doping when the Bosonic band is inside a filled valence band. High-T_c superconductivity sets in when, upon doping, this Bosonic band is depleted. $YBa_2Cu_3O_{6+x}$ would fall into a category in which this band can be depleted roughly by half for $x \lesssim 1$, while in $La_{2-x}Sr_xCuO_4$ it can be depleted entirely. In general the form of $T_c(n)$ in Fig. 3 will vary. Depending on the position of the Bosonic level inside the Fermionic band, it will be assymmetric if this position does not coincide with the band centre. The drop of T_c on one or the other side of the maximum then can be quite abrupt. For the present discussion we have also assumed that this level position is independent of doping. This may not be so, since both the impurity binding energy and the polaronic binding energy can vary not only with doping but also with how the dopant atoms are arranged on the lattice. For all these reasons we can only make here very qualitative comparisons between experiment and theory.

J. Ranninger

5 Conclusions

We have discussed the strong experimental evidence for existence of localized polarons and bipolarons in low-doped insulating samples of high-T_c superconductors and indirect experimental evidence that such polaronic charge carriers continue to exist when doping is increased and the samples become metallic. A characteristic temperature T_{PB} above the superconductivity transition temperature T_c can be assigned to the stability of those bipolarons. Experimentally, T_{PB} can be associated with anomalous changes in the temperature behaviour of the susceptibility [23], the NMR relaxation rate T_1^{-1} [24] and the specific heat [25]. The anomalous behaviour of the specific heat can be interpreted in terms of existence of a pseudo-gap for T below T_{PB}. This pseudo-gap evolved into a true gap upon reducing the temperature towards T_c. In the picture developed in this paper the existence of the pseudo-gap is a consequence of the existence of a Bosonic (bipolaronic) level, which falls inside the band of itinerant electrons.

A proof that T_c is directly related to the number of polaronic charge carriers comes from the doping-dependence of the MIR band observed in optical conductivity, which has been attributed to polaronic transport [26]. It turns out that the spectral weight of this MIR band scales with T_c for different doping rates and that, upon increasing the temperature, for fixed doping, the spectral weight is transferred from this MIR contribution towards the Drude contribution, which provides another indication for the existence of T_{PB}.

In the theoretical picture developed above we do not assume that bipolarons exist in itinerant band states. On the contrary, we envisage them as localized states that exist below a certain characteristic temperature T_{PB}. Owing to overlap of such bipolaronic states with itinerant band states of valence electrons, the initially localized bipolarons acquire itinerant character upon lowering of the temperature. This is due to a precursor effect of Bose–Einstein condensation, which conditions the initially diffusive motion of the Bosons (bipolarons). The type of superconductivity that we obtain is strongly influenced by condensation of the bipolarons, as can be seen from the dependence of T_c on carrier concentration. This might be the reason that, in a Boson–Fermion mixture, so many thermodynamic, electrodynamic and transport properties are very similar to those expected for a purely interacting Bose gas [27].

References

[1] A. S. Alexandrov and J. Ranninger, *Phys. Rev.* B **23** (1981) 1796; *ibid.* **24** (1981) 1164.

[2] M. Marezio and P. D. Dernier, *J. Solid State Chem.* **3** (1971) 340; B. K. Chakraverty and C. Schlenker, *J. Physique* **37** (1976) C4-353; B. K. Chakraverty, M. J. Sienko and J. Bonnerot, *Phys. Rev.* B **17** (1978) 3781; O. F. Schirmer and E. Salje, *J. Phys.* C **13** (1980) L1967.

[3] J. Ranninger and S. Robaszkiewicz, *Physica* B **135** (1985) 468; S. Robaszkiewicz, R. Micnas and J. Ranninger, *Phys. Rev.* B **36** (1987) 180; R. Micnas, J. Ranninger and S. Robaszkiewicz, *Rev. Mod. Phys.* **62** (1990) 113.

[4] C. Taliani *et al.*, *Electronic Properties of High T$_c$ Superconductors and Related Compounds*, ed. H. Kuzmany, M. Mehring and J. Fink, Springer Series of Solid State Science (Berlin) 1990, vol. **99**, p. 280; C. M. Foster *et al.*, *Synthetic Materials* **33** (1989) 171.

[5] D. Mihailovic *et al.*, *Phys. Rev.* B **42** (1990) 7989.

[6] V. N. Denisov *et al.*, *Phys. Rev.* B **48** (1993) 16714.

[7] P. Ruani, private communication.

[8] E. Denisov and B. N. Mavrin, *Sov. Phys. JETP* **75** (1992) 158.

[9] W. Eberhardt, private communication.

[10] M. Knupfer *et al.*, *Phys. Rev.* B **47** (1993) 13944.

[11] J. Mustre de Leon *et al.*, *Lattice Effects in High T$_c$ Superconductors*, ed. Y. Bar-Yam, T. Egami, J. Mustre de Leon and A. R. Bishop, World Scientific (Singapore) 1992, p. 39; A. Bianconi *et al.*, *ibid.* p. 65.

[12] S. J. L. Billinge and T. Egami, *Lattice Effects in High T$_c$ Superconductors*, ed. Y. Bar-Yam, T. Egami, J. Mustre de Leon and A. R. Bishop, World Scientific (Singapore) 1992, p. 93.

[13] T. Egami *et al.*, paper in this volume.

[14] L. E. Rhen *et al.*, *Lattice Effects in High T$_c$ Superconductors*, ed. Y. Bar-Yam, T. Egami, J. Mustre de Leon and A. R. Bishop, World Scientific (Singapore) 1992, p. 27; K. Yamata *et al.*, *ibid.* p. 33.

[15] D. de Fontain *et al.*, *Nature* **343** (1991) 544; J. O. Jorgensen, *ibid.* **349** (1991) 565.

[16] D. Mihailovic and I. Poberaj, paper in this volume.

[17] H. Tolentino *et al.*, *Physica* C **192** (1992) 115.

[18] J. D. Jorgensen *et al.*, *Lattice Effects in High T$_c$ Superconductors*, ed. Y. Bar-Yam, T. Egami, J. Mustre de Leon and A. R. Bishop, World Scientific (Singapore) 1992, p. 84.

[19] J. Ranninger, *Physica Scripta* T **39** (1991) 61.

[20] J. Ranninger and J. M. Robin, 1995, to be published.

[21] Y. J. Uemura *et al.*, *Phys. Rev. Lett.* **62** (1989) 2317.

[22] J. Orenstein *et al.*, *Phys. Rev.* B **36** (1987) 8892; S. Etemad *et al.*, *ibid.* **37** (1988) 3396.

[23] M. Miljhac *et al.*, *Solid State Commun.* **85** (1993) 519.

[24] R. E. Walstedt *et al.*, *Phys. Rev.* B **14** (1990) 9574; M. Horvatic *et al.*, *Physica* C **185/9** (1991) 1139.

[25] J. M. Loram *et al.*, *Phys. Rev. Lett.* **71** (1993) 1740.

[26] J. P. Falck *et al.*, *Phys. Rev.* B **48** (1993) 4043.

[27] A. S. Alexandrov and N. F. Mott, *Supercond. Sci. Technol.* **6** (1993) 215.

6

Formation, phase separation and superconductivity of large bipolarons

DAVID EMIN

Sandia National Laboratories, Albuquerque, NM 87185-0345, USA

Abstract

This paper presents a scenario in which large (multi-site) bipolarons form and give rise to superconductivity. First the physical circumstances in which large bipolarons can form are elucidated. Then several identifying properties of large bipolarons are discussed. Finally, a model of how interactions between large bipolarons lead to their superconductivity is presented. I emphasize the existence of a phonon-mediated intermediate-range attraction between large bipolarons. With attractive interactions between large bipolarons, they can condense into a liquid phase. This liquid is a quantum liquid if the ground state of the interacting large bipolarons is a fluid rather than a solid. This quantum liquid of charged bosons is analogous to the quantum liquid of neutral bosons envisioned for superfluid ^{4}He. As such, the superconductivity of large bipolarons can be understood (or rationalized) in a similar manner to that employed in addressing the superfluidity of liquid ^{4}He.

1 Introduction

This article begins by describing electron–lattice interactions of ionic solids. A long-range electron–lattice interaction results from the dependence of the Coulombic potential energy of a carrier on the positions of the solid's ions. [1] Short-range electron–lattice interactions reflect the sensitivity of the energy of a carrier's local state (e.g., bonding or antibonding state) to the positions of nearby atoms [2,3].

The notions of self-trapping and bipolaron formation are then reviewed. Since a self-trapped carrier can only move when atoms move, the adiabatic approach is employed to discuss polaron and bipolaron formation. That is, the adiabatic treatment obliges electronic carriers to follow the motions of the solid's atoms. Determining the electronic state of a self-trapped carrier is a complicated nonlinear problem because the well that binds a self-trapped carrier depends upon the carrier's state. Nonetheless, by treating the solid as a deformable continuum, scaling theory yields exact results for the nature of self-trapped states in several circumstances [4,5].

Two distinct types of adiabatic polaron are shown to be possible. In multi-dimensional systems with only a short-range electron–lattice interaction (e.g., the deformation potential of covalent systems), a self-trapped carrier always forms a *small* polaron. In these circumstances the self-trapped electronic carrier collapses to a *single* site. By contrast, with the classical long-range electron–lattice interaction of ionic systems, *large* polarons are possible. In such an instance the self-trapped electronic carrier extends over *multiple* sites. The extreme confinement of a small polaron typically leads to its moving by thermally assisted hopping. By contrast, the multi-site extension of a large polaron results in its moving itinerantly.

A bipolaron is formed when two carriers are self-trapped as a pair within a common self-trapping well. Here the possibility of superconductivity resulting from itinerant bipolarons is considered. Presuming that an itinerant bipolaron is necessarily a large bipolaron, the conditions for the formation of a large bipolaron are addressed.

The polaron scaling theory is then extended to treat bipolaron formation [6,7]. Forming a multi-site bipolaron in a multi-dimensional system requires the presence of a long-range component of the electron–lattice interaction. Specifically, the long-range electron–lattice interaction is only sufficiently strong to support a large bipolaron if the ratio of the material's static to optical dielectric constants, $\varepsilon_0/\varepsilon_\infty$, is exceptionally large (>2). Furthermore, the stability of a large bipolaron with respect to dissociation into two large polarons requires the additional presence of a sufficiently strong short-range electron–lattice interaction. However, if the short-range component of the electron–lattice interaction is too strong, then the large bipolaron is unstable with respect to collapsing into a small (severely localized) bipolaron. Thus, large bipolarons can only form within a restricted range of physical parameters. Formation of a stable large bipolaron is most likely in systems with large ratios of the static to optical dielectric constants: $\varepsilon_0 \gg 2\varepsilon_\infty$. The very large values of $\varepsilon_0/\varepsilon_\infty$ found in the insulating parents of superconducting oxides imply that these materials are prime candidates for formation of large bipolarons [8–10].

Self-trapped carriers can manifest themselves by their absorption of photons. In particular, photons are absorbed as self-trapped carriers are raised from the ground states of their self-trapping potential wells. The absorption spectra of large polarons or bipolarons have shapes that differ from those of small polarons or bipolarons [11]. The absorption spectrum that arises from a Franck–Condon transition in which a large bipolaron's self-trapped carrier is liberated is asymmetric. In particular, the absorption due to this photo-ionization rises sharply with increasing photon energy to a maximum and then falls relatively slowly as the wavelength of the final electronic state increasingly falls below the radius of the self-trapped state. By contrast, the absorption spectrum of a small polaron, associated with photon-assisted transfer of the carrier between adjacent sites [12–14], has a complementary asymmetry. The small-polaron absorption rises to a maximum and then falls relatively rapidly [14]. Thus, this paper discusses how single-site and multi-site self-trapped carriers can be distinguished from one another by their absorption profiles.

DC transport of self-trapped carriers requires motion of the atoms associated with self-trapping. For this reason, the low-frequency effective mass of an independent large bipolaron is generally much larger than the mass of a free carrier [7]. Furthermore, the bipolaron's massiveness contributes to the inefficiency with which it is scattered by phonons [15]. Thus, observations of a large effective mass and an exceptionally long scattering time at microwave frequencies in cuprates are consistent with the presence of large bipolarons [8]. However, the large effective mass and long scattering time compensate one another to yield a large-bipolaron mobility that is not clearly distinguishable from that of non-polaronic charge carriers [15].

This paper concludes with a discussion of causes and effects of interactions between large bipolarons [16,17]. Each of the three interactions that are considered reflects distinctive aspects of large bipolarons. First, since large bipolarons move slowly, the Coulomb repulsion between them is reduced by the solid's static dielectric constant. Second, since each large bipolaron has two self-trapped carriers within a single self-trapping well, the Pauli principle leads to a strong short-range repulsion between large bipolarons that opposes their coalescence. Third, since the self-trapped carriers of a large bipolaron are redistributed among sites of their molecular orbitals as their self-trapping potential wells are altered, the lattice stiffness in the vicinity of the bipolaron is reduced. The dependence of the resulting lowering of the zero-point vibrational energy on the spatial distribution of large bipolarons constitutes an intermediate-range phonon-mediated attraction between large bipolarons. Taken together, these three interactions enable large bipolarons to condense from a gaseous state to a liquid phase (when ε_0 is sufficiently large, $\varepsilon_0 \gg 1$) [16,17]. If the ground state of the interacting large bipolarons remains liquid rather than solidifying, it is analogous to the quantum liquid of neutral bosons that yields superfluidity [17]. Thus, considerations of interactions amongst large bipolarons naturally lead to the possibility of their superconductivity.

2 Electron–lattice interactions

A carrier's electron–lattice interaction denotes the dependence of its state on the positions of the solid's atoms. Consider, for example, a tight-binding representation of the state of a carrier in a solid, Ψ. Then the carrier's state is represented as a superposition of local electronic states, ϕ_g, where g denotes the position assigned to each local electronic function: $\Psi = \Sigma_g a_g \phi_g$. The carrier's energy, E, is found by solving the set of coupled equations that link the amplitude for the carrier being at site g, a_g, with the amplitudes for it being at neighboring sites, $g + h$, where h is a vector connecting adjacent sites:

$$(E - \varepsilon_g)a_g = -\sum_h t_{g,g+h} a_{g+h}. \tag{1}$$

In these equations ε_g is the energy of the system when the carrier occupies site g and the electronic transfer energies, $t_{g,g+h}$, provide direct links between adjacent sites.

The electron–lattice interaction may be introduced into this scheme by taking ε_g to depend on atomic positions. The dependence of the local electronic energies on the positions of atoms with which the electron is in direct contact constitutes what is termed the 'short-range' (contact) electron–lattice interaction [2–6]. In a complementary manner, the dependence of the local energy on the positions of the atoms with which its local electronic function does not overlap appreciably is termed a 'long-range' electron–lattice interaction. This long-range electron–lattice interaction arises from the dependence of the potential energy of a carrier in a local state at a given site on the positions of atoms that are not encompassed by the local electronic function of that site [1,3,7].

Much of polaron theory focuses on the long-range Coulomb interactions between a carrier and the ions of a *polar* lattice. The strength of this long-range electron–lattice interaction is often measured by the Fröhlich coupling constant, $\alpha \equiv e^2 [m/(2\omega \hbar^3)]^{1/2}(1/\varepsilon_\infty - 1/\varepsilon_0)$, where ε_∞ and ε_0 are the material's optical and static dielectric constants, respectively [1]. Here, m is the bare effective mass of an electronic carrier and ω is the characteristic optical mode frequency. These long-range electron–lattice interactions are small in covalent materials, in which the electronic polarizabilities are primarily due to electrons: $\varepsilon_\infty \approx \varepsilon_0$. By contrast, the long-range interactions are especially strong in solids with displaceable ions. In these situations $\varepsilon_0/\varepsilon_\infty \gg 1$. Examples of solids with readily displaceable ions include ferroelectric-like materials, layered systems that are doped with intercalated ions, and molecular systems that are doped with ions that reside between the molecular units. Thus, novel superconductivity is frequently suspected in situations in which long-range electron–lattice interactions tend to be especially strong.

Short-range electron–lattice interactions are manifestations of the local chemistry. Models of short-range electron–lattice interactions include the deformation-potential and Holstein models [2,4]. In the deformation-potential scheme for elemental covalent semiconductors the local orbitals are the bonding or antibonding orbitals that link adjacent atoms. The dependence of the bonding and antibonding energies on the interatomic separations constitutes the short-range electron–lattice interaction. In Holstein's molecular-crystal model, the dependence of the energy of a molecular state on the positions of the atoms involved in its molecular orbital constitutes the short-range electron–lattice interaction.

Short-range electron–lattice interactions exist in ionic systems as well as in covalent systems. For example, the strong short-range electron–lattice interaction that causes a hole in KCl to form a small polaron (sometime called the V_K center) [18] reflects the following chemical reaction: hole $+ (Cl)^-$ $+ (Cl)^- \rightarrow (Cl_2)^-$. Oxygen-related states in oxides also often have strong short-range electron–lattice interactions. To understand why this is so, recall that an oxygen atom in free space has an affinity for but one electron. Nonetheless, oxygen atoms in oxide compounds are frequently formally represented as attracting two electrons. In these instances the cations that surround the oxygen anion provide sufficient positive charge that a second electron can be bound in the vicinity of the oxygen atom. Upon moving the cations away from the vicinity of the oxygen atom, the second electron associated with the oxygen dianion is freed from it. In other words, the energy of the outermost electron of an O^{2-} ion depends strongly on the separations between the oxygen anion and the adjacent cations. Thus, the highest-lying oxygen-related state is often associated with a strong short-range electron–lattice interaction.

For large polarons and bipolarons the electronic states extend over multiple sites. Then the continuum approximation, $a_{g+h} \approx a_g + \partial a_g / \partial g \cdot h + (|h|^2/2)\nabla_g^2 a_g$, may be applied. Using this approximation in Eq. (1) yields an effective-mass-type wave equation in which $\varepsilon_g - \Sigma_h t_{g,g+h}$ plays the role of the potential energy of a carrier at site g:

$$Ea_g = \left[-\left(\sum_h t_{g,g+h}|h|^2\right)\nabla_g^2/2 + \left(\varepsilon_g - \sum_h t_{g,g+h}\right)\right]a_g. \qquad (2)$$

The dependence of this effective potential energy on atomic positions constitutes the electron–lattice interaction that is usually studied in the continuum treatment. One sees that dependences of the transfer energies on interatomic separations contribute a short-range electron–lattice interaction. In particular, in the continuum treatment a linear dependence of the transfer energies on interatomic separations (the Su–Schrieffer–Heeger model) [19] is equivalent to the linear dependence of ε_g on a local atomic displacement coordinate envisioned in Holstein's molecular-crystal model [3].

The novel superconductors generally contain some exceptionally displaceable ions. The ionic remnants of dopants (e.g., group II ions in cuprates and alkali ions in Buckyball systems) are often loosely bonded. The interplanar ions of oxide superconductors are extremely displaceable: e.g., $\varepsilon_0/\varepsilon_\infty \gg 1$ for the insulating parents of the cuprate superconductors [8–10]. Thus, strong long-range electron–lattice interactions are expected in the novel superconductors. In addition, significant short-range electron–lattice interactions are expected for oxygen-related states.

3 Self-trapping: general formalism

As a result of its electron–lattice interactions, a charge carrier in a semiconductor may become *self-trapped*. Landau introduced this notion by observing that displacing atoms of a crystal from their carrier-free equilibrium positions can produce a potential well that will bind a charge carrier [20]. Trapping of carriers in this manner is energetically favorable if the lowering of the carrier's energy resulting from its being trapped exceeds the strain energy required to displace atoms so as to produce the trap. This effect is termed *self*-trapping because the carrier is bound in a potential produced by atomic displacements that are themselves stabilized by the carrier occupying the trap. The term *polaron* refers to the quasiparticle comprising a self-trapped carrier and the atomic displacements that bind it. If two carriers are bound together within a common well, then the resultant quasiparticle is termed a *bipolaron*. The bound pair of carriers acting as a unit then behaves as a charged boson.

The self-trapped state is analogous to a conventional trap state (with a time-independent potential) if the carrier circulates in its bound state before the atoms can move significantly. In this circumstance a self-trapped carrier can adiabatically adjust to the relatively slow atomic motion. This situation defines the strong-coupling regime of a polaron [1]. The complementary regime, in which the carrier does not experience a 'rigid' potential, is termed the weak-coupling regime. In this regime the polaronic binding energy and effective mass may be calculated by treating the electron–lattice interaction as a perturbation that couples a free carrier's motion to atomic vibrations.

A mathematical criterion to distinguish between the weak- and strong-coupling regimes is obtained by noting that the distance that an electronic carrier can diffuse in a vibrational period is $[\hbar/(m\omega)]^{1/2}$ [11]. If this distance exceeds the characteristic length scale of a self-trapped state, its 'radius,' then electronic motion is sufficiently rapid to validate the strong-coupling approach. The converse situation defines the domain of the weak-coupling regime. For the classic polaron problem, associated with the long-range Coulomb interactions of a carrier with displaceable ions, this criterion is simply expressed in terms of the Fröhlich coupling constant, α. The weak- and strong-coupling regimes are defined by $\alpha < 1$ and $\alpha > 1$, respectively [1]. Alternatively, this criterion may be written in terms of the binding energy of the self-trapped carrier, E_{bind}. Then the weak- and strong-coupling domains are defined by $E_{bind} < \hbar\omega$, and $E_{bind} > \hbar\omega$, respectively [11].

This work will consider the formation and properties of polarons and bipolarons when the electron–lattice interaction is sufficiently strong to correspond to the strong-coupling regime. In this regime, the standard adiabatic approach may be applied to formation and properties of polarons and bipolarons. In this approach a carrier is presumed to adjust to atomic motions. As a result, a carrier's energy, E_c, is a function of the positions of the atoms in the solid. Furthermore, motion of the solid's atoms is affected by the presence of the carrier. In particular, atomic vibrational motion is governed by a Hamiltonian in which the electronic energy contributes to the atomic potential energy. In the absence of a magnetic field, this Hamiltonian may be approximated by

$$H_{eff} \equiv T_v + V_v + E_c, \tag{3}$$

where the kinetic and potential energies of the atoms of a carrier-free lattice are T_v and V_v, respectively. In the absence of the carrier, minimizing V_v determines the interatomic separations. In addition, the second derivatives of V_v with respect to atomic positions evaluated at the minimum determine the solid's stiffness constants. Thus, introduction of a charge carrier generally results in (1) shifts in equilibrium position of the solid's atoms and (2) changes in the solid's stiffness constants. Both of these effects will be important in the subsequent discussion.

4 Small and large polarons

Studying polaron formation entails determining the ground state of the system comprising a carrier in a solid of displaceable atoms. As will be shown shortly, this task is complicated by the fact that the potential well that binds a self-trapped carrier in the system's ground state itself depends upon the electronic state of that carrier. Thus, self-trapping is a nonlinear phenomenon. In other words, self-trapping involves a feedback mechanism. In particular, increasing the confinement of a self-trapped carrier deepens the potential well that the carrier stabilizes. This effect in turn tends to produce an even more severely confined self-trapped carrier.

The focus of this article will be on self-trapped states that extend over multiple sites. The associated polaronic state is then referred to as *large*. Alternatively, a polaronic state is referred to as *small* if its self-trapped carrier collapses to a single site. In this section the conditions under which a self-trapped carrier will extend over multiple sites rather than collapse to a single site are determined. The properties of large and small polarons are then contrasted with one another.

With our focus on large polarons and bipolarons, it is appropriate to consider a model in which a lattice of displaceable atoms is replaced by a deformable continuum. With the deformation of the continuum about position u being denoted by $\Delta(u)$, the continuum's net strain energy is $V_{\mathrm{v}} = (S/2)\int \mathrm{d}u\, \Delta^2(u)$, where S is the stiffness constant per unit volume. In this model the carrier's Hamiltonian is

$$H_{\mathrm{c}} = T + V(r) = T - \int \mathrm{d}u\, Z(r-u)\Delta(u), \qquad (4)$$

where T is the carrier's kinetic energy and $V(r)$ denotes the potential experienced by a carrier at position r produced by deforming the continuum. Here $Z(r-u)$ describes the electron–lattice interaction. For the classic long-range Fröhlich interaction $Z(r-u) = [(1/\varepsilon_\infty - 1/\varepsilon_0)e^2 S/\pi^3]^{1/2}/|r-u|^2$. For Holstein's short-range interaction $Z(r-u) = A\delta(r-u)$, a contact interaction.

The minimum of the potential energy of the adiabatic Hamiltonian, Eq. (3), is then found to occur when

$$\Delta(u) = \int \mathrm{d}r'\, \frac{|\Psi(r')|^2 Z(r'-u)}{S}. \qquad (5)$$

With this deformation configuration the potential well experienced by the carrier is

$$V(r) = -S^{-1}\int \mathrm{d}r'\, |\Psi(r')|^2 \int \mathrm{d}u\, Z(r'-u)Z(r-u). \qquad (6)$$

In Eq. (6) it is demonstrated that the potential that binds a self-trapped carrier depends upon the state of the carrier, the magnitude and range of the electron–lattice interaction, and the spatial distribution of displaceable atoms (through the domain of the u-integration). Specifically, with only Holstein's short-range interaction

$$V(r) = -\frac{A^2}{S} \int dr' |\Psi(r')|^2 \delta(r'r) = -\frac{A^2}{S} |\Psi(r)|^2, \qquad (7a)$$

and with only the long-range Fröhlich interaction in a *three*-dimensional isotropic medium

$$V(r) = -e^2 \left(\frac{1}{\varepsilon_\infty} - \frac{1}{\varepsilon_0} \right) \int dr' \frac{|\Psi(r')|^2}{|r' - r|}. \qquad (7b)$$

It is generally a formidable task to solve the carrier's wave equation with the potential of Eq. (6). However, a scaling argument can be employed to ascertain fundamental features of the solutions for different electron–lattice interactions and configurations of displaceable atoms [4,5].

The scaling treatment considers the total energy in the adiabatic *limit* in which the vibrational frequencies are set equal to zero. The total energy in the adiabatic limit is then the sum of three contributions. The first contribution is the increase in kinetic energy associated with confining a carrier within its self-trapping well. The second contribution gives the lowering of the potential energy of the carrier due to its confinement within the self-trapping potential well. The final contribution is the strain energy associated with displacing atoms so as to produce the self-trapping potential well. When the carrier's potential energy depends linearly on the displacement coordinate (e.g., Δ in Eq. (4)), the magnitude of the third contribution to the energy is half that of the second contribution to the energy. Thus, these two potential energy terms may be readily combined into a single potential energy term.

The scaling treatment considers how the kinetic and potential energy contributions depend on a dimensionless scaling factor, R, that alters the spatial extent of the self-trapped state. The system's ground state is that for which the energy of this functional, $E(R)$, is a minimum.

With only a short-range (contact) electron–lattice interaction in a homogeneous system of dimensionality d, the energy functional is

$$E(R) = \frac{W}{2R^2} - \frac{E_b}{R^d}. \qquad (8)$$

Here W and E_b are recognized as the electronic bandwidth and small-polaron binding energy, respectively, when R is the radius of the self-trapped state in units of the lattice's fundamental length. For a multi-dimensional system ($d=2$ or 3) there are two distinct minima to this functional. This feature is illustrated in Fig. 1, where $E(R)$ is plotted against R for $d=3$. The minimum that occurs at $R=\infty$, where $E(\infty)=0$, corresponds to the carrier remaining free rather than

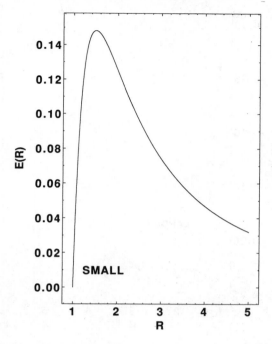

Fig. 1. The energy functional $E(R)$ is plotted against the scaling factor that alters the spatial extent of the self-trapped state, R, for a three-dimensional system with a pure short-range electron–lattice interaction, Eq. (8). The only possible bound state occurs as $R \to 0$, corresponding to the self-trapped state shrinking without limit to form a small polaron.

being self-trapped. Another minimum can occur when the self-trapped carrier shrinks into the smallest possible state commensurate with the atomicity of the solid, R_{min}. This minimum, corresponding to formation of a small polaron, is energetically stable when $E(R_{min}) < 0$. Thus, the only type of polaron that can form with only a short-range electron–lattice interaction in a multi-dimensional system is a small polaron [2,4,5,21].

The situation is qualitatively different for a model that considers only the long-range Fröhlich interaction. Then, in a d-dimensional medium of isotropically displaceable ions, the energy functional is

$$E(R) = \frac{W}{2R^2} - \frac{U}{2R^{4-d}} \left(\frac{1}{\varepsilon_\infty} - \frac{1}{\varepsilon_0} \right).$$ (9)

As illustrated in Fig. 2, for a three-dimensional system ($d = 3$) this functional has a solitary minimum at $R = 2W/[U(1/\varepsilon_\infty - 1/\varepsilon_0)]$, where U is then the Coulomb self-energy of a carrier confined to a single site. The actual radius of

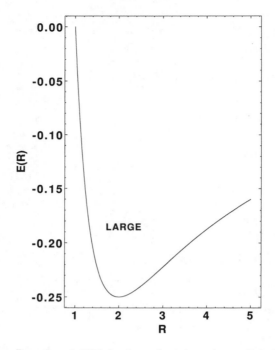

Fig. 2. The energy functional $E(R)$ is plotted against the scaling factor that alters the spatial extent of the self-trapped state, R, for a three-dimensional system with a pure Fröhlich (long-range) electron–lattice interaction, Eq. (9). The only possible bound state occurs at a finite value of R, corresponding to the self-trapped state having a finite radius. In most instances, the self-trapped carrier extends over multiple sites, thereby giving rise to a large polaron.

this Fröhlich polaron is $a_B/(1/\varepsilon_\infty - 1/\varepsilon_0)$, where $a_B \equiv \hbar^2/(m_c e^2)$ is the Bohr radius with the electron's mass being replaced by the electronic effective mass of the carrier, m_c [1]. Thus, even with an exceptionally strong Fröhlich interaction ($\varepsilon_0 \gg \varepsilon_\infty \approx 4$), a Fröhlich polaron will encompass several sites unless the electronic band is so narrow that m_c greatly exceeds the free-electron mass.

Two distinct types of polaronic states are also found in more complicated physical situations. For example, when short- and long-range interactions are treated together, a carrier will either form a large (multi-site) polaron or collapse into a small (single-site) polaron [5]. In this situation, the presence of the long-range component to the electron–lattice interaction tends to facilitate stabilization of a small polaron. Imposition of electronic disorder also fosters small-polaron formation [22].

There are qualitative differences between the transport of small- and large-polarons. These differences exist despite both types of polaron sharing the constraint that their self-trapped carriers only move when atoms alter their

Fig. 3. The coherent motion of a small polaron amongst a row of diatomic molecules is schematically represented. The severely compact self-trapped carrier, hatched region, tunnels between initial and final sites as the atoms, black dots, of these molecules tunnel between positions consistent with the self-trapped carrier's presence.

positions. Transport behaviors of small and large polarons are most simply compared when the temperature is high enough for atomic motion to be treated classically.

In this semiclassical regime the self-trapped carrier of a small polaron moves *incoherently* between single-site localized states as the atoms change positions (semiclassical phonon-assisted hopping). That is, one views the small polaron's self-trapped state being successively destroyed and recreated as it moves between sites. This motion is incoherent because the change in energy of the self-trapped carrier that occurs in the semiclassical hopping process exceeds the electronic transfer energy linking the sites involved in a hop [23–25]. Since the change in the carrier's electronic energy in the semiclassical jump process is a significant fraction of its binding energy and the electronic transfer energy is a fraction of the electronic half-bandwidth, the condition for incoherent motion is similar to that for small-polaron formation, $E_b > W/2$. Indeed, the operational definition of a small polaron is that its self-trapped carrier be sufficiently compact that its semiclassical motion be by hopping.

By contrast, a large polaron moves coherently, with its motion being occasionally interrupted by scattering events. The coherence of a large polaron's motion is a direct consequence of its electronic state extending over multiple sites. That is, the centroid of the self-trapped state can move smoothly between adjacent sites as the self-trapped carrier adjusts to atomic motion. In terms of energies, the motion of a large polaron is coherent because the electronic bandwidth, W, generally exceeds the change of electron energy accompanying intersite motion (a fraction of the self-trapped carrier's binding energy, $3W/(2R^2)$ with $R > 1$). Operationally, observation of high-temperature itinerant transport of a polaron is generally taken to indicate that it is large.

To proceed beyond consideration of semiclassical small-polaron motion, it is to be noted that the translational degeneracy of an ideal lattice ensures that a small-polaron's eigenstates are band-like [23]. However, as depicted in Fig. 3, for a self-trapped carrier to move between adjacent sites, the atomic displacement pattern associated with self-trapping must be transferred between these sites. As a result, the small-polaron bandstates involve collateral tunneling of electrons and atoms. At its greatest (when the self-trapped carrier adiabatically

David Emin

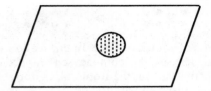

Fig. 4. A schematic representation of a self-trapped carrier confined to a plane. The model of Eq. (10) has the carrier interacting with the atoms of the plane via a short-range (contact) electron–lattice interaction and with ions that surround the plane via the long-range Fröhlich interaction.

follows the atomic motion at absolute zero) the width of a small-polaron band is the product of the vibrational energy, $\hbar\psi$, and a reduction factor (usually about $[(E_b/(\pi\hbar\omega)]^{1/2}\exp(-E_b/\hbar\omega)\ll 1$) associated with atomic tunneling [23].

To maintain the coherence of a small-polaron bandstate in a real solid at absolute zero, the solid's disorder energy must be less than the small-polaron bandwidth. By contrast, maintenance of a large polaron's coherence requires that the disorder energy be less than an energy comparable to the electronic bandwidth. Thus, coherent large-polaron motion generally survives a solid's disorder while coherent small-polaron motion does not.

5 Formation of a large bipolaron

The thrust of this paper is to consider whether bipolarons can produce superconductivity by a mechanism analogous to that which produces the superfluidity of liquid ^{4}He. For such superconductivity the carriers that are analogous to mobile neutral bosons, ^{4}He atoms, are bipolarons that behave as mobile charged bosons. In anticipation of applying this work to materials that are disordered by high densities of charged dopants and structural defects, one requires that these bipolarons robustly resist disorder-induced localization. Since large polarons often form and move itinerantly in doped transition metal oxides [26], it is natural to investigate the conditions under which a large bipolaron, analogous to a large polaron, would form.

Here the formation of a large bipolaron is considered for a model that mimics aspects of the situation found in layered superconductors. As illustrated in Fig. 4, the self-trapped carriers are presumed to exist in a covalent plane embedded within a medium of isotropically displaceable ions. In particular, the carriers sense both (1) atomic displacements within the covalent plane via a short-range electron–lattice interaction and (2) ionic displacements in the surrounding medium via the long-range Fröhlich interaction. Thus, the carriers have their short-range interactions with the atoms of a two-dimensional plane while their long-range interactions are with ions of the three-dimensional medium that surrounds the plane.

With the two carriers of the bipolaron sharing a common state within a common self-trapping well, the energy functional for this model is

$$E_{bi}(R) = \frac{W}{R^2} - 2\left(\frac{1}{\varepsilon_\infty} - \frac{1}{\varepsilon_0}\right)\frac{U}{R} - \frac{4E_b}{R^2} + \frac{U}{\varepsilon_\infty R}. \tag{10}$$

The first term of Eq. (10) is double the corresponding term for a polaron because of there now being two carriers. The second term, arising from the Fröhlich interaction, varies as $1/R$ since the ionic medium is three-dimensional. The third term, arising from the short-range interaction, is proportional to $1/R^2$ since the carriers are confined to a covalent plane. The second and third terms of Eq. (10) are four times the corresponding terms for a polaron because each of the bipolaron's two carriers experiences a potential well that is twice as deep as that for a polaron. The fourth term of Eq. (10) describes the Coulomb repulsion between the two carriers. Similar energy functionals have also been used to study a bipolaron's self-trapping for different arrangements of displaceable atoms [6,7] and when the two carriers of a bipolaron are in a correlated state [27].

To investigate whether a large bipolaron can exist for this model, it is convenient to regroup the terms of Eq. (10):

$$E_{bi}(R) = \frac{W - 4E_b}{R^2} - \left(\frac{1}{\varepsilon_\infty} - \frac{2}{\varepsilon_0}\right)\frac{U}{R}. \tag{11}$$

From Eq. (11) it can be seen that the minimum of this functional can only occur at a finite radius, corresponding to the bipolaron being large, if the coefficient of the term with the weakest R-dependence, namely the term proportional to $1/R$, is negative. Thus, a large bipolaron can only form if $\varepsilon_0 > 2\varepsilon_\infty$.

In addition, for the two carriers' ground state to be that of a large bipolaron, the large bipolaron must be energetically stable with respect to separation into two separate large polarons. This stability question is simply addressed by writing the bipolaron's energy functional in terms of the energy functional of two separate polarons, $2E(R)$:

$$E_{bi}(R) = 2\left[\frac{W}{2R^2} - \left(\frac{1}{\varepsilon_\infty} - \frac{1}{\varepsilon_0}\right)\frac{U}{2R} - \frac{E_b}{R^2}\right] - \frac{2E_b}{R^2} + \frac{U}{\varepsilon_0 R}$$

$$= 2E(R) - \frac{2E_b}{R^2} + \frac{U}{\varepsilon_0 R}. \tag{12}$$

In obtaining Eq. (12) a large portion of the Coulomb repulsion between carriers is offset by the reduction of the Fröhlich potential energy attendant on the two carriers sharing a common potential well.

As shown by Eq. (12), in this treatment, the long-range electron–lattice interaction remains insufficient in itself to produce stable bipolaron formation. Variational treatments that include correlation between the positions of the two electrons find that a large bipolaron's binding can be made slightly greater than that of two separated large polarons for a strong enough long-range electron–lattice coupling strength [28–34]. In these instances, however, the stabilization energy is typically only a very small fraction of a polaron's binding energy.

By contrast, the present work envisions the extra binding provided by the short-range electron–lattice interaction (proportional to E_b) stabilizing a large bipolaron with respect to decomposition into two separate polarons. However, as illustrated in Fig. 5, the bipolaron will collapse to be a small bipolaron if the short-range electron–lattice coupling is too strong. Analysis of this model yields the conditions under which the short-range electron–lattice interaction is strong enough to stabilize a large bipolaron but is not so strong that the bipolaron collapses into a small bipolaron:

$$\frac{4\varepsilon_0/\varepsilon_\infty - 6}{(\varepsilon_0/\varepsilon_\infty)^2 - 2} < \frac{4E_b}{W} < 1. \tag{13}$$

The range of values of $4E_b/W$ bracketed by the inequalities of Eq. (13) widens as $\varepsilon_0/\varepsilon_\infty$ is increased beyond 2.

The model of this paper envisions the most displaceable ions to be those surrounding the plane that contains the carriers. Hence the carriers' Fröhlich interaction with atoms of the carrier-containing plane is ignored. A complementary model, in which the carrier-containing plane contains ions equally displaceable to those that surround the plane, has also been considered [6,7]. Since the carriers then interact with the ions of the plane through both the Fröhlich interaction and the short-range interaction, the energy functionals contain cross terms between the two interactions (c.f. Eq. (6)). While the presence of these cross terms makes that model somewhat more complicated than the model considered here, the findings of the two models are very similar to one another.

Thus, systems in which stable large bipolarons are expected must have large ratios of their static to optical dielectric constants, $\varepsilon_0/\varepsilon_\infty \gg 2$. Structurally stable ionic compounds in which the long-range electron–lattice interactions are relatively strong (e.g., alkali halides) have values of $\varepsilon_0/\varepsilon_\infty$ that only approach 2. However, much larger values of $\varepsilon_0/\varepsilon_\infty$ are measured for the

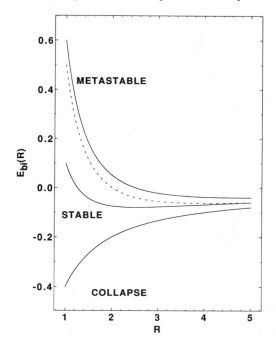

Fig. 5. Energy functionals of a bipolaron, $E_{bi}(R)$, are plotted against R as solid curves. The strength of the short-range component of the electron–lattice interaction increases progressively from zero in the curves labeled as metastable, stable and collapse. The dashed curve is the net energy of two large polarons plotted with the same parameters as those of the metastable bipolaron curve. The physically meaningful energies are those of the minima of these functionals. Upon increasing the strength of the short-range electron–lattice interaction, a large bipolaron can be converted from metastability to stability with respect to separation into two large polarons. However, too strong a short-range electron–lattice interaction causes the large bipolaron to collapse into a small bipolaron.

insulating parents of oxide superconductors [8–10]. These very large values of $\varepsilon_0/\varepsilon_\infty$ (>10) arise from ions that are especially easy to displace. In the cuprate superconductors, these highly displaceable ions surround contiguous CuO_2 layers. Thus, the herein-described model of covalent layers embedded in a three-dimensional medium of displaceable ions is a reasonable representation of the cuprates. The analysis presented here also indicates that it is reasonable to expect large-bipolaron formation in these cases.

6 Some properties of large polarons and bipolarons

Some properties of large bipolarons that distinguish these carriers from conventional carriers and from small bipolarons are discussed in this section.

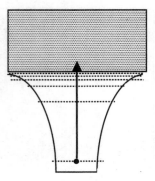

Fig. 6. A large polaron's photoabsorption occurs primarily as a self-trapped carrier is excited from the ground state of the potential well in which it is bound to higher lying states. In this schematic illustration the horizontal dashed lines represent energy levels of the bound and continuum states.

These properties include their infrared absorption, DC effective mass, and very-low-frequency transport. More extensive discussions of these and other properties have been published elsewhere [35,36].

6.1 Polaronic absorption spectra

The absorption spectrum of a polaronic carrier results from photoexcitation of its self-trapped carrier. In the limit of sufficiently strong electron–lattice coupling the photoexcitation can be described within the Franck–Condon approach. In this approach atoms are constrained to remain fixed during the photoexcitation process. The photoabsorption spectrum is then that of a carrier among the corresponding electronic states.

For a large polaron the self-trapped carrier is bound in a hydrogenic-type potential well. As illustrated in Fig. 6, with absorption of a photon a self-trapped electron can be promoted to one of the well's excited levels or to a state in the continuum. Photo-ionizing such a self-trapped carrier gives rise to an absorption spectrum similar to that found for an atom's photoelectric effect [37] or for photo-ionization of a shallow donor of a semiconductor [38]. In particular, absorption associated with photo-ionization begins above the ionization threshold. For the hydrogenic self-trapped state of a large polaron this threshold energy is three times the large polaron's binding energy (c.f., Appendix A of [17]). However, the matrix element for photoabsorption falls as the energy of the final state is increased. This effect is especially rapid when the magnitude of the wavevector of the final electronic state exceeds the reciprocal of the radius of the self-trapped electronic state. As shown in Fig. 7, the

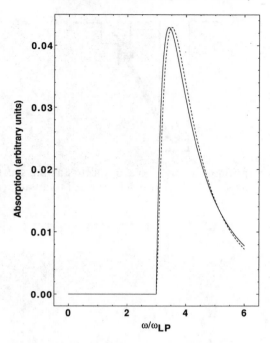

Fig. 7. The absorption coefficients for photo-ionization of three-dimensional (solid curve) and two-dimensional (dashed curve) large polarons are plotted against photon frequency in units of ω_{LP} (the binding energy of a large polaron divided by $\hbar$). For hydrogenic self-trapped states the photo-ionization threshold is three times the large-polaron binding energy, $\omega/\omega_{LP} = 3$.

resulting absorption is generally asymmetric with the high-energy fall-off of the absorption spectrum being more gradual that its rise [35]. The absorption for a large bipolaron is similar in this respect to that for a large polaron.

The absorption spectrum for a small polaron is quite different from that of a large polaron. The relevant electronic levels are those of states that are severely localized at each site of the solid. As illustrated in Fig. 8, the small-polaron absorption results from photo-induced transfer of a carrier from the site at which it is self-trapped to a neighboring site. This absorption is peaked at the difference between the mean electronic energy at an adjacent unoccupied site and the mean electronic energy of a small polaron, $2E_b$ for the Holstein model [12–14]. The absorption between severely localized states is also significantly broadened because atoms often depart from their equilibrium positions. As shown in Fig. 9, the resulting small-polaron absorption band has a qualitatively different shape from that of a large polaron. Thus, the asymmetry of carrier-induced absorptions distinguishes between large- and small-polaronic carriers.

David Emin

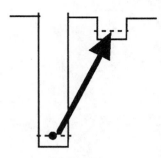

Fig. 8. The photoabsorption of a small polaron occurs as the self-trapped carrier is excited from the site at which it is self-trapped to an adjacent site.

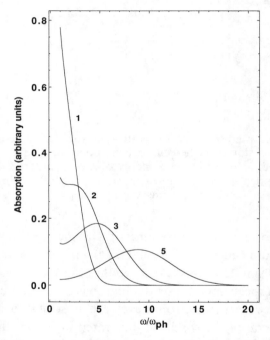

Fig. 9. The $T = 0$ K absorption coefficient of a small polaron in the Holstein model is plotted against photon frequency in units of the characteristic phonon frequency, $\omega/\omega_{\rm ph}$, for the indicated values of $E_{\rm b}/\hbar\omega_{\rm ph}$. For sufficiently strong electron–lattice coupling, an absorption peak is produced at $\hbar\omega = 2E_{\rm b}$.

Production of charge carriers by super-band-gap excitation in insulating parents of cuprate superconductors has been observed to change the material's infrared absorption [39–41] This carrier-induced infrared absorption is asymmetric in the manner expected of large-polaronic carriers. These carrier-induced absorption spectra are similar to those of cuprate superconductors whose carriers are introduced by doping [42]. In some cases, a relatively sharp low-energy peak, suggestive of a carrier's photoexcitation to an excited bound state, is also observed [42]. Thus, the carrier-related infrared absorption spectra in both oxide semiconductors and superconductors are consistent with there being similar large-polaronic carriers in both oxide superconductors and their insulating parents.

6.2 Large-polaronic motion

Motion of a polaron requires motion of the atoms responsible for the carriers' self-trapping. In particular, the mass of an adiabatic large-polaronic carrier, m, depends on the masses of the atoms that must be moved in order to transport the self-trapped carrier: $m = \Sigma_g M_g (\Delta R_g / a)^2$, where ΔR_g is the shift in equilibrium position of the atom of mass M_g at g associated with moving the self-trapped carrier by the lattice constant, a [7,15]. Expressed in other terms, this mass is $m \approx E_p / (\omega R_p)^2$, where E_p is the binding energy of a large polaron of radius R_p [35]. The effective mass of an adiabatic polaron may also be written in terms of the Fröhlich coupling constant, α, and the electronic carrier's mass, m_c, as $m \approx m_c \alpha^4 / 50$ [43,44]. Using this formula for electrons in alkali halides yields effective masses several hundred times the free-electron mass [45]. Since these masses are associated with atomic motion, they can only be observed at frequencies well below those characteristic of atomic motion. Higher frequency measurements tend to measure only the relatively light mass of the electronic carrier, m_c.

A large-polaronic carrier is much heavier and more slowly moving than a free carrier. In particular, the velocity of a large polaron is less than the group velocity of a phonon wavepacket unless the vibrational dispersion is unusually weak. In these instances considerations of conservation of energy and momentum imply that polaron–phonon scattering is dominated by a two-phonon process in which one phonon is absorbed and another is emitted [15]. That is, a large polaron is only weakly scattered as relatively low-momentum phonons are 'reflected' from it. Thus, phonon scattering of a large polaron is unlike that of an electronic carrier in a metal, where the absorption or emission of a phonon greatly affects an electronic carrier's motion. Concomitantly, large polarons are much more effective than electronic carriers in scattering phonons.

Phonons are scattered by a large polaron because its self-trapped charge is redistributed among its sites as atomic positions are altered. This intersite polarization reduces stiffness constants, and hence vibrational frequencies, in the vicinity of the self-trapped carrier from their values in carrier-free regions

[15–17] For this reason, phonons are scattered when they encounter large-polaronic carriers. This scattering is strongest for phonons with small wavevectors, q: $qR_p < 1$ [15].

The scattering rate of a large polaron, $1/\tau$, may be determined by considering the relaxation of a polaron of momentum p. In particular, reflection of a phonon of momentum $\hbar q$ from a large polaron imparts momentum $2\hbar q$ to it. The rate of phonon reflections that slow the polaron exceeds those that accelerate the polaron by $(p/m)\sigma n_q$, where σ is the scattering cross section, about R_p^2, and n_q is the density of phonons of wavevector q. Thus, the momentum relaxation is given by $p/\tau = \langle (2\hbar q)[(p/m)\sigma n_q]\rangle$, where the average is performed over phonon states. For scattering by $q < 1/R_p$ acoustic phonons at high enough temperatures that $n_q \approx k_B T/(\hbar|q|s)$, where s is the velocity of sound, the scattering rate becomes

$$\frac{1}{\tau} \approx \left\langle \left(\frac{2\hbar q}{m}\right) \sigma \left(\frac{k_B T}{\hbar q s}\right)\right\rangle \approx \frac{k_B T}{m s R_p}. \tag{14}$$

Thus, the scattering rate increases with the density of low-energy phonons ($\propto k_B T$) and decreases with the large polaron's effective mass. In other words, the effectiveness with which low-momentum phonons can scatter a relatively high-momentum large polaron falls with the polaron's mass: $\tau \propto m$. For this reason the large polaron's scattering time can be unusually long. Indeed, the long scattering time compensates for the large polaron's heavy mass to yield a sizeable mobility: $\mu = q\tau/m \approx qsR_p/(k_B T)$.

Transport-relaxation in semiconducting cuprates is found to occur in the microwave regime, corresponding to an unusually long relaxation time [8]. Taken together with measurements of the mobility, these results indicate that carriers in these semiconductors have very large effective masses [8]. A long relaxation time and a large effective mass are consistent with motion of independent large polarons.

The similarity between carrier-induced absorptions observed in cuprate semiconductors and superconductors suggests that the carriers are similar in both cases [39–42]. In particular, these carrier-induced absorptions are consistent with the presence of large-polaronic carriers [35]. However, the relaxation times observed in the superconductors' normal states are much shorter than those observed in the corresponding semiconductors [8,42]. These results suggest that large-polaronic carriers move differently in a superconductor's normal state than in a semiconductor [35]. In particular, interactions between large-polaronic carriers in a superconductor's normal state would lead to their moving collectively rather than independently [36].

7 Collective properties of large bipolarons

Interactions between charge carriers can qualitatively affect their motions. In particular, superconductivity and superfluidity are each a property of a collective ground state of interacting particles. Resistanceless flow results from being unable to create excitations that disturb the cooperative flow of interacting particles' collective ground state. A superconductor's Meißner effect manifests the resistance of the collective ground state to magnetic-field-induced rotational flow that would contribute to the system's paramagnetism.

This section considers the sources and effects of interactions between large bipolarons. In particular, the origin of a phonon-mediated attraction between large bipolarons is described. This attractive interaction can enable an imperfect gas of large bipolarons to condense into a liquid. If the ground state of the interacting large bipolarons remains fluid rather than solidifying, then its ground state is a 'quantum liquid' like that of liquid ^{4}He. Such a large-bipolaronic liquid could then exhibit a form of superconductivity analogous to the superfluidity of liquid ^{4}He.

7.1 Interactions between large bipolarons

Being of like sign, large bipolarons exhibit a Coulomb repulsion of one another. However, since the bipolarons are constrained to follow atomic motions, atomic motion is fast enough that polarization due to atomic displacements can affect this Coulomb repulsion. Specifically, the Coulomb repulsion between large bipolarons separated by the distance s is reduced from that of bare charges by the static dielectric constant: $(2e)^2/(\varepsilon_0 s)$, when $s > R_p$ [16,17].

Another effect causes large bipolarons to repel one another strongly when their self-trapped carriers are close enough so that they begin to merge into a quadpolaron. Specifically, with four carriers within a common potential well, the Pauli principle requires that at least two of the four carriers occupy excited states. This promotion energy destabilizes a large quadpolaron with respect to two separate large bipolarons [17,46] In fact, merger of two large bipolarons into a large quadpolaron is energetically very unfavorable. Therefore, large bipolarons strongly repel one another at small enough separations for the wavefunctions of self-trapped carriers of different large bipolarons to overlap with one another. This repulsion may be described as a hard-core repulsion similar to that between ^{4}He atoms: $\varepsilon_R(s) = C_1 E_p \exp(-C_2 s/R_p)$, where C_1 and C_2 are numerical constants [16,17].

A phonon-mediated attraction between large bipolarons also exists [16,17].

This attractive interaction results from the self-trapped charge of a large-polaronic carrier being redistributed among sites in response to a change of atomic positions. This intersite charge redistribution, absent for single-site (small) bipolarons, reduces the stiffness of the solid in the vicinity of a large bipolaron. As a result of this local softening of the lattice, addition of charge carriers can introduce local vibrational modes at large wavevectors, $qR_p > 1$. Indeed, adding charge carriers to cuprates is observed to introduce extra vibrational modes that only exist at large wavevectors, which are called 'ghost modes' [47–49]. In addition, introduction of large-polaronic carriers will generally reduce the frequencies of small-wavevector phonons, $qR_p < 1$. This carrier-induced reduction in phonon frequency is enhanced when the separations between large-polaronic carriers are less than the phonon wavelength, $s < \lambda$, since different softened regions of the solid (e.g., large bipolarons) then affect the phonons coherently [16,17]. Thus, introduction of large-polaronic carriers into a solid reduces its zero-point vibrational energy. Since this energy reduction is enhanced as the separation between large bipolarons is reduced, carrier-induced softening provides a driving force for large bipolarons to amalgamate at low temperatures. In other words, the polarizability of large bipolarons leads to low-temperature phonon-mediated attraction between them. The attractive energy between two large bipolarons is roughly $\varepsilon_A(s) \simeq -\hbar\omega(R_p/s)^4$, for $s > R_p$ [16,17]. The proportionality of this energy to $\hbar\omega$ indicates that the attraction is a quantum-mechanical effect associated with atomic motion.

7.2 Formation of a liquid of large bipolarons

The interaction potential between two large bipolarons at low temperatures is the sum of contributions from their mutual Coulomb repulsion, short-range repulsion and phonon-mediated attraction. However, large bipolarons are only expected to form in multi-dimensional ionic systems when the material's static dielectric constant is especially large. Therefore, it is appropriate to consider the situation in which the Coulomb repulsion between large bipolarons is significantly suppressed by the existence of a very large static dielectric constant. Then, as illustrated in Fig. 10, short-range repulsion can combine with phonon-mediated attraction to produce a minimum in the interaction energy as a function of the separation, s.

Having a minimum in the interaction energy between two large bipolarons ensures that large bipolarons will condense from a gas into a liquid at sufficiently low temperatures. This situation is just that envisioned for a van der

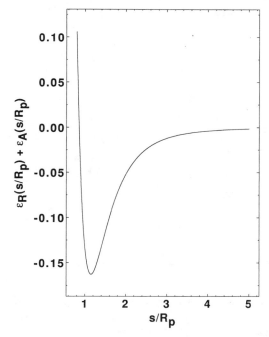

Fig. 10. The interaction energy (in dimensionless units) between two large bipolarons is plotted against their separation, s, divided by their radius, R_p. The short-range repulsion, $\varepsilon_R(s/R_p)$, falls exponentially with s/R_p. The longer-range attraction, $\varepsilon_A(s/R_p)$, diminishes as $(R_p/s)^4$ at large separations.

Waals liquid [50]. The repulsion constant of the van der Waals model, b, then corresponds to the volume of the large bipolaron. The van der Waals attraction constant, a, is roughly the product of the depth of the attractive potential, about $\hbar\omega$, and the volume over which the net interaction is attractive.

Condensation of a gas of large bipolarons into a normal liquid can be described by this van der Waals model [17]. The conditions for condensation that result are shown in the partial phase diagram of Fig. 11 [16,17]. In Fig. 11 the product of the evaporation temperature and the Boltzmann constant, k_B, in units of a/b, is plotted against the product of the global concentration of large bipolarons, c, and b. In addition, a vertical line is drawn at the concentration beyond which large bipolarons are taken to no longer exist: $c = c_m$. In the 'overdoped' regime, $c > c_m$, the atomic displacements responsible for self-trapping of carriers are presumed to interfere sufficiently with one another to preclude self-trapping. Thus, a bipolaronic liquid only exists at sufficiently low temperatures within a temperature-dependent bounded range of overall carrier concentrations.

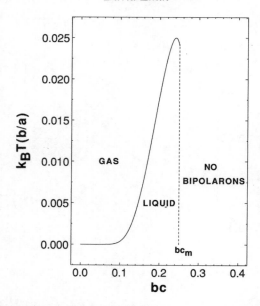

Fig. 11. Large bipolarons will be bound into a liquid only within a limited range of global carrier concentrations, c, and temperatures, T. Here the constants b and a respectively characterize the short-range repulsion and the longer-range attraction of the van der Waals model of a liquid. The solid line gives the temperature above which the liquid of large bipolarons has completely evaporated to a gas of large bipolarons.

When $c > c_m$, the carrier density is too great to permit bipolaron formation.

7.3 Excitations and superconductivity of a large-bipolaronic liquid

Superconductivity can result from large bipolarons if their ground state remains liquid rather that solidifying. In this picture, the large bipolarons are charged bosons that condense into a quantum liquid. The large bipolarons are analogous to the neutral bosons, ^{4}He atoms, whose liquid ground state manifests superfluidity.

A Bose liquid will exhibit resistanceless flow if the requirements of conservation of energy and momentum preclude excitations that would disrupt its ground state's flow. In particular, Landau argues that excitations of a liquid having momentum p and energy $E(p)$ cannot disturb its ground state flowing with a finite velocity, v, if $[E(p)/|p|]_{\text{minimum}} > v > 0$ [51]. Furthermore, requiring gauge invariance for the pure-potential flow of a condensate of charged bosons leads to the London equation that describes the Meißner effect [52]. In addition to their diamagnetic flow, large bipolarons can each manifest an asymmetric Van Vleck paramagnetism as a result of their having a finite size [53]. Thus, just as the liquid ground state of neutral bosons exhibits superfluidity, so the liquid condensate of charged bosons can manifest superconductivity.

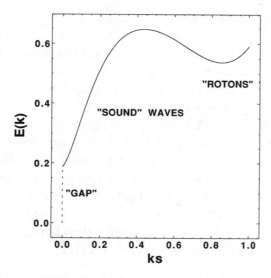

Fig. 12. The excitations energies expected of a bipolaronic liquid, $E(k)$, are plotted against their wavevector, k, up to values of $k \simeq 1/s \simeq 1/R_p$, the large-bipolaron's radius. The maximum energy of this excitation spectrum is expected to be about $\hbar^2/m(R_p)^2 \simeq (\hbar\omega)^2/E_p < \hbar\omega$, a characteristic phonon energy. The long-wavelength excitations of a liquid are sound waves. However, a 'gap' occurs between the ground state and excitations of arbitrarily small k because the carriers carry charge. The non-monotonic excitations sketched near $ks \approx 1$ indicate the possibility of more complicated excitations, akin to the rotons of ^{4}He.

In the absence of interactions between the particles, the system's excitations are just free carriers, $E(p) = p^2/(2m)$. Then, Landau's condition for resistanceless flow cannot be fulfilled since $E(p)/|p| \to 0$ as $|p| \to 0$. However, interactions between carriers can alter the ground states' excitation spectra so as to permit superconductivity. In particular, because of the carriers being charged, the long-wavelength excitations that result for a large-bipolaronic liquid, $E(p) \to \hbar[4\pi(2e)^2 c/(\varepsilon_0 m)]^{1/2} > 0$ as $|p| \to 0$, are compatible with the Landau condition. Of course the Landau condition is also compatible with the long-wavelength excitations of a neutral liquid, namely sound waves: $E(p) \to v_1|p|$ as $|p| \to 0$, where v_1 is the velocity of sound in the liquid. Thus, the excitation spectrum expected for a large-bipolaronic liquid, illustrated in Fig. 12, will generally satisfy the Landau criteria for resistanceless flow. The possibility of a non-monotonic dependence of the excitation energy on p at short wavelengths, analogous to the roton behavior of ^{4}He, is also depicted in the spectrum of Fig. 12.

Resistanceless flow and the Meißner effect are only properties of the bipolaronic liquid's ground state. Nonetheless, finite occupation of this individual state and the resulting superconductivity are possible since bipolarons behave as bosons. The temperature below which finite occupation of the bipolaronic liquid's ground state occurs, namely the liquid's Bose condensation temperature, is the superconducting transition temperature, T_c. However, a Bose liquid's condensate and excitations are unlike those of an ideal Bose gas. In particular, interactions between the condensate's particles drive them from their $k = 0$ momentum state. Penrose and Onsager argue that the number of the condensate's particles that occupy the $k = 0$ state decreases as the liquid's short-range order increases [54]. For example, even at the lowest temperatures, the condensate of superfluid ^{4}He only has $< 10\%$ of its atoms in their $k = 0$ state. In addition, a liquid's excitations possess a collective character that differentiates them from the single-particle excitations of an ideal gas. For this reason, one cannot equate the thermodynamic average for the number of excitations of a Bose liquid with its particle number to obtain the condensation temperature, T_c [55].

Bipolaronic superconductivity cannot occur if the bipolarons' ground state solidifies rather than remains liquid. In the case of superfluid liquid ^{4}He, the atoms' zero-point energies alone preclude their condensation into a solid phase [56]. For bipolarons, the potential produced by the solid's atoms will also tend to frustrate simple solidification of bipolarons causing either liquid formation or multi-phase/multi-domain solidification. However, at some carrier densities, polaronic carriers can order in a manner that is commensurate with the underlying lattice, thereby forming a superlattice.

Evidence of ordering of polaronic carriers, rather than superconductivity, is found in transition metal oxides that have a familial relationship to the cuprate superconductors. Specifically, diffraction studies show polaronic carriers solidifying into superlattices in non-superconducting $La_{2-x}Sr_xNiO_{4+y}$ [57]. This type of solidification may be easier in nickelates than in cuprates because the nickelates' self-trapped carriers are more compact, as indicated by the carrier-induced absorption bands in nickelates having their onset at significantly higher energies than those of cuprates [41]. Nonetheless, solidification of bipolarons may even occur in cuprates, as indicated by the loss of superconductivity when single crystals of $La_{2-x}Sr_xCuO_4$ and $La_{2-x-y}Ba_xTh_yCuO_4$ are doped to yield $\frac{1}{8}$ hole per unit cell [58,59]. In particular, this carrier density is consistent with bipolarons ordering in a manner commensurate with the underlying tetragonal lattice: one bipolaron for each four-by-four superlattice unit, $2/(4 \times 4) = \frac{1}{8}$. The absence of superconductivity when bipolarons condense to an ordered state complements the notion that superconductivity results when the bipolarons' ground state remains liquid. There is even evidence that photocarriers in insulating cuprates condense into droplets and that this liquid phase is a precursor to superconductivity [60–62].

Acknowledgement

This work was performed under the auspices of the USA Department of Energy and was funded in part by the Office of Basic Energy Sciences, Division of Materials Sciences under contract DE-AC04-94AL85 000.

References

[1] H. Fröhlich, *Adv. Phys.* **3**, 325 (1954).
[2] Y. Toyozawa, *Prog. Theor. Phys.* **26**, 29 (1961).
[3] T. Holstein, *Ann. Phys. (N.Y.)* **8**, 325 (1959).
[4] D. Emin, *Adv. Phys.* **22**, 57 (1973).
[5] D. Emin and T. Holstein, *Phys. Rev. Lett.* **36**, 323 (1976).
[6] D. Emin, *Phys. Rev. Lett.* **62**, 1544 (1989).
[7] D. Emin and M. S. Hillery, *Phys. Rev. B* **39**, 6575 (1989).
[8] D. Reagor, E. Ahrens, S.-W. Cheong, A. Migliori and Z. Fisk, *Phys. Rev. Lett.* **64**, 2048 (1989).
[9] G. A. Samara, W. F. Hammetter and E. L. Venturini, *Phys. Rev. B* **41**, 8974 (1990).
[10] G. Cao, J. W. O'Reilly, J. E. Crow and L. R. Testardi, *Phys. Rev. B* **47**, 11510 (1993).
[11] D. Emin, *Phys. Rev. B* **48**, 13691 (1993).
[12] M. I. Klinger, *Phys. Lett.* **7**, 102 (1963).
[13] H. G. Reik and D. Heese, *J. Phys. Chem. Solids* **28**, 581 (1967).
[14] D. Emin, *Adv. Phys.* **24**, 305 (1975).
[15] H.-B. Schüttler and T. Holstein, *Ann. Phys. (N.Y.)* **166**, 93 (1986).
[16] D. Emin, *Phys. Rev. Lett.* **72**, 1052 (1994).
[17] D. Emin, *Phys. Rev. B* **49**, 9157 (1994).
[18] T. G. Castner and W. Känzig, *J. Phys. Chem. Solids* **3**, 178 (1957).
[19] W. P. Su, J. R. Schrieffer and A. J. Heeger, *Phys. Rev. B* **22**, 2099 (1980).
[20] L. D. Landau, *Phys. Z. Sowjetunion* **3**, 644 (1933).
[21] E. I. Rashba, *Opt. Spektrosk.* **2**, 75 (1957).
[22] D. Emin and M.-N. Bussac, *Phys. Rev. B* **49**, 14290 (1994).
[23] T. Holstein, *Ann. Phys. (N.Y.)* **8**, 343 (1959).
[24] D. Emin, in *Electronic and Structural Properties of Amorphous Semiconductors*, edited by P. G. Le Comber and J. Mort (Academic, New York, 1973), pp. 262–328.
[25] D. Emin, *Phys. Rev. B* **43**, 11720 (1991).
[26] A. J. Bosman and H. J. van Daal, *Adv. Phys.* **19**, 1 (1970).
[27] D. Emin, J. Ye and C. L. Beckel, *Phys. Rev. B* **46**, 10710 (1992).
[28] V. L. Vinetskii and M. Sh. Gitterman, *Sov. Phys. JETP* **13**, 1023 (1961).
[29] S. G. Suprun and B. Ya. Moizhes, *Sov. Phys. Solid State* **24**, 903 (1982).
[30] H. Hiramoto and Y. Toyozawa, *J. Phys. Soc. Japan* **54**, 245 (1985).
[31] J. Adamowskii, *Phys. Rev. B* **39**, 3649 (1989).

108 *David Emin*

[32] F. Bassani, M. Geddo, G. Iadonisi and D. Ninno, *Phys. Rev.* B **43**, 5296 (1991).
[33] S. Sil, A. K. Giri and A. Chatterjee, *Phys. Rev.* B **43**, 12642 (1991).
[34] G. Verbist, F. M. Peters, and J. T. Devreese, *Phys. Rev.* B **43**, 2712 (1991).
[35] D. Emin, *Phys. Rev.* B **48**, 13691 (1993).
[36] D. Emin, *Phys. Rev.* B **45**, 5525 (1992).
[37] L. I. Schiff, *Quantum Mechanics*, 2nd Ed. (McGraw-Hill, New York, 1955) pp. 271–5.
[38] E. Burstein, G. S. Pikus and N. Sclar, *Proc. Photoconductivity Conf. Atlantic City* (Wiley, New York, 1956), p. 353.
[39] C. Taliani, R. Sambona, G. Ruani, F. G. Matacotta and K. I. Pokhodya, *Solid State Commun.* **66**, 487 (1988).
[40] Y. H. Kim, C. M. Foster, A. J. Heeger, S. Cox and G. Stucky, *Phys. Rev.* B **38**, 6478 (1988); D. Mihailovic, C. M. Foster, K. Voss and A. J. Heeger, *Phys. Rev.* B **42**, 7989 (1990).
[41] X-X. Bi and P. C. Eklund, *Phys. Rev. Lett.* **70**, 2625 (1993).
[42] K. Kamaras, S. L. Herr, C. D. Porter, N. Tache, D. B. Tanner, S. Etemad, T. Venkatesan, E. Chase, A. Inam, X. D. Wu, M. S. Hedge and B. Dutta, *Phys. Rev. Lett.* **64**, 84 (1990).
[43] L. D. Landau and S. I. Pekar, *Zh. Eksp. Teor. Fiz.* **18**, 419 (1948).
[44] H. Fröhlich, in *Polarons and Excitons* G. G. Kuper and G. D. Whitfield, Eds. (Plenum, New York, 1963) p. 1.
[45] J. Appel, in *Solid State Physics* vol. 21, Ed. F. Seitz, D. Turnbull and H. Ehrenreich (Academic, New York, 1968), p. 215.
[46] D. Emin, in *Physics and Materials Science of High Temperature Superconductors – II*, Ed. R. Kossowsky, B. Raveau, D. Wohlleben and S. Patapis (Kluwer, Dordrecht, 1992) pp. 27–52.
[47] H. Rietschel, L. Pintschovius and W. Reichardt, *Physica* C **162–164**, 1705 (1989).
[48] N. Pyka, W. Reichardt, L. Pintschovius, G. Engel, G. Rossat-Mignod and Y. H. Henry, *Phys. Rev. Lett.* **70**, 1457 (1993).
[49] D. Emin, *Phys. Rev.* B **43**, 8610 (1991).
[50] L. D. Landau and E. M. Lifshitz, *Statistical Physics* (Pergamon, London, 1958) Sec. 73.
[51] L. D. Landau, *Zh. Eskp. Teor. Fiz. Nauk* **59**, 592 (1941).
[52] E. M. Lifshitz and L. P. Pitaevskii, *Statistical Physics, Part 2: Theory of the Condensed State* (Pergamon, Oxford, 1980), Sec. 44.
[53] D. Emin, *Phys. Rev.* B **43**, 2633 (1991).
[54] O. Penrose and L. Onsager, *Phys. Rev.* **104**, 576 (1956).
[55] P. W. Anderson, *Basic Notions of Condensed Matter Physics* (Addison-Wesley, New York, 1984).
[56] R. P. Feynman, *Statistical Mechanics* (Addison-Wesley, New York, 1972) Chap. 11.
[57] C. H. Chen, S-W. Cheong and A. S. Cooper, *Phys. Rev. Lett.* **71**, 2461 (1993); S.-W. Cheong, H. Y. Hwang, C. H. Chen, B. Batlogg, L. W. Rupp, Jr. and S. A. Cooper, *Phys. Rev.* B **49**, 7088 (1994).
[58] Y. Maeno, N. Kakehi, M. Kato and T. Fujita, *Phys. Rev.* B **44**, 7753 (1991).
[59] H. Takagi, B. Batlogg, H. L. Kao, J. Kwo, R. J. Cava, J. J. Krajewski and W. F. Peck, Jr., *Phys. Rev. Lett.* **69**, 2975 (1992).
[60] Y. H. Kim, S.-W. Cheong and Z. Fisk, *Physica* C **200**, 201 (1992).

[61] G. Yu, C. H. Lee, A. J. Heeger, N. Herron, E. M. McCarron, L. Cong, G. C. Spalding, C. A. Nordman and A. M. Goldman, *Phys. Rev.* B **45**, 4964 (1992).
[62] J. B. Goodenough, J.-S. Zhou and J. Chan, *Phys. Rev.* B **47**, 5275 (1993); J.-S. Zhou, H. Chen and J. B. Goodenough, *Phys. Rev.* B **49**, 9084 (1994).

7

Polarons and bipolarons in WO_{3-x} and $YBa_2Cu_3O_7$

E. K. H. SALJE

IRC in Superconductivity and Department of Earth Sciences, Downing Street,
Cambridge CB2 3EQ, UK

Abstract

The physical properties of polarons and bipolarons in WO_{3-x} are reviewed and compared with characteristics of carriers in $YBa_2Cu_3O_7$ and several other high-temperature superconductors, namely $(Ca_{1-x}Y_x)Sr_2(Tl_{0.5}Pb_{0.5})Cu_2O_7$, $Bi_2Sr_2(Ca_{0.9}Y_{0.1})Cu_2O_{8+\delta}$ and $La_2CuO_{4+\delta}$. The fingerprint for (bi)polarons is optical excitations in the spectral near-infrared region. The absorption cross section is drastically reduced in the superconducting phase. The temperature evolution is analysed quantitatively in terms of Bose–Einstein condensation of bipolarons.

1 Introduction

The physics of polarons and bipolarons has recently been reconsidered because it is believed that condensation of bipolarons is closely related to, or may even be the origin of, superconductivity in oxide materials [1–12]. The justification for such belief is based on several experimental observations such as the absence of the Korringa law in the nuclear spin relaxation rate [3], the heat capacity anomaly [4,5] and the softening of phonons above the (pseudo-) gap in the superconducting phase [7,8,13–15]. Few experimental results point directly to the existence of polarons and/or bipolarons in these materials, however. Probably the most direct indication for the existence of such particles stems from observation of their internal excitations in the infrared and visible spectral range. Such excitations were firmly established in WO_{3-x} (in its ε-phase) and related transition metal oxides [16–30]. Similar excitations were recently observed in $YBa_2Cu_3O_7$ and other high-temperature superconductors [31–37]. Although their original discovery by Dewing and Salje [32] was contested on experimental grounds, it is now confirmed that the apparently contradictory result that such excitations were not seen in reflection spectra of $YBa_2Cu_3O_7$ crystals lies in the insensitivity of early reflection measurements and statistical errors introduced by subsequent Kramers–Kronig analysis [38,39]. In fact, existence of an absorption signal is fully compatible with the reflection spectra in [40–44] and new experimental studies clearly confirm the earlier absorption spectra both in powders and in thin films [31].

The importance of these signals lies in the observation that the oscillator strength diminishes when the sample becomes superconducting [31–34]. This result indicates that the particles that produce absorbance at $T > T_c$ are also involved in condensation at $T < T_c$. If these particles are bipolarons then it appears likely that bipolarons are at the origin of superconductivity in $YBa_2Cu_3O_7$ and other high-temperature superconductors.

A very strong indication that bipolarons are, indeed, the carriers in question stems from the close similarity of the excitation spectra in WO_{3-x} and high-temperature superconductors. It is the purpose of this paper to illustrate these similarities and discuss the expected dependence of oscillator strength on the superconducting order parameter. The paper is organized as follows: in part 2 the main results for WO_{3-x} are reviewed and the optical excitations in $YBa_2Cu_3O_7$ and other superconductors are discussed in part 3.

2 Polarons and bipolarons in WO_{3-x}

Tungsten oxide, WO_{3-x}, and related transition metal oxides MoO_{3-x}, $NbO_{2.5-x}$ etc. show properties of charge carrier transport, which can be described neither by that of free electrons nor by diffusive motion of ions [16–30,45–59]. Several mechanisms can exist in one and the same material, e.g. many transition metal oxides show thermal transitions between phases with metallic and insulating properties (MI transitions). Much research has focused on the physical nature of the relevant charge carriers and their transport mechanisms. An almost ideal model compound for such studies is WO_{3-x} for several reasons.

1. Charge carriers can easily be generated or destroyed by chemical reactions or doping of WO_3 with H, V, Mo or implantation/removal of oxygen [24,25,27,53,54,57,60].
2. The carriers are not pinned to structural defects for undoped crystals with reduced oxygen content because point defects are locally compensated by formation of blocks of corner-sharing WO_6 octahedra. These blocks can form extended defect structures (the so-called CS (crystallographic shear) structures). The structural properties of these materials are rather well understood [53,54,60,61].
3. WO_3 undergoes an MI transition from a quasi-metallic phase (δ) to an insulating phase (ϵ) under cooling ($T_{trans} \approx 250$ K). The different transport properties can, therefore, be measured in the same individual crystal [60].

4. Crystal growth of slightly reduced WO_{3-x} ($x \lesssim 0.01$) is very simple and produces thin plates with excellent structural uniformity.
5. The types of carriers can be characterized in a straightforward manner using electron spectroscopy techniques (e.g. XPS). In all well-crystallized WO_{3-x} ($x \leq 0.28$) samples, only two types of carrier are found: those related to local W^{5+} states and quasi-free electrons. The W^{5+} states are indicative of polaronic and/or bipolaronic carriers [57,58].

Formation of polarons of small or intermediate size is specific for some structural phases of WO_{3-x} but not for all. At room temperature (δ-WO_{3-x}) and at elevated temperatures (phases α, β and γ) crystals of WO_{3-x} ($x < 0.001$) behave as poor metals. Their electronic transport properties are characterized by carriers with slightly enhanced mass and Hall coefficients well in agreement with the predictions of almost free carriers with weak electron–phonon coupling (i.e. large polarons) [18,19,24,25,59]. The optical properties are dominated by Drude absorption and a fundamental absorption edge with $E_g = 2.77$ eV. At temperatures close to 250 K, the crystal transforms on cooling into ε-WO_{3-x} [60]. The ε-phase is insulating with a larger gap energy $E_g = 3.05$ eV. The Drude absorption disappears within experimental resolution. Instead, the optical absorption spectra display broad peaks in the near infrared and visible spectral regions [27,28].

2.1 Small polarons in ε WO_{3-x}

We first discuss the characteristics of unpaired single polarons because their physical conceptualization and experimental analysis are substantially easier than those of bipolarons. Single polarons are either generated thermally or, more conveniently, by photoexcitation of bipolarons [19,25]. The fingerprint of single polarons is a strong ESR signal under illumination of ε-WO_{3-x} [30]. The spectral dependence of this photoeffect is shown in Fig. 1. The maximum efficiency occurs near 1 eV; the spectral profile is identical with the excitation spectrum of bipolarons as will be shown below.

Let us now discuss the results of ESR measurements in some detail. The ESR pattern for a magnetic field parallel to the crystallographic b axis (Fig. 2) can be analysed as a superposition of four sets of lines:

a) one set with large hyperfine splitting into two components;
b) two sets with small splitting into two components, each having twice the intensity of the outer set; and
c) one large unsplit central component.

The intensity ratio of the outer components to the central one corresponds to the overlap of the wavefunction with one W nucleus; that of the inner lines

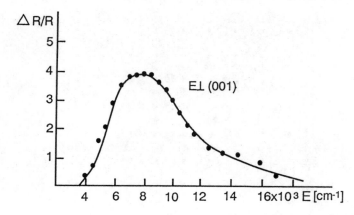

Fig. 1. The energy-dependence of the ESR photoeffect. $\Delta R/R$ is the relative change in ESR signal generated by illumination through splitting of bipolarons into pairs of polarons ($T = 20$ K).

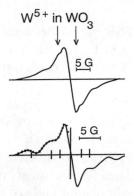

Fig. 2. The ESR signal of W^{5+} in WO_{3-x} at $T = 20$ K for two centres U (top) and L (bottom) showing a sequence of hyperfine splittings (after [4]).

corresponds to the equivalent overlap with two W nuclei. A detailed analysis shows that, in fact, two polaronic signals exist where one (U) has a more extended wavefunction than the other (L).

All the measured hyperfine couplings of centre U are smaller, typically by about 20%, than those of L. Therefore U has a still more extended wavefunction, and the spread by the radii of the d orbitals, as calculated for L, is a lower bound for U. Furthermore, the lines of U are wider than those of L even at 4.2 K, at which hopping is unimportant. In fact, the U lines do not narrow when the temperature decreases below 20 K. The persisting large linewidth is due to unresolved interaction with additional W nuclei, resulting from the larger wavefunction extension. The large wavefunction radius and the corresponding lower binding energy of U are also consistent with the observation that, with

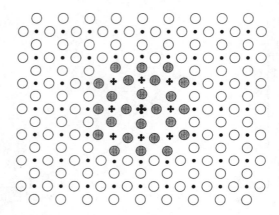

Fig. 3. A sketch of the polaron (U), which extends over nine tungsten sites. Shown is the a–b plane, the extension of the polaron along the c axis is one unit cell, i.e. the polaron is disc-shaped in the a–b plane.

rising temperature, the U ESR signal becomes wider at a faster rate and saturates less easily than that of L. We attribute this to a higher mobility of electrons in U, which causes shorter relaxation times. Near 60 K the lines are broadened beyond detection.

All these observations can consistently be explained if it is assumed that polarons have been identified, characterized by a dominant electron density at the minimum of the self-induced potential well, but still having a large probability density away from the central nucleus. The spread of the wavefunction is larger than that postulated for a small polaron, which is confined to one lattice cell (Fig. 3), but we still cannot classify these electron states as large polarons, since their mobility is too low. From the smallest resolved hyperfine splitting, $A \simeq 5 \times 10^4$ cm^{-1}, we calculate that the electron rests at one site for at least $\tau = \hbar(A\hbar c) \simeq 7 \times 10^{-8}$ s. With the Einstein relation $\mu kT = eD$ and $D \simeq d^2/\tau$ ($d \simeq 4$ Å, W–W distance) one deduces $\mu < 10^{-5}$ cm^2 V^{-1} s. This is much below the limit of validity of a band picture of a large polaron, about 1 cm^2 V^{-1} s, implied by a large polaron model.

We now discuss optical excitations of polarons in WO$_{3-x}$. The Franck–Condon transition with frozen-in phonon degrees of freedom occurs for a single polaron at $2W_p$, where W_p is the polaron self-energy. The shape and position of the absorption band have been anticipated by several authors. The historically first treatment by Reik [62] leads to an approximate profile

$$\alpha \propto \frac{N}{\omega} \exp\left(-\frac{(\hbar\omega - 2W_p)^2}{4W_p \hbar\omega_{phonon}}\right),$$

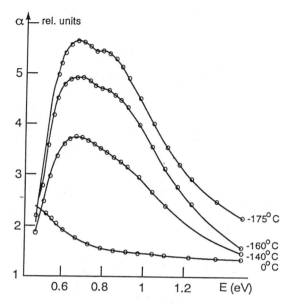

Fig. 4. The absorbance of WO_{3-x} at low temperatures. At 0 °C, WO_{3-x} is in the δ-phase, showing Drude absorbance. Upon cooling the sample into the ε-phase ($T_c = -27$ °C), the polaronic absorption profile appears with two types of polarons (U and L) visible at the lowest temperatures.

where ω_{phonon} is the frequency of the coupled phonon and N is the polaron density. The experimentally observed profile (Fig. 4) shows a maximum at 0.71 eV, which corresponds to $W_p = 0.3$ eV. The phonon energy results from the linewidth and is $\hbar\omega_{phonon} = 0.07$ eV, i.e. in the expected spectral range of phonon modes of the WO_6 octahedra [61]. The linear relationship between α and polaron density N was found to hold for small doping levels whereas high doping levels lead to significant changes in line profiles [27].

The significance of WO_3 for studying optical excitation of polarons lies in its unusual property of possession of a phase transition between the metallic δ-phase without any signature of small or intermediate polarons and the insulating ε-phase, which contains polarons. Optical absorption in the δ-phase is that of Drude-type free electrons, i.e. a weak increase in α with decreasing energy (Fig. 4). In weakly reduced material, one observes absolute values of $\alpha \approx 8$ cm^{-1} at 0.75 eV. The equivalent absorbance is increased by one order of magnitude when the crystal is cooled into the ε-phase. Further cooling increases the absorption cross section. The high absorption cross section of small and intermediate polarons provides experimentalists with an excellent and most sensitive tool for detection of self-trapped carriers. It should also be noted that the relevant matrix element seems to show little dependence on the

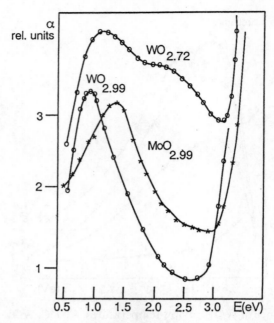

Fig. 5. The absorption spectra of $WO_{2.99}$, $WO_{2.72}$ and $MoO_{2.99}$ showing similar broad absorption profiles near 1 eV.

choice of transition metal in the centre of the polaron. In Fig. 5 the absorption profiles of $WO_{2.99}$, $MoO_{2.99}$ and $WO_{2.72}$ are compared.

2.2 The bipolaron ground state

When WO_3 is cooled in the dark, no ESR signal is observed in the ε-phase. This experimental result shows that all spins are paired in the ground state. The optical absorption shows a signal similar to that of polarons but shifted to higher energies (Fig. 5). The absorption profile is identical, within experimental resolution, to the spectral dependence of the ESR photoeffect in Fig. 1. The close similarity between the spin-paired absorption spectra and those found for the polarons suggests that the optical absorption is due to excitation of spin-paired polarons; e.g. bipolarons, resulting in their splitting into two single polarons. The increased peak energy indicates a lower ground state of the bipolaron with respect to the single polaron ($\Delta W_{p} = \frac{1}{2}\Delta E_{opt} = 0.15$ eV).

The two following models for bipolaron ground states have been discussed:

1. *The on-site bipolaron.* If spin pairing occurs at the same tungsten position then we expect the ground state of all W sites inside the polaron to be W^{4+} whereas the surrounding W sites remain W^{6+}.

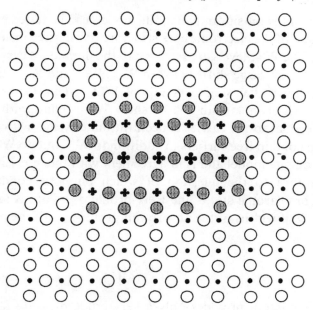

Fig. 6. The model of a bipolaron in ε-WO$_{3-x}$ in the crystallographic *a–b* plane. The extension in the *c* direction is only one unit cell. The deformation cloud contains about 15 tungsten sites.

2. *The inter-site bipolaron.* Spin pairing occurs between sites enclosed by the same phonon 'deformation' cloud, i.e. the total spin vanishes per bipolaron although each W site retains one spin. The valence state is W^{5+}. The spatial extent of the inter-site bipolaron is rather great because it consists of two polaronic wavefunctions with a small elastic contraction due to the common phonon interaction. As shown above, each polaron extends over nine metal sites so that the bipolaron has to exceed this size. The upper limit for its size is twice the polaron size, i.e. 18 sites. If we assume overlap of the three sites at the rim of the polaron wavefunction then we find an inter-site bipolaron configuration in Fig. 6 that contains 15 metal sites. The geometrical size of this bipolaron is 16 Å × 8 Å confined to the *a–b* plane. An alternative arrangement would consist of two polarons on adjacent layers along the crystallographic *c* axis, resulting in a more spherically shaped bipolaron of diameter 10 Å. As the transport properties are highly two-dimensional ($\sigma_{ab}/\sigma_c > 10^2$) [25], it appears that bipolarons are confined to the *a–b* plane, favouring the idea of disc-shaped bipolarons rather than spherical ones. The reason for the disc or pancake shape of polarons and bipolarons in ε-WO$_3$ is presumably not simply the anisotropy of the crystal structure (with larger W–O–W distances along the crystallographic *c* axis compared with those in the *a–b* plane) but the intrinsic d$_{x^2-y^2}$ nature of the W^{5+} orbitals. The role of the crystal structure is to align these orbitals in the *a–b* plane, which greatly enhances the anisotropy of the transport properties compared with a purely structural anisotropy.

The validity of the on-site and inter-site models can be tested experimentally by measurement of the electronic core structure of tungsten inside the polaron cloud [54,55]. The W 4f levels are sensitive to the number of carriers located at the W site and show large shifts between W^{6+}, W^{5+} and W^{4+}. Extensive XPS studies have shown that single crystals of WO_{3-x} ($x < 0.28$) show no W^{4+} signals at all. All crystals with $x < 0.1$ show W^{5+} signals with a $W^{5+}/(W^{6+} + W^{5+})$ ratio equal to the carrier concentration. Crystals with higher carrier concentration contain free carriers in addition to W^{5+} but no W^{4+}. This result rules out the on-site model and is in agreement with the inter-site model.

We now return to the question of which phonons condense into the polaronic or bipolaronic state. The Reik-profile of polarons and bipolarons correlates the half widths of the absorption profile Γ with the phonon energy $\hbar\omega_{phonon}$:

$$\hbar\omega_{phonon} = \frac{\Gamma^2}{(4W_p)}.$$

Although this relation provides only a very crude measure for the phonon energy we may use it to obtain some very revealing estimates. The polaronic absorption near 0.7 eV was correlated with a phonon energy of 0.07 eV (560 cm^{-1}). The broad profile of bipolaronic absorption near 1.1 eV corresponds to a phonon energy $\hbar\omega_{phonon} \approx 0.18$ eV (1484 cm^{-1}). This phonon energy is well outside the range of phonons present in WO_{3-x} ($\hbar\omega < 0.12$ eV) so that the absorption profile cannot be attributed to a single-phonon process. One possibility to resolve this riddle is to consider bipolaron absorption with pairs of phonons emitted. The equivalent single-phonon energy $\frac{1}{2}\hbar\omega_{phonon} = 0.9$ eV is thus well within the energy range of W–O stretching modes. Two-phonon processes have also been observed in resonant Raman scattering experiments when the frequency of the exciting laser light was close to the maximum of the bipolaron absorption [64]. The observed scattering signal at 0.199 eV is close to the effective phonon energy of the bipolaron. The phonon normalization with respect to the bulk material is $(\omega - 2\omega_0)/(2\omega_0) = -4.5\%$, where ω_0 is the frequency of the relevant stretching mode in WO_{3-x}. The negative sign indicates that the phonon softens when W^{6+} is replaced by W^{5+}, the magnitude of the change is characteristic of local structural deformation seen in displacive phase temperatures in framework structures involving changes in oxygen–metal distances of a few percent [63]. We can finally estimate an 'effective polaron and bipolaron mass' for ε-WO_{3-x} in the approximation [64–66] for comparison with other materials and as a rough estimate for the order of magnitude of these quantities

$$m^{**} = m_e^* (1 + \alpha'/6),$$

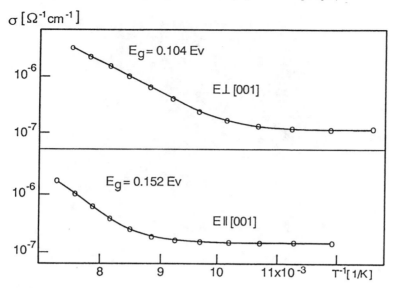

Fig. 7. The conductivity of $\varepsilon - WO_{3-x}$ perpendicular (top) and parallel (bottom) to the crystallographic c-axis.

where α' is the coupling constant

$$\alpha' = 1.25 \, \frac{2W_p}{\pi\hbar\omega_{phonon}}.$$

Using the spectral data the characteristics of the polarons and bipolarons can be summarized as follows.

The polarons have a coupling constant of $\alpha' = 4$, effective mass $m^{**}_{polaron} = 1.7m_e^*$, the spatial extent of a flat disc of diameter 10 Å and a thickness of one atomic layer.

The bipolarons have a larger coupling constant, $\alpha' = 5.6$, and a light effective mass $m^{**}_{bipolaron} = 1.9m_e^*$. The bipolarons extend over elongated discs of size 16 Å × 8 Å and thickness one atomic layer. Their transport is thermally activated at $T \gtrsim 100$ K and tunnelling in a Holstein band appears to occur at lower temperatures (Fig. 7).

2.3 The term diagram and the crab walk of bipolarons in WO_{3-x}

An estimate of the potential barrier between the bipolaron state and two separate polarons can be made from the effective Hamiltonian [55],

$$H_{\text{eff}} = \sum_i \varepsilon_i n_i + \sum{}' V n_j n_i - \lambda \sum{}' x_{ij}(n_i + n_j) + \frac{1}{2} C \sum{}' x_{ij}^2,$$

E. K. H. Salje

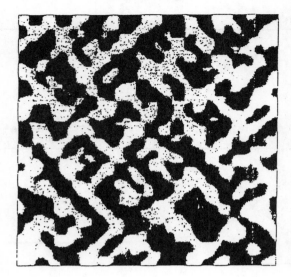

Fig. 8. The result of a computer simulation of an isospin system (one isospin per structural unit) in an elastic medium with bilinear coupling between the isospin and the structural strain. Thermal excitations lead to tweed and stripe patterning (snapshots on a timescale shorter than the time to migrate an isospin variable). Bipolaronic systems with high carrier concentrations are expected to form similar patterns.

where x_{ij} is the strain related to the distance between atoms i and j, C is an elastic constant, ε_i is the site energy, V is the interaction energy and λ describes the bilinear coupling between the strain and the occupancies. The thermodynamic behaviour of this class of Hamiltonians has been explored in great detail and leads to large spatial fluctuations of the occupation number. If C is considered anisotropic, then the fluctuations lead to correlations between the occupancies in a tweed-like or stripe-like pattern (Fig. 8).

Bipolarons are stable for $2\lambda^2 C > V$ with the self-energy

$$E^0_{\text{bipolaron}} = 2E^0_{\text{polaron}} - (2\lambda^2 C - V).$$

The potentials for both bipolarons and polarons can now be developed around the equilibrium points in terms of a general configuration coordinate Q as

$$E_{\text{bipolaron}} = E^0_{\text{polaron}} + aQ^2,$$

$$E_{\text{polaron}} = E^0_{\text{polaron}} + \left(Q - \frac{d}{2}\right)^2,$$

where d is the distance between the relevant cation sites. The optical transition with E_{0p} is then related to the gap between the ground state and the excited state for the same configuration coordinate $Q = Q_0$. Combining this energy with the hopping energy leads to construction of the term diagram in Fig. 9.

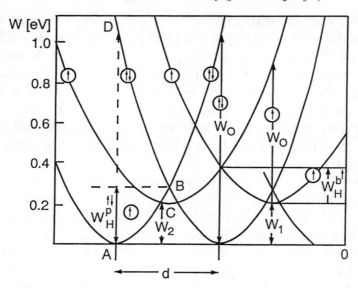

Fig. 9. The term diagram of polarons (single arrow) and bipolarons (double arrows) in ε-WO$_{3-x}$. An optical excitation of a bipolaron is indicated by a broken arrow, the hopping energies W_H are shown by full arrows. An excited bipolaron can decay metastably into pairs of polarons because $W_2 > W_1$.

The potentials of the polaronic states are indicated by a single arrow, and those of bipolarons by double, antiparallel arrows. All potentials are assumed to be fully harmonic. The relevant energies for a transition from one potential well to another are W_H, W_2 and W_1 for conduction processes; E_{0p} is the optical transition energy. The bipolaron state is lowered in energy by W_1 from the single-polaron state. The single-polaron state excited during the photo-effect is stabilized by the energy difference $W_2 - W_1$.

For bipolarons the hopping activation energy W_H is related to the optical transfer energy (neglecting the resonance energy J) by $4W_H = E_{0p}^{bi}$; the experimental values are $E_{0p}^{bi} = 1.1$ eV and $W_H = 0.27$ eV. For single polarons we find experimentally that $E_{0p}^{si} = 0.71$ eV and $W_H^{si} = 0.18$ eV.

In principle, two models are possible for the transport mechanism in the hopping regime. The Heydock model (cited in [55]) considers the limit of strong polaron–polaron coupling without dissociation of the bipolaron during transport. Thus the activation energy, again without the exchange integral J, is $W_H^{bi} = \frac{1}{4}E_{0p}^{bi}$. The alternative model by Chakraverty and Schlenker [55] assumes splitting of the bipolaron into two single polarons, transport of the single polarons, and their subsequent recombination to give the bipolaron. The specific conductivity is

$$\sigma \propto \exp[(-W_2 - W_1/2)/(kT)].$$

W_2 and W_1 are shown in Fig. 9. From the potential functions and the experimental activation energy $E_g = 0.11$ eV, we determine $W_1 = 0.20$ eV and $W_2 = 0.21$ eV. The bipolaronic ground state is therefore 0.20 eV lower than the single-polaronic ground state. The difference $W_2 - W_1$ is 0.01 eV, which is much smaller than the equivalent value 0.09 eV in Ti_4O_7 [55]. This value is comparable with the thermal energy necessary to destabilize the excited polaronic state. The corresponding temperature above which no photo-effect occurs is then $T_0 = (W_2 - W_1)/k \approx 110$ K, which is in good accordance with our experimental results.

The 'dissociation model' is favoured over the Heydock model; the dissociation energy is $W_H^{bi} - W_1 = 0.27$ eV $- 0.20$ eV $- 0.07$ eV. This energy is smaller than that observed in Ti_4O_7 (0.4 eV [55]). Its positive sign confirms that only the dissociation model describes the bipolaron transport correctly.

The transport of bipolarons in an electric field is, thus, a crab walk (Fig. 10). The bipolarons tend to split locally into two polarons. These polarons have higher mobility and lower activation energy than the bipolaron. Pairs of polarons then move and recombine immediately. During dissociation of the bipolaron we see no spin-depairing in the ESR signal. This may indicate that the spins remain correlated and/or that the recombination is virtually instantaneous so that the transient state has to have such a short lifetime that it cannot be observed in an ESR experiment. The two polarons cannot therefore be completely separated in the transient state. Correlations in the transient state follow also from the observation that the recombination time of fully separated polarons in ESR photo-effect measurements allows temperatures to promote transport. The crab walk involves dissociation of the bipolaron only to the extent that spin correlation is preserved and recombination is fast on the time scale of the interatomic transport.

The time scale can be estimated if the mobility measured in Hall experiments is extrapolated to lower temperatures and converted into diffusion coefficients D using the Einstein relation. Typical values for polarons at 120 K are $D = 10^{-3}$ cm^2 s^{-1}. The time to travel one polaron diameter d is then $\tau = 2\ d^2/D = 2\ d^2 e/(\mu k T) = 2 \times 10^{-11}$, i.e. several phonon times. Spin pairing appears to be conserved on this time scale. At lower temperatures the conductivity is independent of temperature with $\sigma \approx 10^{-7} - 10^{-8}$ Ω^{-1} cm^{-1} so that we may assume that the carrier mobility also remains temperature-independent. The time constant increases at 20 K well above the time scale of recombination. If bipolarons were fully split into two non-correlated polarons then their transient states would have lifetimes incompatible with the observed transport properties. In conclusion, the transient state involves crab walk with transient states, which maintain spin correlation, i.e. there is no full decay of the

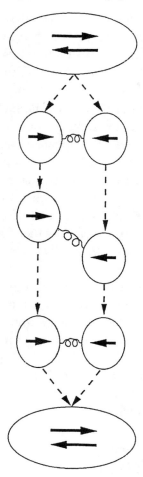

Fig. 10. The 'crabwalk' of a bipolaron during transport in an electric field. The bipolaron separates into two coupled polarons. The first polaron moves in the field, the second follows. The two shifted polarons then recombine to form a bipolaron.

bipolarons into pairs of polarons. This is presumably the reason why WO_{3-x} does not superconduct at the temperatures that such materials as $YBa_2Cu_3O_7$ do.

2.4 The overcrowding effect

Tungsten oxide is an almost ideal material for studying polaron gases. In the preceding part of this review the behaviour of a diluted system of polarons was discussed ($n \approx 10^{18}$–10^{19} cm^{-3}). The following question now arises: what will happen if the polaron density increases? It has been argued that it makes no sense to discuss individual polarons if their density is of the same order of magnitude as the atomic density [27]. More precisely, if bipolarons encompass some ten unit cells, then a higher density of one W^{5+} in ten W^{6+} requires that crystals that have a lower oxygen content than $WO_{2.9}$ should show either substantial overlap between localized wavefunctions or generate additional, delocalized carriers. Careful XPS studies on such materials again show no indication of localized W^{4+} states and transport measurements reveal metallic behaviour. The surplus charges are delocalized fermions, therefore.

This effect that bosonic bipolarons have an upper maximum for the ground state occupancy for geometrical reasons so that fermionic surplus charges are generated if this maximum is exceeded is called the 'overcrowding effect' [27]. Overcrowding is a fundamentally important effect because it mixes fermions and bosons in the same system, which may then show a complex behaviour of the statistical mechanics of its carriers including superconductivity at sufficiently low temperatures.

The change of transport properties and the optical excitations of bipolarons as a function of the carrier concentration are now discussed. In Fig. 11 the Arrhénius plots for the DC conductivity are shown for several chemical compositions with carrier concentrations between 10^{19} and 6×10^{21} cm^{-3}. The stepwise phase transition δ-WO_{3-x}–ε-WO_{3-x} disappears for $n > 10^{20}$ cm^{-3} (i.e. $WO_{2.988}$). Tunnelling transport appears for all insulating materials WO_{3-x} ($x \lesssim 0.1$) at temperatures below about 100 K. A metal–insulator transition is clearly seen when x changes from values below 0.1 (insulating phase) to values above 0.1 (metallic phase).

The nature of the carriers in the metallic phase is now analysed using optical absorption spectroscopy. Absorption spectra of tungsten oxides with four different compositions are shown in Fig. 12. The total bipolaron concentration is determined for these profiles by spectral integration and multiplication with the bipolaron absorption cross section α_0. This value follows from analysis of spectra with low concentration, where strict proportionality between integrated absorbance and carrier concentration is observed. A concentration of $n = 10^{18}$ cm^{-3} leads to an integrated absorbance of bipolarons $\alpha_{total} = 64$ eV cm^{-1}. Increasing bipolaron concentration leads to a proportional increase of α_{total} for concentrations less than 3.7×10^{21} cm^{-3} (equivalent to $WO_{2.9}$). Larger carrier concentrations lead only to a minor increase in total absorbance; additional carriers are not in the bipolaronic ground state. These additional carriers have high mobility and lead to metallic conductivity. It is believed, therefore, that these carriers are (quasi-) free electrons at the bottom of the conduction band.

These results have been confirmed by studies of XPS spectra of highly

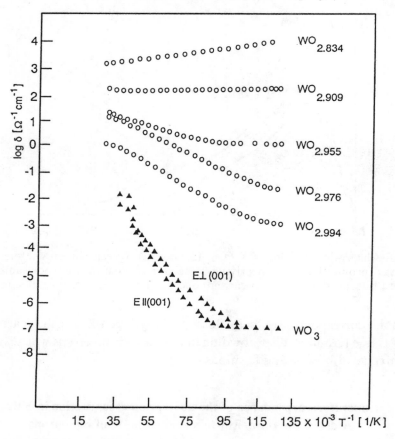

Fig. 11. A conductivity Arrhénius plot of WO$_{3-x}$ at low temperatures. Samples with less oxygen than WO$_{2.9}$ show metallic conductivity. All other samples show hopping conductivity at $T \gtrsim 100$ K and (almost) temperature-independent conductivity at low temperatures.

reduced WO$_{3-x}$. It was found that all samples with carrier concentrations $n < 3.7 \times 10^{21}$ cm^{-3} show W^{5+} states with a density identical to the total carrier concentration, i.e. all carriers are bipolarons for low concentrations. For higher concentrations the W^{5+} concentration increases only slightly but large increases in carrier concentration were found close to the Fermi level. Filling of the conduction band also shifted the Fermi level to higher energies.

The experimental observations in the various phases of WO$_{3-x}$ lead us to the perception that overcrowding is a general phenomenon in polaronic systems. The threshold value for the carrier concentration above which overcrowding occurs depends on geometrical features and the extension of the polaronic wavefunction. In the case of well-localized bipolarons we find that the

        E. K. H. Salje

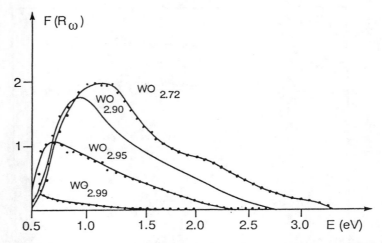

Fig. 12. Absorption profiles of WO_{3-x} for various oxygen contents. Note that the integrated profiles increase linearly with the carrier concentration for samples with $x < 0.1$ but saturate for higher carrier concentrations (i.e. in the metallic regime).

threshold corresponds to their close packing. As the surplus carriers are (almost) free fermions, overcrowding in bipolaronic materials will always lead to a mixture of bosons and fermions.

3 The temperature-dependence of the absorption cross section in $YBa_2Cu_3O_7$: evidence for Bose–Einstein condensation of bipolarons

Polarons and bipolarons in WO_{3-x}, MoO_{3-x}, $NbO_{2.5-x}$ and their chemical mixtures show Frank–Condon transitions in the NIR spectral region. Their peak profiles are very different from those of the Drude absorption of (almost) free carriers in the case of low carrier concentration. Materials with high carrier concentration, e.g. $WO_{2.72}$, show more Drude-like absorption spectra although a quantitative fit with only one type of carrier is not possible. The situation is less clear in high-temperature superconductors. Absorption spectra similar to but slightly broader than those of WO_{3-x} were observed in $YBa_2Cu_3O_{7-\delta}$ [32–34] (Fig. 13) and related Fe- and Zn-doped materials [32], $(Ca_{1-x}Y_x)Sr_2$ $(Tl_{0.5}Pb_{0.5})Cu_2O_7$ [35], $Bi_2Sr_2(Ca_{0.9}Y_{0.1})Cu_2O_{8+\delta}$ and $Ca_2CuO_{4+\delta}$ [36,37]. The peak positions shift somewhat as a function of chemical composition although NIR absorption peaks seem to be a universal feature for all high-temperature superconductors investigated so far.

It is tempting to identify the absorption peaks with excitations of bipolarons. An alternative analysis of the aborption profiles in terms of the Drude model leads to high plasma frequencies ($\omega_p \approx 1.5$–4.3 eV) and large values of

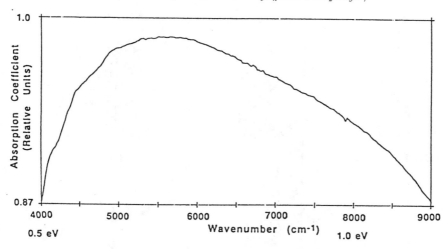

Fig. 13. Absorption spectra of $YBa_2Cu_3O_7$ in the near infrared spectral region.

$\gamma = 1/\tau \approx 0.6\text{–}4.5\,\mathrm{eV}$) whereby these values vary depending on the experimental data and spectral range used for the fit [67–80]. These values are in contradiction with results from Raman spectroscopy [81]. There are several drawbacks of such models, e.g. the resulting effective mass for $\omega_p \approx 1\,\mathrm{eV}$ is $7m_e$, which is rather heavy for the free carriers that underlie the Drude model. If a larger value of ω_p is assumed (i.e. a lighter mass) then the numerical value of γ as determined by fitting of the Drude profile to the experimental data is too large for the Drude model to be self-consistent. Attempts to unravel these inconsistencies were made in the theories of marginal Fermi liquids [82] and nested Fermi liquids [83,84]. Finally, Anderson [85] has analysed the NIR absorption within the RVB idea by assigning different scattering rates to holon–holon and holon–spinon interactions. We conclude that identification of the NIR absorption with excitation of bipolarons, some marginal/nested Fermi liquid or holons seems to be compatible with observed absorption profiles.

We now discuss the question of where the carriers responsible for excitation are localized in the $YBa_2Cu_3O_7$ structure. Doping experiments [32] have shown that incorporation of Fe in Cu–O chains has no influence on the absorption profile. Doping with Zn in the CuO_2 planes, on the other hand, shifts the absorption maximum. The relevant carriers are localized in the CuO_2 planes, therefore, but not in the Cu–O chains. Recent transmission measurements on thin films in which the angle of the incident beam was varied seem to indicate that the dipolar matrix element is also largest for electrical field vectors in the CuO_2 plane and smallest parallel to the crystallographic c axis [86]. It appears, therefore, that the carriers are localized in the CuO_2 plane with dipolar

E. K. H. Salje

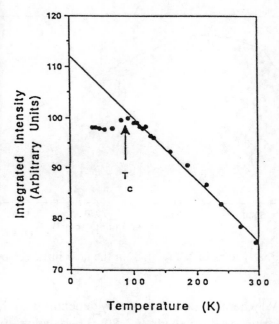

Fig. 14. The temperature-dependence of the absorption cross section near 5000 cm^{-1} in YBa$_2$Cu$_3$O$_7$.

moments inside this plane, i.e. the carriers are subject to the same geometrical constraints as the superconducting pairs in YBa$_2$Cu$_3$O$_7$.

The main significance of the NIR absorption in high-temperature superconductors is its reduction in the superconducting phase [32–34]. In Fig. 14 the results for YBa$_2$Cu$_3$O$_7$ are shown. Cooling a sample below room temperature increases the absorption cross section. This result parallels that for WO$_3$ and other oxides in which we always find an increase in bipolaronic absorption with decreasing temperature. This increase is expected because bipolarons can be thermally excited from the ground state (singlet) into either triplet state (bipolarons with spin $\frac{1}{2} + \frac{1}{2} = 1$) or polarons. With decreasing temperature, such excitations are reduced. The excitation energies for these processes can be calculated directly from the experimental data points in Fig. 14 and Fig. 15 with triplet bipolarons at 440 K above the singlet ground state [87]. This value can be seen as a spin pseudo-gap $E_{gap} \approx 4T_c$, which is compatible with results from neutron scattering studies [88].

Further cooling of samples into the superconducting phase leads to a drastic break in the temperature evolution of the absorption cross section. We quantify this break as an excess reduction of the extrapolated absorbance.

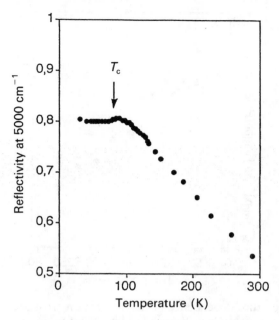

Fig. 15. The temperature evolution of the reflectivity of $YBa_2Cu_3O_7$ at 5000 cm^{-1}.

Ignoring any quantum saturation, we find that superconductivity reduces the absorption coefficient α by some 12–20% at zero temperature compared with the value of the extrapolated normal state. The same behaviour was also found in $(Ca_{1-x}Y_x)\,Sr_2\,(Tl_{0.5}Pb_{0.5})\,Cu_2O_7$ superconductors (Fig. 16). This observation shows that Bose–Einstein condensation does not annihilate the bipolaronic absorption. Incoherent absorption accompanied by emission and absorption of many phonons remains present even at zero temperature. The influence of the superconducting phase transition is restricted to the coherent tunnelling contribution. This latter contribution is small compared with the incoherent absorption which, in turn, does not exist in simple negative U Hubbard Hamiltonians. Such model Hamiltonians cannot therefore explain the temperature-dependence of the NIR aborption in $YBa_2Cu_3O_7$.

Quantitative analysis of the temperature evolution of the absorption cross section in terms of Bose–Einstein condensation of small bipolarons with a Fröhlich Hamiltonian was given by Alexandrov *et al.* [87] and agrees fully with the descriptions and estimates given in this paper. Further discussions on our results in the spirit of the ideas of Kagan and Prokofiev [89] will be published separately.

E. K. H. Salje

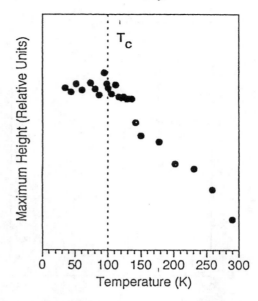

Fig. 16. The temperature-dependence of the absorption cross section in the material $(Ca_{1-x}Y_x)Sr_2(Tl_{0.5}Pb_{0.5})Cu_2O_7$.

References

[1] N. F. Mott, 1990, *Adv. Phys.* **39**, 55.
[2] A. S. Alexandrov, 1987, *JETP Lett. Suppl.* **46**, 107.
[3] A. S. Alexandrov, 1992, *J. Low Temp. Phys.* **87**, 721.
[4] A. S. Alexandrov and J. Ranninger, 1992, *Solid State Commun.* **81**, 403.
[5] N. F. Mott, 1992, *Physica* C **196**, 369.
[6] A. S. Alexandrov and J. Ranninger, 1989, *Physica* C **159**, 367.
[7] H. S. Obhi and E. K. H. Salje, 1992, *J. Phys. Condens. Matter* **4**, 195.
[8] H. S. Obhi and E. K. H. Salje, 1990, *Physica* C **141**, 547.
[9] A. S. Alexandrov, 1992, *Phys. Rev.* B **46**, 14932.
[10] J. Ranninger, 1991, *Z. Phys.* B **84**, 167.
[11] S. Robaszkiewicz, R. Mickas and J. Ranninger, 1987, *Phys. Rev.* B **36**, 180.
[12] D. Emin and M. S. Hillery, 1989, *Phys. Rev.* B **39**, 6575.
[13] H. S. Obhi, E. K. H. Salje and T. Miyatake, 1992, *J. Phys. Condens. Matter* **4**, 10367.
[14] B. Güttler, E. Salje, P. Freemann, J. Blunt, M. Harris, T. Duffield, C. D. Auger and H. P. Hughes, 1991, *J. Supercond. Sci. Technol.* **4**, S70.
[15] B. Güttler, E. Salje, P. Freemann, J. Blunt, M. Harris, T. Duffield, C. D. Auger and H. P. Hughes, 1990, *J. Phys. Condens. Matter* **2**, 8977.
[16] S. K. Dep, 1973, *Phil. Mag.* **27**, 801.
[17] B. W. Faughnan, R. S. Crandall and D. Heyman, 1975, *RCA Rev.* **36**, 177.
[18] E. Salje and G. Hoppmann, 1981, *Phil. Mag.* B **43**, 105.
[19] O. F. Schirmer and E. Salje, 1980, *J. Phys.* C **13**, 1067.
[20] B. W. Faughnan and R. S. Crandall, 1977, *Appl. Phys. Lett.* **31**, 834.
[21] A. Deneuville and P. Gerard, 1978, *J. Electron. Mater.* **7**, 559.

[22] T. Yoshimura, 1985, *J. Appl. Phys.* **57**, 911.
[23] T. Toyoda, 1988, *J. Appl. Phys.* **63**, 5166.
[24] J. M. Berak and M. J. Sienko, 1970, *J. Solid State Chem.* **2**, 109.
[25] R. Gehlig, E. Salje, 1983, *Phil. Mag.* B **47**, 229.
[26] C. Ruscher, E. Salje and A. Hussain, 1988, *J. Phys.* C **21**, 3737.
[27] E. Salje and B. Güttler, 1984, *Phil. Mag.* B **50**, 607.
[28] G. Hoppmann and E. Salje, 1976, *Mater. Res. Bull.* **11**, 1545.
[29] E. Salje and G. Hoppmann, 1981, *Phil. Mag.* B **43**, 145.
[30] O. F. Schirmer and E. Salje, 1980, *Solid State Commun.* **33**, 333.
[31] Y. Yagil and E. K. H. Salje, 1994, *Physica* C **229**, 152.
[32] H. L. Dewing and E. K. H. Salje, 1992, *Supercond. Sci. Technol.* **5**, 50.
[33] H. L. Dewing and E. K. H. Salje, 1992, *J. Solid State Chem.* **100**, 363.
[34] H. L. Dewing, E. K. H. Salje, K. Scott and A. P. Mackenzie, 1992, *J. Phys. Condens. Matter* **4**, L109.
[35] H. L. Dewing and E. K. H. Salje, 1995 in preparation.
[36] C. H. Rüscher, M. Götte, B. Schmidt, C. Quitmann and G. Güntherodt, 1992, *Physica* C **204**, 30.
[37] C. H. Rüscher and M. Götte, 1993, *Solid State Commun.* **85**, 393.
[38] Z. Schlesinger, R. T. Collins, F. Holtzberg, C. Feild, G. Koren and A. Gupta, 1990, *Phys. Rev.* B **41**, 11237.
[39] J. Orenstein, G. A. Thomas, D. H. Rapkine, A. J. Millis, L. F. Schneemeyer and J. V. Waszczak, 1988, *Physica* C **153**, 1740.
[40] J. Orenstein, G. A. Thomas, D. H. Rapkine, C. G. Bethea, B. F. Levine, R. J. Cava, E. A. Rietman and D. W. Johnson, 1987, *Phys. Rev.* B **36**, 729.
[41] Z. Schlesinger, R. T. Collins, D. L. Kaiser, F. Holtzberg, G. V. Chandrashekhar, M. W. Schafer and T. M. Plaskett, 1988, *Physica* C **153–155**, 1734.
[42] F. Lu, C. H. Perry, K. Chen and R. S. Markiewicz, 1989, *J. Opt. Soc. Am.* B **6**, 396.
[43] Y. Watanabe, Z. Z. Wang, S. A. Lyon, D. C. Tsui, N. P. Ong, J. M. Tarascon and P. Barboux, 1989, *Phys. Rev.* B **40**, 6884.
[44] G. A. Thomas, S. L. Cooper, J. Orenstein, D. H. Rapkine, J. V. Waszczak and L. F. Schneemeyer, 1990, *Physica* B **165–166**, 1257.
[45] T. Yoshimura, M. Watanabe, Y. Koike, K. Kiyota and M. Tanaka, 1982, *J. Appl. Phys.* **53**, 7314.
[46] R. Gazzinelli and O. F. Schirmer, 1977, *J. Phys.* C **10**, L145.
[47] D. Davazoglou and A. Donnadieu, 1987, *Thin Solid Films* **147**, 131.
[48] C. J. Raub, A. R. Sweedler, M. A. Jensen, S. Broadston and B. T. Matthias, 1964, *Phys. Rev. Lett.* **13**, 746.
[49] W. A. Kamitakahara, B. N. Harmon, J. G. Traylor, H. R. Shanks and J. Rath, 1976, *Phys. Rev. Lett.* **36**, 1393.
[50] K. L. Ngai and R. Silverglitt, 1976, *Phys. Rev.* B **13**, 1032.
[51] M. Sato, B. H. Grier, G. Shirane and H. Fujishia, 1982, *Phys. Rev.* B **25**, 501.
[52] C. B. Thomas and M. R. Goulding, 1984, *Phil. Mag.* B **49**, 219.
[53] K. Visvanathan and E. Salje, 1981, *Acta Cryst.* A **37**, 449.
[54] E. Salje, R. Gehlig and K. Viswanathan, 1978, *J. Solid State Chem.* **25**, 239.
[55] B. K. Chakraverty and C. Schlenker, 1976, *J. Physique* **10**, C4-353.
[56] C. Ruscher, E. Salje and A. Hussain, 1988, *J. Phys.* C **21**, 4465.
[57] R. Gehlig, E. Salje, A. Carley and M. W. Roberts, 1983, *J. Solid State Chem.* **49**, 318.
[58] E. Salje, A. Carley and W. M. Roberts, 1979, *J. Solid State Chem.* **29**, 237.

[59] W. Sahle and M. Nygren, 1983, *J. Solid State Chem.* **48**, 154.
[60] E. Salje, 1976, *Ferroelectrics* **12**, 215.
[61] E. Salje, 1975, *Acta Cryst.* A **31**, 810.
[62] H. G. Reik and D. Heese, 1967, *J. Phys. Chem. Solids* **28**, 581.
[63] E. K. H. Salje, 1993, *Phase Transitions in Ferroeleastic and Coelastic Crystals*, Cambridge University Press, Cambridge.
[64] T. D. Lee, F. E. Low and D. Pines, 1953, *Phys. Rev.* **90**, 297.
[65] D. C. Langreth, 1965, *Phys. Rev.* A **137**, 760.
[66] D. C. Langreth and L. P. Kadanoff, 1964, *Phys. Rev.* A **133**, 1070.
[67] W. Markowitsch, W. Lang, N. S. Sariciftci and G. Leising, 1989, *Solid State Commun.* **69**, 363.
[68] D. Van der Marel, H.-U. Habermeier, D. Heitmann, W. König and A. Wittlin, 1991, *Physica* C **176**, 1.
[69] J. Ruvalds and A. Virosztek, 1991, *Phys. Rev.* B **43**, 5498.
[70] L. D. Rotter, Z. Schlesinger, R. T. Collins, F. Holtzberg, D. Feild, K. W. Welp, G. W. Crabtree, J. Z. Liu, Y. Fang, K. G. Vandervoort and S. Flesher, 1991, *Phys. Rev. Lett.* **67**, 2741.
[71] W. Ose, P. E. Obermeyer, H. H. Otto, T. Zetterer, H. Lengfellner, J. Kellner and K. F. Renk, 1988, *Z. Phys.* B **70**, 307.
[72] A. C. Nichol, F. L. Pratt, W. Hayer, C. Chen, B. E. Watts and B. M. Wanklyn, 1989, *J. Opt. Soc. Am.* B **6**, 403.
[73] S. Tajima, T. Nakahashi, S. Uchida and S. Tanaka, 1988, *Physica* C **156**, 90.
[74] I. Bozovic, D. Kirillov, A. Kapitulnik, K. Char, M. R. Hahn, M. R. Beasley, T. H. Geballe, Y. H. Kim and A. J. Heeger, 1987, *Phys. Rev. Lett.* **59**, 2219.
[75] Z. Schlesinger, R. T. Collins, D. L. Kaiser, F. Holtzberg, G. V. Chandrashekhar, M. W. Shafer and T. M. Plaskett, 1988, *Physica* C **153–155**, 1734.
[76] R. T. Collins, Z. Schlesinger, T. Holzberg and C. Feild, 1989, *Phys. Rev. Lett.* **63**, 422.
[77] S. L. Cooper, D. Reznik, A. Kotz, M. A. Karlow, R. Liu, M. V. Klein, W. C. Lee, J. Giapiutzakis, D. M. Ginsberg, B. W. Veal and A. P. Paulikas, 1993, *Phys. Rev.* B **47**, 8233.
[78] I. Bozovic, 1990, *Phys. Rev.* B **42**, 1969.
[79] A. Bjorneklett, A. Borg, O. Hunderi and S. Julsrud, 1988, *Solid State Commun.* **67**, 525.
[80] S. Cunsolo, P. Dore, S. Lupi, R. Tripetti, C. P. Varsamis and A. Sherman, 1993, *Physica* C **211**, 22.
[81] R. Gajic, E. K. H. Salje, Z. V. Popovic and H. L. Dewing, 1992, *J. Phys. Condens. Matter* **4**, 9643.
[82] C. M. Varma, S. Schmitt-Rink and E. Abrahams, 1987, *Solid State Commun.* **62**, 681.
[83] J. Ruvalds and A. Virosztek, 1991, *Phys. Rev.* B **43**, 5498.
[84] N. P. Ong, 1991, *Phys. Rev.* B **43**, 193.
[85] P. W. Anderson, 1991, *Phys. Rev. Lett.* **67**, 2992.
[86] Y. Yagil, 1994 personal communication.
[87] A. S. Alexandrov, A. M. Bratkovsky, N. F. Mott and E. K. H. Salje, 1993, *Physica* C **215**, 359.
[88] J. Rossat-Mignod, L. P. Regnault, P. Bourges, C. Vettier, P. Burlet and J. Y. Henry, 1992, *Physica Scripta* **45**, 74.
[89] Yu Kagan and N. V. Prokofiev, 1992 in *Quantum Tunnelling in Condensed Media*, ed. Yu Kagan and A. J. Leggett, Elsevier, Amsterdam.

8

Polaron bands in the far- and mid-infrared spectra of e-doped cuprates

P. CALVANI[†], S. LUPI and P. ROY

Laboratoire pour l'Utilization de Rayonnement Electromagnétique, Université Paris-Sud, 91405 Orsay, France

M. CAPIZZI, P. MASELLI and A. PAOLONE

Dipartimento di Fisica, Università di Roma 'La Sapienza', Piazzale A. Moro 2, I-00185 Roma, Italy

W. SADOWSKI

Faculty of Applied Physics and Mathematics, Technical University of Gdańsk, G. Narutowicza 11/12, 80-592 Gdańsk, Poland

S.-W. CHEONG

AT&T Bell Laboratories, Murray Hill, New Jersey 07974, USA

Abstract

The far- and mid-infrared reflectivity $R(\omega)$ of e-doped single crystals belonging to the family $M_{2-x}Ce_xCuO_{4-y}$ (M = Pr, Nd, Gd; $0 < x < 0.15$; $0 < y < 0.04$) has been studied between 300 and 20 K. In addition to the phonons predicted for the T' structure, $R(\omega)$ shows local modes in the far infrared as well as a broad infrared absorption centered at about 0.1 eV (d or J band). These features depend strongly on both T and y. We have resolved the d band at low T, in samples doped by oxygen vacancies. We demonstrate its polaronic origin by showing that it is made up of intense overtones of the local modes observed in the far infrared. We also find that Ce-doped superconductors ($x > 0.12, y = 0.03$) have the same polaronic structure as the semiconducting ones, partially superimposed on a weak Drude term.

1 Introduction

Since the discovery of high-T_c superconductors (HTSC), infrared reflectivity measurements have been largely employed to investigate both electronic and transport properties of these cuprates [1]. Early spectra already showed several

---

† Permanent address: Dipartimento di Fisica, Università di Roma 'La Sapienza', Piazzale A. Moro 2, I-00185 Roma, Italy

intriguing features common to all HTSC families, which were then attributed to the peculiar properties of the Cu–O plane. Those features are well reproducible and could be studied in greater detail as soon as large single crystals became available. However, their interpretation is still being debated. In the insulating parent compounds of HTSC, one observes phonon modes in the far infrared, and a broad band at high frequencies (from about 1.5 to 2.5 eV). This latter has been unanimously assigned to charge-transfer (CT) transitions between O 2p and Cu 3d orbitals. Nevertheless, the phonon lineshapes are often unusual, as reported for Nd_2CuO_4 (NCO) by several authors [2–5]. Moreover, in La_2CuO_{4+y} the CT peak exhibits a temperature-dependence, which has been interpreted in terms of polaronic effects [6].

When the cuprates are increasingly doped by replacing the lanthanide ion, as in $Nd_{2-x}Ce_xCuO_{4-x}$, spectral weight is transferred from the CT band to lower frequencies [7, 8]. Two new features appear: the mid-infrared (MIR) band at energies ranging from 0.7 eV (for the lowest-T_c materials) to 0.1 eV, and the so-called d or J band at about 0.1 eV [7, 9, 10]. The origin of both of them is controversial. In semiconducting compounds, where the MIR band is well resolved and independent of the free-electron contribution, this band has been attributed to new states created by doping within the CT gap [11, 12]. In metallic cuprates, the MIR contribution gives rise to an anomalous Drude behavior, which has been explained in terms of Holstein processes [13], Holstein-like processes involving magnons [14] or a marginal Fermi liquid [15]. As far as the peak at about 0.1 eV is concerned, it has been attributed to bound charges coupled to both spin and lattice excitations [9, 10] or to polaronic effects [16]. A band at the same energy and having a similar shape has been observed in photoinduced absorption experiments in La_2CuO_{4+y}, $YBa_2Cu_3O_{6.3}$, and $Tl_2Ba_2Ca_{0.98}Gd_{0.02}Cu_2O_8$ [17–21]. This absorption feature has also been attributed to polarons, with Huang–Rhys factors $S \simeq 5$–7 and effective masses $m^* \simeq 10$–15. Increasing experimental evidence points, then, towards an interpretation of the whole infrared spectrum of a class of cuprates in terms of polarons. Up to now, this class is formed by the HTSC parent compounds, slightly doped either chemically or by light irradiation.

From a theoretical point of view, after the early work of Holstein, [22] the influence of polarons on the optical properties of the oxides has been analyzed in several works [23]. More recently, local modes coupled to electrons and holes have been predicted to show up in the normal-state phonon spectra of HTSC for intermediate or large values of the electron–phonon interaction [24, 25]. Moreover, the crucial role that polarons may play in superconductivity has been pointed out by several authors before [26, 27] and after [28–30] the discovery of HTSC. Fundamental support for these theories would come from experimental observation of polarons, and eventually bipolarons, in the optical spectra of metallic cuprates.

Here we study the infrared reflectivity of insulating and metallic single crystals belonging to the family $M_{2-x}Ce_xCuO_{4-y}$ (MCCO, M = Pr, Nd, Gd; $0 < x < 0.15$; $0 < y < 0.04$), with the aim of detecting any signature of polaron formation in these materials. The choice of e-doped cuprates is justified by several arguments. As mentioned above, they often exhibit anomalous phonon lineshapes. The d band, which in photoinduced absorption experiments has been attributed to polarons, is here easily detected [7, 9]. In preliminary tests, we also found that superconducting NCCO (M = Nd) single crystals often show anomalous reflectivities at low energies, suggestive of two different types of carriers. Finally, the reduction process, which triggers all of the above spectroscopic effects, is also a necessary condition for observing superconductivity in these materials. The latter result may hide an intriguing relation between the eventual presence of polarons and the superconducting transition in this class of cuprates.

2 Experiment

Three groups of $M_{2-x}Ce_xCuO_{4-y}$ single crystals have been examined in the present experiment: (i) three Nd_2CuO_{4-y} samples (MN24, MN22, and MN19 in order of increasing oxygen non-stoichiometry); (ii) nearly stoichiometric $Pr_{2-x}Ce_xCuO_{4-y}$ (MP1) and $Gd_{2-x}Ce_xCuO_{4-y}$ (MG1) samples; and (iii) several metallic $Nd_{2-x}Ce_xCuO_{4-y}$ samples, with $0.136 < x < 0.19$. Those among them that had been strongly reduced after growth, like MN15 and MN25, were found to be superconducting with $14\,K < T_c < 20\,K$. The Ce content x was determined by averaging the results of chemical microanalysis at four to seven different points of the sample surface. MN15, MN19, MN20, and MN25 were annealed after growth in a nitrogen atmosphere at 900 °C for 10–15 h and have $y = 0.04 \pm 0.01$ (thermogravimetric datum). MN22 was grown in a reducing atmosphere without further annealing: from its Drude plasma frequency (1200 cm^{-1} against 1700 cm^{-1} for MN19), its oxygen content was found to be intermediate between those of MN19 and MN24. For this latter, as also for MP1 and MG1, an upper limit $y = 0.005$ was fixed by comparison with the oxygen content of samples prepared in the same laboratory by similar procedures [4].

The reflectivity $R(\omega)$ of all samples, relative to gold- or aluminum-plated references, was measured between 300 and 20 K by a Bomem DA3 rapid scanning interferometer. The measurements were repeated in a different laboratory using a Bomem DA8 interferometer, and an excellent reproducibi-

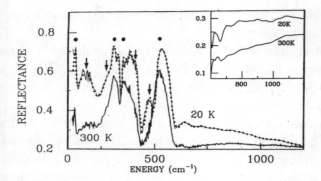

Fig. 1. $R(\omega)$ at different T for the Nd_2CuO_{4-y} single crystal MN22. The fit to a Drude–Lorentz model is reported (dotted line) for the spectrum at 20 K. The inset shows the d-band of the less doped sample MN24. Phonon peaks are indicated by dots, doping-dependent local modes by arrows.

lity of data was found. The light beam, 0.5 mm in diameter, was incident under vacuum on the sample at an angle of about 8°, so that the electric field of the radiation was polarized in the a–b plane. The spectral range extended for most samples from 50 to 30 000 cm^{-1}, with a resolution in the far- and mid-infrared range of 1 cm^{-1}.

3 Results and discussion

3.1 The semiconducting phase of M_2CuO_{4-y} ($M = Pr$, Nd, Gd)

As already mentioned, our starting point is the frequent observation of anomalous lineshapes in the phonon spectra of insulating cuprates, especially at low temperatures [2–5]. To verify whether such anomalies may be related to excess carriers, possibly produced by oxygen non-stoichiometry, we have measured the reflectivities of the three Nd_2CuO_{4-y} single crystals MN24, MN22, and MN19, described in Sec. 2. The clearest results were obtained for the reduced sample MN22, and are shown in Fig. 1. Therein, two groups of spectral features can be distinguished:

1. four bands, which correspond to the four E_u phonons of the T′ structure of NCO ($M = Nd$) [2];
2. four additional peaks, indicated by arrows, which are most clearly seen at 20 K.

The spectral region of the d band is shown in the inset for the less-doped sample MN24. As one can see, it seems to be built up by several oscillators, clearly resolved at low T on its low-frequency side. One can also appreciate its

Table 1. *Peak frequencies ω_p (cm^{-1}) and intensities I ($10^6\,cm^{-2}$) of local modes 1 and 2 and of their overtones and combination bands, as obtained from the best fits to the reflectivities of three Nd_2CuO_{4-y} single crystals at two different temperatures. Sample MN24 is nearly stoichiometric, MN22 is reduced, MN19 is strongly reduced. The absence of a datum indicates that a given oscillator has not been clearly resolved in the corresponding spectrum.*

Mode	Frequency	MN24 I(300 K)	MN24 I(20 K)	MN22 I(300 K)	MN22 I(20 K)	MN19 I(300 K)	MN19 I(20 K)
1	183±5		0.23	0.06	0.37	0.12	0.77
2	285±5		0.30		0.20		0.70
1+1	365±5	0.05	0.05	0.12	0.15	0.08	0.20
1+2	475±7		0.08	0.05	0.16		0.28
2+2	570±2	0.02	0.05				
1+1+2	642±5	0.01	0.05	0.09	0.20	0.05	0.16
1+2+2	713±5	0.02	0.09		0.18	0.18	0.21
1+1+1+2 (or 2+2+2)	783±7		0.34	0.18	0.22		0.19
1+1+2+2	900±30	0.01	0.32	0.20	0.37	0.32	0.34

strong temperature-dependence. Moreover, comparison with the d bands detected in samples MN22 and MN19, not shown in Fig. 1, suggests that its intensity also increases with doping.

In order to get more quantitative information, the spectra of Fig. 1 have been fitted (dots in Fig. 1 on the 20 K spectra) by a Drude–Lorentz complex dielectric function $\tilde\varepsilon(\omega)$ [7]. According to the results of our fits, the E_u oscillators have complex structures, which will not be discussed here. Moreover, their intensities are independent of doping and weakly dependent on temperature.

We shall now focus on the results given by the fit for the additional peaks of type (ii) shown in Fig. 1. In Table 1, the frequencies ω_p and intensities I of such modes are reported for the three NCO samples here considered and for the two extreme temperatures of our experiment. The results confirm that the intensities of type (ii) modes, which are comparable with those of the E_u modes, strongly depend both on doping and temperature. This behavior is particularly evident for those features that are well separated in frequency from the four E_u phonons. By straightforward inspection of Table 1 one finds that all reported frequencies, including those of the modes that build up the d band, can be expressed in terms of overtones or combinations of only two modes, those at 183 and 285 cm^{-1}. The resulting assignment is given in the first column of Table

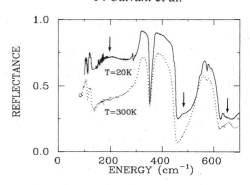

Fig. 2. $R(\omega)$ at 300 K (dotted line) and 20 K (solid line) for the Gd_2CuO_{4-y} single crystal MG1 in the far-infrared region. Doping-dependent local modes are indicated by arrows.

1. As the frequency of the mode increases, large negative shifts with respect to the values computed for overtone or combination bands are observed, as expected for an inharmonic potential.

The strong dependence on doping of type (ii) modes suggests that they are local modes induced by the excess carriers. These latter, in turn, are produced by oxygen vacancies, whose concentration increases from sample MN24 to sample MN19. We also measured, for the sake of comparison, two as-grown samples belonging to the same family, where Nd is replaced by Pr or Gd, respectively. Replacement of Nd by Gd or Pr is known to shift the phonon frequencies through the change in mass and size of the out-of-plane lanthanide [31]. In particular, as one passes from Pr to Nd and Gd, the Cu–O plane stretching mode (the last strong peak on the right-hand side in Fig. 1) is expected to move toward higher frequencies due to a change in Cu–O bond length. This is in fact what is observed in Fig. 2, which shows the reflectivity of Gd_2CuO_{4-y} sample MG1. On the other hand, in Fig. 2 clear structures appear at low T at 180, 480 and 640 cm^{-1} (see arrows), namely at the same frequencies as those of the additional modes detected in Fig. 1 (the narrow features at 280 and 560 cm^{-1} are probably spurious). The insensitivity of the additional mode frequencies to replacement of the lanthanide atom, which is placed out of the Cu–O plane and is known to modulate the Cu–O bond length [31], suggests that these modes do not directly involve atoms out of the Cu–O plane, and also that they are ruled by strong potentials, which can locally determine the Cu–O bond length.

A few oscillators have to be added to the modes directly resolved in the reflectivity spectra of NCO samples, in order to fit their d bands up to 2000 cm^{-1}. Good agreement with data is obtained if the frequencies of these peaks are again chosen to be combinations of the two fundamental ones. The whole d band is therefore found to be made up of modes of type (ii) having frequencies higher than 600 cm^{-1}. Their linewidths are found to follow the power law $\Gamma = B\omega_p^\alpha$, with $B = 1.4 \times 10^{-3}$, and $\alpha = 2.0$ (Γ and the peak frequency ω_p are given in reciprocal centimeters). The linewidths then depend superlinearly on ω_p, as expected for multiphonon processes triggered by electrons [32].

In summary, the whole infrared spectrum of Nd_2CuO_{4-y} up to 2000 cm^{-1} but for the four E_u phonons has the characteristics that one expects for a polaron band. The latter is made up of two local modes of the Cu–O plane (at 183 and 285 cm^{-1}), and of their combinations. The fundamental local modes are largely shifted towards lower frequencies, with respect to the corresponding phonons of the unperturbed lattice. In this framework, the lowest frequency bands $(1,2,1+1,1+2,2+2)$ should correspond to the excess electrons being self-trapped in the lattice distortion. Indeed, recent calculations on Cu–O clusters doped by holes show that the total energy is minimized when the hole resides on a Cu site and the four O nearest neighbors move toward the Cu atom [24]. The lattice distortion gives rise to strongly infrared-active [33] local modes, which will appear as satellite bands of the phonons of the unperturbed lattice [24]. On the other hand, the d band should be given by higher-order transitions allowing the electron to jump from site to site. Such transitions involve lattice relaxation processes, which could explain their large linewidths [34]. As far as the high intensity of type (ii) modes is concerned, one should remember that a similar situation has been found in organic semiconductors, where it has been explained in terms of strong electron–phonon coupling [33].

The polaronic origin of the d band can be confirmed if one uses a one-phonon standard approach, where the polaron band is modeled by a single Gaussian. This allows one to get the strength of the electron–phonon interaction through the Huang–Rhys factor S, and the frequency ω^* of the phonon involved. The d band of sample MN19, which is poorly resolved due to the high concentration of vacancies, has then also been fitted by a single Gaussian, whose parameters have been evaluated as a function of temperature. The linewidth Γ_d is found to depend on T as expected for a polaron band [6, 32], and is given by

$$\Gamma_d = \hbar\omega^* \{8S\ln 2\coth[\hbar\omega^*/(2kT)]\}^{1/2}. \tag{1}$$

S turns out to be 4.3, while $\omega^* = 210$ cm^{-1}. If one now reminds oneself that only two local fundamental modes, 1 and 2, interact with the electron, then one may build up a weighted phonon frequency $\omega^* = (I_1\omega_{p1} + I_2\omega_{p2})/(I_1 + I_2)$. By averaging over the three samples at 20 K, one gets $\omega^* = 237 \pm 10$ cm^{-1}, in excellent agreement with the one-phonon standard polaron approach. This

Table 2. The peak frequencies ω_p *(cm*$^{-1}$ *) of some local modes here observed in* $Nd_{2-x}Ce_xCuO_{4-y}$ *are compared with those of the modes detected by other authors in different samples. Local modes are labeled consistently with their assignment in Table 1. Whenever a polaronic one-phonon model has been applied, the resulting phonon frequency* ω^* *and the Huang–Rhys factor S have been reported.*

Compound	Reference	1	2	1+1	1+2	2+2	1+1+2	ω^*	S
Nd_2CuO_{4-y}	This work	183	285	365	475	570	642	210	4.3
Nd_2CuO_{4-y}	[4]				475				
Gd_2CuO_{4-y}	This work	180			480		640		
$Nd_{1.86}Ce_{0.14}CuO_{3.97}$	This work	160	248	369		580	640		
La_2CuO_{4+y}	[17]		285	398	486	560	640		
La_2CuO_{4-y}	[39]		246	430	495		630		
$YBa_2Cu_3O_{6.3}$	[17]			396/430	520				
$YBa_2Cu_3O_{6.2}$	[21]			435	510				
$YBa_2Cu_3O_{6.3}$	[19]							200	7
$Tl_2Ba_2Ca_{0.98}Gd_{0.02}Cu_2O_8$	[19]	188		403	480	578		200	5.6

model can also explain the temperature-dependence of doping-dependent modes observed here. As already mentioned, their intensity strongly increases with decreasing temperature, until it saturates at $T \approx 200$ K. This is consistent with a small-polaron model [29]. If one enters therein the present experimental values for ω^*, the peak energy of the d band, and the above 'freezing' temperature, then one gets a polaron binding energy $E_b \simeq 2\hbar\omega^* \simeq 450$ cm^{-1}. It may also be worth mentioning that a 'freezing temperature' of about 150 K has been reported for polarons observed by EXAFS in BSCCO [35].

The experimental data reported in the present paper are consistent with previous observations of strong infrared lines induced by excess charges introduced either by chemical or by photon doping (see Table 2). Moreover, structures similar to those shown here for the d band can be distinguished on the top of a mid-infrared (MIR) band (see Fig. 1 in [36]), whose energy is very close to that of the d band observed in this paper [18]. Although these structures were not mentioned or discussed in that paper, the T-dependence of such MIR absorption has been explained [18, 36] in terms of a shake-off polaron band. Incidentally, when a one-phonon model is applied to these data as well as to the mid-infrared band of $YBa_2Cu_3O_{6.3}$, one gets $\omega^* = 200$ cm^{-1} for both compounds [18], in agreement with our present result in Nd_2CuO_{4-y}. One also finds $S = 5.6$ in $Tl_2Ba_2Ca_{0.98}Gd_{0.02}Cu_2O_8$ and $S = 7$ in $YBa_2Cu_3O_{6.3}$, consistent with the present results.

The substantial analogy between the present results in chemically doped Nd_2CuO_{4-y} and those of photoinduced absorption spectra in the above different cuprates shows that the observed phenomena are related to the Cu–O plane and that they are due to extra carriers injected into the plane. Then, oxygen vacancies do not directly produce local modes, and enter the mechanism just in that they provide the excess carriers.

3.2 *The metallic phase of $Nd_{2-x}Ce_xCuO_{4-y}$ $(x > 0.12)$*

As already pointed out, any evidence of polarons in the normal phase of metallic HTSC would provide considerable support for the polaronic models of superconductivity in these materials. As a starting point, we can observe that a band peaked at about 0.1 eV had been introduced in a previous work, to correctly fit the 300 K reflectivity of as-grown metallic samples [7]. Here, we first extend these measurements to reduced $Nd_{2-x}Ce_xCuO_{4-y}$ samples ($y > 0$ and $x > 0.12$), in order to verify whether the above band depends on concentration of oxygen vacancies as observed for the d band in the semiconducting case.

The reflectivity curves of reduced samples are similar to those already published for the as-grown ones [7]. Nevertheless, the former are found to exhibit in the mid-infrared a characteristic dip, whose position approximately corresponds to the low-frequency edge of the d band in semiconducting NCO. In order to compare these spectra with those of as-grown samples [7], $R(\omega)$ has been fitted by a Drude–Lorentz model and the resulting oscillator intensities have been reported in terms of the effective number of carriers

$$n_{eff} = \frac{2m^*V}{\pi e^2} \int_0^\infty \sigma(\omega)\, d\omega. \tag{2}$$

Here, the effective mass m^* is conventionally assumed to be the free-electron mass m_0, and V is the volume of the cell that contains one formula unit (188×10^{-24} cm^3 in the present case). The present procedure of fitting the reflectance data to a Drude–Lorentz model allows one to get an estimate of $\sigma(\omega) = \Sigma_\alpha \sigma^\alpha(\omega)$ for the various α contributions here considered, and therefore to get

$$n_{eff}^{total} = \sum_\alpha n^\alpha_{eff} = \sum_\alpha \frac{2m^*V}{\pi e^2} \int_0^\infty \sigma^\alpha(\omega)\, d\omega. \tag{3}$$

where α = ph (phonons), D (Drude), d (d band), MIR (mid-infrared band).

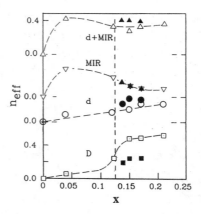

Fig. 3. The spectral weight n_{eff} of the Drude term, the d-band, and the MIR band as a function of Ce concentration x for as-grown (open symbols) and reduced (full symbols) samples of $Nd_{2-x}Ce_xCuO_{4-y}$.

The results of the above procedure are reported in Fig. 3, which shows that, in reduced samples (full symbols), a band centered at 0.1 eV exists, and that it is stronger than in as-grown samples (open symbols). This confirms that this feature has the same nature as the one observed, and resolved, in non-metallic samples. Moreover, the mid-infrared band has nearly the same strength. In turn, in the normal phase of reduced (superconducting) NCCO the Drude contribution is much smaller. This result is intriguing, as it is well known that, among e-doped compounds, only the reduced crystals are superconducting (with $T_c \simeq 20$ K).

To get deeper insight, we have measured the far-infrared reflectivity of the $Nd_{1.86}Ce_{0.14}CuO_{3.97}$ single crystal MN25 ($T_c = 19.6$ K), down to 20 K. Preliminary results are shown in Fig. 4, where several interesting features are evident. At 300 K, the far-infrared $R(\omega)$ is much lower than expected and shows a few resolved structures both in the far infrared and in the d band region. Moreover, a minimum in $R(\omega)$ appears at around 450 cm^{-1}. At 20 K, the lowest normal-phase temperature, $R(\omega) \to 1$ for $\omega \to 0$ with a Drude behavior. The minimum at 450 cm^{-1} strongly deepens, giving rise to a sharp peak in the reflectivity at about 380 cm^{-1}. According to our preliminary analysis, the 20 K spectrum in Fig. 4 results from a Drude term with plasma frequency $\omega_p \simeq 2000$ cm^{-1} and $\Gamma \simeq 100$ cm^{-1}, superimposed on a polaronic band very similar to the one found in the semiconducting samples. Indeed, most oscillators detected for $x = 0$ are also evident in the reflectivity spectrum of the superconductor. Their peak frequencies (Table 2) are in excellent agreement with those observed in the semiconducting phase, but for oscillators 1 and 2, which seem to be shifted

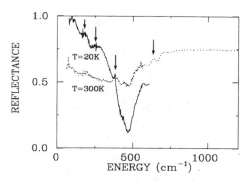

Fig. 4. $R(\omega)$ at 300 K (dotted line) and 20 K (solid line) for the $Nd_{1.86}Ce_{0.14}CuO_{3.97}$ sample MN25 with $T_c = 19.6$ K. Local modes are indicated by arrows.

towards lower energies. Further work is now in progress, to verify the reproducibility and significance of this result.

4 Conclusion

The results presented here can be summarized as follows. In reduced $M_{2-x}Ce_xCuO_{4-y}$ samples ($y > 0$), the far-infrared spectra show a series of temperature-dependent structures, in addition to the extended modes of the T' lattice, which we attribute to local modes associated with the introduction of oxygen vacancies. The intensity of these local modes, which account for a considerable amount of far-infrared absorption, can be explained only by assuming that they are coupled to the excess electrons produced by vacancies. The intensity of the d band, which is peaked at about 0.1 eV, also increases with doping and decreases with temperature. This band is here first partially resolved, and shown to be the superposition of strong overtones of the local modes detected in the far infrared. The intensities of the observed overtones again require the existence of a medium-strong coupling between electrons and lattice vibrations. The peak widths increase with peak frequencies ω_p according to a quadratic law. The ensemble of these results strongly points towards the existence of small polarons in chemically e-doped cuprates. The local modes in the far infrared originate from bound electron-vibration states, while the d band would correspond to shake-off processes, which allow the polaron to move from site to site.

We also observe the above local modes and the d band in the normal phase of a superconducting $Nd_{2-x}Ce_xCuO_{4-y}$ sample. For the first time in a HTSC, the Drude optical absorption has been directly resolved from all extra-Drude contributions in the spectra. Our results then yield direct evidence that polarons survive in the metallic phase, and that two different kinds of carriers exist in e-doped compounds. The former give rise to a Drude term, which in such low-T_c compounds has low plasma frequencies; the latter are self-trapped and produce polaronic absorption spectra. Such a conclusion can be extended to hole-doped HTSC, if one considers that the minimum in Fig. 4 corresponds to that detected [37] at 440 cm^{-1} in the optical conductivity of $YBa_2Cu_3O_7$, and interpreted in terms of strong electron–phonon coupling [38]. We then conclude that polarons are present in HTSC together with nearly free (Drude) carriers. They will have major effects on the transport, and possibly the superconducting properties, of these materials.

References

[1] T. Timusk and D. B. Tanner, *Infrared Properties of High T_c Superconductors*, in *Physical Properties of High Temperature Superconductors*, edited by D. M. Ginsberg (World Scientific, Singapore, 1989), pp. 339–407, and references therein.

[2] E. T. Heyen, G. Kliche, W. Kress, W. König, M. Cardona, E. Rampf, J. Prade, U. Schröder, A. D. Kulkarni, F. W. de Wette, S. Piñol, D. McK. Paul, E. Morán, and M. A. Alario-Franco, *Solid State Commun.* **74**, 1299 (1990).

[3] P. Calvani, M. Capizzi, S. Lupi, P. Maselli, M. Virgilio, A. Fabrizi, M. Pompa, W. Sadowski, and E. Walker, *J. Less-Common Metals* **164–165**, 776 (1990).

[4] G. A. Thomas, D. H. Rapkine, S.-W. Cheong, and L. F. Schneemeyer, *Phys. Rev.* B **47**, 11 369 (1993).

[5] M. K. Crawford, G. Burns, G. V. Chandrashekhar, F. H. Dacol, W. E. Farneth, E. M. McCarron III, and R. J. Smalley, *Phys. Rev.* B **41**, 8933 (1990).

[6] J. P. Falck, A. Levy, M. A. Kastner, and R. J. Birgenau, *Phys. Rev. Lett.* **69**, 1109 (1992).

[7] S. Lupi, P. Calvani, M. Capizzi, P. Maselli, W. Sadowski, and E. Walker, *Phys. Rev* B **45**, 12470 (1992).

[8] T. Arima, Y. Tokura, and S. Uchida, *Phys. Rev.* B **48**, 6597 (1993).

[9] G. A. Thomas, D. H. Rapkine, S. L. Cooper, S.-W. Cheong, and A. S. Cooper, *Phys. Rev. Lett.* **67**, 2906 (1991).

[10] G. A. Thomas, D. H. Rapkine, S. L. Cooper, S.-W. Cheong, A. S. Cooper, L. F. Schneemeyer, and J. V. Waszczak, *Phys. Rev.* B **45**, 2474 (1992).

[11] I. Terasaki, T. Nakahashi, S. Takebayashi, A. Maeda, and K. Uchinokura, *Physica* C **165**, 152 (1990).

[12] J. Lorenzana and L. Yu, *Phys. Rev. Lett.* **70**, 861 (1993).

[13] S. L. Cooper, G. A. Thomas, J. Orenstein, D. H. Rapkine, M. Capizzi, T. Timiusk, A. J. Millis, L. F. Schneemeyer, and J. V. Waszczak, *Phys. Rev.* B **40**, 11358 (1993).

[14] C. X. Chen and H. B. Schüttler, *Phys. Rev.* B **43**, 3771 (1991).

[15] C. M. Varma, P. B. Littlewood, S. Schmitt-Rink, E. Abrahams, and A. E. Rukenstein, *Phys. Rev. Lett.* **63**, 1996 (1989).

[16] J. P. Falck, A. Levy, M. A. Kastner, and R. J. Birgenau, *Phys. Rev.* B **48**, 4043 (1993).

[17] Y. H. Kim, S.-W. Cheong, and Z. Fisk, *Phys. Rev. Lett.* **67**, 2227 (1991); Y. H. Kim, A. J. Heeger, L. Acedo, G. Stucky, and F. Wudl, *Phys. Rev.* B **36**, 7252 (1987).

[18] D. Mihailovic, C. M. Foster, K. Voss, and A. J. Heeger, *Phys. Rev.* B **42**, 7989 (1990).

[19] D. Mihailovic, C. M. Foster, K. F. Voss, T. Mertelj, I. Poberaj, and N. Herron, *Phys. Rev.* B **44**, 237 (1991).

[20] J. M. Ginder, M. G. Roe, Y. Song, R. P. McCall, J. R. Gaines, E. Ehrenfreund, and A. J. Epstein, *Phys. Rev.* B **37**, 7506 (1988).

[21] C. Taliani, R. Zamboni, G. Ruani, F. C. Matacotta, and K. I. Pokhodyna, *Solid State Commun.* **66**, 487 (1988).

[22] T. Holstein, *Ann. Phys. (N. Y.)* **8**, 325 (1959).

[23] H. G. Reik, in *Polarons in Ionic Crystals and Polar Semiconductors*, edited by J. Devreese (North-Holland, Amsterdam, 1972), and references therein.

[24] K. Yonemitsu, A. R. Bishop, and J. Lorenzana, *Phys. Rev. Lett.* **69**, 965 (1992); *Phys. Rev.* B **47**, 8065 (1993).

[25] J. Mustre de Leon, I. Batistic, A. R. Bishop, S. D. Conradson, and S. A. Trugman, *Phys. Rev. Lett.* **68**, 3236 (1992).

[26] A. Alexandrov and J. Ranninger, *Phys. Rev.* B **23**, 1796 (1981).

[27] B. K. Chakraverty, *J. Physique* **42**, 1351 (1981).

[28] T. M. Rice and F. C. Zhang, *Phys. Rev.* B **39**, 815 (1989).

[29] D. Emin, *Phys. Rev. Lett.* **62**, 1544 (1989).

[30] S. Robaszkiewicz, R. Micnas, and J. Ranninger, *Phys. Rev.* B **36**, 180 (1987).

[31] S. Tajima, T. Ido, S. Ishibashi, T. Itoh, H. Eisaki, Y. Mizuo, T. Arima, H. Takaji, and S. Uchida, *Phys. Rev.* B **43**, 10496 (1991).

[32] C. P. Flynn, *Point Defects and Diffusion*, Clarendon Press, Oxford, 1972, p. 239–68.

[33] M. J. Rice, L. Pietronero, and P. Brüesch, *Solid State Commun.* **21**, 757 (1977).

[34] G. D. Mahan, *Many-Particle Physics*, second edition (Plenum Press, New York and London, 1990), p. 553.

[35] A. Bianconi and M. Missori, in *Phase Separation in Cuprate Superconductors*, ed. by E. Sigmund and A. K. Müller (Springer, Berlin, 1994), p. 316.

[36] C. M. Foster, A. J. Heeger, G. Stucky, and N. Herron, *Solid State Commun.* **71**, 945 (1989).

[37] K. Kamaras, S. L. Herr, C. D. Porter, N. Tache, D. B. Tanner, S. Etemad, T. Vankatesan, E. Chase, A. Inam, X. D. Wu, M. S. Hegde, and B. Dutta, *Phys. Rev. Lett.* **64**, 84 (1990).

[38] T. Timusk, C. D. Porter, and D. B. Tanner, *Phys. Rev. Lett.* **66**, 663 (1991).

[39] A. V. Bazhenov, A. V. Gorbunov, K. B. Rezchikov, T. N. Fursova, A. A. Zakharov, and M. B. Tsetlin, *Physica* C **214**, 45 (1993).

9

Electron–phonon interaction of non-equilibrim carriers in the photoinduced state of $YBa_2Cu_3O_{7-\delta}$

D. MIHAILOVIĆ and I. POBERAJ
J. Stefan Institute, University of Ljubljana, Ljubljana, Slovenia

Abstract

Using picosecond pulses we excite a large number of carriers in $YBa_2Cu_3O_{7-\delta}$ and drive the system through an insulator-to-metal transition. As we do this, we study the carrier interaction with the apical O and the 340 cm^{-1} planar O buckling vibrations by counting the number of non-equilibrium phonons that the carriers generate as they relax towards equilibrium. Counting is done directly by photoexcited anti-Stokes Raman scattering (PEARS). Carrier transport is found to be thermally activated; presumably the carriers relax by hopping between localized states. The in-plane and chain activation energies are found to be very different, suggesting that different carriers are involved: the chain carriers form polarons, which are strongly coupled to apical O vibrations, while the planar holes, which have a smaller activation energy, are suggested to interact less strongly with the lattice, especially in the metallic state.

1 Introduction

One may ask a very simple question about charge carrier transport in a cuprate superconductor: are the carriers moving through the crystal as in an extended band, or are they hopping from site to site? Given that the structure of these materials is composed of two distinct parts – the CuO_2 planes plus charge reservoirs – one may perhaps also wonder whether the behaviour of carriers in the two parts is different. In order to answer such questions, we need a microscopic probe, which, it is to be hoped, might tell us something about both the type of interaction between the carriers and the symmetry and location of the interaction. In this paper we suggest that the question may be answered by counting the phonons that are scattered (or released) as carriers move through the crystal. Phonon counting can be done effectively by using Stokes/anti-

Stokes Raman spectroscopy. The spectra of different phonons can be measured, so that the strength of coupling of the carriers to phonons either in different parts of the unit cell (e.g. planes or chains) or of different symmetry can be determined.

Although there are many more standard techniques, which can in principle answer the posed questions, in many cases the results have so far been inconclusive and the method of carrier transport in copper oxides is still not clear. For example, optical conductivity measurements, especially in comparison with photoinduced absorption measurements, suggest the existence of polarons in these materials [1]. On the other hand, the often observed linear temperature-dependence of the resistivity of the cuprate materials and the most recent observation of Mathiessen's rule for in-plane substituted Zn resistivity (where to a first approximation the zinc doping just adds a temperature-independent term to the scattering lifetime) suggest that the holes in the CuO_2 planes are behaving in a fashion consistent with simple band theory [2]. Measurements of photoconductivity in $YBa_2Cu_3O_{7-\delta}$ for different doping levels in the range $0.6 < \delta < 1$ when compared with the optical conductivity, $\sigma(\omega)$, imply fairly conclusively that the energy levels in the 'gap' between the planar Cu antibonding band (or upper Hubbard band) and the hybridized planar Cu $d_{x^2-y^2}$–O$p_{x,y}$ band is filled with localized states [3]. These experiments suggest that the materials can be described as Fermi glasses in a large part of their phase diagram. In $YBa_2Cu_3O_{7-\delta}$ at approximately $\delta = 0.6$, the Fermi level crosses the mobility edge and an insulator-to-metal transition occurs. Whether the localized states are due to Anderson localization, i.e. O disorder, or due to polarons is not clear from those experiments. However, since there is relatively little evidence for any significant changes in the (static) disorder of the CuO_2 planes with doping, either (i) it is the disorder outside the planes that is important, or (ii) the localized states in the CuO_2 planes form because of carrier self-trapping, or (iii) both.

In order to be able to observe the interaction of the carriers with phonons we need to excite the charge carriers from equilibrium. This is easily done by photoexcitation using a short laser pulse. If the timescale of the experiment is such that phonons scattered from the carriers have not yet decayed or dispersed, then there will be a transient measurable non-equilibrium population of phonons, which we can detect by recording Raman spectra with the same short laser pulses that we used to excite the system. Clearly we are assuming that the initial transfer of energy from the photoexcited carrier to lattice vibrations is faster than the phonon lifetime, which we believe is a reasonable assumption. From the point of view of the present investigation it is also important to note that an insulator-to-metal transition, similar to the one

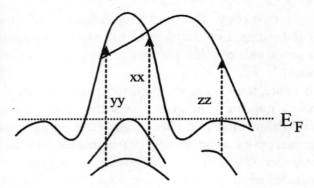

Fig. 1. A schematic diagram of the excitation process is shown in the band picture. The bands shown are simplified from various band structure calculations and are discussed in detail in the text. The arrows represent the vertical transitions at 2.33 eV.

that occurs upon doping with O, apparently occurs also upon photoexcitation with short pulse laser pulses [4]. This means that, by varying the laser excitation intensity, we can obtain information on how the carrier–lattice coupling changes with doping density.

2 The excitation process

The laser pulses used in the present experiments were 1.5 ps long from a frequency-doubled, pulse-compressed mode-locked Nd:YAG laser. The experimental details have already been described elsewhere [5]. The photon energy of 2.33 eV is sufficient to excite carriers into a number of different possible states. The best data on what states are available in this energy range are from optical ellipsometry measurements or from resonant Raman experiments [6], where the experimental cross-section has been compared with the calculated LDA band structure. In the metallic compound $YBa_2Cu_3O_{7-\delta}$ with $\delta \simeq 0$, where these calculations work quite well, the assignment of the transitions can be taken to be fairly reliable. We show a schematic diagram of the possible vertical exitations in $YBa_2Cu_3O_7$ in Fig. 1.

There are two possible excitations at 2.33 eV:

1. for incident light polarization in the x or y direction (i.e. along the crystal a or b axes), the inital states are primarily the planar $O(2)\ p_y$–$Cu(2)\ d_{xy}$–$O(3)\ p_x$ planar non-bonding bands, while the final states are in the hybridized plane band $(Cu(2)\ d_{z^2-y^2}$–$O(2)\ p_x$–$O(3)\ p_y)$ and
2. for polarization in the z direction (along the crystal c axis), the initial states are in the $O(2)\ p_z$–$O(3)\ p_z$–$Cu(2)\ d_{3z^2-r^2}$–$O(4)\ p_z$–$O(1)\ p_z\ z$ orbital hybridized bands, while the final state is in the chain band $(Cu(1)\ d_{x^2-y^2}$–$O(1)\ p_y$–$O(4)\ p_z)$.

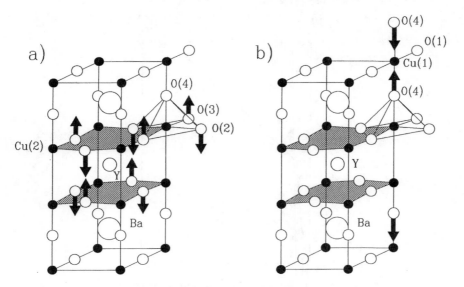

Fig. 2. The labels of the different ions are shown, together with the eigenvectors of the two phonons investigated: (a) the in-plane O(2)/O(3) plane buckling mode at $340\,\mathrm{cm}^{-1}$ and (b) the apical O at $470\,\mathrm{cm}^{-1}$ vibration. The $340\,\mathrm{cm}^{-1}$ vibration couples predominantly to carriers in the CuO_2 planes, while the apical O vibration is coupled predominantly to the Cu(1) ion in the chains.

(The nomenclature of the ions is shown in Fig. 2). In the insulating material with $\delta = 1.0$ (i.e. with *no* chain O sites filled), the states involved in the excitation are less clear, but since the experiments were performed with samples that had a significant O concentration ($\delta < 0.75$), we will not concern ourselves here with the pure O_6 material.

The experimental method of 'phonon counting', which is described here, has to some extent been utilized in semiconductor physics, particularly in determining the carrier energy loss rates and LO phonon lifetimes in direct-gap semiconductors like GaAs for example. Because of its importance from the point of view of device applications, the energy loss rate in semiconductors has been extensively calculated and verified by experiments and is thus well understood [7]. We can use these results to test whether band-like behaviour is also observed in materials like $YBa_2Cu_3O_{7-\delta}$.

3 The experimental results

The excitation photon flux that is available in our experiments is in the range 10^{11}–10^{14} photons cm^{-2}. Although the excited carrier density cannot be calculated accurately unless we know the carrier lifetime, we estimate the

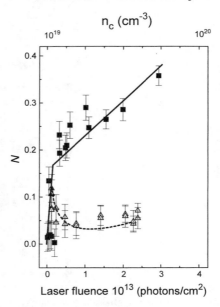

Fig. 3. The phonon occupation numbers as a function of photoexcited flux density. In the metallic state the coupling of carriers to the two phonons is significantly reduced. The 340 cm^{-1} planar phonon is apparently significantly less coupled in the metallic state than the apical O phonon.

photoexcited carrier density at maximum fluence to be near 10^{20} cm^{-3}. This compares with a carrier density at optimal doping ($\delta = 0$) of approximately 3×10^{21} cm^{-3}. We discuss the interaction of carriers with two different vibrations, which are well observed in the Raman spectra: the B_{1g}-like plane buckling O(2)/O(3) c-axis out-of-phase vibration at 340 cm^{-1} and the A_g symmetry apical or bridging O(4) c-axis vibration at 470 cm^{-1} (see Fig. 2). (We are experimentally restricted by the bandpass filter of our apparatus to the range above 150 cm^{-1}, so we cannot measure the behaviour of the lower lying vibrations.) The occupation numbers for the two vibrations measured on the 1.5 ps timescale are shown in Fig. 3 as a function of laser fluence, Φ. Both phonon occupation numbers rise rapidly from their equilibrium values with photoexcitation, indicating that we are indeed creating a non-equilibrium population and that there is a sufficiently large phonon relaxation bottleneck for us to be able to study the effects of non-equilibrium carrier–lattice coupling.

At a photon fluence of approximately 10^{13} photons cm^{-2} a discontinuity is observed for both phonons, which we attribute to the doping-induced I–M transition. We can be confident of this suggestion for a number of reasons: (i) at a similar value of photon fluence an onset of metallic behaviour has been observed in measurements of transient photoconductivity [4] and (ii) we observe a change in reflectivity of the sample at the same point as the discontinuity, signifying a change in dielectric constant. The occupation number of the plane-buckling vibration is observed to show a rapid fall in the metallic phase of the material, whereas the apical O vibration shows only a reduction in the slope of N versus Φ. Unfortunately it is not easy to interpret quantitatively the results in Fig. 3, but qualitatively, the reduction of N in the metallic state signifies a reduction in the e–p coupling strength, assuming that there is no significant reduction in phonon lifetimes with doping. (This latter assumption is justified by the absence of any significant systematic change in vibrational linewidths as a function of doping.) The planar vibration shows a particularly large change in coupling strength. We will discuss the nature of the e–p coupling in the next section, when it becomes clear that the e–p interaction is actually between the lattice vibrations and localized carriers.

4 The relaxation process

So far we have presented no data on the type of coupling between photoexcited carriers and lattice vibrations. From the temperature-dependence of the phonon occupation number we are able to determine fairly unambiguously that the interactions we are studying are between the vibrations and carriers in localized electronic states. In general, carrier relaxation in solids proceeds partly by energy loss to the lattice and partly by radiative recombination. The radiative recombination channel appears to be very weak in the cuprates, as there is very little luminescence observed in these materials. The fact that in $YBa_2Cu_3O_{7-\delta}$ radiative recombination is not observed – except for some weak hot luminescence and a relatively weak luminescence band at around 1 eV – indicates that the photoexcited electron and hole wavefunctions must have very little overlap. The most likely reason is that the electrons and holes remain spatially separated: the electrons presumably confined to the chains, and the holes in the CuO_2 planes. The dominant relaxation process appears to proceed non-radiatively.

In Fig. 4 we show the temperature-dependence of relaxation in the photo-induced metallic state of $YBa_2Cu_3O_{7-\delta}$. We see that the phonon occupation number for both vibrations shows activated behaviour, where at low temperatures we see no excess vibrational occupation with strong increase in N with increasing temperature. The activation energies for the 340 and 470 cm^{-1} phonons are 30 and 64 meV respectively (given by the slope in the insert to Fig. 4).

First, let us examine the temperature-dependence that we would expect were the photoexcited carrier relaxation within a semiconductor-like CuO_2 band. As

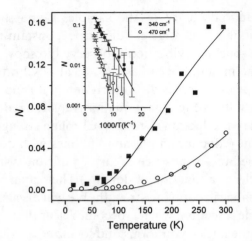

Fig. 4. The temperature-dependence of the phonon occupation number implies activated carrier hopping. The insert shows the same data in a $\log N$ versus $1/T$ plot. The occupation number due to ambient temperature has been subtracted for every data point.

already mentioned in the introduction, theoretical description of the energy relaxation has been dealt with in detail by Seeger [7]. The most important result is that, for carriers that are significantly out of equilibrium, the energy relaxation rate R_E is essentially temperature-independent. This is because the energy relaxation rate is proportional to the rate of phonon absorption plus phonon emission, the sum of which is independent of lattice temperature. The energy relaxation rate becomes dependent on the ambient lattice temperature only when the carriers are cooled to the bottom of the band, such that they are nearly in equilibrium with the lattice. (We can be sure that we do not have the latter situation since the phonon occupation numbers are very high: the excess occupation number of 0.3 implies a vibrational temperature in excess of 700 K for the apical O mode.) We are thus dealing with a case that apparently cannot be described by band theory.

We propose that the non-equilibrium occupation numbers that we observe are a result of carrier relaxation towards equilibrium through hopping between localized states. The process is indicated schematically in Fig. 5. Every time a carrier makes a hop it requires lattice activation energy from its localized state. Since the process is one of energy relaxation from a non-equilibrium state, in the hopping process the carrier releases excess energy to lattice vibrations, which we observe. The phonon occupation number is thus directly proportional to the relaxation rate of the photoexcited carriers and so the activation energies that we measure åre the *carrier* hopping activation energies.

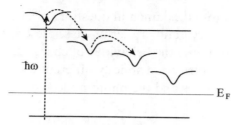

Fig. 5. A diagram schematically showing the photoexcited charge carrier relaxation process via localized states.

The activation energy for in-plane carrier hopping appears to be significantly smaller than for the carriers that are coupled to the apical O vibrations, which suggests that we are dealing with two different types of carriers: one type predominantly coupled to vibrations in the planes, the other – most probably chain carriers – predominantly coupled to the apical O vibrations. Although the planar carriers may also be coupled to the apical O vibrations, judging by their lower activation energy, they are clearly in shallower potentials than are the chain carriers.

The origin of the localized states that we observe can be either disorder or self-trapping, or most likely, both. In addition, localization due to spin coupling to the antiferromagnetic background should be considered as well, especially for the planar carriers. Polaronic states in $YBa_2Cu_3O_{7-\delta}$ have been considered in detail by Ranninger [8]. In this picture, relaxation proceeds predominantly via relaxation of Frank–Condon excited polaronic states of the O(4)–Cu(1)–O(4), O(4)–O(1)–Cu(1)–O(4) and O(4)–O(1)–Cu(1)–O(1)–O(4) chain clusters. The relaxation process in this case, just as in the case of disorder-localized states (which would arise due to O disorder in the interstitial O(1) and O(4) positions), proceeds via apical O phonon emission. The model does not so far explain the activated behaviour of the carriers coupled to the plane-bucking $340\ cm^{-1}$ phonons. (In principle the cause of localization of these carriers could be quite different, e.g. the antiferromagnetic spin background.)

5 Conclusion

In the experiments outlined in this paper we believe that we have conclusively shown that carrier transport in the photoexcited 'metallic' state of $YBa_2Cu_3O_{7-\delta}$ proceeds via activated hopping. The relaxation process clearly cannot be described by a band model in which the energy relaxation rate is independent of temperature. Although many features in the data still have to be understood in detail, it appears that the experiments have succeeded in answering one of the more fundamental questions about the cuprates: a band picture is not satisfactory, and a Fermi glass description is much more appropriate in understanding the behaviour of charge carrier transport in the cuprates. Moreover, we observe distinctly different coupling between planar and chain vibrational modes and the photoinduced carriers.

References

[1] D. Mihailović *et al., Phys. Rev.* B **42**, 7989 (1990).
[2] J. R. Cooper, private communication.
[3] G. Yu *et al., Phys. Rev.* B **48**, 7545 (1993).
[4] G. Yu *et al., Phys. Rev.* B **45**, 4964 (1992).
[5] D. Mihailović and I. Poberaj, *J. Phys. Chem. Solids* **54**, 1315 (1993).
[6] E. T. Heyen *et al., Phys. Rev.* B **45**, 3037 (1992); E. T. Heyen *et al., Phys. Rev. Lett.* **65**, 3048 (1990).
[7] K. Seeger, *Semiconductor Physics*, Springer Series in Solid State Sciences (1985).
[8] J. Ranninger, *Solid State Commun.* **85**, 929 (1992).

10

Experimental evidence of local lattice distortion in superconducting oxides

T. EGAMI[1,2], W. DMOWSKI[1,2], R. J. McQUEENEY[1,3],
T. R. SENDYKA[1,2], S. ISHIHARA[4], M. TACHIKI[4],
H. YAMAUCHI[5], S. TANAKA[5], T. HINATSU[6] and S. UCHIDA[6]

1) *Laboratory for Research on the Structure of Matter, University of Pennsylvania,*
Philadelphia, PA 19104-6272, USA
2) *Department of Materials Science and Engineering*
3) *Department of Physics*
4) *Institute for Materials Research, Tohoku University, Sendai 980, Japan*
5) *International Superconductivity Technology Center, Superconductivity Research Laboratory,*
Tokyo 135, Japan
6) *Department of Applied Physics, University of Tokyo, Tokyo 103, Japan*

Abstract

If lattice polarons exist in high-temperature superconducting oxides then there must be evidence of local lattice distortion associated with polarons. While the distortions are dynamic and subtle, making direct observation difficult, there are numerous indications that some anomalous local deviations from the crystallographic lattice structure exist in superconducting oxides. Based largely upon the results of pulsed neutron scattering measurements, we present an argument in favor of the presence of local lattice distortions consistent with lattice polarons. A few implications of the observation in relation to other physical properties are discussed.

1 Introduction

Even though polarons have been known for a long time, direct experimental observation of lattice distortions associated with them is surprisingly scarce, largely because the density of polarons is usually low and consequently the lattice distortion is small on average, making observation very difficult. While some observations of lattice distortion associated with polarons have been made for low-dimensional organic conductors in which the periodic lattice distortion (Peierls distortion) can be regarded as an array of localized polarons [1], there are very few such reports for oxides [2]. Moreover, most known cases of polarons are heavy, small polarons, while in high-temperature superconducting (HTSC) oxides the presence of mobile large polarons is suspected. For

155

those reasons, local lattice distortion has been observed so far mostly by non-traditional methods of structural study, while the crystallographic community has largely been skeptical. In this paper we discuss why observation is difficult, whether there is sufficient experimental evidence to support the presence of polarons in high-temperature superconducting oxides or not, and the implications of these observations.

2 Experimental methods to observe local lattice distortion

2.1 Crystallographic methods

The intensity of X-rays or neutrons scattered by a sample is given by

$$I(Q) = |F(Q)|^2$$

$$F(Q) = \frac{1}{\sqrt{N\langle f(Q)\rangle}} \sum_i f_i(Q) e^{iQ \cdot R_i}, \tag{1}$$

where Q is the momentum transfer in the scattering ($Q = 4\pi \sin\theta/\lambda$), $f(Q)$ is the atomic scattering factor, $\langle \ldots \rangle$ denotes compositional average, R_i is the position of the ith atom,

$$R_i = \langle R_i\rangle + u_i, \tag{2}$$

where u_i is the displacement and $\langle R_i\rangle$ is the average crystallographic atomic position. By Fourier-transforming the displacement

$$u_i = \frac{1}{\sqrt{N}} \sum_q u_q e^{-iQ \cdot R_i} \tag{3}$$

the intensity becomes

$$I(Q) = I_B(Q) + I_D(Q) + \ldots, \tag{4}$$

where

$$I_B(Q) = \frac{e^{-2BQ^2}}{N\langle f(Q)\rangle^2} \sum_{i,j} f_i(Q) f_j(Q) e^{iQ \cdot (R_i - R_j)}, \tag{5}$$

$$I_D(Q) = \frac{1}{N\langle f(Q)\rangle^2} \sum_{i,j,q} (Q \cdot u_q)^2 f_i(Q) f_j(Q) \langle e^{i(Q-q) \cdot (R_i - R_j)}\rangle. \tag{6}$$

The first term describes the Bragg diffraction peaks at the reciprocal lattice points, while the second term describes the diffuse scattering intensity observed in between the Bragg peaks. The factor e^{-2BQ^2} is the Debye–Waller factor, where B is called the thermal factor,

$$B = \tfrac{1}{2}\langle u^2 \rangle = \tfrac{1}{2}(\langle u_{\text{phonon}}^2 \rangle + \langle u_{\text{strain}}^2 \rangle). \tag{7}$$

Here $\langle u_{\text{phonon}}^2 \rangle$ is the thermal amplitude for phonons, and $\langle u_{\text{strain}}^2 \rangle$ is the average amplitude of displacement due to strains.

However, displacements due to strain are often rather small, and difficult to differentiate from the thermal amplitude. In such a case more careful examinations are required. For instance the Debye–Waller approximation

$$\langle e^{iQ \cdot u_i} \rangle = e^{-2BQ^2} \tag{8}$$

assumes random Gaussian distribution of u_i. However, if the local displacements are not random, such as in the case of a local double-well potential, then this approximation breaks down. Instead, if u_i is equal to $\pm u$, assuming that u is parallel to Q,

$$\langle e^{iQ \cdot u_i} \rangle = \cos(Qu), \tag{9}$$

which shows a slow oscillatory Q dependence. However, in conventional crystallographic analysis, data are usually not taken up to high enough Q values to observe the oscillation. Then, for small values of Q ($Qu \ll 1$), Eq. (8) is

$$\cos(Qu) = 1 - \tfrac{1}{2}(Qu)^2 + \dots \approx e^{-(Qu)^2/2}, \tag{10}$$

which cannot be differentiated from the Debye–Waller form. For this reason, the effects of all displacements including those in the double-well potential are usually swept into the Debye–Waller factor in the conventional analysis. Only in non-crystallographic methods such as EXAFS or PDF analysis is $I(Q)$ determined up to high enough Q values to differentiate local displacements from phonons. Thus it is not surprising that the crystallographic community has been less inclined to recognize the presence of local displacements in superconducting oxides.

In equation (6) the lattice sum requires $K = Q - q$ to be the reciprocal lattice vector. Diffuse scattering is caused by any deviation from perfect periodicity, including phonons (thermal diffuse scattering, TDS), local strains and other defects. Diffuse scattering intensity is usually low, and it is not easy to differentiate diffuse scattering arising from local strains from the TDS. This will be discussed later.

2.2 The EXAFS method

In the extended X-ray absorption fine structure (EXAFS) method the structural information is contained in the quantum interference of the outgoing and incoming wavefunctions of the photo-excited electrons, which modifies the X-ray absorption coefficient. A classical analogy is back-scattering diffraction of photo-excited electrons. This method has the particular advantage that it is capable of describing the distribution of atomic distances (PDF) from a specific element, for instance Cu. It also determines the structure factor up to large values of Q ($=2k$), typically 30 Å. On the other hand, an important part of the structure factor, up to k about 3 Å^{-1} or $Q=6$ Å^{-1}, cannot be observed by EXAFS because of interference by band effects (XANES), and is left out of the analysis. This often renders the coordination number and sometimes even the near-neighbor distances determined by the EXAFS method less than reliable. By using highly textured grain-oriented samples and polarized X-rays, a directional PDF can be determined by EXAFS. However, small amounts of misoriented grains can produce misleading results. Thus the results of EXAFS studies are excellent indicators of the presence of anomaly, but they are not necessarily quantitatively reliable.

A number of X-ray absorption (EXAFS and XANES) studies suggest that, in HTSC oxides, the behavior of the lattice at the onset of superconductivity may be anomalous [3], and some atoms, notably the apical oxygen atoms, are locally in the double-well potential [4]. However, this has been particularly controversial, since other structural measurements did not agree with this observation of such double-well potential behavior, as we will discuss below.

2.3 PDF analysis of pulsed neutron scattering data

The premise of crystallographic structural analysis is periodicity of the structure. Consequently only the Bragg peaks are considered for the analysis, and diffuse scattering intensities are discarded as a part of the background. On the other hand, in the method of atomic pair-distribution function (PDF) analysis no such assumption is made. PDF analysis has been widely used to study liquids and glasses [5, 6], but has only rarely been used to study crystalline solids. The PDF, $\rho(r)$, is obtained by Fourier-transformation of the normalized scattering intensity, or the total structure factor, $S(Q) = I(Q)/\langle f(Q) \rangle^2$,

$$\rho(r) = \rho_0 + \frac{1}{2\pi^2 r} \int [S(Q) - 1] \sin(Qr) Q \, dQ, \tag{11}$$

where ρ_0 is the average number density of atoms and describes the probability of two atoms being separated by a distance r. It is a one-dimensional quantity averaged over all angles. In obtaining the PDF not only the Bragg diffraction intensity but also diffuse scattering intensities are included.

For a long time the experimental limitation on the range of Q has made it impractical to apply this method to well-ordered materials. In glasses and liquids $S(Q)$ attenuates quickly to unity beyond $Q = 20$–25 Å^{-1}, because structural disorder makes the structure incoherent at large Q. In crystals, on the other hand, $S(Q)$ persists up to larger values of Q. Fortunately, even in a perfect crystal, thermal and quantum mechanical positional disorder due to lattice vibrations renders the structure factor $S(Q)$ incoherent for very large values of Q, as represented by the Debye–Waller factor discussed above. Usually, when $S(Q)$ is determined up to Q of 35–45 Å^{-1}, the Fourier-transformation can be calculated without serious termination errors [7]. With the advent of synchrotron based sources such as the pulsed neutron source or the synchrotron radiation source, neutrons and X-rays of sufficiently short wavelengths are now available, so that $S(Q)$ can be determined up to these high ranges of momentum transfer.

The main advantage of this technique is that *correlated atomic displacements* can be studied directly. For instance, if two atoms are oscillating in-phase, then, while the Debye–Waller factor records the amplitude of each atom, the PDF shows no oscillation at all because the interatomic distance does not change with time. On the other hand, if they are oscillating out-of-phase, then the PDF shows an amplitude twice as great as does the Debye–Waller factor. Thus, by properly modeling the PDF, correlated atomic motion can be detected. Furthermore, by virtue of collecting data up to high values of Q, anharmonic displacements such as formation of double-well states can be detected directly.

One of the problems of the PDF method is that, in spite of the large Q ranges over which data are collected, some spurious oscillations remain. They are due to statistical as well as systematic errors. Methods to evaluate the statistical errors have been developed [7], and are used in the modeling process so that real effects can be separated from spurious effects. Another problem is that, unless isotopic substitution is used, only the total PDF, not resolved for correlations among the elements, is obtained. However, in most cases the structure can be identified by a careful modeling process. Yet another difficulty often mentioned is the uniqueness of the model. This problem, however, is common to any structural study, conventional or otherwise. In the case of the PDF method, this is due mainly to general lack of experience with this technique; psychological barriers remain with respect to the reliability of the technique as a method of structural determination. This problem can be eased only by continued success of the method and its thereby increased acceptance by the scientific community.

3 Evidence of local distortion

3.1 Crystallographic evidence

The first indication of anomalous lattice distortion was already seen early in 1987 in the Rietfeld analysis of the structure of $La_{2-x}Sr_xCuO_4$ (LSCO), for which thermal factors were found to be larger than expected for phonons [8].

Since then many HTSC oxides have been found to have large thermal factors [9]. For instance the apparent amplitude $\langle u^2 \rangle^{1/2}$ is as much as 0.35 Å for oxygen in the Tl–O plane of $Tl_2Ba_2CaCu_2O_8$ [10]. Since the condition for melting according to Lindeman's rule corresponds to about 0.15 Å in thermal amplitude, this apparent amplitude is certainly anomalous. In this case we found by PDF analysis that this is due to rearrangement in the Tl–O plane [11].

However, large thermal factors are often seen as a consequence of lattice defects, including the mixed ion effect. When two kinds of ions with different ionic radii are occupying crystallographically equivalent sites, the atomic size difference produces local strain, and contributes to the thermal factor. Thus thermal factors are small for stoichiometric compounds. Indeed simple unmixed compounds such as $YBa_2Cu_3O_7$ (YBCO) show rather small thermal factors. For instance Sullivan *et al.* concluded from their single-crystal X-ray diffraction study that there is no evidence for apical oxygen double-well behavior in $YBa_2Cu_3O_7$ [12]. However, in our view too much attention has been focused just on the apical oxygen, while much more is going on in the compound. For instance, the chain oxygen atoms in this compound show anomalous amplitude, possibly similar to those in $YBa_2Cu_4O_8$. In addition this compound is optimally doped as far as the charge carrier concentration is concerned, and represents a special case, while underdoped HTSC compounds are likely to show anomalous lattice behavior, as will be discussed below.

3.2 EXAFS results

As we mentioned above, several EXAFS measurements show double-well features as well as anomalous changes near the superconducting transition temperature. For instance, double-well behavior has been reported for Bi–Sr–Ca–Cu–O (BISCO) [13] and Tl–Ba–Ca–Cu–O [14] in addition to YBCO(123) samples. However, these results need to be seen with some caution. In order to separate the in-plane Cu–O distance from the out-of-plane distances, grain-oriented powder samples were used in measurements. However, if each powder particulate contains more than one crystal grain, then these particles will be misoriented during the orientation process by the applied magnetic field, and the in-plane and out-of-plane signals become mixed up. This can produce a false double-well pattern. In addition, in the case of BISCO the modulation in the Bi–O plane resulting in superlattice behavior produces complex local structure with more than one atomic distance. In Tl(2212), complex atomic rearrangement in the Tl–O plane can also affect other parts of the structure. Indeed, in YBaCuO, according to Stern *et al.*, only if the oxygen content is reduced is a double-well feature observed [15]. Thus the evidence is still

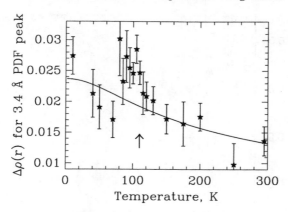

Fig. 1. The temperature-dependence of the PDF peak height at 3.4 Å for $Tl_2Ba_2CaCu_2O_8$ determined by pulsed neutron scattering. Vertical bars represent statistical error. The solid line is the normal temperature dependence calculated from the phonon density of states measured by neutron inelastic scattering. The arrow indicates T_c [17].

conflicting, and it is important to carefully separate the effect of structural disorder or complexity from the intrinsic effect possibly resulting from the particular electron–lattice interaction.

3.3 *Pulsed neutron scattering*

Since 1987 we have been studying the local structure and its temperature dependence by using PDF analysis for a number of HTSC oxides [11, 16–19]. All the HTSC oxides that we examined showed in one way or another in the PDF marked deviations from the crystallographic structure. However, the mode of displacement is not identical, and varies from one system to another. In some cases the displacements are merely the consequence of static or quasi-static disorder unrelated to superconductivity, such as in the case of Tl–O plane rearrangement in Tl(2212) [11].

3.3.1 *Evidence of anomalous temperature-dependence*

Anomalous temperature-dependence of PDF peak height was observed for a number of compounds. Among them the observation made for one peak of Tl(2212) [17] is the best and most unambiguous case. As shown in Fig. 1, the height of the peak at 3.4 Å changes anomalously near the superconducting transition temperature of 110 K shown by an arrow. Here the solid line is the normal temperature-dependence calculated from the phonon density of states determined by inelastic neutron measurement. The deviation of the data points

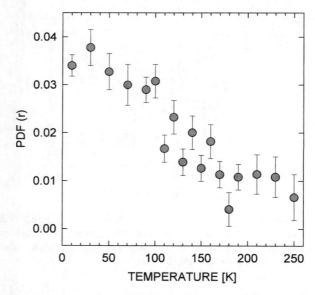

Fig. 2. The anomalous temperature dependence of the PDF peak height at 3.43 Å for
YBa$_2$Cu$_4$O$_8$ determined by pulsed neutron scattering.

from the expected normal behavior near T_c can be concluded with 99.999%
statistical confidence. If one attempts to interpret this change in terms of the
change in phonon frequencies, then the magnitude of change in PDF corresponds to the change in phonon frequency by as much as 20% or so. While some
changes in lattice dynamics are expected at T_c, the changes usually are much
smaller. For YBCO(123) the phonon frequencies observed by Raman scattering [20], which are consistent with the results of calculation according to the
Eliashberg theory [21], amount only to about 2%. Since the change in PDF
peak height cannot be explained in terms of change in phonon frequency, it is
appropriate to call this change an anomaly. In other compounds the anomaly
near T_c is often found, but it tends to be weaker than in the case of Tl(2212).

In the case of YBa$_2$Cu$_4$O$_8$ (124) the anomalous change was observed at
temperatures significantly above T_c. As shown in Fig. 2 and Fig. 3, while the
peak at 3.88 Å corresponding to a and b translation shows normal temperature-
dependence, other peaks including those at 2.97, 3.43 and 3.66 Å showed rapid
decrease in peak height, and almost completely disappeared at temperatures
above 200 K. It is very interesting to compare the observed temperature
dependence with the temperature-dependencies of the Cu NMR spin–lattice
relaxation time, T_1 [22], the Cu Knight shift and the resistivity divided by T [23].
Below around 160 K ($T_c = 80$ K), $1/T_1$ precipitously decreases with lowering of
temperature. This decrease in $1/T_1$ corresponds closely to the change in height

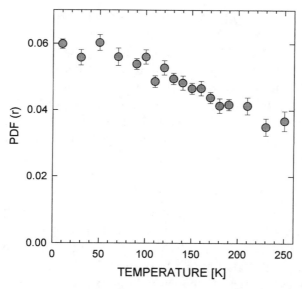

Fig. 3. The normal temperature-dependence of the PDF peak height at 3.88 Å for $YBa_2Cu_4O_8$ determined by pulsed neutron scattering.

of some peaks of the PDF of the (124) compound. Since the nuclear spin relaxation rate, $1/T_1$, is proportional to the imaginary part of the spin susceptibility of the system, $\chi''(q=0, \omega \to 0)$, this behavior is usually interpreted as indicating an opening up of the spin gap. Indeed, neutron inelastic scattering measurement detects a decrease in $\chi''(q, \omega)$ below this temperature in $La_{2-x}Sr_xCuO_4$ and $YBa_2Cu_3O_{7-\delta}$ [24]. This opening of the spin-gap is a consequence of the increase in the Cu spin–spin correlation, which, in the spin fluctuation theory [25, 26], is considered to provide the mechanism for high-temperature superconductivity. However, the present results suggest that the spin-gap may be at least in part structural in origin, and that this phenomenon is also a part of the lattice effect as discussed below.

Thus the evidence for anomalous temperature-dependence appears unmistakable. What is not yet clearly established, however, is its implication. In particular, in pulsed neutron scattering measurements the issue of neutron scattering dynamics complicates matters. There is evidence that the observed anomaly is dynamic in nature. This comes from the fact that the intensity of the anomaly depends upon the setting angle of the neutron detector. The data used in computing the PDF were obtained by the time-of-flight method. In this method the neutron detector has no energy discrimination. However, because of the neutron dynamics (the Placzek shift), only neutrons with a certain range of energy transfer contribute to the PDF. For instance, those neutrons that are slowed down too much by giving a large amount of their energy to the sample will not be coherent with higher energy neutrons. The detector angle determines the energy range. This dynamic aspect is being investigated now by inelastic neutron scattering, performed at the Rutherford–Appleton Laboratory using the MARI spectrometer. A similar experiment carried out earlier for YBCO(123) showed pronounced changes in lattice dynamics near T_c [27]. Exactly what is changing with temperature is often difficult to determine when the changes are small, but in some cases it can be identified clearly by structure modeling as discussed below.

3.3.2 Evidence for local displacement

Significant local deviations from the average structure have been seen in all HTSC compounds. While in some cases they are merely the consequences of a mixed ion effect or defects, in YBCO(123, 124) they must be intrinsic. The PDFs (solid lines) of $YBa_2Cu_4O_8$ (124) and $YBa_2Cu_3O_7$ (123) are compared with the PDFs calculated for the crystallographic structure (dotted lines) in Fig. 4. The calculated PDFs were convoluted by a Gaussian function to represent thermal vibrations. The amplitude of thermal vibration assumed ($\langle u_x^2 \rangle^{1/2} = 0.048$ Å) is consistent with the phonon dispersion in these solids. The measured PDFs agree reasonably well with the calculated PDFs up to about 4 Å, but beyond 4 Å they show significant departures from the calculated PDFs. The amplitude of oscillations in the measured PDFs is less than that in the calculated PDFs beyond 4 Å, suggesting that larger values of $\langle u_x^2 \rangle$ have to be assumed to account for the part of the PDF beyond 4 Å. This implies that there are collective atomic displacements of correlation length about 4 Å. Similar amounts of deviation are seen for the (124) and (123) compounds. In order to quantify the degree of deviation, we define an agreement factor

$$A = \left(\int_{r_1}^{r_2} (\rho_{exp}(r) - \rho_{calc}(r))^2 \, dr \Big/ \int_{r_1}^{r_2} \rho_0^2 \, dr \right)^{1/2}, \tag{12}$$

where $\rho_{exp}(r)$ is the experimentally determined PDF, $\rho_{calc}(r)$ is the calculated PDF, and r_1 and r_2 define the range over which the A-factor is calculated. The A-factor is a real-space version of the crystallographic R-factor. The A-factor evaluated for the range 2.17–9 Å is 0.0866 for the (124) and 0.0887 for the (123) compound.

The PDF, however, is a one-dimensional correlation function averaged over all angles. In order to deduce a three-dimensional structure, modeling or structural refinement has to be carried out. This can be done by using the A-factor. What we do is to modify the structural model so that the A-factor is

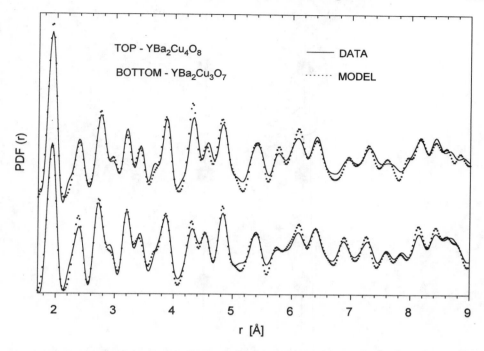

Fig. 4. Neutron PDFs of $YBa_2Cu_4O_8$ (above) and $YBa_2Cu_3O_7$ (below). Solid lines represent the experimental PDFs, and dotted lines are PDFs calculated for the crystallographic structure.

minimized, for instance starting with the crystallographic structure. We call this procedure real-space structural refinement, as opposed to the usual process of reciprocal space refinement in which an experimental $S(Q)$ is compared with a theoretical $S(Q)$. This can be done by the so-called simulated annealing, using the Monte Carlo method and the genetic algorithm [11, 17]. We can also minimize the A-factor with respect to a small number of controlled variables when the choices of these variables are clear [19].

Such a modeling process suggested that, in the (124) compound, the chain oxygen (O4) atoms are significantly displaced from the atomic position in the average structure. At low temperatures (below 100 K) the O4 atoms are displaced by about 0.1 Å in the x-direction, which is perpendicular both to the chain and to the c-axis. The displacements are not random, but are highly correlated as shown in Fig. 5, within a domain of about 7 Å × 15 Å, with the long dimension perpendicular to the chain. Since Cu atoms in the chain are not much displaced, these collective displacements of oxygen atoms break the mirror symmetry, and create local electrical polarization as in ferroelectric solids. At temperatures above 190 K, however, the correlations in displace-

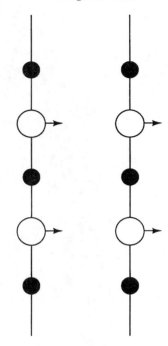

Fig. 5. The pattern of displacement for chain oxygen (O4) atoms. Open circles are
oxygen atoms and closed circles are Cu atoms, both in the chain.

ments are lost. Thus the anomalous temperature-dependence shown in Fig. 2 is
due to formation of such ferroelectric micro-domains at low temperatures.
Implications of such displacements will be discussed later.

Similar studies carried out for $Tl_2Ba_2CaCu_2O_8$, $La_{2-x}Sr_xCuO_4$, (Ba_6K_4) BiO_3, and $Nd_{1.835}Ce_{0.165}CuO_4$ revealed several common features of the atomic displacements in HTSC oxides [18].

1. Significant displacements of oxygen atoms, often in excess of 0.1 Å in magnitude, are observed. The displacements of other atoms are smaller in magnitude, at least by a factor of two.
2. Displacements are not random, but locally well correlated, particularly at low temperatures, often resulting in new, well-defined peaks in the PDF.
3. Not all the atoms are displaced. Nearly half of the atoms are located at the average crystallographic sites.
4. At low temperatures, well-defined displacements occur collectively in a small domain, of size 6–15 Å. Thus the total system appears to be made of micro-domains of two phases, with and without local atomic displacement.
5. Displacements are often observed along the c-axis.
6. The dynamics of the displacements change near or above T_c. The oxygen–oxygen correlation seems to behave particularly anomalously with temperature.

Since we observed only the nuclear positions by neutron scattering, these observations do not directly prove the presence of polarons. However, a case can be made for the following reasons.

1. The local displacements suggest involvement of charge fluctuations. In the case of YBCO(124), the local ferroelectric polarization in the micro-domain indicates charge inhomogeneity. In the case of Tl(2212), $La_{2-x}Sr_xCuO_4$ and $Nd_{2-x}Ce_xCuO_4$ the close oxygen pairs with the interatomic distance reduced from the expected 3.2 Å to 3.0 Å suggest trapped holes [18, 19].
2. The density of the microdomains suggested by their size is comparable to the density of doped charge carriers. Thus each micro-domain cannot be assoiciated with more than one or two charge carriers.

Thus, while the evidence is rather indirect and circumstantial, it is possible, or even likely, that these microdomains represent polarons or bipolarons. If this is the case then the spin-gap temperature corresponds to the bipolaron formation temperature.

3.4 X-ray diffuse scattering

As discussed above, local displacements of atoms produce diffuse scattering of X-rays, neutrons or electrons. Such measurements, however, are very difficult, because their intensity is often quite low, and can be masked by thermal diffuse scattering. The sole reported study of diffuse scattering from HTSC solids concerns the effect of oxygen vacancies [28]. In order to observe atomic displacement, however, the measurement needs to be carried out at larger Q values, since as shown in equation (6) the diffuse scattering intensity is proportional to $(Qu)^2$. We studied the diffuse scattering from a single crystal of $La_{1.85}Sr_{0.15}CuO_4$ in the neighborhood of the (0, 0, 18), (2, 0, 18), (0, 0, 14), (1, 1, 14), and (2, 0, 14) Bragg peaks. The measurements were made at the synchrotron wiggler beamline X-25 of the National Synchrotron Light Source of the Brookhaven National Laboratory. The results were compared with the calculated thermal diffuse scattering (TDS) intensity. The TDS was calculated using a shell model by the Karlsruhe group determined to reproduce the phonon dispersion measured by neutron inelastic scattering [29].

The intensity of a scan along $(\xi, \xi, 17.85)$ is shown in Fig. 6 and Fig. 7, for $T = 300$ K and 10 K. In each case the solid line shows the calculated TDS intensity. It is clear that the intensity at $T = 300$ K is well accounted for by the calculated TDS, but at $T = 10$ K the TDS intensity amounts to only one third or so, and there is a substantial additional intensity. This additional, or extra, diffuse scattering (XDS) increases in intensity below 200 K, and saturates

T. Egami et al.

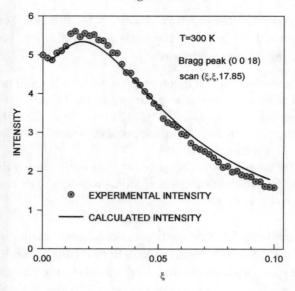

Fig. 6. The diffuse scattering along $(\xi, \xi, 17.85)$ at $T = 300$ K. Circles are data points, and the solid curve is the calculated thermal diffuse scattering (TDS) due to phonons. The calculated intensity is scaled to obtain best fitting.

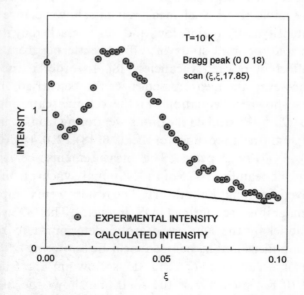

Fig. 7. The diffuse scattering along $(\xi, \xi, 17.85)$ at $T = 10$ K. Circles are data points, and the solid curve is the calculated thermal diffuse scattering due to phonons, with the normalization constant set for 300 K. Note the difference in the vertical scale.

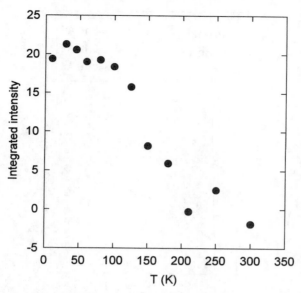

Fig. 8. The temperature-dependence of the integrated intensity of extra diffuse scattering.

around 100 K, as shown in Fig. 8, perhaps accidentally in a very similar way to the PDF peak height shown in Fig. 2. The XDS intensity depends upon the deviation from the nearest Bragg peak, q, roughly as $1/|q|^4$ as shown in Fig. 9. Note that the TDS intensity and the intensity of diffuse scattering due to point defects are proportional to $1/|q|^2$. Therefore the XDS is not due to phonon softening or point-like inclusions. Comparison of intensities around different Bragg peaks suggests that the polarization of displacements, u, causing XDS is transverse, or $u \perp q$. The XDS intensity is not isotropic, but has a strong angular dependence as shown in Fig. 10. This peculiar angular dependence suggests that this scattering is not due to artifacts such as finite resolution of the X-ray optics allowing the tail of the Bragg peak to be seen. The problem of resolution was also checked by a high-resolution measurement using a crystal analyzer. While the implication of this observation is not yet clear, this experiment establishes that there are local distortions at an intermediate length scale, which change anomalously with temperature. The symmetry of the XDS, being strong in the (11) direction in the x–y plane, is in accordance with the electronic susceptibility, as we will discuss below.

3.5 Observation by other methods

Two recent reports appear to be consistent with the local lattice distortion described above. One is the observation by scanning tunneling microscope

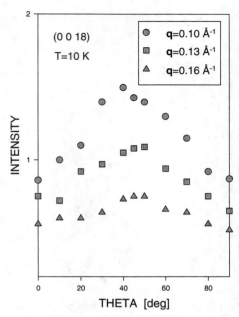

Fig. 9. The dependence of the diffuse scattering intensity on q, indicating q^{-4} dependence.

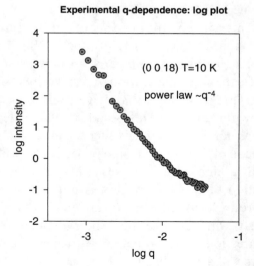

Fig. 10. The dependence of the diffuse scattering intensity at $(h, k, 18)$ with constant $q = (q_x^2 + q_y^2)^{1/2}$ on the angle in the x–y plane, θ.

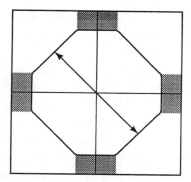

Fig. 11. A schematic representation of the Fermi surface. The shaded area is the extended saddle point, and the arrow indicates nesting.

(STM) of inhomogeneous charge distribution on the surface of YBCO(123), which appears like a local charge density wave (CDW) [30]. Usually CDWs or polarons are mobile, but at the surface they could be pinned by surface irregularity. The other is an observation by electron diffraction on Tl(2212) [31]. In addition, a recent high-resolution imaging study by transmission electron microscopy (HRTEM) shows some micro-domains [32].

4 Consistency with measurements of other physical properties

4.1 Electron band structure

If the charge carriers are in the polaronic state, then there must be a signature in the electronic band structure, while the band structure observed by the photoemission experiments shows well-defined dispersion and appears to be well accounted for by the LDA band structure calculations. This point is often raised against the polaron argument. However, the observed band structure is *different from the LDA band structure in detailed but important ways*. In particular, the saddle point in the X-point in the Cu–O plane (the M-point in some structures, but always in the direction of the Cu–O bond) is supposed to be more than 50 meV below the Fermi level according to the LDA calculations, but in all of the HTSC solids other than Nd–Ce–Cu–O it is within 20 meV of the Fermi level, and is very much more extended than in calculations [33], as is schematically presented in Fig. 11. In order to explain this phenomenon Andersen *et al.* pointed out that the oxygen displacements, which reduce the oxygen–oxygen distance just as observed by pulsed neutron scattering, would raise the energy of the saddle point [34]. Thus the presence of local distortion is consistent with the observed and calculated band structure.

Moreover, the extended saddle point is consistent with the localization tendency. The size of the extended saddle point over which the energy is nearly constant should be inversely related to the size of the large polaron. This comes to be about 10 Å × 20 Å, quite consistent with the size of the micro-domains suggested by the pulsed neutron PDF study. Furthermore, the shape of the Fermi surface shown in Fig. 11 is consistent with the diffuse scattering observed. Because of the linear edge parallel to the (11) direction, there will be a high susceptibility in this direction. Indeed the calculated generalized dielectric constant (susceptibility) shows strong intensities in that direction [35]. The observed strong intensity of XDS in the (11) direction may be related to this square Fermi surface.

4.2 Phonon lifetime

Local lattice distortions should scatter phonons and make the phonon lifetime shorter. So far there have not been many reports supporting this, but recent observations are pointing precisely in this direction. Careful measurement of thermal conductivity showed that the phonon contribution to thermal conductivity is anomalously small, indicating strong phonon scattering [36]. Neutron inelastic scattering shows rather broad phonon peaks indicative of strong anharmonic effects [37], although more theoretical studies are needed to quantify this point.

4.3 Transport properties

The possibility of the observed electronic transport properties being consistent with the idea of polarons has been extensively discussed by Emin [38], and more recently by Alexandrov, Bratkovsky and Mott in terms of Anderson localization [39]. Here we note in addition that the anomalous temperature-dependence of the Hall coefficient recently noted by Hwang *et al.* [40] may also be consistent with the presence of polarons.

4.4 Magnetic neutron scattering

The spin-gap observed by magnetic neutron scattering [24] is regarded as important supporting evidence for the spin-fluctuation theory [25, 26]. However, the neutron result can also be interpreted in terms of polarons. Localization of charge carriers as polarons disrupts Cu spin correlations within the polarons. The Cu spin is momentarily depolarized whenever the polaron passes over the Cu atom, so that the lifetime of Cu spin correlation is controlled by the polaron dynamics. The average time between departure of a polaron from a Cu site and arrival of another polaron there is given by $\tau_s = d/v$, where d is the mean distance between polarons and v is the polaron velocity, and this defines the lifetime of local spin correlation. Therefore an inelastic neutron scattering event with the energy transfer larger than $\hbar\omega_g = \hbar/\tau_s$ experiences good spin correlation, and thus neutrons are strongly scattered. However, when the energy transfer is less than $\hbar\omega_g$, the spin correlation appears weak and the neutron scattering intensity is lower, resulting in the spin-gap behavior. If we assume that $d = 10$ Å and $v = 5000$ m s^{-1} (the velocity of sound), then $\hbar\omega_g$ is equal to 20 meV, as observed for LSCO [24]. When the carrier concentration is reduced, d is increased, and thus $\hbar\omega_g$ is expected to be reduced, as is observed [24]. When the energy transfer $\hbar\omega$ is less than $\hbar\omega_g$, the scattering intensity is inversely related to the number of polarons that would arrive during the scattering time, $\tau = 1/\omega$, which is $(4\pi/3)(v\tau)^3$. Thus the scattering intensity should be proportional to ω^3, as is observed. Also the width of the magnetic peak in Q should be reciprocally proportional to $v\tau$, and thus proportional to ω, again as is observed. Since the Cu spin–spin exchange interaction ($J \simeq 100$ meV) and the phonon energy ($\hbar\omega_{max} \simeq 80$ meV) are comparable in magnitude, spin dynamics is expected to have significant interplay with lattice dynamics. This point warrants further careful studies.

5 The nature of electron–lattice interaction in HTSC solids

5.1 The effect of electron correlation

The effects of strong electron correlation on the magnetic properties of transition metal oxides are widely known and well studied, but relatively less attention has been given to its effects on the lattice. Earlier we have shown that, at the crossover point between the Mott state and the ionic state, the electron–lattice interaction is strongly enhanced by the electron correlation effect [41]. Let us consider two atoms, A and B, with electron energy levels E_A and E_B ($E_A - E_B = \Delta$). If the electron repulsion energy for two electrons to occupy each state is U, then the system can have three ground state configurations. When Δ is strongly negative and $|\Delta| > U$, two electrons will occupy the A atom level. If $\Delta < U$, one electron will be on each atom. When $\Delta > U$, both electrons will be on the B atom level. The situation with $|\Delta| < U$ can be described as the Mott state, in the sense that the repulsion energy U decides the electron configuration. In the limit of $\Delta = 0$ this state is the usual Mott state. $\Delta = U$ signals the crossover condition. If we introduce a hopping term in between the A and B levels then the effect of modulating the hopping integral is maximum at the crossover

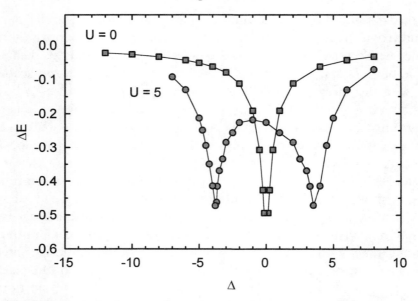

Fig. 12. The energy gain due to Peierls distortion in the two-band Hubbard model, for $t = 1$, $\delta t = 0.15$, $U = U_A = U_B = 0$ (Peierls instability of non-interacting electrons), and $U = 5$, as a function of Δ.

point. We calculated the effect of Peierls distortion (dimerization) for the two-band Peierls–Hubbard model by exact diagonalization in one dimension. The Hamiltonian is given by

$$H = t \sum_{i,j,\sigma} (c^+_{A,i,\sigma} c_{B,j,\sigma} + c^+_{B,j,\sigma} c_{a,i,\sigma}) + \Delta \sum_{i,j,\sigma} c^+_{A,i,\sigma} c_{A,i,\sigma} + U_A \sum_i n_{A,i,\uparrow} n_{A,i,\downarrow}$$
$$+ U_B \sum_j n_{B,j,\uparrow} n_{B,j,\downarrow}. \tag{13}$$

The energy gain due to Peierls distortion, which modulates the hopping integral to $t \pm \delta t$ was calculated for a ring of eight atoms (four A and four B atoms) as shown in Fig. 12. Here the case of $\Delta = U = 0$ corresponds to the usual Peierls instability of a half-filled band. For the case of $U > 0$, the effect of Peierls distortion is largest at the two values of Δ corresponding to the crossover points. In between the two crossover points the system is in the bond-order Mott state. Over a fairly wide range of the values of Δ the energy gain is significant, showing that the effect of lattice distortion is enhanced by U.

Our preliminary calculations show that, even when the charge density is away from half-filling, a similar response to the lattice distortion near the crossover point occurs. In that case, however, the wavelength of the distortion to which the system responds strongly, λ, is not $2a$, but $2a/n$, where n is the charge density. Thus, by this distortion, charge carriers are divided into groups of two, forming bipolarons. Note that the wavevector of distortion, $2\pi/\lambda$, corresponds to the usual Fermi surface nesting condition. In the regular band case, Fermi surface nesting produces lattice distortion and CDW, thus creating the superlattice and new Brillouin zones, so that the electronic states are partitioned in the new Brillouin zone, two electrons per unit cell. In the case in which electron repulsion is dominant, the CDW is accompanied by the bond-order wave (BOW). Thus strong electron correlation promotes bipolaron formation, at least in one dimension.

It is important to note that this enhancement occurs due to charge transfer between A and B atoms induced by lattice distortion. Thus a large part of the energy gain is carried by the electron–electron interaction energy. Therefore this effect cannot be described in terms of the conventional deformation potential picture, but requires description of the heavy dressing of phonons by electrons due to electron–electron interaction. Near the crossover point a small lattice deformation can induce a large electronic effect due to this enhancement. If this crossover point corresponds to the charge concentration required to produce maximum T_c, then at that point the lattice deformation necessary to produce polarons is at its minimum. As the charge density is reduced and the condition departs from the crossover point, larger lattice deformation becomes necessary. This agrees with our observation that, for optimally doped $YBa_2Cu_3O_7$, the lattice distortion is small, whereas in underdoped (124) and reduced (123) compounds the distortions are more easily observable. This, furthermore, is consistent with the isotope effect, which is at its minimum when T_c is at its maximum. At the crossover point the polarons are almost electron-polarons, since most of the screening is achieved by transferred charges induced by small lattice deformation.

5.2 The two-component model

The Fermi surface depicted in Fig. 11 clearly shows two kinds of carriers; one in the extended saddle points, which are nearly localized, and the other on the more strongly dispersed parts of the Fermi surface, which produce nesting. Fermi surface nesting is known to produce short-range magnetic ordering in LSCO and possibly in YBCO, but there is so far no evidence of CDWs. We have searched for evidence of CDWs near the (π, π) point of LSCO by X-ray scattering, but were not able to find any diffuse scattering intensity that could be indicative of CDW. There were reports of observing CDWs by electron diffraction, but that appears to be a radiation-induced effect. Therefore the fast dispersing carriers must be delocalized and nearly free. Thus, if the carriers in the extended saddle points are nearly localized and forming bipolarons while free carriers hybridize with the local bipolarons, then we have an ideal case of two-component superconductivity [42, 43]. It is further appealing that the superconducting gap is at its maximum in the direction of the saddle point, suggesting that the coupling driving force for superconductivity comes from the carriers on the extended saddle point.

In this picture, at low concentrations of carriers the polarons are strongly localized and only weakly interacting through hybridization with free carriers. Thus the polaron mass is large, making the Bose condensation temperature, T_{BE}, low; while the bipolaron formation temperature or spin-gap temperature, T_g, is high. Upon increasing the carrier concentration, the bipolaron binding energy becomes reduced due to increased screening. Consequently the bipolaron formation temperature is reduced, while the Bose condensation temperature is increased because of the reduced mass. When T_{BE} and T_g cross each other T_c is maximized. To summarize, to the left of the optimum concentration at which T_c is maximum, or in the underdoped state, bipolarons are formed above T_c, which is the bipolaron Bose condensation temperature. To the right of the optimum concentration, or in the overdoped state, bipolarons overlap each other, and the system becomes a normal phonon-mediated strong-coupling superconductor. Thus the lattice displacements are large and relatively easily seen in the underdoped state, but in the overdoped state the atomic displacements associated with the Cooper pairs are essentially phonon-like, and difficult to differentiate from thermal phonons. Our observations of the lattice distortion are largely consistent with this picture. Larger atomic displacements are seen in YBCO(124) and YBCO(123) with reduced oxygen content, which are underdoped, than in fully oxygenated YBCO(123). However, more thorough studies of concentration-dependence need to be carried out.

It is interesting to note that a recent photoemission study of overdoped BISCO showed a temperature-dependence of the superconducting gap strongly suggestive of a two-component model [44]. In the Γ–M direction (parallel to the Cu–O bond) the gap is maximum with the value of $2\Delta/(kT_c)$ = 4.6 at $T \ll T_c$, while in the Γ–X direction (45° away) the gap is smaller but non-zero at low temperatures and decreases quickly with increasing temperature. The result first of all contradicts the pure $d(x^2 - y^2)$ wave scenario, and at the same time the single anisotropic s-wave scenario. It is completely consistent with the picture of a large gap at the saddle point (the M point) and the induced superconductivity along the rapidly dispersing Fermi surface including the X point. When the carrier concentration is reduced, the gap in the X-direction is expected to become smaller because of reduced hybridization, and the gap anisotropy would appear d-like, consistent with various reports of d-like anisotropy of the gap [45, 46]. It is, furthermore, possible that the d-wave is actually involved. For instance, as for the symmetry of the wavefunction of the bipolarons, the BCS-like phonon mean-field would result in the s-wave, but if the coupling is local and mediated by a specific mode of local lattice distortion, then it may be possible that the d-wave state occurs even in the lattice mechanism. However, this scenario requires further study.

6 Conclusions

If the lattice polarons exist in the HTSC solids, then the lattice distortions associated with them should be experimentally observable, since the density of the polarons is supposed to be reasonably high. However, conventional tools for structural analysis have not provided evidence of such lattice distortion. On the other hand, unconventional methods such as EXAFS and PDF analysis of pulsed neutron scattering data suggest evidence of local distortion, which could be reconciled with the existence of polarons. This contrast is not all that surprising, since polarons are aperiodic and dynamic, thus their effects are not easily captured by conventional crystallographic techniques, which presume periodicity. Surveying the available experimental observations, it appears that we can make a convincing case for the presence of dynamic local lattice distortions in HTSC solids. What we are still lacking, however, is a direct connection between observed distortions and charge carriers; the connections are all circumstantial. Carrying out an experiment that would establish a firm connection is obviously the most crucial step in establishing the relevance of polarons to the HTSC phenomena.

Acknowledgements

This work was supported by the National Science Foundation through grants DMR93-00728 and DMR91-20668. Discussions with A. R. Bishop, M. Onellion, E. A. Stern, S. D. Conradson, A. Bianconi, C. Humphreys, J. Etheridge, E. Mele, S. K. Sinha, D. E. Moncton, S. L. J. Billinge, Y. Bar-Yam, E. Kaldis, M. Arai, K. Yamada, A. L. de Lozanne, M. Salamon and J. Goodenough are gratefully acknowledged.

References

[1] C. R. Fincher, Jr., C.-E. Chen, A. J. Heeger, A. G. MacDiarmid and J. B. Hastings, *Phys. Rev. Lett.* **48**, 100 (1982).
[2] M. Marezio, D. B. McWhan, P. D. Dernier and J. P. Remeika, *J. Solid State Chem.* **6**, 213 (1973).

[3] S. D. Conradson and I. D. Raistrick, *Science* **243**, 1340 (1989).
[4] J. Mustre-de Leon, S. D. Conradson, I. Batistic and A. R. Bishop, *Phys. Rev. Lett.* **65**, 1765 (1990).
[5] B. E. Warren, *X-Ray Diffraction* (Addison-Wesley, 1969).
[6] H. P. Klug and L. E. Alexander, *X-Ray Diffraction Procedures for Polycrystalline and Amorphous Materials*, 2nd ed. (John Wiley and Sons, 1968).
[7] B. H. Toby and T. Egami, *Acta Cryst.* A **48**, 336 (1992).
[8] J. D. Jorgensen, H. B. Schuttler, D. G. Hinks, D. W. Capone, II, K. Zhang, M. B. Brodsky and D. J. Scalapino, *Phys. Rev. Lett.* **58**, 1024 (1987).
[9] K. Yvon and M. Francois, *Z. Physik* B **76**, 413 (1989).
[10] P. Bordet, J. J. Capponi, C. Chaillout, J. Chenavas, A. W. Hewat, E. A. Hewat, J. L. Hodeau, M. Marezio, J. L. Tholence and D. Tranqui, *Physica* C **156**, 189 (1988).
[11] W. Dmowski, B. H. Toby, T. Egami, M. A. Subramanian, J. Gopalakrishnan and A. W. Sleight, *Phys. Rev. Lett.* **61**, 2608 (1988).
[12] J. D. Sullivan, P. Bordet, M. Marezio, K. Takenaka and S. Uchida, *Phys. Rev.* B **48**, 10638 (1993).
[13] A. Bianconi, S. Della Longa, M. Missori, I. Pettiti and M. Pompa, in *Lattice Effects in High T_c Superconductors*, eds. Y. Bar-Yam, T. Egami, J. Mustre-de Leon and A. R. Bishop (World Scientific, 1992) p. 65.
[14] H. Yamaguchi, S. Nakajima, Y. Kuwahara, H. Oyanagi and Y. Syono, in *Advances in Superconductivity V*, eds. Y. Bando and H. Yamauchi (Springer-Verlag, 1993) p. 73.
[15] E. A. Stern, private communication.
[16] T. Egami, W. Dmowski, J. D. Jorgensen, D. G. Hinks, D. W. Capone, II, C. U. Segre and K. Zhang, *Rev. Solid State Sci.* **1**, 101 (1987).
[17] B. H. Toby, T. Egami, J. D. Jorgensen and M. A. Subramanian, *Phys. Rev. Lett.* **64**, 2414 (1990).
[18] T. Egami and S. J. L. Billinge, in *Advances in Superconductivity V*, eds. Y. Bando and H. Yamauchi (Springer-Verlag, 1993) p. 67.
[19] S. J. L. Billinge and T. Egami, *Phys. Rev.* B **47**, 14386 (1993).
[20] B. Friedl, C. Thomsen and M. Cardona, *Phys. Rev.* B **46**, 5757 (1990).
[21] R. Zeyher and G. Zwicknagl, *Solid State Commun.* **66**, 617 (1988).
[22] T. Machi, I. Tomeno, T. Miyatake, N. Koshizuka, S. Tanaka, T. Imai and H. Yasuoka, *Physica* C **173**, 32 (1992).
[23] B. Bucher, P. Steiner, J. Karpinski, E. Kaldis and P. Wachter, *Phys. Rev. Lett.* **70**, 2012 (1993).
[24] G. Shirane, *Physica* C **185–189**, 80 (1991).
[25] T. Moriya, Y. Takahashi and K. Ueda, *J. Phys. Soc. Japan* **59**, 2905 (1990).
[26] H. Monien, P. Monthoux and D. Pines, *Phys. Rev.* B **43**, 275 (1991).
[27] M. Arai, K. Yamada, Y. Hidaka, S. Itoh, Z. A. Bowden, A. D. Taylor and Y. Endoh, *Phys. Rev. Lett.* **69**, 359 (1992).
[28] X. Jiang, P. Wochner, S. C. Moss and P. Zschack, *Phys. Rev. Lett.* **67**, 2167 (1991).
[29] L. Pinchovius, N. Pyka, W. Reichardt, A. Yu. Rumiantsev, N. L. Mitrofanov, A. S. Ivanov, G. Collin and P. Bourges, *Physica* C **185–189**, 156 (1991).
[30] H. L. Edwards, J. T. Markert and A. L. de Lozanne, *Phys. Rev. Lett.* **69**, 2967 (1992); *J. Vac. Sci. Technol.* in press (1994).
[31] Y. Koyama, S.-I. Nakamura, Y. Inoue and T. Ohno, *Phys. Rev.* B **46**, 5757 (1992).

[32] J. Etheridge and C. Humphreys, private communication.
[33] K. Gofron, J. C. Campusano, H. Ding, R. Liu, A. A. Abrikosov, D. D. Koelling, B. Dabrowski and B. W. Veal, unpublished.
[34] O. K. Andersen, O. Jepsen, A. I. Liechtenstein and I. I. Mazin, *Phys. Rev. Lett.* B **49**, 4145 (1994).
[35] H. Krakauer, W. Pickett and R. E. Cohen, *Phys. Rev.* B **47**, 1002 (1993).
[36] M. Salamon, private communication.
[37] H. A. Mook, B. C. Chakoumakos, M. Mostoller, A. T. Boothroyd and D. McK. Paul, *Phys. Rev. Lett.* **69**, 2272 (1992).
[38] D. Emin, in *Lattice Effects in High T_c Superconductors*, eds. Y. Bar-Yam, T. Egami, J. Mustre-de Leon and A. R. Bishop (World Scientific, 1992) p. 377.
[39] A. S. Alexandrov, A. M. Bratkovsky and N. F. Mott, *Phys. Rev. Lett.* **72**, 1734 (1994).
[40] H. Y. Hwang, B. Batlogg, H. Takagi, H. L. Kao, J. Kwo, R. J. Cava, J. J. Krajewski and W. F. Peck, Jr., *Phys. Rev. Lett.* **72**, 2636 (1994).
[41] T. Egami, S. Ishihara and M. Tachiki, *Science* **261**, 1307 (1993); S. Ishihara, T. Egami and M. Tachiki, *Phys. Rev.* B **49**, 8944 (1994); S. Ishihara, M. Tachiki and T. Egami, *Phys. Rev.* B **49**, 16123 (1994).
[42] Y. Bar-Yam, *Phys. Rev.* B **43**, 359 (1991); *ibid.* **43**, 2601 (1991).
[43] R. Micnas, J. Ranninger and S. Robaszkiewicz, *Rev. Mod. Phys.* **62**, 113 (1990).
[44] J. Ma, C. Quitmann, R. J. Kelly, H. Berger, G. Margaritondo and M. Onellion, unpublished.
[45] W. N. Hardy, D. A. Bonn, D. C. Morgan, Ruixing Liang and Kuan Zhang, *Phys. Rev. Lett.* **70**, 3999 (1993).
[46] Z.-X. Shen *et al.*, *Phys. Rev. Lett.* **70**, 1553 (1993).

11

The Hall effect due to small polarons and conduction in narrow energy bands

LIONEL FRIEDMAN

National Research Council Fellow
Rome Laboratories, Hanscom AFB, MA 01731, USA

Abstract

The small polaron has proved useful in understanding the transport properties of such low-mobility solids as oxides, glasses, and amorphous semiconductors. Polarons and bipolarons are of interest in high-T_c superconductors. I will first briefly review the basic mechanism for the Hall effect found in the localized regime where transport is due to multi-phonon-assisted transitions between localized small polaron states. The temperature-dependence of the Hall mobility will be reviewed for the non-adiabatic, adiabatic and three- and four-site cases. I will then indicate how the magnetic phase factors in the localized regime give the conventional magnetic Lorentz force in a description of polaron band motion or of purely electronic bands of narrow width. This narrow-band regime is more relevant to the normal state of high-T_c materials in which carrier motion is itinerant. I will then survey experimental evidence for the Hall effect due to small polarons and in the narrow-band regime for several materials and conclude with an example of the Hall effect in the normal state of the cuprate superconductors taken from David Emin.

1 The basic mechanism of the Hall effect in the localized regime

The model used is a straightforward two-dimensional generalization of the molecular crystal model of Holstein [1]. (This case admits only a small-polaron and free-particle solution and no large-polaron solution.) Briefly, the model consists of a site occupied by diatomic molecules with fixed centres of gravity and orientation but variable internuclear separation so that each acts like an Einstein oscillator with fixed frequency, ω_0. The oscillators are subject to weak coupling giving rise to dispersion of the vibrational frequencies. A single charge carrier resides on a molecule and its electronic energy is taken to be a linear function of the internuclear separation of the occupied molecule-ion. Motion of the carrier is formulated in a generalized tight-binding approach as follows.

The total wavefunction of the system, dependent on the electron coordinate r and the internal coordinates x_g of the individual sites, is expanded in the basis of the local electronic wavefunctions

$$\psi = \sum_g a_g(\ldots x_g \ldots)\phi^{(0)}(r-g, x_g). \tag{1}$$

In the presence of a constant magnetic field H, an appropriate set of local basis functions are [2]

$$\phi_{(H)}(r-g) = \exp\left(-i\frac{e}{2hc}(H \times g)\cdot r\right)\phi^{(0)}(r-g), \tag{2}$$

where the x_g-dependence of the local wavefunctions is neglected, and the latter are taken to be spherically or cylindrically symmetrical about the magnetic field direction. This results in a phase modification of the transfer integrals $J_{g,g'}$ governing intersite transport:

$$J_{g,g'}^{(H)} = J_{g,g'}^{(0)}\exp\left(-i\frac{e}{2hc}H\cdot(g' \times g)\right). \tag{3}$$

It is to be noted that any arbitrariness due to the phase of the local states or to the choice of gauge cancels in the final results, which are always proportional to the sum of phase factors about a closed path. This is shown immediately below.

The lowest order two-site jump rate w_2, proportional to the absolute square of $J_{g,g'}$ is clearly unaffected. However, the phase does appear in the next higher order jump process, in which the carrier moves to a final site both directly and indirectly via a third intermediate site, if permitted by the site geometry. (The three-site arrangement also holds for spatially random centres, for example in impurity conduction.) The corresponding first- and second-order transition amplitudes, 90° out of phase for $H=0$, are shifted in phase by an amount proportional to H both in magnitude and sign. The resulting interference term provides the Hall current. The probability per unit time for the above event is governed by an energetic coincidence of the electronic energies of the three involved sites, i.e. equal displacements of the x_g terms. The probability of the three-site coincidence event is less than that of two uncorrelated coincidences, $\varepsilon_3 < 2\varepsilon_2$. In the jump perturbation (non-adiabatic) regime $J < h\omega_0$ and in the limit of vanishing vibrational dispersion it is found that

$$2\varepsilon_2 > \varepsilon_3 = (\tfrac{4}{3})\varepsilon_2, \tag{4}$$

i.e. the three-site coincidence is more likely than two uncorreleated two-site coincidences.

The magnetic-field-dependent three-site jump rate is

$$w_3 \approx J^3\, e^{-\varepsilon_3/(kT)}, \tag{5}$$

and since the two-site rate is

$$w_2 \approx J^2\, e^{-\varepsilon_2/(kT)}, \tag{6}$$

the Hall angle and Hall mobility are given by [3]

$$\Theta_{\rm H} \approx \mu_{\rm H} \approx w_3/w_2 \approx J e^{(-1/3)\varepsilon_2/(kT)}. \tag{7}$$

So, even in the three-site case in the jump perturbation (non-adiabatic) regime, the activation energy of the Hall mobility is less than that of the conductivity mobility, i.e. it shows a milder temperature-dependence. Calculations by Emin [4] for the four-site case in the non-adiabatic regime gave a Hall mobility even more temperature-independent and even decreasing with increasing temperature at high temperatures. This is true *a fortiori* in the adiabatic [5] regime ($J > \hbar\omega_0$) in which the electron follows the lattice motions, and for which there is a J-dependent lowering of the hopping (barrier) energies,

$$\mu_{\rm D} \approx \frac{ea^2\psi_0}{kT} \exp\left(-\frac{\varepsilon_2 - J}{kT}\right), \qquad \mu_{\rm H} \simeq \frac{ea^2\omega_0}{kT} F(T)\exp\left[-\frac{1}{kT}\left(\frac{\varepsilon_3}{3} - J\right)\right]. \tag{8}$$

Regarding the sign of the Hall effect, an argument due to Holstein is as follows. The transfer term of the electron Hamiltonian

$$H_{\rm e} = -\sum_{i,j} J_{ij}^{(e)} C_j^{(e)*} C_i^{(e)} \tag{9}$$

can be written in terms of hole creation and annihilation operators as

$$C_i^{(h)} = C_i^{(e)*}, \tag{10}$$

$$H_{\rm e} = -\sum_{i,j} J_{ij}^{(h)} C_j^{(h)*} C_i^{(h)}, \tag{11}$$

where the transfer integral for a hole is

$$J_{ij}^{(h)} = -[J_{ij}^{(e)}]^*. \tag{12}$$

It follows that the sign of the Hall effect for holes relative to that for electrons in the same energy band is given by the factor $-(-1)^n$, where n is the number of sites in the closed path. Thus, for $n=3$, this relation states that the sign of the Hall effect for holes is the same as that for electrons, namely negative (*n*-signed). It is to be emphasized that this argument is applicable only to the same energy band and to spherically symmetrical s-like orbitals. This was later extended by Emin [7] to consideration of the absolute (rather than the relative) sign of the Hall effect, and to orbitals other than simple s-states.

2 The Hall effect in the narrow band regime

The point that I wish to make here is that the basic magnetic phase factors responsible for the Hall effect in the local regime give the classical magnetic Lorentz force term in a Boltzmann equation description in the band regime [8]. In the small-polaron-band regime at low temperatures $(T < \frac{1}{3}\theta_D)$, the translational invariance of the electron–lattice system requires that the basic state be plane-wave combinations of localized small-polaron states characterized by wave vector σ

$$a_{\sigma,\ldots,N_k\ldots} = e^{i g \cdot \sigma} \chi_{g,\ldots,N_k\ldots}, \tag{13}$$

where $\chi_{g,\ldots,N_k\ldots}$ is the vibrational state when the carrier is localized on site g; here the transitions are diagonal in the $(\ldots N_k \ldots)$. Then, by forming a classical wavepacket as a superposition of states with a narrow range of wave vectors

$$\sum_{\sigma,\ldots,N_k\ldots} C_{\sigma,\ldots,N_k\ldots}\, e^{i g \cdot (\sigma - \sigma_0)} \chi_{g,\ldots,N_k\ldots} \tag{14}$$

one finds, on averaging over the vibrational quantum numbers, that $f_\sigma = |\langle c_\sigma \rangle|^2$ obeys a Boltzmann equation of standard form

$$\frac{\delta f_\sigma}{\delta t} = \frac{e}{h}\,[E + (v \times H)] \cdot \frac{\delta f_\sigma}{\delta \sigma} - \frac{f_\sigma - f_\sigma^{(0)}}{\tau_\sigma}, \tag{15}$$

where E is an applied electric field and a phenomenological relaxation term has been added.

It is found that, for lattices with four-site geometry (SC and FCC),

$$R_4 = R_N, \tag{16}$$

where

$$R_N = 1/(nq) \tag{17}$$

is the 'normal' magnitude of the Hall coefficient and where the sign is that normally expected. However, for sites with three mutually nearest neighbours,

$$R_3 = R_N x\, kT/J, \tag{18}$$

where J is the relevant bandwidth and the sign is anomalous. Thus $(R_3)^{-1} \approx 1/T$, and the sign anomalies are the same as those found in the localized regime.

The reason for this deviation from the normal result is the fact that the energy bandwidths are not large relative to the thermal energy $k_B T$. This may be seen as follows. In the derived Boltzmann equation we set

$$f_\sigma = f_\sigma^{(0)} - \left(\frac{\delta f_\sigma^{(0)}}{\delta E_\sigma}\right) \phi_\sigma \qquad (19)$$

For the Hall current,

$$j_y = \sigma_{xy} E_x, \qquad (20)$$

we obtain for the Hall conductivity

$$\sigma_{xy} = \frac{e^3 \tau}{k_B T} H \frac{2}{(2\pi)^2} \int d\sigma_x \int d\sigma_y (v_y^2 M_{xx}^{-1} - v_x v_y M_{xy}^{-1}) \exp\left(-\frac{E_\sigma}{k_B T}\right), \qquad (21)$$

where $M_{xy}^{-1} = h^{-2}(\delta^2 E_\sigma/\delta\sigma_x\delta\sigma_y)$ are the elements of the inverse effective mass tensor. For $E < k_B T$, negative mass states are strongly weighted, resulting in a deviation from the normal result obtained for wide energy bands on the basis of a constant band edge effective mass. The same approach was applied to conduction in narrow purely electronic energy bands. Examples will be given below.

3 Experimental results

3.1 The hopping regime

Experimental results on the Hall effect in hopping and in narrow-band semiconduction are rare. The original interest in hopping conduction was for hole conduction in lithium-doped NiO, which was of interest to Heikes and co-workers of Westinghouse Research Laboratories as a potential thermoelectric material. The first reported work was by Bosman and Crevecoeur [9] of the Philips Laboratory, Eindhoven. They found equal activation energies for both resistivity and thermopower, implying an unactivated mobility. They suggested narrow-band semiconduction at high temperatures and impurity conduction at low temperatures. However, a difference in the two activation energies was found for undoped high-purity NiO by Keem *et al.* [10], suggesting small-polaron motion. The same was found for MnO by Crevecoeur and de Wit [11]. Another even earlier observation of small-polaron formation was in orthorhombic sulphur by Spear and co-workers [12]. They found a thermally activated drift mobility from time-of-flight measurements, and a pressure-dependence [13] of the mobility consistent with pressure-induced modulation of the transfer integral J in the non-adiabatic regime.

The most promising possibility for observing the Hall effect in hopping conduction was thought to be for AC impurity conduction in doped, compensated semiconductors at cryogenic temperatures. Carriers are transferred from an occupied to an empty impurity site assisted by absorption or emission of a single acoustic phonon. There were two serious attempts to measure a Hall signal, first by Pollak and Amitay [14], and some years later by Pollak and Klein [15]. Pollak and Amitay attempted to measure the AC Hall effect in Ge and Si at low temperatures as predicted by Holstein [16], but obtained a negative result. Inclusion of spatial correlations of the impurities reduced their corrected theoretical value below the sensitivity of their apparatus for their Si sample, but not for their Ge sample. Klein used p-Ge doped with Ga acceptors to 1×10^{15} cm^{-3}. His upper bound for the Hall conductivity was a factor of 37 below the theoretical estimate and a factor of six below his correction to the theoretical value. Kastner [17] reported measurement of the DC Hall effect in uncompensated Si:As samples on the insulating (as well as on the metallic) side of the metal–insulator transition, where variable-range hopping was observed. In this case, the Hall effect arises at three-site configurations at the intersections of percolation paths and has been treated theoretically [18–20].

The only definitive result for the Hall effect due to thermally activated hopping of small polarons is by Nagels [21] for slightly reduced $LiNbO_3$. He measured the conductivity, thermopower and Hall mobility as functions of temperature. The activation energy of the Hall mobility was found to be one third that of the difference in activation energies between the conductivity and thermopower, in agreement with the result quoted above for the non-adiabatic case and three-site geometry.

3.2 The band regime

The first example is that of the Hall effect due to polaron-band motion at low temperatures previously described. The anomalies here have the same dependence on site geometry as in the hopping regime, as previously noted. The same considerations can be applied to narrow energy bands (bandwidths $< kT$) of purely electronic origin. Some examples will be given.

The first examples are the organic molecular crystals [22] anthracene and napthaline (aromatic hydrocarbons) for which polaronic localization on the individual molecules was assumed to be negligible. This is due to delocalization of the carrier over the large planar molecule and was confirmed by comparing the energy of the molecule with that of the molecule-ion with altered bond lengths. Owing to the low symmetry of the crystal structure (base-centred monoclinic), the Hall coefficient is anomalous, $R_H \simeq (kT/J_{eff})$ and hole conduction is predicted to give an n-signed Hall effect. The only measurement of this class of organics superconductors was on Cu-phthalocyanine [23], which is structurally similar to the aromatic hydrocarbons. This mobility T^{-n} was unactivated, consistent with the above picture, and the sign was undetermined.

Finally, I will conclude with an example of the Hall effect for large bipolarons in the normal state of the cuprates taken from David Emin [24]. The quasi-particle dispersion is given by

$$E(k) = h\omega_{\rm p}[(kc/2)\coth(kc/2)]^{1/2}, \tag{22}$$

where c is the lattice constant normal to the layers. Unlike a symmetrical energy band, this dispersion shows a point of inflection at small k ($\ll \pi/a$), where a is the in-plane lattice constant and $2/c \ll \pi/a$. Since the plasma energy $h\omega_{\rm p} \simeq 0.01$–$0.1$ eV, the band is 'narrow' and thermal excitation of negative mass states above the inflection point will be important. As carriers are thermally excited above the inflection point, the Hall number, $n_{\rm H} \equiv 1/(qR_{\rm H}) \simeq T$. However, at sufficiently high temperatures $kT >$ bandwidth, the rise of $n_{\rm H}$ with T will saturate and fall, in agreement with the results presented earlier. The negative-mass states also result in a sign reversal of the themoelectic power [24] with increasing temperature above $T_{\rm c}$.

References

[1] T. Holstein, *Ann. Phys (N.Y.)* **8**, 343 (1959).
[2] L. Friedman and T. Holstein, *Ann. Phys. (N.Y.)* **21**, 494 (1963); T. Holstein and L. Friedman, *Phys. Rev.* **165**, 1019 (1968).
[3] T. Holstein, *Phys. Rev.* **124**, 1329 (1961).
[4] D. Emin, *Ann. Phys. (N.Y.)* **64**, 336 (1971).
[5] D. Emin and T. Holstein, *Ann. Phys. (N.Y.)* **53**, 439 (1969).
[6] T. Holstein, *Phil. Mag.* **27**, 225 (1973).
[7] D. Emin, *Phil. Mag.* **35**, 1189 (1977).
[8] L. Friedman, *Phys. Rev.* **131**, 2455 (1963).
[9] A. G. Bosman and C. Crevecoeur, *Phys. Rev.* **144**, 763 (1966).
[10] J. E. Keem, J. M. Honig and L. L. Van Zandt, *Phil. Mag.* B **7**, 537 (1978).
[11] C. Crevecoeur and H. J. deWit, *Solid State Commun.* **6**, 295 (1968).
[12] A. R. Adams and W. E. Spear, *J. Phys. Chem. Solids* **25**, 1113 (1964); Gibbons and W. E. Spear *ibid.* **27**, 1917 (1966).
[13] F. K. Dolezalek and W. E. Spear, *J. Non-Cryst. Solids* **4**, 97 (1970).
[14] M. Amitay and M. Pollak, *J. Phys. Soc. Japan 21 Suppl.,* 549 (1966).
[15] R. S. Klein, *Phys. Rev.* B **31**, 2014 (1985).
[16] T. Holstein, *Phys. Rev.* **124**, 1329 (1961).
[17] D. W. Koon and T. G. Kastner, *Phys. Rev.* **41**, 12054 (1990).
[18] H. Bottger and V. V. Bryksin, *Phys. Status Solidi* B **80**, 569 (1977).
[19] L. Friedman and M. Pollak, *Phil. Mag.* B **38**, (1978); *ibid.* **44**, 487 (1981).
[20] M. Gruenewald, H. Mueller, P. Thomas and D. Wuertz, *Solid State Commun.* **38**, 1011 (1981).

[21] P. Nagels, R. Callaerts and M. Denayer, *Proc. 5th Int. Conf. on Amorphous and Liquid Semiconductors*, p. 867 (1974).

[22] L. Friedman, *Phys. Rev.* **133**, A1668 (1964).

[23] G. Heilmeier, G. Warfield and S. Harrison, *Phys. Rev. Lett.* **8**, 8 (1962).

[24] D. Emin, in *Lattice Effects in High Temperature Superconductors*, ed. Y. Bar-Yan (World Scientific, New York, 1992).

12

Static and dynamic conductivity of untwinned Y₁Ba₂Cu₄O₈: gaps or condensation?

P. WACHTER, B. BUCHER and R. PITTINI

Laboratorium für Festkörperphysik, ETH Zürich, 8093 Zürich, Switzerland

Abstract

We report on experimental evidence for different electronic phases in the superconductor $YBa_2Cu_4O_8$: simple metallic Cu–O chains and highly correlated CuO_2 planes. $YBa_2Cu_4O_8$ is a genuine untwinned compound; hence, we were able to determine the anisotropy of the resistivity along the a and b directions. Along the b direction (chain and plane conduction channels), a normal metallic temperature behavior of the chain dominates. For the a direction (only the plane conduction channel), the resistivity reveals unconventional behavior with a kink at 160 K, becoming linear at higher temperatures. Further, we present results of the dynamical (optical) conductivity of the CuO_2 plane as a function of frequency and temperature. The frequency-dependence of the optical conductivity is consistent with a model of ferromagnetic polarons in an antiferromagnetic matrix. This is further confirmed by a depolarization experiment, which is sensitive to crystal regions with different spin polarizations. The temperature behavior of the thermal occupation of the ground state is in agreement with a real-space condensation.

1 Introduction

From the beginning [1] the high-T_c superconductors (HTSC) have been very challenging systems. Regarding their solid state chemistry, the defect structure with its wide variability in stoichiometry gives rise to homogeneity problems: the samples may show a *chemical* phase separation of different oxygenated regions. So, proper preparation of homogeneous samples is crucial. Concerning the physics, the normal as well as the superconducting state show unusual features. In addition, as we will show at least for $YBa_2Cu_3O_7$ (123) or $YBa_2Cu_4O_8$ (124), one has to take into account the coexistence of different *physical* phases. The Cu–O chains behave as a normal Fermi liquid, whereas

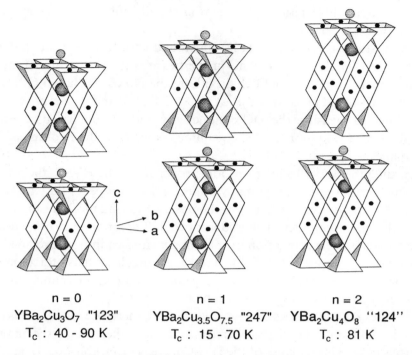

n = 0   n = 1   n = 2
$YBa_2Cu_3O_7$ "123" $YBa_2Cu_{3.5}O_{7.5}$ "247" $YBa_2Cu_4O_8$ "124"
T_c : 40 - 90 K T_c : 15 - 70 K T_c : 81 K

Fig. 1. The crystal structure of the $n=0$, 1, 2 members of the homologous series $Y_2Ba_4Cu_{6+n}O_{14+n}$. The picture of the $n=0$ member shows two unit cells. For the $n=1$ structure, only one half of the unit cell is depicted. The $n=2$ member is represented by just one unit cell. Large spheres are Ba, medium are Y and small are Cu.

the CuO_2 planes exhibit a strongly correlated behavior. These two subsystems are connected by the apex oxygen of the pyramids and it is an interesting task to resolve the role of the apex oxygen with respect to charge transfer, plane – chain interaction switch etc.

In Fig. 1 we show the crystal structures of $Y_2Ba_4Cu_{6+n}O_{14+n}$ with $n=0$, 1, 2, which can be synthesized by adding various amounts of CuO to the 123 compound. In the 123 compound there is one Cu–O chain between the CuO_2 planes, in the 123.5 (or 247) compound with $n=1$ there are alternating single and double chains between the planes and in the 124 compound with $n=2$ there are always double chains between the CuO_2 planes. In the presence of additional CuO, even under ambient pressure, the 123 transforms completely into the 124 compound, which suggests that the latter is the stable phase [2].

To our knowledge, $YBa_2Cu_4O_8$ (124) is the only naturally stoichiometric HTSC (Fig. 1) i.e. it is not possible to vary the oxygen content because of the high coordination number of the oxygen inside the double chains [2]. This is one great advantage of the 124 system as we do not have to worry about *chemical* phase separation.

The well-studied $YBa_2Cu_3O_7$ (123) superconductor has the disadvantage of being twinned. Thus, the 'a–b plane' properties are rather a mixture of those of the CuO_2 plane and the Cu–O chain (along the b axis). $YBa_2Cu_4O_8$ (124) is naturally untwinned (and so is $Y_2Ba_4Cu_7O_{15}$ (247) [3]) and it thus provides the opportunity to measure the DC conductivity and the polarized reflectivity along the a direction alone, i.e. the genuine response of the two-dimensional CuO_2 plane.

Regarding the normal state transport properties of the HTSC, the question of doping is very relevant. Optimally doped and overdoped $YBa_2Cu_3O_7$ (123) contains too many carriers to reveal the basic scattering mechanism; a Fermi liquid behavior dominates. The underdoped 123 compound is non-stoichiometric and chemical segregation or inhomogeneities cannot be excluded. The 124 compound is thus a gift from nature inasmuch as it is stoichiometric, untwinned and underdoped so that the genuine static and dynamic conductivity of the low-doped CuO_2 planes can be determined.

The oxides are prototypes of ionic crystals and, therefore, one expects also a polaronic nature of the charge carriers. On the other hand, the magnetic properties of the CuO_2 planes of the HTSC have been established to be crucial for superconductivity. Considering these two features for the HTSC, one is tempted to anticipate magnetic, polaronic charge carriers: magnetic polarons. The magnetic polarons could be ferromagnetic polarized regions in the short-range-ordered antiferromagnetic CuO_2 plane, as has been proposed by Wachter and Degiorgi [4], Wachter et al. [5], de Jongh [6, 7], Alexandrov and Krebs [8] and Mott [9, 10]. A physical manifestation of such magnetic polarons is diffusive motion instead of a propagating one. The dynamical (optical) conductivity to be presented is indeed consistent with the hypothesis of magnetic polarons.

2 Experiments and results

We have measured the static and dynamic (optical) transport properties of $YBa_2Cu_4O_8$ (124). In the DC resistivity for current along the a (plane) and b (plane and chain) directions, respectively, strong anisotropy is found [11] (Fig. 2(a),(b)). For the b direction one expects two conduction channels: the plane and the chain. Assuming a model of shunted conduction channels along the b direction (plane and chain) [11, 12], one can separate out the chain contribution alone, which is shown in Fig. 2(a) (thin curve). The a direction, on the other hand, represents the intrinsic conductivity of the interesting CuO_2 plane alone.

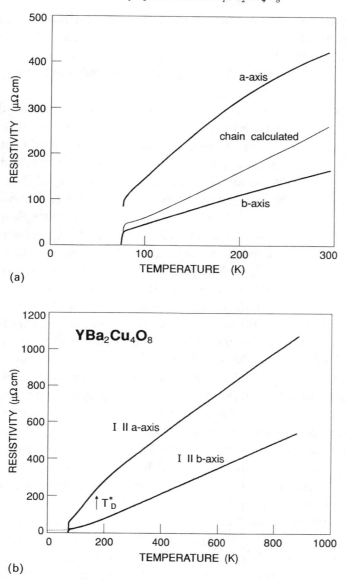

(a)

(b)

Fig. 2(a) Resistivity of a YBa$_2$Cu$_4$O$_8$ single crystal for the two plane directions (Montgommery corrected): along the b axis (lower curves) and the a axis (upper curve). Also shown is the calculated resistivity of the chain alone (thin curve) exploiting the model of shunted resistors along the b direction[11,12] (b) The resistivity of a YBa$_2$Cu$_4$O$_8$ single crystal along the a and b axes in an extended temperature range.

Only a single conduction channel exists for the a direction (plane). It is remarkable that the conductivity of the b axis is practically dominated by the chain contribution along the b axis, which is an important hint for the reflectivity measurements, inasmuch as for $E \parallel b$ mainly a reflectivity due to the free carriers in the chain is expected.

The two conduction channels (plane and chain) show a completely different temperature-dependence. The resistivity of the b axis reveals a Bloch–Grüneisen signature as expected for a normal metal (see dotted fit in Fig. 2(b)), whereas that of the plane shows an unusual kink at about 160 K and above 400 K up to 900 K a linear behavior and quite a high value [13] (Fig. 2(b)). The intersection of a linear extrapolation towards zero temperature gives for both channels a negative value of the resistivity, indicating high sample quality and, more important, that a nonlinear resistivity has to be assumed at low temperature. These results are in accordance with measurements of the spin-lattice relaxation time T_1 of $YBa_2Cu_4O_8$ by Zimmermann *et al.* [14]. They found a normal metallic relaxation for the chain Cu(1) down to 4 K with no change below T_c; condensation of the free carriers in the chain is not observed at Cu(1). However, they found a magnetic relaxation mechanism for Cu(2) of the plane and claimed opening of a spin gap below 130 K in the naturally underdoped $YBa_2Cu_4O_8$. On the other hand, the kink in the plane resistivity is at around 160 K [13] and Hall effect measurements confirmed the existence of a spin gap around $T_D^* \approx 160$ K [13]. As the spins are frozen below T_D^* the scattering channel closes and hence the resistivity should decrease faster with decreasing temperature in accordance with observation (Fig. 2(a),(b)). Therefore, the unusual charge transport of the HTSCs must be attributed to the spin dynamics of the CuO_2 plane. It should be mentioned that also in the 123 compound at 5 K a spin gap of about 28 meV (320 K) has been measured by inelastic neutron scattering [15].

The polarized optical reflectivity of the 124 compound has been measured between 4 meV and 12 eV (unpolarized down to 1 meV) between room temperature and 6 K. The polarized reflectivity spectrum between 0.2 and 3.3 eV has already been published in [3] for 300 K and between 0.05 and 5.5 eV at 300 K by Kircher *et al.* [16]. Two plasma edges near 1.2 and 1.7 eV are seen at 300 K for light polarized parallel to the a and b axes, respectively. In Fig. 3 we show the reflectivity of the 124 compound at different temperatures with the light polarized along the a or b axes only in the far infrared (FIR). For light polarized along the b axis only very little structure is seen, because in this direction we have the plane and chain contributions, and, as we have seen in Fig. 2, the chain contribution dominates. It is thus not very relevant to measure the reflectivity of HTSCs on polycrystalline material, with unpolarized light or

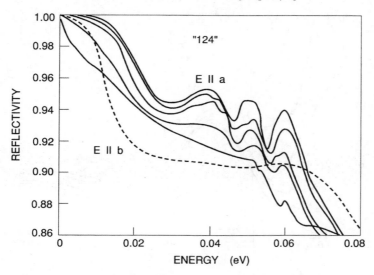

Fig. 3. The near normal reflectivity of $YBa_2Cu_4O_8$ ($T_c = 72$ K) for $E_{light} \parallel$ the a axis. The temperatures are, from top to bottom: 6, 30, 50, 65 and 95 K. The dashed line is for E_{light} $\parallel$ the b axis at 6 K.

on twinned single crystals, since the chain contribution with its normal Drude behavior dominates the spectrum. Single crystals of the 124 compound have a T_c of 72 K; polycrystalline material has T_c of 81 K [2]. The reason is slight incorporation of Al from the Al_2O_3 crucibles into the single crystals during high-temperature and high-pressure crystal growth. Crystal growth has recently become possible in Y_2O_3 crucibles and now also the single crystals have a T_c of 81 K [17].) In 124 the compound below T_c of 72 K an additional plasma edge develops around 20 meV. At energies 28, 44 and 54 meV an absorption takes place, which emerges with decreasing temperature below T_c.

We expect that the experimentally determined reflectivity curves for the 123, 247 (123.5) and 124 compounds are similar when the T_c values are the same. In [3] we have shown that, in samples of the untwinned 247 compound we can vary T_c between 30 and 70 K, where the lower T_c is related to a smaller carrier concentration. This enabled us to see phonon structures that are hardly detected with a high T_c and a large carrier concentration, i.e. for ideally or overdoped samples. These phonons are near 50 meV in energy and again we expect similar types of phonons for all HTSCs. On the other hand, we have not seen a charge gap structure in the 247 (123.5) compound [3]. In thermomechanically prepared untwinned very small single crystals of the 123 compound Schlesinger [18] claimed to have seen a charge gap in the superconducting state at 60 meV and, for smaller energies, a 100% reflectivity. Unfortunately the

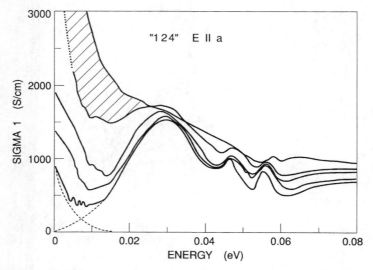

Fig. 4. The optical conductivity $\sigma_1^a(\omega)$ of the 124 compound for the *a* direction. The temperatures are, from top to bottom: 95, 65, 50, 30 and 6 K. The dashed lines are estimates of a separation of σ_1^a into bound and free carrier contributions at 6 K.

spectral range did not go very much below 50 meV. We show in Fig. 3 that (similarly to the case of the 247) there are phonons near 50 meV whose reflectivity increase may simulate a gap structure when one does not measure very much below this energy. The calibration of 100% reflectivity is also difficult. In Fig. 3 a self-checking of the measurement can be seen when the reflectivity still rises for lower photon energies than a proposed gap and even develops structure. A measurement performed with the same equipment as above, but on superconducting Rb_3C_{60}, shows a 100% reflectivity line between 70 and 10 cm^{-1} for a T_c of 29 K [19].

On the polarized reflectivity (in the range 4 meV to 12 eV), we performed a Kramers–Kronig (K–K) transformation to evaluate the optical constants. The dynamical (optical) conductivity σ_1^a (along the *a* direction) is depicted in Fig. 4. With lower temperature, the spectral weight becomes smaller, i.e. there is condensation to the superconducting state and the missing spectral weight is represented by a delta function at $\omega = 0$.

3 Discussion

3.1 Frequency-dependent scattering

The conductivity in the mid and near IR is not Drude-like ($\sigma \propto \omega^{-2}$). Especially the CuO_2 plane reveals a strong spectral weight at 1 eV, compared with that at

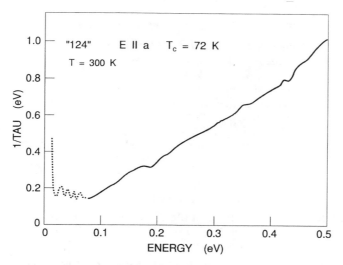

Fig. 5. The frequency-dependent scattering rate $\Gamma^*(\omega) = 1/\tau$ for the a direction of $YBa_2Cu_4O_8$.

zero frequency. An approach to the IR conductivity, in contrast to interband transitions, is to assume an inelastic scattering of the free carriers in the mid IR by a non-phonon excitation spectrum (electron–spin, electron–electron etc.). The underlying excitations need not be infrared active themselves; but, the interactions can be seen via an unusual conductivity of the carriers involved. The optical conductivity in the mid infrared at 300 K can be described as

$$\sigma(\omega) = \frac{ne^2}{m^*} \frac{1}{\Gamma^*(\omega) - i\omega} \tag{1}$$

with $m^*(\omega) = m(1 + \lambda)$ a renormalized mass and $\Gamma^*(\omega) = \Gamma/(1 + \lambda)$ a renormalized scattering rate. From $\sigma_1(\omega)$ and $\sigma_2(\omega)$ (obtained via the K–K relation) we can derive the frequency-dependence of $\Gamma^*(\omega)$ and it is shown in the mid infrared in Fig. 5.

An astonishing linearity comes into being between 0.08 and 0.5 eV for a sample of the 124 compound at 72 K. For $YBa_2Cu_{3.5}O_\delta$ (247) we have even found [3] a linearity up to 0.8 eV. The linearity of $\Gamma^*(\omega)$ has been thought of as evidence for a nearly localized Fermi liquid (a marginal Fermi liquid), a phenomenological model proposed by Varma *et al.* [20]. They predicted the universal relation $\Gamma^*(\omega) \propto \hbar\omega + k_B T$, i.e. a static conductivity ($\omega \to 0$) reciprocal in temperature and a $\Gamma^*(\omega)$ linear in frequency ω at constant temperature as observed in Fig. 5. However, there is no correlation of the optical scattering rate and T_c as listed in Table 1 when using the relation $\Gamma^* \propto k_B T_c$. Therefore, the interaction responsible for this scattering seems not to be related to the

P. Wachter et al.

Table 1. *The strength of the frequency-
dependent scattering rate* $(\Gamma^*(\omega) = A\hbar\omega)$

Material	T_c (K)	A
$YBa_2Cu_3O_7$ [18]	91	0.6
$YBa_2Cu_{3.5}O_{7.5-x}$ [3]	72	1.1
$YBa_2Cu_{3.5}O_{7.5-x}$ [3]	30	1.1
$YBa_2Cu_4O_8$	81	1.8

interaction resulting in superconductivity. The scattering rate of the static conductivity also does not relate to the T_c values for the various compositions.

3.2 Spin-flips, magnetic polarons and bipolarons

The HTSCs differ in two aspects from a common superconducting metal. First, the bonding is at least partly of ionic nature. Second, the CuO_2 planes show short-range-ordered antiferromagnetically correlated regions of the magnetic moments of the Cu^{2+} ions. Regarding the ionic bonding of oxides, one should expect a polaronic nature of the charge carriers. On the other hand, Mott [21] has pointed out that spin holes in an antiferromagnetic matrix polarize the surroundings, resulting in a ferromagnetic region as sketched in the top of Fig. 6. In the HTSCs the spin holes are the consequence of the charge holes [4, 5] in the Cu $3d^9$ matrix introduced by the high affinity of oxygen ions, for which filling of the p shells is energetically favored (only oxygen is able to do this and in fact no HTSC has been found with S or Se as the anion). The charge hole is thus a $3d^8$ state (resulting in mixed or even intermediate valence $3d^9$ and $3d^8$ states [22]) and it has been shown by Wachter and Degiorgi [4, 5] and Ospelt *et al.* [23] that the $3d^8$ state is a non-magnetic low-spin state, i.e. a singlet or a spin hole. Another argument based mainly on XPS data e.g. that of Eskes and Sawatzky [24] assumes that the charge hole sits at an oxygen ion, creating a $2p^5$ state, which is also a spin $\frac{1}{2}$ state just as is the $3d^9$ state of Cu. (This p state would be much easier to achieve with S or Se anions, but has never been found.) Both assumptions use a very strong ionic model, but we know that some covalency also exists. Brandow [25] has re-analyzed the XPS data [24] and comes to the conclusion that the hole has a 60% probability of being at the Cu site and 40% of being at the oxygen site. Independently of the model that one accepts, Mott [9, 10] has shown that the spin at the oxygen site also creates a ferromagnetic spin polaron in the antiferromagnetic Cu matrix.

Hence, as a first heuristic approach to the HTSCs, one may anticipate

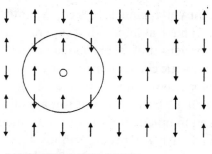

MAGNETIC POLARON

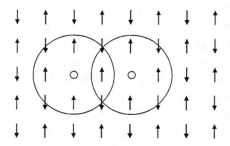

MAGNETIC - BOUND BIPOLARON

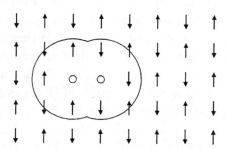

MAGNETIC - BOUND BIPOLARON

Fig. 6. The model of ferromagnetic polarons and bipolarons in an antiferromagnetic matrix. The binding energy results from competing effects such as Jahn–Teller splitting, Coulomb repulsion and gained exchange energy.

magnetic polarons in the CuO_2 planes. Indeed, Wachter and Degiorgi [4, 5] and similarly de Jongh [6, 7] have assumed that superconductivity in the HTSCs is based on a Bose-like condensation of exchange-coupled non-magnetic bipolarons. In the antiferromagnetic cluster matrix, magnetic polarons can combine to form bipolarons because they gain exchange energy and elastic energy [26]: only seven broken magnetic bonds for a ferromagnetic bipolaron (middle part of Fig. 6) or six broken magnetic bonds for a non-magnetic bipolaron (lower part of Fig. 6) instead of eight broken bonds for two independent single polarons (see Fig. 6). As argued by Wachter and Degiorgi [4, 5], the binding mechanism of the bipolarons is a combined effect of gained magnetic exchange energy (about 0.1 eV per magnetic bond as estimated from T_N of the antiferromagnetic insulator $YBa_2Cu_3O_6$), lattice energy (due to a Jahn–Teller effect), Coulomb repulsion and Hund's rule energy. The latter entity is related to the probability of finding the Cu^{3+} $3d^8$ configuration in the low- or high-spin configuration. There are about 50 Cu^{3+} compounds (nearly all oxides) known in the literature, the most significant being the insulator $NaCuO_2$ [27, 28], which are all diamagnetic (except for K_3CuF_6), which can only be achieved with a non-magnetic, low-spin $3d^8$ state (a combination of a $3d^9$ state of Cu with a $2p^5$ state of oxygen invariably results in a metal). In other words, nature favors the low-spin state and the expenditure of Hund's rule energy in formation of a spin-hole ferromagnetic Cu polaron is maybe not too much.

Formation of ferromagnetic polarons is only conceivable in an antiferromagnetic matrix, be it of long- or short-range order. Thus the magnetic polarons and also bipolarons come into existence only below the Néel temperature T_N, but not necessarily at T_N, since thermal disorder might be large. What is the Néel temperature of the antiferromagnetic clusters in the 124 compound? The highest T_N in insulating $YBa_2Cu_3O_6$ is 400 K, for $YBa_2Cu_3O_{6.35}$ it is 300 K [15]. Single crystals of CuO_2 have T_N of 212 K [29]. In the superconducting compositions of the 123 compound a pseudo-spin gap exists far above T_c with an energy of 130–150 K [15]. For the 124 composition no inelastic neutron scattering of similar kind is known, but, if we take the onset of deviation from the high-temperature straight line relation of the resistivity in Fig. 2(a), (b) then we find it to be near 200 K, with the fully opened spin gap near 130 K. We thus use the beginning of the opening of the spin gap near 200 K as the Néel temperature of the antiferromagnetic clusters in the 124 compound.

We thought of performing a critical experiment to detect ferromagnetic polarons. The possibility of having untwinned single crystals of the 124 compound permits an optical depolarization measurement. Linearly polarized light is reflected from the crystal with incident polarization direction 45° to crystal axes *a* and *b*. This choice of angle represents an optimum between a

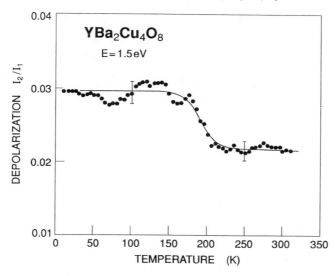

Fig. 7. The temperature-dependence of the depolarization of $YBa_2Cu_4O_8$ single crystals due to ferromagnetic microregions. The continuous line is a guide to the eye.

relatively high contribution of the planes (the *a* axis) to the measured signal, and a high value of absolute reflected light intensity (the *b* axis), which is necessary to obtain high accuracy in ellipsometric measurement of the depolarization. Before reaching the analyzer the light beam passes a Faraday modulator, where only the azimuth of the polarization is modulated and not the ellipticity. As a consequence the modulated light is no longer elliptically polarized. With a crossed analyzer, and in the absence of magnetic field, the measured intensity at 300 K is quite low (about 2% of the incident intensity). However, on cooling below 300 K a certain degree of depolarization sets in at about 200 K and below, which cannot be reduced by setting the analyzer to any other angle. The main change in depolarization occurs between 200 and 130 K (Fig. 7). An antiferromagnetic lattice, even when only present in clusters, cannot depolarize the reflected light because there is no net magnetization in the clusters. However, individual ferromagnetic microregions (magnetic polarons, ferrons, ferromagnetic bipolarons) with independent and random magnetization directions will cause local Kerr rotations, right- and left-handed with respect to the original incident polarization direction. As a consequence the analyzer cannot extinguish the reflected light to the same degree as in the absence of ferromagnetic microregions: depolarization sets in. (The scattering and depolarizing entities can in reality be much smaller than the wavelength of light as also experienced by the air molecules that cause partial polarization of the light of the sky.)

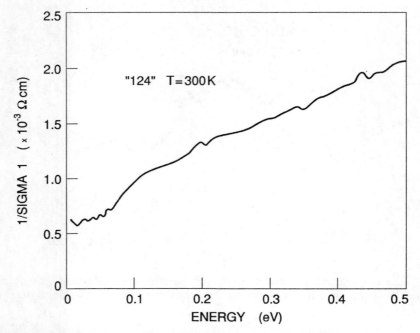

Fig. 8. The inverse optical conductivity of $YBa_2Cu_4O_8$ at 300 K. In the energy region $\omega > J \approx 0.1$ eV, we have found $\sigma \propto \omega^{-1}$, as expected for diffusive motion of polarons.

That the onset of depolarization is at about 200 K (see Fig. 7) and not at T_c is very significant and we thus expect the Néel temperature of the antiferromagnetic clusters in the 124 compound also to be at this temperature, and simultaneously the birth of magnetic polarons. With further decreasing temperature the alignment of the spins in the antiferromagnetic as well as in the ferromagnetic regions follows roughly a magnetization function (rather than a spin correlation function) as seen in Fig. 7. At and below T_c no significant further change in depolarization is observed. At T_c we expect only condensation of the bipolarons. At the moment we do not want to over-interpret our results until further experimental depolarization results are available.

The transport behavior of the polarons should be a diffusive motion in contrast to a propagating one. A theoretical prediction of this spin-flip diffusive motion has been made by Rice and Zhang [30]. The optical conductivity for the energy range above the magnetic exchange energy J (about 0.1 eV) has been calculated to be inversely proportional to the frequency: $\sigma \propto \omega^{-1}$. There is a good agreement with our experiment on the 124 compound (Fig. 8). Thus, the unconventional optical behavior in the mid IR energy range above 0.1 eV could be understood in terms of diffusive motion of magnetic polarons throughout the CuO_2 plane.

3.3 Superconductivity

With the f-sum rule, it is possible to calculate the amount of condensed carriers in the superfluid phase. For the superconducting state, we have proposed [3] that the superconducting carriers be described by the London model. Thus the optical conductivity of the condensed carriers should be given by $\sigma_1^{\text{sup}}(\omega) = B\delta(\omega)$ where $B = \pi n_s e^2/(2m^*) = \frac{1}{8}\omega_{\text{ps}}^2$ with n_s the density of superconducting carriers and m^* their effective mass; ω_{ps} is called the plasma frequency of the condensed carriers. In the superconducting state, an absorption manifests itself by a dip in the reflectivity as seen in our measurements (Fig. 3). The consequence of these absorptive dips is the peaks in optical conductivity at 28, 44 and 54 meV shown in Fig. 4. In Fig. 4 we observe also the temperature-dependence of the secondary plasma edge near 20 meV, which exhibits a strong temperature-dependence below T_c. In fact the missing area (hatched, e.g. between the curves for 95 and 65 K in Fig. 4), which corresponds to a loss of spectral weight, is compensated by an increase in the delta function at $\omega = 0$, which cannot be observed experimentally. This is the consequence of the f-sum rule [3]. At 6 K the lowest curve in Fig. 4 still clearly shows some non-superconducting carriers since a small plasma edge remains. This is a consequence of the experimental fact that the reflectivity never reaches exactly 100% except for $\omega = 0$. A further consequence of this is that there is no low-energy edge of the absorption peak centered at 28 meV, which could be taken as a charge gap of the superconducting state. An estimated decoupling of the plasma and absorptive parts is shown by dashed lines in Fig. 4. The existence of non-superconducting carriers in the CuO_2 plane at 6 K, i.e. at only $(1/12)T_c$ (the same 6 K at which one measured an absolute zero line for the optical conductivity of Rb_3C_{60} with a T_c of 29 K [19]) with energies much less than $k_B T_c$ is incompatible with s-wave pairing. A d-wave pairing, on the other hand, with a gap not extending over the whole Fermi surface, i.e. a pseudo-gap, is compatible with the experimental findings, but the size of the pseudo-gap cannot be larger than about 10 meV as estimated from Fig. 4, even permitting excessive experimental error margins. It is also clear from Fig. 4 that a charge gap of about 60 meV, as proposed by Schlesinger for the untwinned 123 compound [18] is impossible in the 124 compound. On the other hand, Bose condensation of non-magnetic bipolarons is gapless [9] and thus agrees with the experiment. In this connection it is not important whether the *non-magnetic bipolarons* consist of two magnetic *polarons* in the sense of two low-spin $3d^8$ Cu states as proposed by Wachter and Degiorgi [4] or Wachter *et al.* [5], or of two magnetic *polarons* in the sense of two $2p^5$ states that couple with opposite spin as proposed by Mott [9]. The non-magnetic bipolarons can be visualized as in the lower part of Fig. 6.

However, the fact that, at 6 K, there exist still uncondensed carriers (see Fig. 4) prompts us to assume that there are either still unpaired magnetic polarons as in the upper part of Fig. 6, or ferromagnetic bipolarons as in the middle part of Fig. 6, which as non-bosonic ones would not condense (the number of spins in the polarization cloud of the middle part of Fig. 6 is not well defined, for the shown size of the cloud a total spin of $\frac{7}{2}$ would result). The depolarization experiment, which does not exhibit significant changes below T_c, would rather be in favor of the existence of ferromagnetic entities down to absolute zero.

The question remains of how to interpret the absorption peaks at 28, 44 and 55 meV, which seem to emerge from the conductivity spectrum of Fig. 4 below T_c. At least the peak at 55 meV seems to be present already above T_c and can easily be associated with excitations of magnetic polarons or bipolarons or even phonons. The lower energy peaks we propose to be associated with the condensed superfluid phase below T_c. Similar examples can be observed even in the H_2O system, in which intramolecular vibrations remain the same above and below the condensation to the liquid, but in the condensed liquid phase lower energy excitations like surface waves are present. Another example may be superfluid ^4He. We also want to remark that, in inelastic neutron scattering in the 123 compound [15], one observes (spin flip) transitions below T_c with similar energies, namely about 30 and 41 meV. In such a complex system of antiferromagnetic clusters with incorporated magnetic or non-magnetic bipolarons, it is of course easy to observe spin-conserving (optical) and non-spin-conserving (neutrons) transitions with similar energies.

We now want to show the analysis of the thermal occupation of the superfluid ground state. In Fig. 9 the f-sum rule has been used to determine the superconducting fraction of the free carriers. The thermal occupation has been normalized to the value at 6 K, which can be calculated to be between 0.25 and 0.4 for various stoichiometries of the 247 compound [3], but in Fig. 9 is set to be unity, using the hatched areas for different temperatures in Fig. 4. If we postulate a Bose condensation of the exchange-coupled bipolarons, then the thermal occupation should follow the empirical two-fluid approximation: $n_s \propto 1 - (T/T_c)^4$ [31]. The experimental results of Fig. 9 (full line) show good consistency with the concept of a Bose condensation. However, it has been shown [32] that classical BCS superconductors like Sn can also have an exponent near 4 in the condensate fraction of the superfluid but only in the extreme anomalous limit of the coherence length ξ_0 being much larger than the London penetration depth λ_L. For the HTSCs the coherence length is much smaller than λ_L so that in a BCS model of the HTSCs an exponent near 4 in the condensate fraction is hardly possible. Instead, in the London limit of the BCS model, we would obtain an exponent of 3 [32] which is shown by the dashed

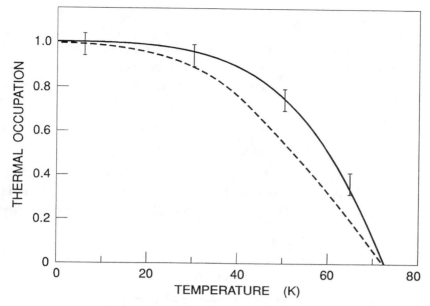

Fig. 9. Thermal occupation of the superfluid ground state as deduced from *f*-sum calculations. The two-fluid model (upper plot) of a Bose condensation gives good agreement with theory[30,31]. The lower curve represents the BCS theory for the HTSCs.

curve in Fig. 9, but the experimental results disagree with the concept of the BCS theory for the HTSCs and instead promote the concept of a Bose condensation of non-magnetic bipolarons. It should be mentioned that in the bismuth HTSCs an exponent of 4 has also been found [33].

4 Conclusion

The stoichiometric system $YBa_2Cu_4O_8$ (124) shows no chemical phase separation. The intrinsic properties of the CuO_2 planes have been probed by the static and dynamic conductivity. Within the framework of the hypothesis of magnetic polarons and condensed non-magnetic bipolarons normal as well as superconducting features can be understood. There is no signature of a gap in the density of states in the optical spectra, excluding s-wave pairing. A d-wave pairing is not excluded but a pseudo-gap cannot be larger than 10 meV. The unusual optical conductivity, manifested in a proportionality to ω^{-1}, is a consequence of diffusion of holes through the CuO_2 plane, which is compatible with the concept of magnetic polarons. The depolarization experiment indicates the birth of ferromagnetic microregions (such as magnetic polarons or bipolarons) at about 200 K. The temperature-dependence of the secondary plasma edge near 20 meV below T_c may be regarded as the consequence of Bose-condensation of non-magnetic bipolarons. Thus the two-fluid model of the superconducting state with a superfluid of condensed bipolarons in the planes obtains a high degree of probability. Some excitation energies in the superfluid state could be obtained, the exact nature of which remains to be determined in future work.

Acknowledgement

The authors are most grateful to Dr L. Degiorgi for critically reading the manuscript and fruitful discussions. The technical help of J. Müller and H. P. Staub is gratefully acknowledged. Single crystals of the 124 compound were grown by Dr J. Karpinski.

References

[1] J. G. Bednorz and K. A. Müller, *Z. Phys.* B **64**, (1986), 189.
[2] J. Karpinski, S. Rusiecki, E. Kaldis, B. Bucher and E. Jilek, *Physica* C **160**, (1989), 449.
[3] B. Bucher, J. Karpinski, E. Kaldis and P. Wachter, *Phys. Rev.* B **45**, (1992), 3026.
[4] P. Wachter and L. Degiorgi, *Solid State Commun.* **66**, (1988), 211.
[5] P. Wachter, L. Degiorgi and E. Kaldis, *Proc. 1st. Int. Conf. Metallurgy and Materials Science of W–Ti–RE–Sb*, Vol. 2, (1988), 1178, Fu Chongyue ed. (Int. Acad. Publ.).
[6] L. J. de Jongh, *Solid State Commun.* **65**, (1988), 963.
[7] L. J. de Jongh, *Physica* C **161**, (1989), 631.
[8] A. S. Alexandrov and A. B. Krebs, *Sov. Phys. Usp.* **35**, (1992), 345.
[9] N. F. Mott, *Physica* C **205**, (1993), 191.
[10] N. F. Mott, *J. Phys. Condens. Matter* **5**, (1993), 3487.
[11] B. Bucher, J. Karpinski, E. Kaldis and P. Wachter, *Physica* C **167**, (1990), 324.
[12] B. Bucher, J. Karpinski, E. Kaldis and P. Wachter, *J. Less Common Metals* **164 & 165**, (1990), 20.
[13] B. Bucher, P. Steiner, J. Karpinski, E. Kaldis and P. Wachter, *Phys. Rev. Lett.* **70**, (1993), 2012.
[14] H. Zimmermann, M. Mali, D. Brinkmann, J. Karpinski, E. Kaldis and S. Rusiecki, *Physica* C **159**, (1989), 681.
[15] J. Rossat-Mignot, L. P. Regnault, P. Bourges, C. Vettier, P. Burlet and J. Y. Henry, *Physica* B **186–188**, (1993), 1.
[16] J. Kircher, M. Cardona, A. Zibold, H.-P. Geserich, E. Kaldis, J. Karpinski and S. Rusiecki, *Phys. Rev.* B **48**, (1993), 3993.
[17] J. Karpinski, private communication.
[18] Z. Schlesinger, *Phys. Rev. Lett.* **65**, (1990) 801 (for the *a* direction of the untwinned 123 compound).
[19] L. Degiorgi, P. Wachter, G. Grüner, S. M. Huang, J. Wiley and R. B. Kaner, *Phys. Rev. Lett.* **69**, (1992), 2987.

[20] C. M. Varma, P. B. Littlewood, S. Schmitt-Rink, E. Abrahams and E. Ruckenstein, *Phys. Rev. Lett.* **63**, (1989), 1996.

[21] N. F. Mott, *Metal–Insulator Transition*, Taylor & Francis Ltd, London, 1974.

[22] B. H. Brandow, *J. Phys. Chem. Solids* **54**, (1993), 1137.

[23] M. Ospelt, J. Henz, E. Kaldis and P. Wachter, *Physica* C **153–155**, (1988), 159.

[24] H. Eskes and G. A. Sawatzky, *Phys. Rev. Lett.* **61**, (1988), 1415.

[25] B. Brandow, *J. Solid State Chem.* **88**, (1990), 28.

[26] A. S. Alexandrov and J. Ranninger, *Phys. Rev.* B **23**, (1981), 1796.

[27] K. Hestermann and R. Hoppe, *Z. Anorg. Allg. Chem.* **367**, (1969), 261.

[28] L. Degiorgi, E. Kaldis and P. Wachter, *Physica* C **153–155**, (1988), 657.

[29] F. Marabelli, G. B. Parravicini and P. Wachter, *Solid State Commun.* **86**, (1993), 131.

[30] T. M. Rice and F. C. Zhang, *Phys. Rev.* B **39**, (1989), 815.

[31] D. R. Tilley and J. Tilley, *Superfluidity and Superconductivity*, Adam Hilger Ltd, Bristol, 1986.

[32] B. Mühlschlegel, *Z. Phys.* **155**, (1959), 313.

[33] D. B. Romero, C. D. Porter, D. B. Tanner, L. Forro, D. Mandrus, L. Mihaly, G. L. Carr and G. P. Williams, *Phys. Rev. Lett.* **68**, (1992), 1590.

13

The near infrared and optical absorption of high-T_c superconductors using powders

C. H. RÜSCHER

Institut für Mineralogie der Universität Hannover, Welfengarten 1, 30169 Hannover, Germany

Abstract

The near infrared and optical absorption of several high-T_c superconductors is investigated using the KBr powder method. The spectra show broad NIR absorption peaks. Examples of absorption spectra obtained using transmission data of thin film work for $Ba_{1-x}K_xBiO_3$ (BKB) and single-crystal measurements for $Bi_2Sr_2Ca_{1-x}Y_xCu_2O_{8+\delta}$ (BSCC) are given for comparison. They confirm the results of the KBr spectra. An example of similar investigations on NiO also shows good agreement over the whole spectral range under investigation. These facts provide some evidence that the line profiles deduced by the KBr technique may indicate a correct absorption line shape. A possible explanation of the NIR line profile is discussed.

1 Introduction

The absorption of various high-T_c systems measured using Kubelka-Munk [1, 2] or standard KBr powder techniques [3–7] is known to possess broad peaks in the near infrared spectral range. Similar features have earlier been discussed in connection with a small-polaron absorption mechanism in the non-stoichiometric compounds of systems like TiO_{2-x} [8], WO_{3-x} [9], $NbO_{2.5-x}$ [10, 11] and the ternary compounds $Nb_xW_{1-x}O_y$ [12]. For the superconductors of the system $YBa_2Cu_3O_{7-\delta}$, $La_2CuO_{4+\delta}$ and $Bi_2Sr_2CaCu_2O_{8+\delta}$ [3–5] it has been shown that the NIR absorption cross-sections increase approximately linearly with decreasing temperature down to the superconducting transition temperatures (T_c). Below T_c the slope decreases to smaller values. Dewing *et al.* [3, 4] have explained this effect as evidence for Bose–Einstein condensation of small bipolarons for $YBa_2Cu_3O_7$. Alexandrov *et al.* [13] later showed that the conductivity sum rule for small bipolarons largely predicts the temperature variations observed in the NIR absorption experiments. On the other hand, any confirmation of the bipolaron theory of high-T_c superconductivity, which is related to the NIR absorption feature, should also quantify its line shape. So far line profile analyses have been carried out for the KBr spectra of the BSCC and BKB family of compounds [6, 7]. They make evident that the NIR line profiles closely follow predictions for small-polaron absorption in disordered systems as described by Bryksin [14]. However, there can be restrictions against quantitative analysis of powder spectra because of possible grain size effects and single-crystal work is highly desirable. Thus KBr spectra of examples of high-T_c superconductors (and families thereof) will be given together with a few examples of new single-crystal and thin film work for their confirmation.

2 Concepts of the KBr method and results

The idea of using the KBr powder method for investigation in the near infrared (NIR) and into the visible (VIS) spectral range of high-T_c superconductors follows its use as a standard method for investigation of phonon-absorption. Its disadvantages are possible grain size effects together with Mie or Rayleigh scattering [15], which can occur if the wavelength becomes smaller than the typical grain sizes to be used ($1 - 0.1$ µm). The loss in information concerning the anisotropy or related effects [16] may be only slightly unfavourable, especially if absorption occurs exclusively strongly only for one or two (nearly) isotropic crystallographic directions.

An advantage and necessity of the powder experiments is to dilute the sample with the (KBr) matrix by use of particles of small grain size. Therefore, ceramics and powders with high absorption coefficients may be used as starting materials. For samples of the high-T_c families concentrations of 0.05 mol% are typically used. The spectra are plotted as

$$\text{absorption} = -\ln(I/I_0) \tag{1}$$

with I/I_0 given by the intensity of transmitted light of the sample plus KBr over that of KBr alone.

Typical examples of spectra of compounds of the systems $YBa_2Cu_3O_{7-\delta}$, $La_{2-x}Sr_xCuO_4$ and $La_2CuO_{4+\delta}$ are shown in Fig. 1. It is observed that the high-T_c superconductors of each family show a pronounced and structureless absorption peak in the NIR together with considerably overlapping structure towards higher energies. It may be noted that there are systematic changes in the spectra with change of dopant. This is also observed for $YBa_2Cu_3O_{7-\delta}$ (not shown here, compare [1, 2]), where a strong reduction in NIR absorption occurs together with appearance of a new peak at about $13\,000$ cm^{-1} and an

C. R. Rüscher

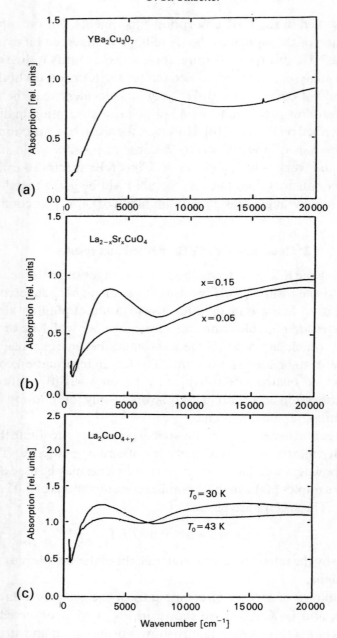

Fig. 1. Typical absorption spectra for superconducting members of different high-T_c families obtained by the KBr method.

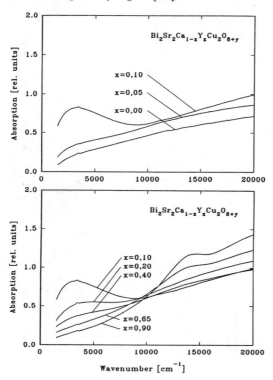

Fig. 2. Absorption spectra for superconducting ($x < 0.6$) and non-superconducting ($0.6 < x < 1$) samples of a series of compounds of the $Bi_2Sr_2Ca_{1-x}Y_xCu_2O_{8+y}$ system. For this system, the maximum in T_c is about 90 K and is given for samples with $x \approx 0.1$–0.2.

increase in total intensity towards higher wavenumbers for increasing δ. It may also be noted that the temperature-dependence [3–5] of the NIR absorption of the high-T_c superconductors is uniform, as mentioned in the introduction. However, because of the strong superposition of different absorption mechanisms and possible scattering effects of light, it is difficult to separate the various contributions for their interpretation. For this purpose it is very profitable to measure series of samples, where chemical components influence the optical absorption by substitutions or by change in composition, e.g. as indicated in Fig. 1(b) and Fig. 1(c).

Examples for which separation into different contributions is very obvious are given for the systems $Bi_2Sr_2Ca_{1-x}Y_xCu_2O_{8+\delta}$ (Fig. 2) and $Ba_{1-x}K_xBiO_{3-\delta}$ (Fig. 3). For the BKB system (Fig. 3) the spectra are dominated by the NIR contribution, which again becomes considerable towards the superconducting members of the family ($x > 0.2$). For the BSCC system (Fig. 2) three major

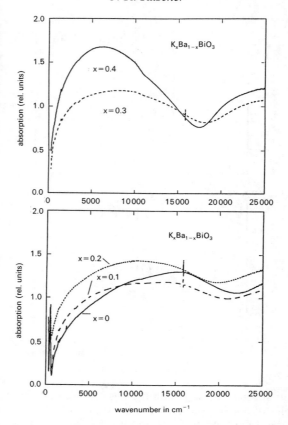

Fig. 3. Absorption spectra for the system $K_x Ba_{1-x} BiO_3$. The spectra of the superconducting members are shown in the upper graph ($T_c \approx 29.5$ and 30.3 K for the $x = 0.3$ and 0.4 samples, respectively).

contributions may be extracted, which certainly depend on the variation in Ca/Y ratio. These are the following.

1. The increase in absorption intensity towards higher wavenumbers, which decreases with increasing Ca content, i.e. towards the high-T_c members of the family ($0 < x \lesssim 0.6$).
2. The disappearance of the superimposed peak at about 13 000 cm^{-1} for $x \lesssim 0.6$.
3. The increase of the NIR peak with decreasing x to the maximum for $x = 0.1$. For compounds with $x < 0.1$ the NIR peak largely reduces and there appears only a flat increase towards higher wavenumbers. It may be noted that, for compounds with $x < 0.1$, T_c decreases again.

3 Comparison with single-crystal and thin film work

In earlier work on the NIR absorption of $YBa_2Cu_3O_7$ [2–4], thin film spectra were also investigated. The spectra closely coincide with absorption spectra

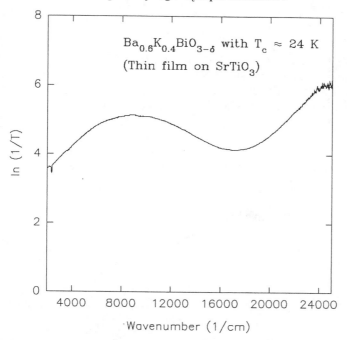

Fig. 4. Absorption spectra of a thin film sample of Ba$_{0.6}$K$_{0.4}$BiO$_{3-\delta}$.

obtained using the KBr method. A similar good agreement can be given here for an example of the BKB system. Shown in Fig. 4 is the as-measured absorption spectrum ($-\ln(T)$, T = transmission) of a thin film of composition Ba$_{0.6}$K$_{0.4}$BiO$_{3-\delta}$. This film has a T_c onset of 24 K, measured using a DC SQUID, with a broadening within about 8 K towards the plateau at lower temperature. Comparing the line shape with those given for the BKB system in Fig. 3, very close agreement is obtained for the sample with $x = 0.3$ over the whole spectral range shown. It can well be understood that the thin film sample shows closer agreement with the $x = 0.3$ sample rather than with the $x = 0.4$ one, considering the possibility of deviations in oxygen stoichiometry.

Comparison for the BSCC system is given in Fig. 5. Absorption spectra calculated using the transmission data of thin sections of single-crystal slices of composition Bi$_2$Sr$_2$Ca$_{1-x}$Y$_x$O$_{8+\delta}$ with $x = 1$ and $x = 0.6$ are shown. The latter x value has been estimated on the basis of the line shape, which is obtained from the reflectivity curve (not shown). The reflectivity increases from about 10% at 8000 cm^{-1} to about 50% at 3000 cm^{-1}, which is typical for crystals of this composition. For the crystal with $x = 1$ a completely flat reflectivity with $R \approx 10\%$ is observed. Below 800 cm^{-1}, phonon structures are observed. These crystals can easily be thinned (a–b slices) and measured using a microscope technique in transmission and reflection (at the same crystal position) with

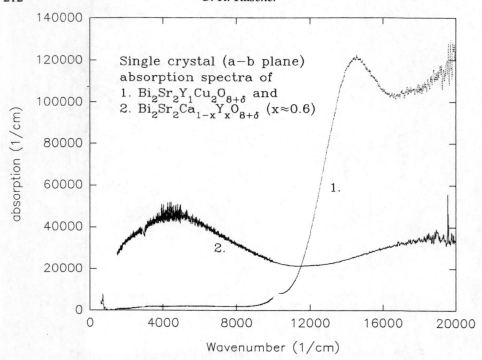

Fig. 5. Single-crystal absorption spectra for the BSCC system shown.

typical spot sizes of about 50–80 μm. Details of this procedure will be given elsewhere [17]. Interference patterns in the reflectivity (not shown here) were used to estimate the thicknesses, which are about 0.1–0.3 μm, for determination of the absolute absorption. However, the absolute value for the crystal with $x = 1$ might be less certain above $10\,000$ cm^{-1}, because a thinner section had to be used. In any case, the main features as pointed out for the series of KBr measurements on the BSCC system (Fig. 2) can well be inferred from the results shown in Fig. 5, by comparing the line profiles. On crossing against the metallic samples of the system (decreasing x), the $13\,000$ cm^{-1} peak disappears and the total increase in absorption above about $10\,000$ cm^{-1} decreases and becomes washed out, while the NIR peak strongly increases in intensity, together with a shift of the peak maximum towards lower values. It may be noted that the series of KBr spectra starts with an (already doped) $x - 0.9$ sample, which implies that the features above $10\,000$ cm^{-1} have already flattened out. Thus, the line profiles of the single-crystal absorption spectra and the KBr spectra suggest tentative agreement for the BSCC system.

To finish this section a typical KBr spectrum obtained for NiO is shown in Fig. 6. The peak at about 7000 cm^{-1} and the structures around $14\,000$ cm^{-1}

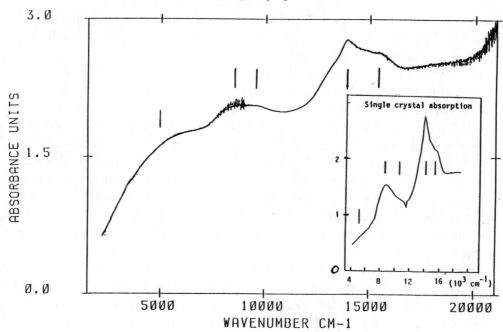

Fig. 6. KBr absorption spectra of a (green) NiO sample. In the inset the spectrum obtained for a single crystal is shown. The horizontal bars indicate marked features in both spectra (see text).

closely resemble the line profiles known for single-crystalline NiO (e.g. [18]). For the sake of better comparison, a single-crystal spectrum is shown in the inset of Fig. 6. The peaks at 7000 and 14 000 cm^{-1} have been explained in terms of Ni d–d transitions, while the broad feature in the background, i.e. with a knee at about 5000 cm^{-1} is at present unspecified. This could be due to some stoichiometric deviation from NiO, which has still to be investigated. It may be noted that similar features are also observed in Fe-rich (highly distorted) biotites, which has been explained in terms of $Fe^{2+} + Fe^{3+}$ hopping [19]. Above about 18 000 cm^{-1} the absorption increases further, as is also known for the single-crystal behaviour (not shown, compare [18]). Thus, the line shape, including any sharp structure as seen in the single-crystal spectra, is detected in the KBr experiment for NiO as well.

4 Discussion: possible explanation of the NIR absorption using the Bryksin model

The examples given in Section 2 and the comparative work outlined in Section 3 show that the KBr method can be useful also for studying line profiles of absorption in the NIR and VIS spectral range. This may help to understand the nature of the NIR absorption of high-T_c superconductors beside or in addition to single-crystal or thin film work. As yet explanations of the origin of the uncommon NIR line shape of the single-crystal reflectivity of high-T_c super-conductors are still controversial, e.g. [20–24]. Possible explanations of the observed behaviour follow the frequency-dependent scattering rates or the addition of a Lorentz oscillator with a Drude free-carrier description. At least for the BKB high-T_c system a rather broad absorption structure around 7000–10 000 cm^{-1} is clearly observed, using standard Kramers–Kronig analysis of single-crystal reflection data [25]. A possible Drude contribution, i.e. an effective plasma edge in the reflectivity line shape, is below about 0.1 eV in this system. This might indicate that heavy quasi-particles are relevant for the drift contribution to the frequency-dependent conductivity (or a small number of effective free carriers).

Coming back to the KBr spectra, the BKB system might, therefore, be a good candidate for line profile analysis, because competition with Drude-like effects can be ruled out for the NIR peak. In addition, because of the cubic structure, any anisotropy effect can be ruled out. A tentative description of the absorption profiles for the BKB system has been proposed in [7] to follow the Bryksin model of small-bipolaron absorption in disordered systems. This is given for the BKB system by (compare [26, 27])

$$\sigma' \approx \exp\left(\frac{-\mu^2}{\Gamma^2} - \frac{(w - 8E_a - 2V_c)^2}{16E_a kT}\right) \tag{2}$$

for $w < 8E_a - 2V_c - 8E_a \mu kT/\Gamma^2$,

$$\sigma' \approx \exp\left(\frac{-\mu^2}{2\Gamma^2} - \frac{(w - 8E_a - 2V_c - \mu)^2}{16E_a kT + 2\Gamma^2}\right) \tag{3}$$

for $8E_a - 2V_c - 8E_a \mu kT/\Gamma^2 < w < 8E_a - 2V_c + 8E_a \mu kT/\Gamma^2 + 2\mu$ and

$$\sigma' \approx \exp\left(\frac{-(w - 4E_a)^2}{16E_a kT + 4\Gamma^2}\right) \tag{4}$$

for $w > 8E_a - 2V_c + 8E_a \mu kT/\Gamma^2 + 2\mu$, where E_a and V_c are the small-polaron hopping energy and on-site repulsion in the two-site model, respectively; Γ is the Gaussian width of localized states; and μ is the filling level with respect to the maximum in the distribution of localized states.

Some results are illustrated in Fig. 7, for the sake of better comparison. The parameters used are given in the caption. It can be seen that the description turns out to follow well the NIR line profiles, and resolves another absorption

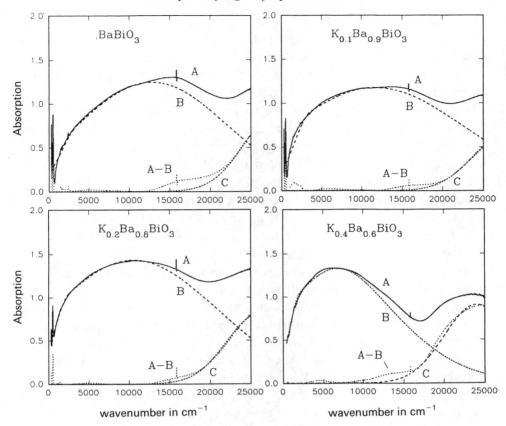

Fig. 7. Four examples of absorption spectra of BKB compounds (A) compared with calculated spectra (B) with the use of Eq. (2)–(4), and difference spectra (A − B). C denotes a fit with a Gaussian line shape to the difference spectrum (details are discussed in [7]). The parameters derived vary within the limits as follows (in cm^{-1}):

Material	E_a	V_c	μ	Γ
$BaBiO_{3-\delta}$	3100–4375	10650–15810	8400–9500	8800
$Ba_{0.9}K_{0.1}BiO_{3-\delta}$	3000–3500	10350–12250	7200–7700	11000
$Ba_{0.8}K_{0.2}BiO_{3-\delta}$	1875–3000	6300–10500	7500–7700	10000
$Ba_{0.6}K_{0.4}BiO_{3-\delta}$	1600–2100	5300–7150	4200–4800	6700

(The scaling factors to Eq. (2)–(4) are within 1.4–2.2).

contribution above about 10 000 cm^{-1}. The parameters, although largely tentatively, may also indicate systematic changes as a function of x. For further details of the description procedure and also a comparison of similar results obtained for the BSCC system I refer to [6, 7], where this has been done in detail. Here I shall just point out that the description works very well, which does not exclude other possibilities. However, that use of the Bryksin model turns out to be relevant for the NIR absorption implies very much the presence of localized states, i.e. polarons or bipolarons. It may be noted that a certain increase in polaron absorption intensity together with a change in the line profile on the low-energy side ($w < 8E_a - 2V_c - 8E_a\mu kT/\Gamma^2$) of the peak as a function of decreasing temperature is also to be expected for the small-polaron assumption. However, this effect is mainly included in the pre-factors to Eq. (2)–(4), which are not considered here. For the BKB system the temperature-dependence of the NIR absorption is also unknown at present.

Finally, the possible origin of the distribution of localized states itself needs to be considered. The spread of levels could be due to a certain amount of disorder, introduced by the non-stoichiometry in oxygen, and/or to the variation in site energies by substitution effects. Other more 'intrinsical' effects like that of correlation could also be considered. Such broad distributions as obtained for the BSCC ($\Gamma \approx 2000$ cm^{-1}) and for BKB ($\Gamma \approx 7000$–$10\,000$ cm^{-1}) are common e.g. for the $NbO_{2.5-x}$ and WO_{3-x} [9–12] system. One remarkable difference for $NbO_{2.5-x}$ and WO_{3-x} is that one starts to dope polarons into empty sites and the system becomes gradually metallic, while for the bismuth-ates and cuprates the situation is the opposite, with all sites in principle being filled in the undoped case. Considering the Bryksin model, this implies a filling level of about 60–90% in the case of the highly conducting bismuthates and cuprates and only 10–30% of occupied localized states in the $NbO_{2.5-x}$ and WO_{3-x} compounds, which also possess high conductivity. In my opinion this can provide for T_c values of about 70 K in the former and up to 10 K in the latter cases, considering the occupied localized states to be bipolarons with the possibility of bipolaron condensation and disregarding other competing effects (e.g. pinning, charge ordering, coexistence of polarons and band states etc.).

Acknowledgement

The results presented were obtained with financial support of the BMFT (F.+ E 13N5738).

References

[1] C. H. Rüscher, *Proc. 2nd USSR–FRG Bilateral Seminar in Tallinn* 1989, pp. 124–8.

[2] N. Rüffer, C. H. Rüscher, J. Erxmeyer and K. Schulze, in *Supraleitung und Tieftemperaturtechnik* 1991 (VdI Verlag), pp. 127–31.

[3] H. L. Dewing, E. K. H. Salje, K. Scott and A. P. Mackenzie, *J. Phys. Condens. Matter* **4**, (1992) L109.

[4] H. L. Dewing and E. K. H. Salje, *Supercond. Sci. Technol.* **5**, (1992) 50.

[5] C. H. Rüscher and M. Götte, *Solid State Commun.* **85**, (1993) 323.

[6] C. H. Rüscher, M. Götte, B. Schmidt, C. Quittmann and G. Güntherodt, *Physica* C **204**, (1992) 30.

[7] C. H. Rüscher, A. Heinrich and W. Urland, *Physica* C **219**, (1994) 471.

[8] V. N. Boglomolov, E. K. Kudinov, D. N. Mirlin and Yu. A. Firsov, *Sov. Phys. Solid State* **9**, (1968) 1630.

[9] E. Salje and B. Güttler, *Phil. Mag.* B **43**, (1984) 607.

[10] C. H. Rüscher, E. Salje and A. Hussain, *J. Phys. C: Solid State Phys.* **21**, (1988) 3737.

[11] C. H. Rüscher, *Physica* C **200**, (1992) 129.

[12] C. H. Rüscher, E. Salje and A. Hussain, *J. Phys. C: Solid State Phys.* **21**, (1988) 4465.

[13] A. S. Alexandrov, A. M. Bratkovsky, N. F. Mott and E. K. H. Salje, *Physica* C **215**, (1993) 359.

[14] V. V. Bryksin, *Sov. Phys. Solid State* **24**, (1982) 627.

[15] M. Born and E. Wolf, *Principles of Optics*, 1980, 6th ed. (Pergamon Press), p. 633.

[16] J. Orenstein and D. R. Rapkine, *Phys. Rev. Lett.* **60**, (1988) 968.

[17] C. H. Rüscher, C. Haas, G. Wiegers and S. van Smaalen, 1995, in preparation.

[18] R. Newman and R. M. Shrenko, *Phys. Rev.* **114**, (1959) 1507.

[19] C. H. Rüscher and S. Gall, 1995, in preparation.

[20] K. F. Renk, H. Eschrig, G. Schreiber, J. Keller, J. Schützmann and W. Ose, *Physica* C **165**, (1990) 1.

[21] L. D. Rotter, Z. Schlesinger, R. T. Collins, F. Holtzberg and C. Field, *Phys. Rev.* B **43**, (1991) 13 102.

[22] I. Bozovic, *Phys. Rev.* B **42**, (1990) 1969.

[23] I. Terasaki, T. Nakahaski, J. Z. Liu, Y. Fang, K. G. Vandervoort and S. Flesher, *Phys. Rev. Lett.* **67**, (1991) 2741.

[24] A. Zibold, M. Dürrler, A. Gayman, H. P. Geserich, N. Nücker, V. M. Burlakov and P. Müller, *Physica* C **193**, (1992) 171.

[25] S. H. Blanton, R. T. Collins, K. H. Kelleher, L. D. Rotter, D. G. Hinks and Y. Zheng, *Phys. Rev.* B **47**, (1993) 996.

[26] V. V. Bryksin, V. S. Voloshin and A. V. Raitsev, *Sov. Phys. Solid State* **25**, (1983) 820.

[27] V. V. Bryksin and V. S. Voloshin, *Sov. Phys. Solid State* **26**, (1984) 1429.

14

Polaronic theory of mid-infrared conductivity: a numerical cluster study

A. S. ALEXANDROV[1], V. V. KABANOV[2] and D. K. RAY[2]

[1]*IRC in Superconductivity, University of Cambridge, Madingley Road,*
Cambridge CB3 OHE, UK
[2]*Laboratoire des Propriétés Mécaniques et Thermodynamiques des Matériaux,*
CNRS Université Paris-Nord, 93430 Villetaneuse, France

Abstract

The observed characteristics of mid infrared (MIR) spectra in doped semiconductors are discussed. These characteristics were explained by Reik and co-workers on the basis of hopping motion of small polarons from a localized site to a neighbouring localized site. The success and limitations of this model are pointed out. Emin, on the other hand, showed the importance of large polarons for the conductivity. The recently observed features of MIR spectra in high-T_c cuprates are then summarized. The low-frequency peak in many cuprates with frequency 0.1–0.2 eV has been ascribed by many investigators to polaronic origin. We have undertaken in the present work numerical studies of polaronic conductivity in the two-site and four-site cluster model by diagonalization of the dynamical matrix. Broadening of the phonon spectra due to damping has been taken into account by considering a small but finite phonon lifetime. For intermediate and strong coupling, a number of peaks in the optical conductivity appear due to bound states with different numbers of phonons. We have also studied the importance of Hubbard U by calculating the optical conductivity as a function of U with two electrons in a two-site model. The experimental results of MIR spectra for the cuprates can be better understood on the basis of the present calculations.

1 Introduction

It was Landau who first introduced the idea of polarons for explaining the F centres in NaCl as due to self-trapping of electrons [1]. Polarons are the quasiparticles formed by the accompanying self-consistent polarization field and are generated due to the dynamical electron–phonon interaction. As a consequence there is extra scattering of the charge carriers, phonon energies are renormalized and the charge carriers are heavy [2].

The temperature- and frequency-dependences of the hopping mobility due to the polarons have been calculated to understand the optical conductivity in doped semiconductors [2] and it has been found that the hopping motion of the polarons gives temperature- and frequency-dependences quite different from those for ordinary electronic conduction [3].

Studies on the polarons showed that these can be of two types – large and small – depending on the radius of the polarons $r \gtrless a$, the inter-lattice distance. The range and strength of the electron–lattice interaction determine the type of polarons formed. Before we discuss the polaronic contributions to the conductivity, we point out the principal features of MIR spectra in the doped semiconductors measured during the last 30 years in systems such as TiO_2 doped with Nd and Li [4], p- and n-type $LaCO_3$ [5], $Sr\,TiO_{3-y}$ [6] and WO_3 [7].

The principal features of the optical spectra in the mid-infrared region are: (i) a maximum is observed in $\sigma(\omega)$ as a function of ω in addition to the Drude peak, which has a maximum at $\omega = 0$; (ii) the line shape is strongly asymmetric and (iii) the magnitude of $\sigma(\omega_m)$, where ω_m is the frequency at which the maximum of $\sigma(\omega)$ occurs, in general decreases with increasing temperature. Reik and co-workers [5, 8] attempted to explain these results in terms of hopping mobility of small polarons. However, detailed comparison between theory and experiment showed that good fitting of the spectral shape is obtained only for $\omega < \omega_m$. The experimental asymmetry is more pronounced than that given by Reik's calculations. On the other hand, Reik's calculations give more pronouced temperature-dependence of $\sigma(\omega)$ than the experimental results. Better agreement with experimental results has been obtained by Emin, who estimated the contribution of large polarons $\sigma(\omega)$ by using the Fermi golden rule [9] for excitation of a self-trapped carrier from the ground state to the free carrier state. His results for the large polarons are as follows.

(a) The frequency-dependence of $\sigma(\omega)$ shows a sharp rise at photon energy ω exceeding $3E_p/T$ for both 2D and 3D cases where E_p is the ground state energy of the polarons.

(b) The curves are strongly asymmetric. The values of $\sigma(\omega)$ are much higher for $\omega > \omega_m$ than for $\omega < \omega_m$, in agreement with experimental results.

(c) $\sigma(\omega)$ does not depend on temperature.

In the background of this experimental and theoretical work on MIR spectra in non-magnetic and non-superconducting semiconductors, work has been undertaken on the high-T_c cuprates to see whether polaronic theory holds good in these interesting systems. The observed results are as follows.

(a) In cases like those of $Nd_2 CuO_{4-x}$, $YBa_2 Cu_3O_{6+x}$ and $LaCuO_{4+x}$ [10] two principal peaks have been observed. The one close to 0.2 eV is rather weak and the other with

$\omega \simeq 0.7$ eV broad and relatively strong. In $La_{2-x} Sr_x Cu O_{4-y}$ crystals [11] the low frequency peak shifts to lower values with increase of x. Similarly the shift of the peak to lower energy with increase of T_c has been observed in $Tl_2Ba_2Ca_2O_8$, $YBa_2Cu_3O_7$ and in $La_{2-x}Sr_xCuO_3$ by Mihailović *et al* [12]

(b) Experiments on $La_{1.8}Sr_{0.2}CuO_{4+\delta}$ and $La_{1-x}Sr_xNiO_{4+\delta}$ done by Xiang-Xui Bi and Eklund [13] showed larger asymmetry of $\sigma(\omega)$ than given by Reik's theory.

(c) From the observed temperature-dependence of $\sigma(\omega)$ in $La_2CuO_{4+\delta}$, Falck *et al.* [14] confirmed the polaronic origin of the spectrum at 0.13 eV.

(d) The reflectivity spectra in Nd_2CuO_{4-y} single crystals showed temperature dependent superstructures in the far infrared at frequencies that are multiples of those of the local modes [15]. The correlation of these modes with phonons supports a polaronic origin of the low-frequency spectra in the cuprates.

(e) The high-frequency peak (at 0.7 eV) in $YBa_2Cu_3O_{7-\delta}$ has been studied by Dewing and Salje [16] as a function of temperature. These results have been interpreted by Alexandrov *et al.* [17] as a manifestation of Bose–Einstein condensation of small bipolarons.

On the basis of these experimental results we can conclude that the low-frequency peak in the MIR spectra of the cuprates is of polaronic origin. The magnitudes of ω_m, the low oscillator strength of the observed peaks, the nature of the observed asymmetry in the shape of the spectrum, the temperature-dependence of the observed spectra, the variation in broadening of the spectrum with temperature and the correlation of the superstructure with phonons in Nd_2CuO_{4-y} strongly favour a polaronic origin for these spectra. By comparison of the theories of Reik and Emin with the experimental results it seems that the polarons are of intermediate size, arising from moderate electron–phonon coupling. An analysis of this peak by Millis and Shraiman [18] showed that dielectric polarons rather than spin waves are responsible for this peak.

We have recently studied the conditions of polaron formation within the framework of the Holstein model [19, 20] both in the adiabatic and in the non-adiabatic limit. The characteristics of the polarons as a function of coupling strength have been studied.

Here we use this model to calculate the optical conductivity for a two-site and four-site cluster with variation of the size of the polarons and these are presented in Section 2. We next consider in Section 3 the importance of Hubbard U for the case of two electrons in the cluster. We obtain the asymmetry and the phonon superstructure corresponding to the two-site bound states of polarons. These numerical results are then compared with the experimental ones for the cuprates. Details of these results for the intermediate polarons will be useful for understanding some aspects of optical conductivity in cuprates in contrast to the analytical theories developed earlier for small and large polarons.

2 Numerical results with the two-site and four-site models of optical conductivity

The multi-site Hamiltonian in the presence of electron–phonon interaction and Hubbard Coulomb correlation is given by

$$\mathcal{H} = -t\sum_{\langle i,j \rangle} C_{i\sigma}^{+} C_{j\sigma} + g\Omega \sum_{i} C_{i\sigma}^{+}(d_i + d_i^{+}) + \Omega \sum_{i} d_i^{+} d_i + U\sum_{i} U_{i\uparrow} n_{i\downarrow}, \qquad (1)$$

where t is the hopping integral for nearest neighbours, C_i and d_i are the electron and phonon operators, Ω is the single optical phonon frequency considered here, g is a dimensionless constant and $i, j \leq n$, $n = 2$ and 4. The last term in (1) is the Coulomb repulsion when two electrons are present. The polaronic level shift E_p is given by $E_p = g^2 \Omega$. For diagonalizing the Hamiltonian (1) matrix elements between antisymmetrical electronic states are to be considered. The dimension of the matrix is, thereby, reduced and exact diagonalization can be easily done with 50 phonon modes for $N = 2$ with good energy convergence [20]. For $N = 4$, we used 20 phonon states for each mode and diagonalization is done following the Lanczos procedure, which has been discussed in detail by Wagner *et al.* [22].

We can then express $\sigma(\omega)$ in terms of the eigenvalues E_n and eigenstates α_n^i of this Hamiltonian as

$$\sigma(\omega) \sim \frac{t^2}{2\omega} \sum_{n=1}^{N_m} \frac{\eta}{\omega^2 - (E_m - E_n)^2 + \eta^2} \left(\sum_{m=1} (\alpha_1^{2m-1}\alpha_N^{2m} - \alpha_1^{2m}\alpha_N^{2m-1}) \right)^2. \qquad (2)$$

Here η is a parameter signifying the phonon broadening of the lines. For $\eta \to 0$, the phonon spectrum is a δ-function. The shape of $\sigma(\omega)$ does not depend on the value of η if η is large for the two-site model. We calculate $\sigma(\omega)$ for different values of $\lambda = g^2 \Omega/(zt)$ (where z is the coordination number) in the adiabatic limit and compare these with those given by Reik's formula:

$$\sigma(\omega) = \frac{t^2}{2E_p T} \frac{1 - \exp(-\beta\omega)}{\omega} \exp\left(-\frac{(\omega - 2E_p)^2}{8E_p T} \right). \qquad (3)$$

These results are shown in Fig. 1 of [20]. The main points of this comparison are as follows.

1. Reik's formula agrees with our numerical results in the strong-compling small-polaron limit.

2. For the case of intermediate coupling the shape of $\sigma(\omega)$ is more asymmetric than that given by Reik's formula.
3. Additional superstructure appears for different spectral weights of the states with different numbers of phonons for intermediate and small polarons. Calculations with four sites show additional asymmetry in the shape of $\sigma(\omega)$ due to hopping to more distant sites.

3 The effect of Hubbard U on $\sigma(\omega)$

We have calculated $\sigma(\omega)$ for polarons in the presence of intra-site Hubbard U by considering two electrons in a two site cluster. For small values of U, the ground state is of bipolaronic type [19] in agreement with the analytical results obtained earlier by Bryksin and Voloshin [23]. We get the main peak at $\omega = 4E_p - U$. We have studied the evolution of $\sigma(\omega)$ with increasing U for $\omega = 0.2t$. For $U = 0$ we find one peak in $\sigma(\omega)$ in the high-energy region $\omega = 4E_p$. With increasing U, this peak shifts to the lower value of ω and the peak shows more asymmetry. Further increase of U destroys the bipolaronic ground state and we get an additional peak in $\sigma(\omega)$, which corresponds to excitation from the unpaired ground state to the bonding and antibonding paired states. We also get additional phonon superstructure in all the spectra. These results are shown in Fig. 5 of [20].

4 Conclusion

We have thus presented our results on the optical conductivity $\sigma(\omega)$ for a finite-size cluster with two and four electrons. We have got good agreement with results obtained from Reik's formula in the strong-coupling limit. In the intermediate-coupling limit the results differ both in the position of the peak and in the asymmetry of the spectral shape. When we increase N to 4, the asymmetry was found to increase, particularly for the intermediate-coupling case. We obtained phonon superstructure in $\sigma(\omega)$ in the intermediate and strong-coupling cases. We have also studied the effect of Hubbard U on $\sigma(\omega)$ for two electrons in a two-site cluster. In the weak-U limit, the on-site bipolarons are formed and dominate the optical properties. For larger U, these bipolarons dissociate into polarons with the appearance of a new peak. The experimental results on $\sigma(\omega)$ spectra in cuprates are in better agreement with the intermediate-size polarons [13] rather than small polarons (Reik's case) and large polarons (Emin's case). It seems that the temperature-dependence of the polaronic peak is a crucial test of the size of the polarons. For small polarons the temperature-dependence is strong whereas large polarons show temperature-independent spectra [9]. We intend to study the temperature-dependence of the conductivity on the basis of our cluster model for the intermediate polarons in the future.

Acknowledgement

The authors are grateful to Sir Nevill Mott for helpful and stimulating discussions.

References

[1] L. D. Landau, *Phys. Z. Sowjet Union* **3**, 664 (1933).
[2] J. Appel in *Solid State Physics* edited by Frederick Seitz, David Turnbull and Henry Ehrenreich vol. 21 (Academic Press, New York, 1968) p. 193.
[3] T. Holstein, *Ann. Phys. (N.Y.)* **8**, 343 (1959).
[4] V. N. Bogomolov and D. N. Morlin, *Phys. Stat. Sol.* **27**, 443 (1968); E. K. Kudinov, D. N. Morlin and Yu. A. Firsov, *Sov. Phys. Solid State* **11**, 2257 (1970).
[5] R. Muhlstroh and H. G. Reik, *Phys. Rev.* **162**, 703 (1967).
[6] W. S. Baer, *Phys. Rev.* **144**, 734 (1966).
[7] O. F. Schirmer and E. Salje, *Solid State Commun* **33**, 333 (1980).
[8] H. G. Reik, *Z. Phys.* **203**, 346 (1967); H. G. Reik in *Polarons in Ionic Crystal and Polar Semiconductors*, edited by J. Dereese (North-Holland, Amsterdam, 1972).
[9] D. Emin, *Adv. Phys.* **24**, 3058 (1975); *Phys. Rev. B* **48**, 13691 (1993).
[10] G. A. Thomas, R. H. Rapkine, S. L. Cooper, S.-W. Cheong, A. S. Cooper, L. F. Schneemeyer and J. V. Waszczak, *Phys. Rev.* **45**, 2474 (1992); S. L. Cooper, G. A. Thomas, A. J. Millis, P. E. Sulewski, J. Orensteion, D. Rapkine, S. W. Cheong and P. L. Trevor, *Phys. Rev. B* **42**, 14785 (1990).
[11] S. Uchida, T. Ido, H. Takagi, T. Arima, Y. Tokura and S. Tajima, *Phys. Rev. B* **43**, 7942 (1991).
[12] D. Mihailović, C. M. Foster, K. Voss and A. J. Heeger, *Phys. Rev. B* **42**, 7989 (1990).
[13] Xiang-Xui Bi and P. C. Eklund, *Phys. Rev. Lett.* **70**, 2625 (1993).
[14] J. P. Falck, A. Levy, M. A. Kastner and R. J. Birgeneau, *Phys. Rev. B* **48**, 4043 (1993).
[15] P. Calvani, M. Capizzi, P. Maselli, A. Palone, S. Lupi, P. Roy, W. Sadowski and E. Walker to be published.
[16] H. L. Dewing and E. K. H. Salje, *Supercond. Sci. Technol.* **5**, 50 (1992).
[17] A. S. Alexandrov, A. M. Bratkovsky, N. F. Mott and E. K. H. Salje, *Physica* C **215**, 359 (1993).
[18] A. L. Millis and B. L. Shraiman, *Phys. Rev. B* **46**, 14834 (1992).
[19] A. S. Alexandrov, V. V. Kabanov and D. K. Ray, *Phys. Rev. B* **49**, 9915 (1994).
[20] A. S. Alexandrov, V. V. Kabanov and D. K. Ray, *Physica* C **224**, 247 (1994).
[21] J. Ranninger and U. Thibblin, *Phys. Rev. B* **45**, 773 (1992).
[22] J. Wagner, W. Hanke and D. J. Scalapino, *Phys. Rev. B* **43**, 10517 (1991).
[23] V. V. Bryksin and V. S. Voloshin, *Sov. Phys. Solid State* **26**, 1429 (1984).

15

Electromagnetic properties of local pair superconductors

S. ROBASZKIEWICZ, R. MICNAS and T. KOSTYRKO

Institute of Physics, A. Mickiewicz University, Grunwaldzka 6, 60-780 Poznań, Poland

Abstract

Recent experimental findings concerning the temperature-dependence of the penetration depth and strong empirical evidence for a universal 3D $X–Y$ critical behaviour in several classes of high-T_c superconductors suggest that Bose condensation of 'weakly charged' bosons can be a driving mechanism for the phase transition in extreme type II superconductors. We explore the occurence of analogous behaviour in a simple model of local electron pairs, which is that of hard-core charged bosons on a lattice. We examine the electromagnetic properties of the model in the superfluid phase as a function of boson concentration and density–density interaction. In the low-density limit, with the use of the exact scattering length, we present new results for the ground state characteristics. The relevance of the obtained results to interpretation of experimental data on high-T_c oxides is discussed.

1 Introduction

Some recent experimental results concerning temperature-dependence of the London penetration depth and empirical evidence for a universal 3D $X–Y$ critical behaviour in several families of high-T_c superconductors indicate that Bose condensation can be a driving mechanism for the phase transition in these extreme type II superconductors [1–3]. Motivated by these experimentally established universal trends, we examine the electromagnetic properties of the effective pseudo-spin model of the local pair superconductor, equivalent to a hard-core charged Bose gas on a lattice, for the case of arbitrary electron concentration n. From linear response theory, the current response is obtained and expressions for the paramagnetic and diamagnetic kernels are evaluated in the RPA (random phase approximation) and SWA (spin-wave approximation). The evolution of the London penetration depth λ at $T = 0$ K as a function of electron concentration and inter-site interaction is determined for various lattice structures. In the low-density limit the analytical formulae for $\lambda^{-2}(0)$ to the order of $n^{5/2}$ are obtained for cubic lattices, with the interaction expressed in terms of the exact two-particle scattering amplitude. For the short-range Coulomb interaction the universal feature of $[\lambda(0)/\lambda(T)]^2$ in the considered model is the T^{d+1} behaviour in the $T \to 0$ limit ($d > 2$) and the X–Y critical behaviour close to T_c. In the intermediate temperature regime $\lambda(T)$ depends on the electron density and the interactions and reflects the location of the crossover from the weak (dilute) to the strong (dense) coupling limit. We compare theoretical results with experimental data for the penetration depth in several families of high-T_c oxides.

2 Hard-core charged bosons in the magnetic field

We describe a system of hard-core charged bosons (bipolarons) on the lattice in the magnetic field by the following Hamiltonian [1]:

$$\mathcal{H} = -\frac{1}{2} \sum_{jm} J_{jm} (\rho_j^+ \rho_m^- e^{i\Phi_{jm}} + \text{H.c.}) + \sum_{jm} K_{jm} \rho_j^z \rho_m^z - \bar{\mu} \sum_j (2\rho_j^z + 1) + \text{constant}. \quad (1)$$

The boson creation/annihilation operators $\rho_j^\pm$, and ρ_j^z satisfy the commutation rules of the $s = \frac{1}{2}$ operators. In the above

$$\Phi_{jm} = -\frac{2e}{\hbar c} \int_{R_j}^{R_m} A(r) \, dr, \quad (2)$$

where $A(r)$ denotes the vector potential at r.

Using the RPA method one obtains the quasi-particle energy spectrum for the superfluid state (SS), when $\langle \rho^x \rangle \neq 0$ and the inter-site repulsion is short-ranged:

$$E_k = R[(\varepsilon_k^0)^2 + V_k \varepsilon_k^0 \sin^2 \theta]^{1/2} \overset{k \to 0}{\sim} \hbar s |k|, \quad (3)$$

where $s = 2\langle \rho^x \rangle Ja(1 + K/J)^{1/2} \sqrt{Z}/\hbar$ denotes the velocity of sound and

$$R \equiv 2(\langle \rho^x \rangle^2 + \langle \rho^z \rangle^2)^{1/2}, \qquad \cos^2 \theta = (n-1)^2/R^2,$$
$$A_k = R\varepsilon_k^0 + B_k, \qquad \varepsilon_k^0 = J_0 - J_k, \qquad B_k = \frac{1}{2} R V_k \sin^2 \theta,$$
$$V_k = J_k + K_k.$$

In the above, $n = 2\bar{n}$, where $\bar{n}$ is the boson concentration, a is the lattice spacing and Z is the coordination number. The length of pseudo-spin R is given as a solution of the equation

$$1 = \frac{R}{N} \sum_k \frac{A_k}{E_k} \coth\left(\frac{E_k}{2k_{\mathrm{B}}T}\right). \tag{4}$$

The expectation value of the Fourier transform of the total current operator can be obtained from the linear response theory as

$$\mathscr{I}_\alpha(\boldsymbol{q},\omega) = \frac{c}{4\pi} \sum_\beta [K_{\alpha\beta}^{\mathrm{dia}}(\boldsymbol{q},\omega) + K_{\alpha\beta}^{\mathrm{para}}(\boldsymbol{q},\omega)] A_\beta(\boldsymbol{q},\omega), \tag{5}$$

where the diamagnetic part of the kernel in the RPA is given by

$$K_{\alpha\beta}^{\mathrm{dia}}(\boldsymbol{q},\omega) = -\delta_{\alpha\beta}4\pi \frac{2\bar{e}^2}{Z\hbar^2 c^2}\left[\sin^2\theta \frac{R^2 J_0}{4} + \sin^2\theta \frac{R}{4N}\sum_k J_k \frac{B_k}{E_k}\coth\left(\frac{E_k}{2k_{\mathrm{B}}T}\right)\right.$$
$$\left. + (1 + \cos^2\theta)\frac{R}{4N}\sum_k J_k \frac{A_k}{E_k}\coth\left(\frac{E_k}{2k_{\mathrm{B}}T}\right)\right]. \tag{6}$$

The paramagnetic part evaluated in the RPA, in the limit: $\omega \to 0$, $\boldsymbol{q} \to 0$ (for $n \neq 1$) reads

$$K_{\alpha\beta}^{\mathrm{para}}(0,0) = -2\delta_{\alpha\beta}4\pi \left(\frac{2\bar{e}JR\cos\theta}{\hbar c}\right)^2 \frac{1}{N}\sum_k \sin k_\alpha \sin k_\beta \frac{\partial f}{\partial E_k}. \tag{7}$$

The London penetration depth can be expressed in terms of the total kernel as

$$\lambda^{-2}(T) = -K_{\alpha\alpha}(T) = \frac{4\pi\bar{e}^2 n_{\mathrm{s}}(T)}{m^* c^2} \tag{8}$$

($\bar{e} = 2e$), defining the superfluid number density versus effective mass ratio n_{s}/m^*. One can show by means of integration by parts of Eq. (7) that the paramagnetic part of the kernel reduces exactly to the diamagnetic one for vanishing superconducting order parameter, thus proving disappearance of the Meißner effect above T_c. In the low-density limit ($\tilde{n} \ll 1$) and neglecting zero-point quantum corrections, we obtain for $K_{\alpha\beta}$ per unit volume

$$K_{\alpha\beta} = \frac{4\pi\bar{e}^2\tilde{n}}{m^* c^2}\left[\delta_{\alpha\beta} - \frac{\hbar^2}{\tilde{n}m^*}\int \frac{\mathrm{d}^d p}{(2\pi)^d} p_\alpha p_\beta \left(-\frac{\partial f}{\partial E_k}\right)\right] \tag{9}$$

$$\tag{10}$$

($m^* = \hbar^2/(2Ja^2)$, $\tilde{n} = n/(2a^3)$), justifying use of the Landau quasi-particle formula [4].

For $T \to 0$, $K_{\alpha\beta}^{\mathrm{para}} \to 0$, $n_{\mathrm{s}}(0) = \tilde{n}$ for $\tilde{n} < \frac{1}{2}$ and

$$\lambda^{-2}(0) = -K_{\alpha\alpha}^{\mathrm{dia}}(0) = \frac{4\pi\bar{e}^2\tilde{n}}{m^* c^2}. \tag{11}$$

Note that $n_s(0)$ for $\tilde{n} > \frac{1}{2}$ is $(1 - \tilde{n})/a^3$ due to the electron–hole symmetry of the model Eq. (1). The results of numerical evaluation of Eq. (11) for the square and SC lattices for different values of K/J are presented in Fig. 1(a),(b). For $K/J \leq 1$ the maximum of λ^{-2} is placed near $n = 1$. The zero-point quantum corrections are strongest for the 2D lattice and, as for the absolute value of λ, they are largest in the high-density limit, $|n - 1| \sim 0$.

At $T = 0$ K, for $K/J > 1$ the inter-site density–density interactions between the local pairs stabilize the CDW (charge density wave) phase if $n = 1$ and the M (CDW plus SS) phase for $n_c < n < 1$ and $1 < n < 2 - n_c$, where n_c is the critical concentration dependent on K/J [6]. In such a case the maximum of λ^{-2} is placed near $|n_c - 1| < 1$ and, with decreasing $|n - 1|$, λ^{-2} increases first, then goes through a maximum near a border between the SS and M phases and goes to zero as $|n - 1| \to 0$.

3 Low-density expansion for 3D lattices

In the low-density limit one can derive the penetration depth within a systematic expansion exploiting exact two-body scattering amplitude [5, 6] as

$$\frac{t_0}{2J_0} = \frac{A}{1 + A(C - 1)}, \tag{12}$$

where $A = 1 + K/J$, $J_0 = JZ$ and C is a constant dependent on a lattice structure ($C_{SC} = 1.51638$, $C_{BCC} = 1.3932$, $C_{FCC} = 1.3446$). At $T = 0$ K we obtain

$$\frac{1}{\lambda^2(0)} = \frac{\alpha J_0}{4} \left\{ -n(2 - n) - n^2(C - 1) \left[2\left(\frac{t_0}{2J_0}\right) + \left(\frac{t_0}{2J_0}\right)^2 (2C - 3) \right] \right.$$
$$\left. + \frac{\gamma Z^{3/2}}{6\pi^2} n^{5/2} \left[5\left(\frac{t_0}{2J_0}\right)^{3/2} - 3\left(\frac{t_0}{2J_0}\right)^{5/2} \right] \right\} + \mathcal{O}(n^3), \tag{13}$$

where $\alpha = -8\pi \bar{e}^2 Z/(\hbar^2 c^2 a)$, $\gamma = v_0/a^3$ and v_0 is the volume of the unit cell. The low-T expansion of $\lambda(T)$ expressed in terms of the scattering amplitude reads

$$\frac{n_s}{\tilde{n}} = \left(\frac{\lambda(0)}{\lambda(T)}\right)^2 = 1 - \eta T^4, \tag{14}$$

where

$$\eta = \frac{2\pi^2 \hbar}{45 s m^* \tilde{n}} \left(\frac{k_B}{\hbar s}\right)^4, \tag{15}$$

$$\hbar^2 s^2 = J_0 J a^2 n(2 - n) \left(\frac{t_0}{2J_0}\right) \left[1 + \left(\frac{t_0}{2J_0}\right)^{3/2} \frac{2\gamma}{\sqrt{2\pi^2}} Z^{3/2} n^{1/2} \right]. \tag{16}$$

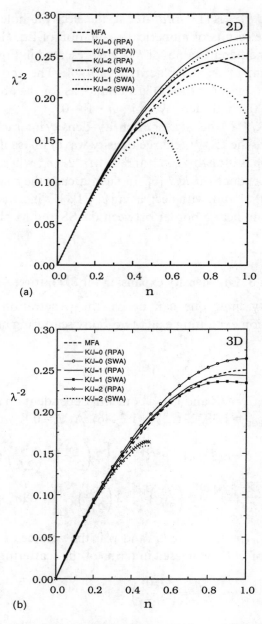

Fig. 1. The concentration-dependence of the inverse square of the London penetration depth at $T = 0$ for several values of K/J from RPA, SWA and mean-field approximation (MFA) ($\hbar^2 c^2 a/(8\pi \bar{e}^2 J) = 1$): (*a*) 2D square lattice and (*b*) SC lattice. The plots are symmetric with respect to $n = 1(\bar{n} = 1/2)$.

In the low-density limit we are also able to calculate the thermodynamic critical field H_c, which to the lowest order in n is given by

$$\frac{H_c^2(0)}{8\pi} = E_0^N - E_0 = \frac{J_0}{4}\left(\frac{t_0}{2J_0}\right)n^2 + \mathcal{O}(n^{5/2}), \qquad (17)$$

where E_0^N is the ground state energy of the normal phase. Using the Ginzburg–Landau relations between λ, H_c and $\xi(0)$

$$\xi(0) = \frac{\Phi_0}{2\pi\lambda H_c\sqrt{2}}, \qquad (18)$$

where Φ_0 is the flux quantum, and expressions (11 and 17) for λ and H_c, one obtains the following estimations for $\xi(0)$, H_{c2} and H_{c1}:

$$\xi^2(0) \simeq \frac{a^2}{Z}\left(\frac{t_0}{2J_0}n(2-n)\right)^{-1},$$

$$H_{c2} \simeq \frac{\phi_0}{2\pi\xi(0)^2},$$

$$H_{c1} \simeq \frac{\ln\kappa}{\kappa}H_c \sim \frac{\ln\kappa}{\lambda^2}, \qquad \kappa = \frac{\lambda(0)}{\xi(0)}. \qquad (19)$$

4 Universal behaviour and superfluid characterics

At low T the superfluid density has the phonon form as in the standard case of a weakly interacting Bose gas [7]:

$$1 - \frac{n_s}{\tilde{n}} = 1 - \left(\frac{\lambda(0)}{\lambda(T)}\right)^2 \sim T^{d+1}. \qquad (20)$$

At higher temperatures, in the low-density limit the k^2 term in the excitation spectrum, Eq. (3), becomes dominant and $n_s(T)$ is approximately given by

$$\frac{n_s(T)}{\tilde{n}} \approx 1 - \left(\frac{T}{T_c}\right)^{d/2}, \qquad (21)$$

where $T_c \approx T_B$ for $n \ll 1$ and T_B is the Bose–Einstein transition temperature ($T_B = 3.31\hbar^2(\tilde{n})^{2/3}/(k_B m^*)$). Formula (21) is valid for $T_0 < T < T_c$ (but below the critical regime), where T_0 is the crossover temperature and $k_B T_0 \approx \bar{V}_0 n_0$ ($\bar{V}_0 = 2(J_0 + K_0)$, $n_0 = \langle\rho^x\rangle^2$).

Near T_c there are fluctuation corrections arising from critical phenomena characteristic of X–Y $d=3$ model type behaviour. They occur in a critical

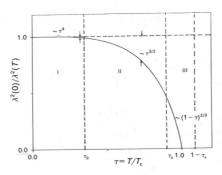

Fig. 2. A schematic representation of the temperature-dependence of the London penetration depth for model (1) and short-range density–density interaction, in 3D.

region of width $|(T/T_B)-1| \leq n^{1/3}\bar{V}_0/(2J_0)$ and cause a shift in transition temperature from T_B to T_c as well as a change in the critical exponents from their mean-field values to the usual scaling exponents inside this region. In particular, we have

$$\left(\frac{\lambda(0)}{\lambda(T)}\right)^2 = \left(\frac{1-T}{T_c}\right)^\nu, \qquad \nu \approx \tfrac{2}{3} \text{ for } d=3. \tag{22}$$

In Fig. 2 we present a schematic representation of the temperature-dependence of the London penetration depth for model (1) in the dilute limit. There are three different regimes: a low-T 'phonon' region (I) for $\tau = T/T_c < \tau_0 = T_0/T_c$, an intermediate 'free particle' region (II) for $\tau_0 < \tau < \tau_x \sim 1 - \tau_0$ and a critical region (III) for $\tau_x < \tau < 1 - \tau_x$ with the X–Y model type critical behaviour. With increasing $\bar{n}$ and/or $\bar{V}_0$ regions I and III will expand at the cost of suppressing region II. Thus the universal features of $(\lambda(0)/\lambda(T))^2$ in the considered model are the T^4 behaviour in the $T \to 0$ limit and the 3D X–Y critical point behaviour close to T_c. The temperature-dependence in the intermediate T region depends on $\bar{n}$ and $\bar{V}_0$ and reflects the location of the crossover from the dilute to the dense limit in terms of the exponent x in the formal expression

$$\left(\frac{\lambda(0)}{\lambda(T)}\right)^2 = 1 - \left(\frac{T}{T_c}\right)^x. \tag{23}$$

The temperature-dependence of the penetration depth observed in high-T_c cuprates appears to be bounded by the behaviour of the two-fluid model and the dilute Bose gas [9, 8, 3]. The compounds with high T_c turn out to be closer to two-fluid behaviour ($x \sim 4$ in Eq. (23)), while the systems with reduced T_c (lower electron concentration) exhibit crossover to the dilute gas behaviour ($x \sim \tfrac{3}{2}$).

In Fig. 3 we show T_c/T_c^m versus $(\lambda^m(0)/\lambda(0))^2$ for several values of K/J

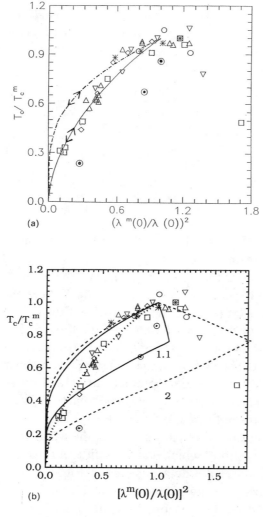

(a)

(b)

Fig. 3. T_c/T_c^m versus $[\lambda^m(0)/\lambda(0)]^2$ for the hard core charged Bose gas on a lattice. $\lambda^m(0)$ corresponds to the maximum transition temperature, T_c^m.

(a) $K/J=1$. Solid and dotted–dashed lines denote the RPA and MFA results for a SC lattice, respectively. $T_c(n)$ is taken from [6]. The arrows up (down) denote behaviour in the underdoped (overdoped) regime as $|n-1|$ decreases from 1 to 0.

(b) $K/J>1$. Solid and dashed lines denote the MFA results for $K/J=1.1$ and 2, respectively. The dotted line is the RPA for a SC lattice in the underdoped regime. The experimental data for T_c/T_c^m and the zero-temperature penetration depth $\lambda_\parallel^m(0)/\lambda_\parallel(0)$ are from [3] and [9]. ($\bigcirc$) $Tl_2Ba_2Ca_2Cu_3O_{10}$, $Tl_{0.5}Pb_{0.5}Sr_2Ca_2Cu_3O_9$, $Bi_{2-x}Pb_xSr_2Ca_2Cu_3O_{16}$; ($\Diamond$) $Y_{1-x}Pr_xBa_2Cu_3O_{6.97}$; ($\triangle$) $YBa_2Cu_3O_x$; ($\triangledown$) $La_{2-x}Sr_xCuO_4$; ($\bigstar$) $Bi_2Sr_2Ca_{1-x}Y_xCu_2O_{8+\delta}$($\square$) $LaMo_6Se_8$, $PbMo_6S_8$, $SnMo_6S_4Se_4$, $SnMo_6S_7Se$, $SnMo_6S_1Se_7$,$LaMo_6S_8$,$PbMo_6S_4Se_4$; and ($\odot$) $Tl_2Ba_2CuO_{6+\delta}$

Table 1. *Summary of results. The $T_0 < T < T_c$ results are only for low concentration and beyond the critical regime.*

Quantity	$0 < T < T_0$		$T_0 < T < T_c$	
	$d = 3$	$d = 2 + \varepsilon$	$d = 3$	$d = 2 + \varepsilon$
$\langle \rho^x(0) \rangle - \langle \rho^x(T) \rangle$	T^2	T	$T^{3/2}$	T
$E(T) - E(0)$	T^4	T^3	$T^{5/2}$	T^2
$C(T)$	T^3	T^2	$T^{3/2}$	T
$\rho_s - \rho_s(T)$	T^4	T^3	$T^{3/2}$	T
$1 - (\lambda(0)/\lambda(T))^2$				
$H_{c1}(0) - H_{c1}(T)$	T^4	T^3	$T^{3/2}$	T

evaluated in the MFA and RPA, where T_c^m corresponds to the maximum critical temperature in the $0 < \bar{n} < 1$ interval and $\lambda^m(0)$ corresponds to T_0^m. Notice the very substantial K/J-dependence of the lower branch of the wing (corresponding to the case of overdoping). The K/J-dependence of the upper branch is much weaker (even in the MFA) [10]. For $K/J \to 1$ the lower part of the wing sticks to the upper one (compare Fig. 3(a) and Fig. 3(b)). Thus, only systems with suppressed inter-site repulsion ($K/J \simeq 1$) are expected to maintain the universality of the behaviour in the underdoped and overdoped regimes (cf. Fig. 3(a)). For $K/J > 1$ the theory predicts an almost universal T_c versus λ^{-2} behaviour in the underdoped regime (upper parts of the wings in Fig. 3(b), $\lambda^m/\lambda < 1$) and deviations from universality for systems in the overdoped regime: the shape and extension of the lower part of the wing strongly depend on the inter-particle interaction, which can be different for various families of materials. The experimental data depicted in Fig. 3 compare well with these predictions. In particular, notice the data on $Tl_2Ba_2CuO_{6+\delta}$ extending for the first time into the strongly overdoped regime [9].

Finally, we have summarized in Table 1 the temperature-dependences of $\langle \rho^x \rangle$, internal energy E, specific heat C, superfluid mass density $\rho_s = m n_s$, as well as $\lambda(T)$ and the lower critical field H_{c1}, for a $d = 3$ lattice and for a quasi-2D layered structure ($d = 2 + \varepsilon$).

Acknowledgements

This work received financial support from the Committee of Scientific Research (KBN, Poland, project numbers 2 P3 02 057 04 and 200 11 91 01). R.M. would like to thank T. Schneider for discussions.

References

[1] R. Micnas, J. Ranninger and S. Robaszkiewicz, *Rev. Mod. Phys.* **62**, 113 ('1990).

[2] A. S. Alexandrov and N. F. Mott, *Supercond. Sci. Technol.* **6**, 215 (1993).

[3] T. Schneider and H. Keller, *Int. J. Mod. Phys.* B **8**, 487 (1994); T. Schneider *et al., Physica* C **216**, 432 (1993).

[4] A. L. Fetter, *Ann. Phys. (N.Y.)* **60**, 464 (1970).

[5] R. T. Whitlock and P. R. Zilsel, *Phys. Rev.* **131**, 2409 (1963).

[6] R. Micnas and S. Robaszkiewicz, *Phys. Rev.* B **45**, 9900 (1992).

[7] P. B. Weichmann, *Phys. Rev.* B **38**, 8739 (1989).

[8] J. F. Annett, N. Goldenfeld and S. Ren, *J. Low Temp. Phys.* **89**, 197 (1992); J. F. Annett, N. Goldenfeld and S. R. Ren, *Phys. Rev.* B **43**, 2778 (1991).

[9] Ch. Niedermayer *et al., Phys. Rev. Lett.* **71**, 1764 (1992).

[10] An improved analysis of the phase diagram of (1), using the Bethe–Peierls–Weiss approximation for CDW ordering and the RPA for superconducting order, provides a very weak K/J-dependence for the upper part of the T_c/T_c^m versus $(\lambda^m(0)/\lambda(0))^2$ plot, even for large K/J.

16

Electron–hole asymmetric polarons

J. E. HIRSCH

Department of Physics, University of California, San Diego, La Jolla, CA 92093-0319, USA

Abstract

In small-polaron models the hopping amplitude for a carrier from a site to a neighboring site is reduced due to 'dressing' by a background degree of freedom. Electron–hole symmetry is broken if this reduction is different for a carrier in a singly occupied site and one in a doubly occupied site. Assuming that the reduction is smaller in the latter case, the implication is that a gradual 'undressing' of the carriers takes place as the system is doped and the carrier concentration increases. A similar 'undressing' will occur at fixed (low) carrier concentration as the temperature is lowered, if the carriers pair below a critical temperature and as a result the 'local' carrier concentration increases (and the system becomes a superconductor). In both cases the 'undressing' can be seen in a transfer of spectral weight in the frequency-dependent conductivity from high frequencies (corresponding to non-diagonal transitions) to low frequencies (corresponding to diagonal transitions), as the carrier concentration increases or the temperature is lowered repetively. This experimental signature of electron–hole asymmetric polaronic superconductors as well as several others have been seen in high-temperature superconducting oxides. Other experimental signatures predicted by electron–hole-asymmetric polaron models remain to be tested.

1 The physics of high-T_c oxides

From the beginning of the high-T_c era there have been indications that small polarons may play an important role in the physics of these materials [1–10]. Among the workers that have not completely abandoned the Fermi liquid framework for this problem most would agree that the physics of the normal state may be described in terms of heavily dressed quasi-particles [11]. There is little agreement, however, concerning the physical origin of this quasi-particle

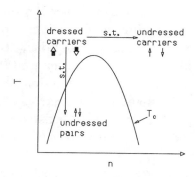

Fig. 1. A schematic depiction of the physics of electron-hole asymmetric polaron systems. Heavily dressed quasi-particles at low concentrations undress as the temperature is lowered or as the carrier concentration increases. s.t. denotes spectral weight transfer, from high to low frequencies (in the direction of the arrows). Below the curve labeled T_c the system is superconducting. The initial rise in T_c versus n is due to the increasing number of carriers.

dressing, with proposed explanations ranging from strong electron–phonon interactions [1–6] to electron–spin interactions (magnetic polarons) [7–9] to electron–electron interactions (electronic polarons) [10].

When a high-T_c oxide is doped increasingly with holes the normal state becomes less 'strange' [11], suggesting that the quasi-particles evolve from being heavily dressed to being lightly dressed, or undressed. In other words, a gradual undressing of carriers occurs as the carrier density increases. Direct evidence for this process is seen in optical absorption in the normal state: a transfer of spectral weight in the frequency-dependent conductivity occurs, from high-energy excitations (in the 1–3 eV range) to mid-infrared and lower energy (intra-band) excitations, as the carrier concentration increases [12, 13]. Furthermore, the band effective mass as inferred, for example, from resistivity is seen to decrease as the carrier concentration increases.

When a low-carrier-concentration system becomes superconducting the 'local' carrier concentration around a given carrier will also increase, due to pairing, particularly if the pair wavefunction has a short spatial extent (coherence length). Thus one may expect that an 'undressing' similar to what takes place in the normal state upon doping would occur at a fixed (low) carrier concentration as the temperature is lowered and the system becomes superconducting. Furthermore, assuming that it is this 'undressing' that drives the superconducting transition, one would conclude that superconductivity should disappear at high carrier concentration [14] because carriers are already undressed in the normal state. This physics is qualitatively depicted in Fig. 1.

The physics just described arises in electron–hole-asymmetric polaron models. Electron–hole asymmetry is a generic property of solids and in particular arises naturally in various models of small polarons, including the Holstein model [15] when slightly generalized. When electron–hole asymmetry is introduced in these models a new pairing mechanism arises and superconductivity can occur under conditions on the interaction parameters in the models that are vastly less restrictive than in their electron–hole-symmetric counterparts. Put another way, possibly the most effective way to suppress superconductivity in a generic model of small polarons is to make it electron–hole symmetric.

There are many characteristic features of such electron–hole-asymmetric polaron models. Several are seen in high-T_c oxide superconductors, others are still experimentally untested. In addition there are likely to be many other phenomena predicted by these models that have yet to be elucidated and quantified. Because electron–hole asymmetry can arise in any small-polaron model, whether electronic, magnetic or electron–phonon, we hope that workers on these problems will be interested in exploring its consequences further.

2 Electron–hole-asymmetric small polarons

A small-polaron model describes the propagation of a carrier (electron or hole) that carries with it a 'cloud' describing the deformation of a local background degree of freedom by the carrier [15–17]. This local degree of freedom may be an ionic coordinate, as in Holstein's model, an electronic degree of freedom (an electronic polaron) or a spin degree of freedom (a magnetic polaron). In the simplest realization there is one background degree of freedom associated with each lattice site. In a tight-binding description, the possible ground states of a lattice site with different numbers of carriers are then

$$|0\rangle|0\rangle \equiv |0\rangle\rangle, \tag{1a}$$

$$|\uparrow\rangle|1\rangle \equiv |\uparrow\rangle\rangle, \tag{1b}$$

$$|\downarrow\rangle|1\rangle \equiv |\downarrow\rangle\rangle, \tag{1c}$$

$$|\uparrow\downarrow\rangle|2\rangle \equiv |\uparrow\downarrow\rangle\rangle. \tag{1d}$$

On the left-hand side of Eq. (1) the first ket describes the electronic state and the second the background ground state: $|n\rangle$ denotes the background state when there are n carriers at the site. On the right-hand side we have defined a 'composite state' of electron and background state in its ground state and denote it with a double bracket.

Coherent motion of these small polarons occurs when the background degree of freedom remains in the ground state when the carrier hops from site to site (diagonal transitions) [15]. It is assumed that the background state instantaneously relaxes from the ground state corresponding to the old number of carriers to that corresponding to the new number of carriers as hopping occurs (the anti-adiabatic limit). The possible hopping amplitudes for a carrier are then, depending on the number of carriers at the two sites involved in the hopping process,

$$|\uparrow\rangle\rangle|0\rangle\rangle \xleftrightarrow{t_0} |0\rangle\rangle|\uparrow\rangle\rangle, \tag{2a}$$

$$|\uparrow\rangle\rangle|\downarrow\rangle\rangle \xleftrightarrow{t_1} |0\rangle\rangle|\uparrow\downarrow\rangle\rangle, \tag{2b}$$

$$|\uparrow\downarrow\rangle\rangle|\downarrow\rangle\rangle \xleftrightarrow{t_2} |\uparrow\rangle\rangle|\uparrow\downarrow\rangle\rangle, \tag{2c}$$

where the hopping amplitudes are obtained from the 'bare' hopping amplitude t by multiplying by appropriate overlap matrix elements of the background states:

$$t_0 = t\langle 0|1\rangle^2, \tag{3a}$$

$$t_1 = t\langle 0|1\rangle\langle 1|2\rangle, \tag{3b}$$

$$t_2 = t\langle 1|2\rangle^2. \tag{3c}$$

Our basic assumption is that

$$\langle 0|1\rangle \neq \langle 1|2\rangle \tag{4}$$

in general. Eq. (4) defines what we mean by electron–hole-asymmetric polarons. It implies that the effective model describing the dressed quasi-particles is not electron-hole symmetric even if the 'bare' model involving only the carriers is: since $t_0 \neq t_2$, the hopping amplitude for a single electron and a single hole will be different. We assume for definiteness that the carriers in Eq. (1) are electrons, and that

$$\langle 0|1\rangle > \langle 1|2\rangle \tag{5}$$

so that holes are heavier than electrons. If the converse of Eq. (5) were to hold, then the subsequent discussion would still apply, provided that the words 'electron' and 'hole' were interchanged.

The assumption Eq. (5) implies that the deformation of the background state induced by the first electron is smaller than that induced by the second electron

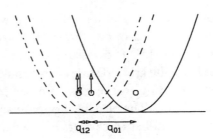

Fig. 2. A schematic depiction of potential wells for the electron–hole-asymmetric Holstein model. The full, dashed and dash–dotted lines show the oscillator potential when the site has zero, one and two holes respectively (indicated by vertical arrows on the circles centered at the respective equilibrium positions). q_{01} and q_{12} denote the difference in equilibrium positions when the site has zero and one hole, and one and two holes, respectively; $q_{01} > q_{12}$. In the electron–hole-symmetric model $q_{01} = q_{12}$.

on a site. Conversely, the deformation induced by the first hole on a site is larger than that induced by the second hole. In a Holstein model for small polarons, this occurs if the equilibrium position of the oscillator changes by a different amount in adding the first and the second carrier to a site. This is shown schematically in Fig. 2. Generalized Holstein models with this property will be discussed in Sect. 5.

The hopping amplitude for a single hole in such a system is t_2. For a system with hole density n per site the average hopping amplitude is

$$t(n) = t_2(1 - n) + t_1 n \equiv t_2 + n \Delta t, \tag{6a}$$

$$\Delta t = t_1 - t_2, \tag{6b}$$

since n is the probability that another hole of opposite spin will occupy one of the two sites involved in the hopping process. Eq. (6) is correct only to lowest order in n but is sufficient for our purpose since we will be interested only in systems with low carrier concentration.

The conductivity sum rule states that the integrated optical absorption for intra-band processes is given by

$$\int_0^{\omega_m} d\omega \, \sigma_1(\omega) = \frac{\pi e^2 n}{2m^*} \tag{7}$$

with σ_1 the frequency-dependent real part of the conductivity (per site), n the number of carriers per site and m^* the effective mass. ω_m is a high-frequency cut-off that allows only for intra-band processes. The effective mass in a tight-binding model is

$$m^* = \hbar^2/(2ta^2) \qquad (8)$$

with a the lattice spacing (we assume a simple hypercubic lattice). In our case, the hopping amplitude Eq. (6) varies with carrier concentration so that the integrated optical absorption

$$\int_0^{\omega_m} d\omega\, \sigma_1(\omega) = \frac{\pi e^2 a^2 t(n) n}{\hbar^2} \qquad (9)$$

increases faster than linearly with carrier concentration. The extra spectral weight in the intra-band spectrum arises from a decrease in the spectral weight of non-diagonal transitions: hopping processes in which the background degrees of freedom end up in excited states rather than the ground state.

Thus, an essential feature of electron–hole-asymmetric polarons is that, as a function of doping, a transfer of spectral weight in the frequency-dependent conductivity from non-diagonal to diagonal transitions occurs. Such behavior is seen in high-T_c oxides doped with holes [12, 13]: the intra-band spectral weight increases more rapidly than linearly with hole concentration, and the absorption in the visible range (1–3 eV) decreases correspondingly. This indicates that the carriers gradually undress as the hole concentration increases.

3 Superconductivity from electron–hole-asymmetric polarons

The hopping amplitude for a hole polaron of spin σ between sites i and j can be written as

$$t_{ij}^\sigma = t_2 + \Delta t\,(n_{i,\,-\sigma} + n_{j,\,-\sigma}) \qquad (10)$$

with $n_{i,\,-\sigma}$ the number of holes of opposite spin at site i and Δt given by Eq. (6b). Eq. (10) ignores the possibility of holes of opposite spin being present at both sites i and j, but this will be unimportant at low hole concentration. A Hamiltonian that describes the low-energy physics of these small polarons is then

$$H = -\sum_{\langle ij \rangle / \sigma} t_{ij}^\sigma (c_{i\sigma}^\dagger c_{j\sigma} + \text{h.c.}) + U\sum_i n_{i\uparrow} n_{i\downarrow} + V\sum_{\langle ij \rangle} n_i n_j \qquad (11)$$

with U and V repulsive on-site and nearest-neighbor interactions [18]. For a dilute system of hole carriers, this Hamiltonian can be accurately studied within the BCS framework [19], and leads to superconductivity if the parameters satisfy the condition

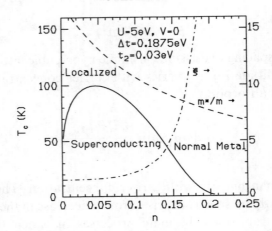

Fig. 3. Critical temperature versus hole concentration (solid line) for parameters given in the figure. Also shown is the coherence length (dash–dotted line) and the effective mass enhancement (dashed line) versus hole concentration.

$$\frac{\Delta t}{t_2} > \left[\left(1 + \frac{U}{2zt_2} \right) \left(1 + \frac{V}{2t_2} \right) \right]^{1/2} - 1 \qquad (12)$$

with z the number of nearest neighbors per site. If the single-hole hopping amplitude t_2 is very small compared with t_1 then we have $t_2 \ll \Delta t$ and Eq. (12) simplifies to

$$\Delta t > \left(\frac{UV}{4z} \right)^{1/2}. \qquad (13)$$

The physics leading to superconductivity is gain of kinetic energy: when polarons pair, their mobility increases and the energy is lowered. The mobility of a bound pair can be calculated explicitly and is found to be always larger than half the single-particle mobility [20]. In other words, a polaron pair is lighter than the sum of its individual components.

In Fig. 3 is shown the critical temperature resulting from this model for a typical set of parameters. It also shows the pair coherence length and the single-particle effective mass in the normal state (Eq. (6a) and Eq. (8)). The coherence length increases monotonically with hole doping, and the single-particle effective mass decreases. Thus a cross-over occurs from a strong-coupling regime at low hole doping to a weak-coupling regime at high hole doping, for which conventional behavior in all properties should be seen. In particular, the BCS gap ratio attains the weak-coupling value 3.53 for large hole doping and is larger for low hole doping [19].

The quasi-particle energy in this model is given by the usual form

$$E_k = [(\varepsilon_k - \mu)^2 + \Delta_k^2]^{1/2} \tag{14}$$

with μ the chemical potential and ε_k the single-particle band energy. The function Δ_k is only a function of band energy and of the form

$$\Delta_k = \Delta_m \left(\frac{-\varepsilon_k}{D/2} + c \right) \equiv \Delta(\varepsilon_k), \tag{15}$$

with

$$D = 2zt(n) \tag{16}$$

the bandwidth (that increases with hole doping) and Δ_m and c parameters obtained from solution of the BCS equations. Analytic forms for these as well as for the critical temperature can be found in the weak- and strong-coupling limits [19].

The quasi-particle energy can be rewritten as

$$E_k = [a^2(\varepsilon_k - \mu - v)^2 + \Delta_0^2]^{1/2}, \tag{17a}$$

$$a = \left[1 + \left(\frac{\Delta_m}{D/2} \right)^2 \right]^{1/2}, \tag{17b}$$

$$\Delta_0 = \Delta(\mu)/a, \tag{17c}$$

$$v = \frac{1}{a} \frac{\Delta_m}{D/2} \Delta_0. \tag{17d}$$

In Fig. 4 is shown the behavior of the quasi-particle energy, gap function and coherence factors

$$u_k^2 = \frac{1}{2} \left(1 + \frac{\varepsilon_k - \mu}{E_k} \right), \tag{18a}$$

$$v_k^2 = \frac{1}{2} \left(1 - \frac{\varepsilon_k - \mu}{E_k} \right), \tag{18b}$$

versus band energy. Because the minimum in the quasi-particle energy is shifted from μ to $\mu + v$, various characteristic features arise in tunneling and photo emission experiments [19]. In particular, tunneling characteristics exhibit an asymmetry of universal sign, and a thermoelectric voltage (also of universal sign) appears across a tunnel junction when the quasi-particles on both sides obey different distribution functions [21]. These effects should be experimentally observable.

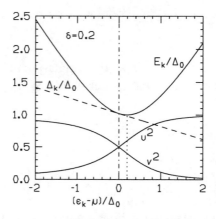

Fig. 4. Quasi-particle energy, gap function and coherence factors versus hole band energy for gap slope $\delta = \Delta_m/(D/2) = 0.2$. The vertical dot–dashed and dotted lines indicate the positions of the chemical potential μ and of the quasi-particle energy minimum $\mu + \nu$, respectively.

4 Bose decondensation versus pair unbinding

In electron–hole-symmetric bipolaron models of superconductivity, transition to the normal state generally occurs through Bose decondensation, and bound pairs still exist above T_c [22]. More generally, it is assumed that such a scenario always occurs when the pair coherence length is short [23]. However, this is not so in the models considered here. Qualitatively, the coherence length is given approximately by

$$\xi \simeq t_2/\varepsilon_b \qquad (19)$$

with ε_b the pair binding energy and t_2 the single-particle hopping amplitude. The Bose condensation temperature is in the range in which center of mass excitations of the pairs start to play a significant role and is proportional to the pair hopping amplitude

$$T_c^{\mathrm{Bose}} \propto t_p \qquad (20)$$

while the BCS pair unbinding transition is proportional to the pair binding energy

$$T_c^{\mathrm{BCS}} \propto \varepsilon_b. \qquad (21)$$

Thus, even in the regime of short coherence length ($\xi < 1$), one may have

$$T_c^{\mathrm{BCS}} \ll T_c^{\mathrm{Bose}} \qquad (22)$$

provided that the pair hopping amplitude is larger than the single-particle amplitude

$$t_p \gg t_2. \tag{23}$$

While Eq. (23) cannot be valid for electron–hole-symmetric polarons (in fact, the opposite occurs) it is an essential feature of the models considered here. In particular, in the strong-coupling regime ($t_2 \ll \Delta t$) the pair binding energy and pair hopping amplitude are given by [19]

$$\varepsilon_b = 2zt_p - V, \tag{24a}$$

$$t_p = \frac{2\Delta t^2}{U - V} \tag{24b}$$

so that the pair binding energy can be arbitrarily small for a finite value of the pair hopping amplitude.

In other words, in the models discussed here in the regime of low hole concentration the pairs are of small spatial extent and yet highly mobile and weakly bound. When the temperature is raised, they will dissociate well before the temperature range in which center of mass excitations of the pairs would have started to play a role (cf. Eq. (22)). This scenario is rather different from what occurs in electron–hole-symmetric polaron models and should be evident experimentally, e.g. by a sharp break in the Knight shift as the temperature is raised across the critical temperature. Only at extremely low hole concentration does a cross-over to a Bose-condensation description occur in the models considered here [24].

5 Models for electron–hole-asymmetric polarons

Because electron–hole symmetry is not an exact symmetry of solids, any microscopic model for description of physical reality should, in principle, contain electron–hole symmetry-breaking terms, which should only be discarded after establishing that they are irrelevant for the physics of interest. Here we discuss how electron–hole symmetry-breaking terms arise in small-polaron models in a natural way.

5.1 Generalized Holstein models

The site Hamiltonian for a conventional Holstein model is given by [15]

$$H_i = \frac{p_i^2}{2M} + \frac{1}{2} Kq_i^2 + \alpha q_i n_i + U n_{i\uparrow} n_{i\downarrow} \tag{25}$$

with (p_i, q_i) canonical coordinates of a vibrational degree of freedom of mass M and force constant K. This model is electron–hole-symmetric; however, various modifications of it to include physical effects that occur in nature will turn it electron–hole-asymmetric [25], namely the following.

1. Allow for a variation of the electron–phonon coupling constant α with site occupation [26]. A possible parametrization is

$$\alpha \rightarrow \alpha + \frac{\alpha'}{2}(n_i - 1). \tag{26}$$

This is equivalent to allowing for dependence of the on-site repulsion U on the phonon coordinate q_i.

2. Allow for a variation of the stiffness K with site occupation

$$K \rightarrow K(n_i). \tag{27}$$

3. Allow for a variation in ionic mass with site occupation, $M \rightarrow M(n_i)$. (This effect is likely to be small.) Both this as well as effect 2 imply variation of the vibration frequency $\omega_0 = (K/M)^{1/2}$ with site occupation.

4. Anharmonic effects [27, 28]: the potential energy is modified to

$$\tfrac{1}{2}Kq^2 \rightarrow \tfrac{1}{2}Kq^2 + \beta q^4 \tag{28}$$

$(\beta > 0)$.

The overlap matrix element of the oscillator ground state wave function with n and n' carriers at the site is given by

$$\langle n|n'\rangle = \left(\frac{2(a_n a_{n'})^{1/2}}{a_n + a_{n'}}\right)^{1/2} \exp\left(-\frac{a_n a_{n'}}{2(a_n + a_{n'})}(q_n - q_{n'})^2\right) \tag{29}$$

with

$$q_n = -\frac{\alpha(n)}{K(n)}n + \frac{4\beta\alpha(n)^3}{K(n)^4}n^3, \tag{30a}$$

$$a_n = \frac{(K(n)M(n))^{1/2}}{\hbar}. \tag{30b}$$

If any of the above-listed situations occur,

$$\langle 0|1\rangle \neq \langle 1|2\rangle \tag{31}$$

and electron–hole asymmetry results. Consideration of the physical nature of each of these effects leads to the conclusion [25] that the *sign* of the symmetry breaking is always such that, in the electron representation, $\langle 0|1\rangle > \langle 1|2\rangle$, as assumed in the previous sections. Furthermore, the *magnitude* of the symmetry-breaking effects needed to satisfy the condition for superconductivity Eq. (12) is estimated to be rather modest: of order 10% for cases 1 or 2, and less than 1% for case 4, when these effects are considered separately. (In reality, they are likely to act together.) The magnitude of polaronic band narrowing in the regime in which the criterion for superconductivity is satisfied is much smaller than in the conventional electron–hole-symmetric Holstein model [22].

For the particular case 1, the usual Lang–Firsov canonical transformation [29] is easily generalized. New boson creation and annihilation operators and new fermion operators are introduced through the relations

$$\bar{a}_i = a_i + g\left(n_i + \frac{\alpha'}{\alpha}\, n_{i\uparrow}n_{i\downarrow}\right), \tag{32a}$$

$$\bar{c}_{i\sigma} = X_{i\sigma}c_{i\sigma} \tag{32b}$$

with

$$X_{i\sigma} = \exp\left[g\left(1 + \frac{\alpha'}{\alpha}\, n_{i,-\sigma}\right)(a_i^\dagger - a_i)\right], \tag{33a}$$

$$g = \left(\frac{\varepsilon_b}{\hbar\omega_0}\right)^{1/2}, \tag{33b}$$

$$\varepsilon_b = \alpha^2/(2K). \tag{33c}$$

Here, $\bar{a}_i$ is the new phonon destruction operator, $\bar{c}_{i\sigma}$ the new fermion destruction operator and ε_b the polaron binding energy. The site Hamiltonian is diagonal in terms of the new operators:

$$H_i = \hbar\omega_0(\bar{a}_i^\dagger \bar{a}_i + \tfrac{1}{2}) + U_{\text{eff}}\bar{n}_{i\uparrow}\bar{n}_{i\downarrow} - \varepsilon_b\bar{n}_i \tag{34}$$

with [30]

$$U_{\text{eff}} = U - \frac{1}{K}\left(\alpha^2 + 2\alpha\alpha' + \frac{\alpha'^2}{2}\right) \tag{35}$$

the effective on-site interaction. The inter-site hopping operator

$$H_t = -t\sum_{\langle ij\rangle/\sigma}(c_{i\sigma}^\dagger c_{j\sigma} + \text{H.c.}) \tag{36}$$

is rewritten in terms of the new operators as

$$H_t = -t\sum_{\langle ij\rangle/\sigma}(\bar{X}_{i\sigma}^\dagger \bar{X}_{j\sigma}\bar{c}_{i\sigma}^\dagger \bar{c}_{j\sigma} + \text{H.c.}), \tag{37a}$$

$$\bar{X}_{i\sigma} = \exp\left[g\left(1 + \frac{\alpha'}{\alpha}\,\bar{n}_{i,-\sigma}\right)(\bar{a}_i - \bar{a}_i^\dagger)\right], \tag{37b}$$

and the Hamiltonian

$$H = \sum_i H_i + H_t \tag{38}$$

describes the full dynamics of the problem in terms of the new 'polaron operators' $\bar{c}_{i\sigma}$ and the new phonon operators that refer to vibrations around the equilibrium position for given electronic occupation of the site

$$\bar{q}_i = q_i + \frac{1}{K}(\alpha n_i + \alpha' n_{i\uparrow} n_{i\downarrow}). \tag{39}$$

The situation here is somewhat more complicated than the usual case because the 'dressing operators' Eq. (37b) depend on fermion in addition to boson operators. The coherent hopping amplitude resulting from zero-phonon processes is obtained by taking the expectation value of Eq. (37b) in the zero-phonon states, yielding the matrix elements Eq. (29), which for this case are

$$\langle \bar{X}_{i\sigma}\rangle_0(\bar{n}_{i,-\sigma}=0) = \langle 0|1\rangle = \exp\left(-\frac{\alpha^2}{4Kh\omega_0}\right), \tag{40a}$$

$$\langle \bar{X}_{i\sigma}\rangle_0(\bar{n}_{i,-\sigma}=1) = \langle 1|2\rangle = \exp\left(-\frac{(\alpha+\alpha')^2}{4Kh\omega_0}\right). \tag{40b}$$

In the zero-phonon subspace the Hamiltonian Eq. (38) takes the form Eq. (11), without the V term. Nearest-neighbor interaction will arise from direct Coulomb interaction as well as from virtual hopping processes [31].

5.2 Electronic tight-binding models

In the derivation of a tight-binding model from first principles, a variation of the hopping amplitude with site occupation arises from the following two effects.

1. An off-diagonal matrix element of the Coulomb interaction between Wannier orbitals at nearest-neighbor sites [32, 33].
2. The 'orbital expansion' that occurs when a second electron is added to an orbital that already has an electron, arising due to the strong on-site Coulomb repulsion [34, 10, 35]. In other words, higher energy atomic orbitals become partially occupied. Because of this the hopping amplitudes to or from such a doubly occupied site will be different from those of a singly occupied site.

The second effect in particular is a polaronic effect: when an electron leaves a doubly occupied site the second electron relaxes to a new orbital configuration, and an overlap matrix element modulates the hopping amplitude of the first electron. A simple way to describe this physics is by using a tight-binding model with two orbitals per site [36].

First principles calculations bear out these qualitative considerations and yield estimates of the variation of hopping amplitude with occupation [35, 36]. It is found that, whenever the hopping amplitudes show substantial variation with site occupation, it is the hole hopping amplitude that is smaller.

5.3 The spin-fermion Hamiltonian

A Hamiltonian describing the interaction of electrons with local spin $\frac{1}{2}$ degrees of freedom:

$$H_i = (Vn_i - \omega_0)\sigma_z + \Delta\sigma_x^i + Un_{i\uparrow}n_{i\downarrow} \tag{41}$$

has been used in a variety of contexts: to describe excitonic superconductivity [37], coupling to anharmonic apex oxygen motion in high-T_c oxides [27], and conduction of holes through anions [10] to list just a few. This Hamiltonian gives rise to small polarons like those of the Holstein model, and, except for the special case $V = \omega_0$, is electron–hole-asymmetric. For the cases in which this Hamiltonian has been used, there is no physical reason that would determine that $V = \omega_0$.

To illustrate the enlargement of the region in parameter space in which superconductivity can occur in the presence of electron–hole asymmetry, we plot in fig. 5 for the Holstein model the effective mass enhancement

$$\frac{m^*}{m} = \frac{1}{\langle 1|2\rangle^2} = \exp\left(-\frac{\alpha^2}{2K\hbar\omega_0}\right) \tag{42}$$

versus effective interaction

$$U_{\text{eff}} = U - \frac{\alpha^2}{K} \tag{43}$$

for various values of the oscillator frequency ω_0. For the electron–hole-symmetric case the condition for superconductivity is $U_{\text{eff}} < 0$; as the frequency decreases the effective mass enhancement that exists when the condition for superconductivity is met (intersection of straight lines with the left-hand vertical axis) becomes very large. For the electron–hole-asymmetric case, instead the condition for superconductivity Eq. (12) becomes (for $V = 0$)

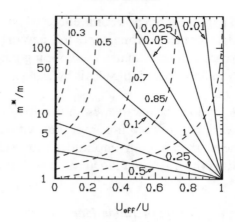

Fig. 5. The parameter range in which superconductivity can occur in the Holstein model. The straight solid lines give the values of the effective mass enhancement as a function of the effective on-site repulsion (Eq. (43)) for various values of the oscillator frequency divided by the on-site repulsion: $\hbar\omega_0/U = 05$, 0.25, 0.1, 0.05, 0.025 and 0.01 (numbers next to the solid lines). In the electron–hole-symmetric case the superconducting boundary is $U_{eff}/U = 0$ (thick vertical line); superconductivity occurs to the left of that line. The dashed lines give the criteria for superconductivity in the asymmetric case, for various values of the overlap $\langle 0|1 \rangle$ (numbers next to dashed lines); superconductivity occurs in the region above a given dashed line for that value of $\langle 0|1 \rangle$.

$$\frac{m^*}{m} = \left[\langle 0|1 \rangle^2 - \frac{U}{2zt}\frac{U_{eff}}{U} \right]^{-1}. \tag{44}$$

The dashed lines show the condition Eq. (44) for various values of the overlap $\langle 0|1 \rangle$ between the ground states of the oscillator with one and two holes, assuming for definiteness that $U/(2zt) = 1$. Above a given dashed line, superconductivity can occur for that value of $\langle 0|1 \rangle$. It can be seen that, as $\langle 0|1 \rangle$ increases, the region in parameter space in which superconductivity occurs is greatly enlarged, and in particular the mass enhancement required for given $\hbar\omega_0/U$ is much smaller than in the electron–hole-symmetric case. The optimal situation occurs for $\langle 0|1 \rangle = 1$, for which adding a second hole does not change the equilibrium position of the oscillator with respect to that when it had one hole. For example, for that particular case, if the oscillator frequency were $\hbar\omega_0/U = 0.05$ then superconductivity would occur for $m^*/m \simeq 6$, while in the electron–hole-symmetric case it occurs only for $m^*/m \simeq 22\,000$.

In summary, a variety of models of small polarons are naturally extended to describe electron–hole-asymmetric polarons. These and other models can be used to describe polaronic effects arising from various physical processes such as electron–phonon, electron–electron or electron–spin interactions (even the

Holstein model can be used to describe electronic effects if the excitation energy ω_0 is an electronic energy scale). Superconductivity is much easier to achieve if electron–hole asymmetry is present. The essential point is that the electron–hole-symmetric special cases of these models that are commonly studied represent approximations that are questionable in the light of the fact that electron–hole symmetry-breaking terms do occur in nature and significantly modify the physics of these models.

6 Coupling of low- and high-energy physics

The conductivity sum rule for a system governed by the tight-binding Hamiltonian Eq. (11) yields [38]

$$\int_0^{\omega_m} d\omega\, \sigma_1^{\delta\delta}(\omega) = \frac{\pi e^2 a_\delta^2}{2\hbar^2} \langle -T_\delta \rangle. \tag{45}$$

Here, δ denotes a principal direction in the crystal lattice (which is assumed to be hypercubic). a_δ is the lattice spacing, and ω_m is a high-frequency cut-off. T_δ is the kinetic energy in direction δ:

$$T_\delta = -\sum_{i,\sigma} t_{i,i+\delta}^\sigma (c_{i+\delta,\sigma}^\dagger c_{i\sigma} + \text{H.c.}) \tag{46}$$

with the hopping amplitude given by Eq. (10). The expectation value of T_δ is

$$\langle T_\delta \rangle = -t(n)\sum_{i,\sigma} \langle c_{i+\delta,\sigma}^\dagger c_{i\sigma} + \text{H.c.} \rangle$$
$$-2\Delta t \sum_{i,\sigma} [\langle c_{i\sigma}^\dagger c_{i,-\sigma}^\dagger \rangle \langle c_{i,-\sigma} c_{i+\delta,\sigma} \rangle + \text{H.c.}] \equiv \langle T_\delta^t \rangle + \langle T_\delta^{\Delta t} \rangle. \tag{47}$$

As the system becomes superconducting, the anomalous expectation values in the second term of Eq. (47) become non-zero and the kinetic energy decreases (the first term in Eq. (47) is essentially unchanged when T is lowered below T_c). Thus, the integrated optical absorption Eq. (45) increases. Quantitative examples are given in [39]. This extra spectral weight goes into the δ-function response at zero frequency that determines the London penetration depth, together with the spectral weight coming from the decrease in optical absorption at frequencies below the superconducting gap [40].

What is the origin of this extra spectral weight? It does not come from any energy scale associated with the Hamiltonian Eq. (11). However, we should remember that Eq. (11) is a low-energy effective Hamiltonian that only accounts for the diagonal transitions in the background degrees of freedom. The conductivity sum rule reminds us that there is also the possibility of high-

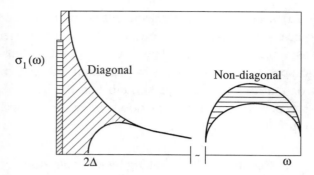

Fig. 6. A schematic depiction of spectral weight redistribution in the frequency-dependent conductivity as the system goes superconducting. Both the 'missing areas' arising from intra-band (diagonal) transitions (diagonally hatched) and from non-diagonal transitions (horizontally hatched) contribute to the δ-function at zero frequency that determines the London penetration depth.

energy non-diagonal transitions in the background states, and a decrease in the weight of those transitions should account for the extra spectral weight that appears at low energies. The situation is shown schematically in Fig. 6. The extra spectral weight is given by

$$\delta A_{\mathrm{h}} = \frac{\pi e^2 a_\delta^2}{2\hbar^2} \langle -T_\delta^{\Delta t} \rangle. \tag{48}$$

The frequency-dependent conductivity can be written in a spectral representation (at zero temperature) as

$$\sigma_1^{\delta\delta}(\omega) = \pi \sum_M \frac{|\langle 0|J_\delta|M\rangle|^2}{E_M - E_0} \delta\left(\omega - \frac{E_M - E_0}{\hbar}\right), \tag{49}$$

where E_M is the energy of the mth excited state and J_δ is the current operator in the tight-binding model. We denote by $|n_i^m\rangle$ the mth excited state of the background degree of freedom at site i when there are n_i electrons at the site. For an optical transition in which an electron hops from site i to site j the matrix elements involving the background degrees of freedom are

$$R_{ij}^{mm'} = \langle n_i | n_i'^m \rangle \langle n_j | n_j'^{m'} \rangle, \tag{50}$$

where n_i and n_i' are site occupation numbers before and after the transition, and we assume that initially the background degrees of freedom are in their ground state ($|n_i\rangle = |n_i^0\rangle$). The optical absorption will be proportional to the square of these matrix elements. The 'diagonal' transition that contributes to the intra-band spectral weight is R_{ij}^{00}. By completeness we have

$$\sum_{mm'} |R_{ij}^{mm'}|^2 = 1. \tag{51}$$

Now, for a single hole hopping between two sites

$$R_{ij}^{mm'}(0) = \langle 1|2^m\rangle\langle 2|1^{m'}\rangle \tag{52}$$

while for a hole hopping when there is another hole present at one of the two sites

$$R_{ij}^{mm'}(1) = \langle 1|2^m\rangle\langle 1|0^{m'}\rangle \tag{53}$$

By our assumption Eq. (5) we have then

$$R_{ij}^{00}(0) < R_{ij}^{00}(1) \tag{54}$$

so that by completeness

$$\sum_{(m,m')\neq(0,0)} |R_{ij}^{mm'}(0)|^2 > \sum_{(m,m')\neq(0,0)} |R_{ij}^{mm'}(1)|^2. \tag{55}$$

Eq. (55) implies that the contribution from hopping of one hole to the high-frequency optical absorption (non-diagonal transitions) is larger when it is isolated than when it is in the presence of another hole, while the converse occurs for low-frequency intra-band absorption (diagonal transition) (Eq. (54)). The physics involved is simply the Franck–Condon principle that explains the distribution in intensity in absorption bands in molecules [41], and is depicted schematically in Fig. 7. Thus, as holes pair or as the hole concentration increases through doping, a transfer of spectral weight from high frequencies to low frequencies occurs.

In particular for the generalized Holstein models the matrix elements are given by

$$\langle n|n'^m\rangle \equiv G_{0m}(g_{nn'}) = \frac{(2g_{nn'})^{m/2}e^{-g_{nn'}}}{(m!)^{1/2}}, \tag{56a}$$

$$g_{nn'} = \ln\left(\frac{1}{\langle n|n'\rangle}\right). \tag{56b}$$

For a transition in which the final states at sites i and j are m and m', we can sum over all possible contributions with $m+m'=M$ so that the total energy of the background final states is $\hbar\omega_0 M$:

$$\sum_{m+m'=M} |R_{ij}^{mm'}|^2 = \sum_{m+m'=M} G_{0m}^2(g_{n_in_i})G_{0m'}^2(g_{n_in_i}) = G_{0M}^2(g_{n_in_i}+g_{n_in_j}). \tag{57}$$

Thus, for a single hole hopping between sites i and j, the contribution to the

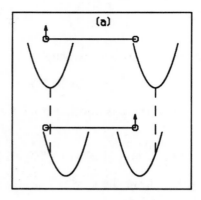

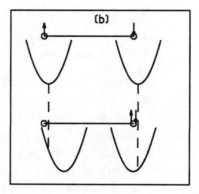

Fig. 7. Initial and final states for a hole undergoing an optical transition to a neighboring site in the absence (a) and in the presence (b) of another hole. The relative positions of the potential curves for the background degrees of freedom are shown. Vertical transitions in the background degree of freedom, that carry large spectral weight, are shown as dashed lines; note that they occur at a lower total energy for the case of paired holes than for isolated holes.

optical absorption at frequency $\hbar\omega_0 M$ is $G_{0M}^2(2g_{12})$, when another hole is present it is $G_{0M}^2(g_{01} + g_{12})$ and when two other holes are present it is $G_{0M}^2(2g_{01})$. In Fig. 8 are shown these probabilities for one case. The effect of both doping and pairing is to decrease the number of transitions in which a single hole is present at the two sites involved relative to transitions with two holes present. Thus the total spectral weight gets an increasing contribution from Fig. 8(b) relative to that from Fig. 8(a). To calculate the relative contributions of the various hole configurations in the superconducting state, the probability of two holes being at the same or nearest-neighbor sites can be obtained from the superconducting pair wave function [19].

Note also that, in addition to the $M = 0$ peak (the intra-band contribution),

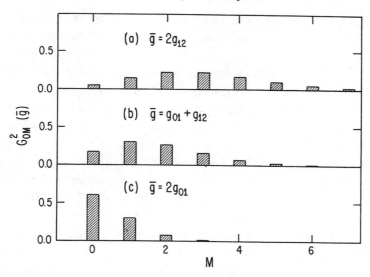

Fig. 8. Matrix elements for transition probabilities in a generalized Holstein model, $G^2_{0M}(\bar{g})$, versus M. $\hbar\omega_0 M$ is the energy of the final state, with ω_0 the oscillator frequency. Cases (a), (b) and (c) correspond to an isolated hole, and a hole in the presence of one and two other holes respectively. Note the shift in spectral weight towards lower frequencies as the hole concentration increases, and in particular the increase in the ground-state to ground-state (diagonal) transition probability at $M = 0$. $g_{nn'}$ is given by Eq. (56b), and for this example $\langle 0|1 \rangle = 0.88$ and $\langle 1|2 \rangle = 0.47$.

other low-frequency peaks also grow in going from (a) to (b) to (c). This may relate to the growth of the mid-infrared feature in high-T_c oxides that is seen under doping [12, 13]. While the simple Holstein model should not be expected to reproduce the details of the optical spectra observed in high-T_c oxides, we believe that it may capture the qualitative features, with the excitation energy ω_0 being of electronic origin and of magnitude a fraction of an electron-volt.

We have also performed explicit calculations of the various contributions to the optical absorption in the spin-fermion model Eq. (41) [42]. While the shift in relative weight of the various contributions upon pairing is clearly illustrated by these results, it was not possible to establish quantitatively the equivalence of the extra spectral weight at low frequencies as given by Eq. (48) and the part of the optical absorption arising from non-diagonal transitions in Eq. (49). The reason is that that model, and tight-binding models in general (including the Holstein model), do not conserve total oscillator strength.

7 Discussion

Within the renormalization group theory of critical phenomena [43] a perturbation is 'relevant' if it changes qualitatively the physics of the system, even if the magnitude of the perturbation is small. We have tried to argue here that electron–hole symmetry-breaking perturbations are 'relevant', in that sense, for the physics of polaronic systems. Because there is no underlying symmetry reason that would exclude such perturbations in real systems, it is necessary to consider them, in particular with respect to the physics of superconductivity.

Superconductivity arises in electron–hole-asymmetric polaronic systems due to the lowering of kinetic energy that occurs when carriers pair. The resulting superconductor exhibits a number of characteristic features associated with its low-energy physics: an isotropic s-wave state, strong dependence of the critical temperature on carrier concentration, an energy-dependent BCS gap function, branch imbalance, positive pressure-dependence of T_c, sensitivity to non-magnetic disorder, tunneling asymmetry, cross-over from strong to weak coupling as the carrier concentration increases, etc. The transition from the superconducting to the normal state occurs through pair unbinding rather than Bose decondensation even in the regime in which the coherence length is short. Many of these characteristic features have been seen in high-T_c oxide superconductors.

The most characteristic feature of these systems, however, is the remarkable coupling between low- and high-energy physics. When the superconducting gap opens up, which is a low-energy phenomenon, changes in optical absorption are predicted to occur at high frequencies, unrelated to the energy scale of the superconducting gap. This signature of the underlying physics has been seen experimentally [44, 45]. The detailed dependences on temperature and carrier concentration of the effect that have been predicted [39] have, however, not yet been experimentally tested. Furthermore, the related experimental prediction [20] that the observed London penetration depth should be shorter than expected from the normal state effective mass and from the low-frequency missing area in the conductivity remains open to experimental verification.

The model does not yield a precise prediction from first principles of the high-frequency range in which changes in conductivity should occur, because the microscopic physics of the underlying polaronic processes has not yet been entirely clarified. However, the phenomenology predicted by the model is unambiguous: the frequency range in which changes in optical conductivity should occur when the system goes superconducting is the same range in which changes in normal state conductivity upon doping should occur. Experiments then indicate that the main range in high-T_c oxides in which a decrease in absorption should be seen upon pairing is in the visible (1–3 eV). Additionally, shift of spectral weight to lower frequencies should also be seen from the mid- and near-infrared region. For a given frequency in that range, however, this could result in either an increase or a decrease in intensity depending on the doping level.

Experiments on high-T_c oxides also show a variation of optical absorption at high frequencies with temperature in the normal state [44, 45]. Within the class of models considered here we have found that variation of the number of carriers with temperature in the normal state is likely to occur in the high-T_c oxide structures [46], which may account for this as well as other anomalous normal state properties. Other observations in high-T_c oxides such as variations in the gap magnitude may be accounted for by generalizing the models discussed here to describe more than one band [47] and by existence of disorder [48].

In summary, superconductivity in electron–hole-asymmetric polaron systems is driven by 'undressing' of the heavily dressed polaronic carriers that occurs upon pairing. It parallels the same phenomenon occuring in the normal state as the carrier concentration is increased. Dilute carriers become increasingly 'free' as the local carrier concentration increases, and are able to pay the price of extra Coulomb repulsion that arises when they pair in order to achieve this 'freer' state. However, if they are sufficiently 'free' in the unpaired state (which is the normal state at higher carrier concentration) then the incentive to develop pairing correlations disappears. The same basic physics would apply to any Fermi liquid system in which the quasi-particle dressing is a strong function of the local carrier concentration, even if the carriers are not strictly speaking 'polaronic' in the conventional sense. If this is indeed the essential physics underlying the phenomenon of superconductivity in high-T_c oxides then the possibility that it plays an essential role in other superconductors as well should not be excluded.

Acknowledgements

The author is grateful to UCSD for support and to F. Marsiglio and S. Tang for collaboration.

References

[1] J. G. Bednorz and K. A. Müller, *Z. Phys.* B **64**, 189 (1986).
[2] A. S. Alexandrov, *Physica* C **158**, 337 (1989).
[3] B. K. Chakraverty, D. Feinberg, Z. Hang and M. Avignon, *Solid State Commun.* **64**, 1147 (1987).
[4] D. J. Scalapino, R. T. Scalettar and N. E. Bickers, in *Novel Superconductivity*, ed. S. A. Wolf and V. Z. Kresin (Plenum, New York, 1987) p. 475.
[5] Y. Suzuki, P. Pincus and A. J. Heeger, *Phys. Rev.* B **44**, 7127 (1991).
[6] J. Ranninger, *Z. Phys.* B **84**, 167 (1991).
[7] H. Kamimura, S. Matsuno and L. Saito, *Solid State Commun.* **67**, 363 (1988).
[8] J. Appel, P. Hertel and F. Richter, *Physica* C **162**, 1499 (1989).
[9] N. F. Mott, *Adv. Phys.* **39**, 55 (1990).

[10] J. E. Hirsch and S. Tang, *Phys. Rev.* B **40**, 2179 (1989); *Solid State Commun.*
     69, 987 (1989); J. E. Hirsch, *Phys. Lett.* A **134**, 451 (1989).
[11] K. Levin *et al.*, *Physica* C **175**, 449 (1991) and reference therein.
[12] S. Uchida *et al.*, *Phys. Rev.* B **43**, 7942 (1991).
[13] S. L. Cooper *et al.*, *Phys. Rev.* B **41**, 11605 (1990); *Phys. Rev.* B **45**, 2549 (1992);
     Phys. Rev. B **47**, 8233 (1993).
[14] J. B. Torrance *et al.*, *Phys. Rev. Lett.* **61**, 1127 (1988).
[15] T. Holstein, *Ann. Phys. (N.Y.)* **8**, 325 (1959).
[16] G. D. Mahan, *Many Particle Physics* (Plenum, New York, 1981) Ch. 6.
[17] H. G. Reik, *Polarons in Ionic Crystals and Polar Semiconductors* (North
     Holland, Amsterdam, 1972), p. 686.
[18] The interaction Δt can in fact be substantially larger than given by Eq. (6b),
     due to contributions from virtual processes to high-energy states: J. E.
     Hirsch and F. Marsiglio, *Phys. Rev.* B **41**, 2049 (1990).
[19] J. E. Hirsch and F. Marsiglio, *Phys. Rev.* B **39**, 11515 (1989); *Physica* C **162–
     164**, 591 (1989); F. Marsiglio and J. E. Hirsch, *Phys. Rev.* B **41**, 6435
     (1990); *Physica* C **165**, 71 (1990); *Physica* C **171**, 554 (1990); J. E. Hirsch,
     Physica C **158**, 326 (1989); *Physica* C **161**, 185 (1989); *Physica* C **182**, 277
     (1991).
[20] J. E. Hirsch and F. Marsiglio, *Phys. Rev.* B **45**, 4807 (1992).
[21] J. E. Hirsch, *Phys. Rev. Lett.* **72**, 558 (1994).
[22] A. S. Alexandrov and J. Ranninger, *Phys. Rev.* B **23**, 1796 (1981); A. S.
     Alexandrov, J. Ranninger and S. Robaszkiewcz, *Phys. Rev.* B **33**, 4526
     (1986).
[23] R. Micnas, J. Ranninger and S. Robaszkiewcz, *Rev. Mod. Phys.* **62**, 1131
     (1990).
[24] J. E. Hirsch, *Physica* C **179**, 317 (1991).
[25] J. E. Hirsch, *Phys. Rev.* B **47**, 5251 (1993); *Phys. Lett.* A **168**, 305 (1992).
[26] P. Pincus, *Solid State Commun.* **11**, 51 (1972).
[27] K. A. Müller, *Z. Phys.* B **80**, 193 (1990).
[28] A. Bussmann-Holder and A. R. Bishop, *Phys. Rev.* B **44**, 2853 (1991).
[29] I. G. Lang and Y. A. Firsov, *Sov. Phys. JETP* **16**, 1301 (1963).
[30] Note that the expression for U_{eff} given in the second paper in [25] is erroneous.
[31] G. Beni, P. Pincus and J. Kanamori, *Phys. Rev.* B **10**, 1896 (1974).
[32] J. Hubbard, *Proc. Roy. Soc. London* A **276**, 238 (1963).
[33] S. Kivelson, W. P. Su, J. R. Schrieffer and A. Heeger, *Phys. Rev. Lett.* **58**, 1899
     (1987).
[34] A. Zawadowski, *Phys. Scripta* T **27**, 66 (1989); *Phys. Rev.* B **39**, 4682 (1989).
[35] J. E. Hirsch, *Phys. Rev.* B **48**, 3327 (1993); *Phys. Rev.* **48**, 9815 (1993); *Chem.
     Phys. Lett.* **171**, 161 (1990).
[36] J. E. Hirsch, *Phys. Rev.* B **43**, 11400 (1991); *Phys. Rev.* B **48**, 3340 (1993).
[37] W. A. Little, *Phys. Rev.* A **134**, 1416 (1964); R. Bari, *Phys. Rev. Lett.* **30**, 790
     (1973).
[38] P. F. Maldague, *Phys. Rev.* B **16**, 2437 (1977).
[39] J. E. Hirsch, *Physica* C **199**, 305 (1992).
[40] R. A. Ferrell and R. E. Glover, *Phys. Rev.* **109**, 1398 (1958); M. Tinkham and
     R. A. Ferrell, *Phys. Rev. Lett.* **2**, 331 (1959).
[41] G. Herzberg, *Molecular Spectra and Molecular Structure* (Van Nostrand,
     Princeton, 1950) p. 194.
[42] J. E. Hirsch, *Physica* C **201**, 347 (1992).

[43] S. K. Ma, *Modern Theory of Critical Phenomena* (Benjamin, Reading, 1976).
[44] H. L. Dewing and E. K. H. Salje, *Supercond. Sci. Technol.* **5**, 50 (1992).
[45] I. Fugol, V. Samovarov, A. Ratner and V. Zhuravlev, *Solid State Commun.* **86**, 385 (1993); *Physica C* **216**, 391 (1993).
[46] J. E. Hirsch and F. Marsiglio, *Physica C* **195**, 355 (1992).
[47] J. E. Hirsch and F. Marsiglio, *Phys. Rev. B* **43**, 424 (1991).
[48] F. Marsiglio, *Phys. Rev.* **45**, 956 (1992); K. I. Wysokinski, *Physica C* **198**, 87 (1992); J. E. Hirsch, *Physica C* **194**, 119 (1992).

17

On the nature of the superconducting state in high-T_c cuprates

T. SCHNEIDER

IBM Research Division, Zürich Research Laboratory, CH-8803 Rüschlikon, Switzerland

Abstract

We provide strong evidence that cuprate superconductors and uncharged superfluids like ^{4}He share the universal 3D x–y properties in the fluctuation-dominated regimes. The universal relations between critical amplitudes and T_c, supplemented by the empirical phase diagram (doping-dependence of T_c) also imply – in agreement with recent kinetic-induction measurements on $La_{2-x}Sr_xCuO_4$ films and muon spin resonance (µSR) data – unconventional behavior of the penetration depth in the overdoped regime. The evidence of uncharged superfluid of 3D x–y behavior is completed in terms of the asymptotic low-temperature behavior of the penetration depth, because the sound-wave contribution of the uncharged superfluid accounts remarkably well for the experimental data. The dominant role of 3D x–y fluctuations, implying tightly bound and interacting pairs even above T_c, together with the doping-dependent specific heat singularity point uniquely to Bose condensation of hard pairs on a lattice as the mechanism that drives the transition from the normal to the superconducting state.

Considerable debate has arisen over the nature of superconductivity and the symmetry of the order parameter in high-T_c superconductors [1–9]. In view of the fact that thermal fluctuations reflect the structure of the order parameter and that these extreme type II materials exhibit pronounced fluctuation effects [10–17], we discuss in this paper the use of thermal fluctuations to elucidate the nature of the superconducting state.

The organization of this paper is as follows. First we provide strong evidence that cuprate superconductors and uncharged superfluids like ^{4}He share the universal three-dimensional (3D) x–y properties in the fluctuation-dominated regimes. Second, we explain how the universal relations between critical amplitudes and T_c, supplemented by the empirical phase diagram (doping-dependence of T_c), imply – in agreement with recent kinetic-induction measurements on $La_{2-x}Sr_xCuO_4$ films and μSR data – unconventional behavior of the penetration depth in the overdoped regime. The evidence of uncharged superfluid or 3D or x–y behavior will be completed in terms of the asymptotic low-temperature behavior of the penetration depth because the sound-wave contribution of the uncharged superfluid accounts remarkably well for the experimental data. The remainder of the paper addresses the implications. In particular, the evidence of strong 3D x–y fluctuations, implying tightly bound and interacting pairs even above T_c, together with the doping-dependent specific heat singularity point uniquely to Bose condensation of hard pairs on a lattice as the mechanism that drives the transition from the normal to the superconducting state.

Measurements of various properties, including the magnetic penetration depth [13,14,16] and specific heat [12–15, 17], strongly suggest that the fluctuations of the order parameter in the Meißner phase and close to the transition temperature are essentially those of a neutral superfluid or three-dimensional (3D) x–y model. The associated complex scalar order parameter field

$$\Psi(r) = |\Psi(r)| \exp(i\varphi(r)) \tag{1}$$

represents in a superconductor the macroscopic wave function of singlet pairs of charge $2e$. The term $|\Psi(r)|$ is related to the superfluid number density and $\varphi(r)$ denotes the phase. Introducing the magnetic penetration depth λ and the correlation length ζ, and describing the decay of the fluctuations in the modulus $|\Psi(r)|$ of the order parameter, the uncharged behavior of the cuprates can be understood in terms of the dimensionless charge, $\tilde{e} = 1/\kappa = \xi/\lambda$, which determines the strength of the screening. In cuprate superconductors, this charge is small, because $\lambda \gg \xi$ [11, 13, 14]. In an uncharged superfluid or 3D x–y model the specific heat diverges close to T_c as [13, 14, 18]

$$C^\pm = \frac{A^\pm}{\alpha}|t|^{-\alpha} + B^\pm \approx \frac{A^\pm}{\alpha} + B^\pm - A^\pm \ln|t|, \qquad t = 1 - \frac{T}{T_c}, \qquad A^+ \approx A^-. \tag{2}$$

Here we have used the fact that the critical exponent α is very small. The term $B^\pm$ accounts for the background. To illustrate the evidence of 3D x–y critical behavior, we reproduce in Fig. 1 the specific heat measurements of polycrystalline $DyBa_2Cu_3O_{7-x}$ [15]. In the temperature range $6 > -\ln|t| > 4$ the data are remarkably consistent with Eq. (2), and indeed are seen to fall on nearly parallel branches with finite slope. The upper branch corresponds to $T < T_c$ and the lower one to $T > T_c$ (Fig. 1b) so that $B^- + A^-/\alpha > B^+ + A^+/\alpha$, reflecting the mean-field background behavior. Apparently, these measurements are fully consistent with the 3D x–y critical point behavior given by Eq. (2). Moreover,

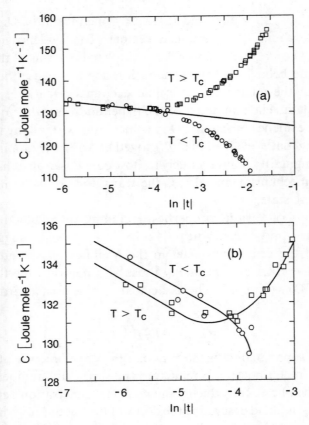

Fig. 1. Specific heat versus $\ln |t|$ of $DyBa_2Cu_3O_{7-x}$ taken from [11]. Reprinted from [12].

the resulting value for $A^+ \approx A^-$ can be used to determine the correlation volumes for the phase, $(\xi_{\|0}^\varphi)^2 \xi_{\perp 0}^\varphi$, and amplitude, $(\xi_{\|0})^2 \xi_{\perp 0}$, fluctuations via the universal relations [13, 14, 18]

$$R_\varphi^3 = A^-(\xi_{\|0}^\varphi)^2 \, \xi_{\perp 0}^\varphi \approx (0.8)^3, \qquad R^3 = A^+(\xi_{\|0})^2 \, \xi_{\perp 0} \approx (0.3)^3. \qquad (3)$$

Here '$\|$' denotes a direction parallel and '$\perp$' perpendicular to the layers. The relations between the length scales and the effective pair masses are [12–14]

$$\frac{\xi_{\|0}^\varphi}{\xi_{\perp 0}^\varphi} = \left(\frac{\lambda_{\|0}}{\lambda_{\perp 0}}\right)^2 = \frac{M_\|}{M_\perp}, \qquad \frac{\xi_{\|0}}{\xi_{\perp 0}} = \left(\frac{M_\perp}{M_\|}\right)^{1/2}. \qquad (4)$$

Given the experimental estimates of $A^+ \approx A^-$ and the mass anisotropy $M_\|/M_\perp$, Eq. (3) and Eq. (4) can be used to evaluate correlation volumes and correlation length amplitudes. Because pronounced fluctuation effects are tied to small correlation volumes, the estimates listed in Table 1 clearly reveal the

Table 1. *Measured values of T_c and experimental estimates for $A^+ \approx A^-$ [12,14]. The correlation volumes for the phase and amplitude fluctuations were derived from the universal relations (3), the correlation length amplitudes with the aid of Eq. (4) and $M_\perp/M_\| \approx 25$ for $YBa_2Cu_3O_{7-\delta}$ [14].*

Material	T_c (K)	$1/A^- \approx 1/A^+$ (Å³)	$(\xi^\varphi_{\|0})^2\xi^\varphi_{\perp0}$ (Å³)	$(\xi_{\|0})^2\xi_{\perp0}$ (Å³)	$\xi^\varphi_{\|0}$ (Å)	$\xi^\varphi_{\perp0}$ (Å)	$\xi_{\|0}$ (Å)	$\xi_{\perp0}$ (Å)
DyBa$_2$Cu$_3$O$_7$	91.4	1130	579	30	2.9	72	3.7	0.7
YBa$_2$Cu$_3$O$_7$	91.7	1790	916	48	3.3	83	6.2	1.2
^{4}He	2.18	600	307	16	6.7		2.5	

importance of fluctuations. Indeed, the correlation volumes of DyBa$_2$Cu$_3$O$_7$ and YBa$_2$Cu$_3$O$_7$ are comparable to that in superfluid helium and are on the order of the unit-cell volume. Thus, in analogy to superfluid helium, fluctuation effects, exposing the universal 3D behavior, are essential and experimentally accessible. Moreover, a mean-field treatment of the phase transition is not appropriate.

Another quantity of interest is the magnetic penetration depth. In extreme type II superconductors it is simply related to the superfluid mass density $\rho_{\|,\perp} = M_{\|,\perp}n_s$, where n_s is the number density of the superfluid pairs, in terms of the London relation

$$\frac{1}{\lambda_{\|,\perp}^2(T)} = \frac{16\pi\rho_{s,\|,\perp}(T)e^2}{M_{\|,\perp}^2c^2} = \frac{16\pi n_s(T)e^2}{M_{\|,\perp}c^2}. \tag{5}$$

Approaching the 3D x–y critical point from below, the superfluid mass density tends to zero as $\rho_{s,\|,\perp} \propto t^{-\nu}$, with $\nu \approx \frac{2}{3}$ [13, 14], so that, according to Eq. (5)

$$\frac{1}{\lambda_{\|,\perp}^2(T)} = \frac{1}{\lambda_{\|,\perp,0}^2}t^\nu. \tag{6}$$

This behavior is nicely confirmed by the experimental data shown in Fig. 2 [16], which yield $\nu \approx \frac{2}{3}$ and $(\lambda_\|(T=0)/\lambda_{\|0})^2 \approx 1.5$. In this context it is important to recognize that, at the 3D x–y critical point transition temperature, phase correlation length, specific heat and superfluid mass density or, equivalently, the London penetration depth, are not independent of one another. Indeed, there are the universal relations [13, 14]

$$k_BT_c = \frac{\Phi_0^2}{16\pi^3}\frac{\xi^\varphi_{\|,\perp0}}{\lambda_{\|,\perp0}^2}, \qquad (k_BT_c)^3A^-\frac{M_\perp}{M_\|} = \left(\frac{\Phi_0^2}{16\pi^3}\right)^3\frac{R_\varphi^3}{\lambda_{\|0}^6}, \tag{7}$$

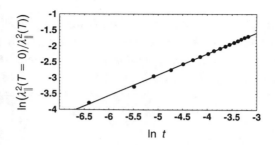

Fig. 2. $2\ln(\lambda_\parallel(T=0))/(\lambda_\parallel(T))$ versus $\ln t = \ln[1-(T/T_c)]$ for $YBa_2Cu_3O_{7-\delta}$, using data from [16]. The straight line represents the fit to $(\lambda_\parallel(T=0)/\lambda_\parallel(T))^2 = (\lambda_\parallel(T=0)/\lambda_{\parallel 0})^2 t^\nu$ yielding $\nu \approx 2/3$ and $(\lambda_\parallel(T=0)/\lambda_{\parallel 0})^2 \approx 1.5$.

connecting T_c with the amplitudes of penetration depth, $\lambda_{\parallel 0}$ and phase correlation length, $\xi_{\parallel,\perp 0}^\varphi$ or specific heat, A^-. The latter expression follows from Eq. (3) and Eq. (4). Because A^-, $M_\perp/M_\parallel$ and $\lambda_{\parallel 0}$ can be measured independently, relation (7) can be used to check the consistency of the estimates for the critical amplitudes. As an example, we evaluate the amplitude of the phase correlation length with the aid of Eq. (7) and the λ measurements shown in Fig. 2 and compare it with the value derived from the specific heat anomaly. The data shown in Fig. 2 yield $\lambda_{\parallel 0}^2 \approx 0.67\lambda_\parallel^2(T=0)$ and with $T_c \approx 92.4$ K [16], $\lambda_\parallel(T=0) = 1817$ Å [5] and the universal relation (7), we find $\xi_{\parallel 0}^\varphi \approx 3.24$ Å, which is close to the value listed in Table 1. Thus, the evidence of 3D $x-y$ behavior is consistent, because independent measurements and various universal relations lead essentially to the same estimate for the phase correlation amplitude $\xi_{\parallel 0}^\varphi$.

Another important implication of the universal relations (3) and (7) emerges from the fact that the properties of cuprate superconductors depend on doping. Empirically, this dependence is well described by [19, 20]

$$\bar{T}_c = \frac{T_c}{T_c^m} = 1 - \beta(x-x_c)^2 = A\rho(1-\rho), \tag{8}$$

with $x_c \approx 0.16$ and $\beta \approx 82.6$. Here x is the concentration of holes per CuO_2 in the CuO_2 layer, while T_c^m denotes the transition temperature for optimum doping ($x = x_c$). The fraction of mobile holes and hole concentration are related by

$$2\rho = 1 \pm \sqrt{\beta(x-x_c)}. \tag{9}$$

In the underdoped ($\rho \to 0$) and overdoped ($\rho \to 1$) limit, T_c tends to zero and the system is expected to approach the 3D quantum $x-y$ critical point or, equivalently, the $3+1 = 4$D classical $x-y$ critical point [21], at which the mean-field treatment becomes exact up to logarithmic corrections, and $A^\pm$ vanishes. Thus A^- is expected to adopt its maximum value at T_c^m and to decrease as T_c

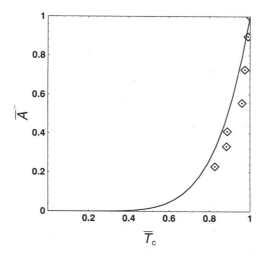

Fig. 3. $\bar{A}^- = A^- / A_m^-$ versus T_c / T_c^m for $YBa_2Cu_3O_{7-\delta}$ according to [17]. A_m^- is the value of A^- for optimum doping and T_c^m the corresponding value of T_c. The curve corresponds to Eq. (10).

tends to zero in the underdoped ($\rho \to 0$) and overdoped ($\rho \to 1$) limit. Experimental evidence of this scenario comes from the specific heat measurements of underdoped $YBa_2Cu_3O_{7-x}$ [17]. The data are shown in Fig. 3 together with the Ansatz

$$\bar{A}^- = \frac{A^-}{A_m^-} \approx (\bar{T}_c)^6, \tag{10}$$

interpolating between the limits $\rho \to 0$ and $\rho \to 1$. Combining Eq. (5) and Eq. (7) with Ansatz (10) we obtain

$$\bar{T}_c = \left(\frac{\bar{M}_\parallel}{\bar{M}_\perp} \right)^{1/9} (\bar{\lambda}_{\parallel 0}^2)^{-1/3} = 4\rho(1-\rho), \tag{11}$$

where

$$\bar{M}_{\parallel,\perp} = \frac{M_{\parallel,\perp}}{M_{\parallel,\perp}^m}, \qquad \bar{\lambda}_{\parallel,\perp} = \frac{\lambda_{\parallel,\perp}}{\lambda_{\parallel,\perp}^m}. \tag{12}$$

The index m refers to the respective values for optimum doping. In Eq. (10) there are also exposed the doping-dependence of the effective mass, mass anisotropy and penetration depth, namely

$$\bar{\lambda}_{\parallel 0}^2 = \bar{M}_\parallel \frac{n_s^m}{n_s} = \left(\frac{\bar{M}_\parallel}{\bar{M}_\perp} \right)^{1/3} [4\rho(1-\rho)]^{-3}. \tag{13}$$

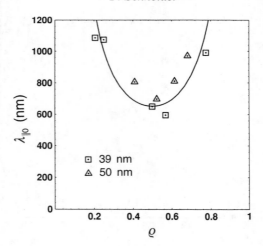

Fig. 4. Kinetic inductance estimates for $\lambda_{\parallel 0}$ and $\lambda_{\parallel}(T=0)$ in La$_{2-x}$Sr$_x$CuO$_4$ films plotted in terms of $\lambda_{\parallel 0}$ and $\lambda_{\parallel}(T=0)$ versus ρ[23]. The solid curve corresponds to Eq. (13) with
$$\bar{M}_\perp/\bar{M}_\parallel = 1.$$

A crucial implication of the 3D x–y universality, supplemented by the empirical doping-dependence of T_c, is the divergence of the penetration depth amplitudes in both the underdoped and the overdoped limits. This behavior differs drastically from the standard picture, in which the effective mass M is assumed to be independent of doping, so that $\lambda^2 \propto M_\parallel/n_s$ decreases monotonically with increasing superfluid density n_s [22].

The doping dependence of $\lambda_{\parallel 0}$ has recently been studied in La$_{2-x}$Sr$_x$CuO$_4$ films [23]. The kinetic inductance data, covering both the underdoped and overdoped regimes, are shown in Fig. 4 in terms of $\lambda_{\parallel 0}$ versus ρ. The solid curve is a fit to Eq. (13) with $\bar{M}_\parallel/\bar{M}_\perp \approx 1$. Thus the standard picture [22] fails, while the 3D x–y critical point behavior, supplemented by the empirical doping-dependence of T_c, describes the experimental data remarkably well. This behavior is further confirmed by the μSR data shown in Fig. 5 in terms of T_c versus $\bar{\sigma}_\parallel(T=0) \propto 1/\lambda_\parallel^2(T=0)$. The term $\sigma(0)$ is the μSR relaxation rate, which is proportional to the square of the inverse penetration depth. The solid curve corresponds to Eq. (11) with $\bar{M}_\parallel/\bar{M}_\perp \approx 1$ and the arrows denote the behavior in the underdoped ($\uparrow, 0 \leq \rho \leq \frac{1}{2}$) and overdoped ($\downarrow, \frac{1}{2} \leq \rho \leq 1$) regimes as ρ increases from 0 to 1. Deviations from this behavior are not unexpected in the low-T_c regime, because the effective mass ratio, $\bar{M}_\parallel/\bar{M}_\perp$, entering Eq. (11) depends on doping (Eq. (13)). Better agreement requires that $\bar{M}_\parallel/\bar{M}_\perp$, entering Eq. (11) depends on doping (Eq. (13)). Better agreement requires that $M_\parallel/M_\perp$ increases with reduced T_c as observed in YBa$_2$Cu$_3$O$_{7-\delta}$ [24]. More importantly, the experimental data of overdoped Tl$_2$Ba$_2$CuO$_{6+\delta}$ [25, 26] and

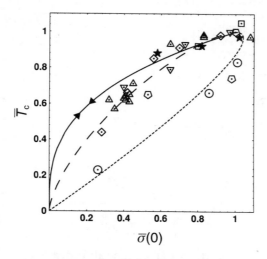

$$\bar{\sigma}(0)$$

Fig. 5. $\bar{T}_c$ versus $\bar{\sigma}(T=0)=(\lambda_\parallel^m(T=0))^2/\lambda_\parallel^2(T=0)$ taken from [5–7,24–26]. The solid curve corresponds to Eq. (10) with $\bar{M}_\parallel/\bar{M}_\perp \approx 1$ and the arrows denote the underdoped $(\uparrow, 0 \leq \rho \leq 1/2)$ and overdoped $(\downarrow, 1/2 \leq \rho \leq 1)$ regimes. The dashed lines are a guide to the eye and sketch the effect of the doping-dependence of the effective mass ratio, $\bar{M}_\parallel/\bar{M}_\perp$. Data taken from [13,32]. ($\square$) $Tl_2Ba_2Ca_2Cu_3O_{10}$, $Tl_{0.5}Pb_{0.5}Sr_2Ca_2Cu_3O_9$, $Bi_{2-x}Pb_xSr_2Ca_2Cu_3O_{16}$; ($\diamond$) $Y_{1-x}Pr_xBa_2Cu_3O_{6.97}$; ($\triangle$) $YBa_2Cu_3O_x$; ($\triangledown$) $La_{2-x}Sr_xCuO_4$; ($\bigstar$) $Bi_2Sr_2Ca_{1-x}Y_xCu_2O_{8+\delta}$; ($\bigcirc$) $Tl_2Ba_2CuO_{6+\delta}$; and ($\varhexagon$) $Ca_{1-x}Y_xSr_2Tl_{0.5}Pb_{0.5}Cu_2O_7$.

$Ca_{1-x}Y_xSr_2Tl_{0.5}Pb_{0.5}Cu_2O_7$ [27] confirm the essential implication of Eq. (11), namely, the reversed evolution of underdoped ($\uparrow$) and overdoped ($\downarrow$) systems. The dashed lines in Fig. 5, resembling the outline of a fly's wing, are guides to the eye and illustrate the effect of the doping-dependent anisotropy, $\bar{M}_\parallel/\bar{M}_\perp$. The upper part of the outline describes the underdoped and the lower one the overdoped system. Indeed, the experimental points of overdoped $Tl_2Ba_2CuO_{6+\delta}$ and $Ca_{1-x}Y_xSr_2Tl_{0.5}Pb_{0.5}Cu_2O_7$ are close to the outline of the lower part of the wing, while the remaining data, referring to underdoped and mildly overdoped materials, are close to the upper branch.

The implications of 3D x–y behavior are not restricted, however, to critical phenomena and the associated phase diagram in the (T_c, x) plane. If the fluctuations of the order parameter Ψ are indeed essentially those of a neutral superfluid, than this scenario also implies the existence of sound waves, which dominate the low-temperature behavior [28–30]. As far as measurable low-temperature properties are concerned, the magnetic penetration depth appears to be particularly suitable. It is directly accessible and not, as in the case of the specific heat, subject to large background subtractions. Although the temperature-dependence of λ has been studied rather extensively [6–9, 14] only the study of Sun *et al.* [5] extends it to sufficiently low temperatures ($T < 4.2$ K).

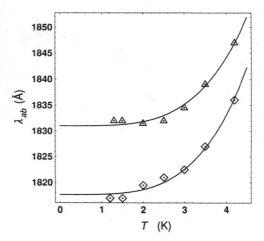

Fig. 6. The temperature-dependence of the penetration depth λ_{ab} for $YBa_2Cu_3O_{7-\delta}$ ($\diamond$) and $Y_{1-x}Pr_xBa_2Cu_3O_{7-\delta}$ ($x=0.1$) ($\triangle$), taken from [5]. The solid line is a fit to Eq. (15). From [33].

They measured the penetration depth using planar Josephson junctions (Pb/ insulator/$Y_{1-x}Pr_xBa_2Cu_3O_{7-\delta}$) tunneling into the c-axis and an applied magnetic field H parallel to the a–b plane. Neglecting screening effects and adopting a macroscopic description, where the order parameter is a complex scalar, the maximum supercurrent that can be sent through the junction is given by [31]

$$I_{max}(\Phi)=I_c\left|\frac{\sin(\pi\Phi/\Phi_0)}{\pi\Phi/\Phi_0}\right|, \qquad \Phi=HY(\lambda+\lambda_{Pb}+d), \qquad \Phi_0=\frac{hc}{2e}, \quad (14)$$

where d denotes the barrier thickness, λ and λ_{Pb} the penetration depths of $Y_{1-x}Pr_xBa_2Cu_3O_{7-\delta}$ and Pb, respectively. $I_c=YZJ_c$ is the maximum critical current and J_c the critical current density over area YZ of the junction. The 'diffraction pattern' has been verified experimentally and the resulting estimates of the penetration depth are shown in Fig. 6 together with fits to the leading sound wave contribution to the penetration depth,

$$\frac{\lambda(T)}{\lambda(0)}=1+AT^4. \qquad (15)$$

Apparently, the experimental data are remarkably consistent with a T^4-dependence and yield the values $\lambda(0)=1831$ Å, $A=3.29\times10^{-5}$ K^{-4} for $Y_{1-x}Pr_xBa_2Cu_3O_{7-\delta}$ and $\lambda(0)=1818$ Å, $A=2.78\times10^{-5}$ K^{-4} for $YBa_2Cu_3O_{7-\delta}$. To establish the link between Eq. (15) and the sound waves in the superfluid, we adopt the standard effective Hamiltonian of an uncharged superfluid in the hydrodynamic regime [28–30] and calculate the superfluid mass density ρ_s entering the London penetration depth (Eq. (5)). For this purpose, it is convenient to consider a superfluid moving with velocity $\mathbf{v}$ [28–30]. The distribution function of the superfluid moving as a whole is then obtained from the distribution function $n(E)$ for the gas at rest by replacing the excitation energy $E(p) = \hbar\omega(p)$ by $E = E(p) - \hbar\mathbf{p}\cdot\mathbf{v}$. Thus the total momentum of the quasi-particle gas is

$$\mathbf{P} = \sum_p \hbar p n(E(p) - \hbar\mathbf{p}\cdot\mathbf{v}) \approx \sum_p \hbar\mathbf{p}(\hbar\mathbf{p}\cdot\mathbf{v})\left(-\frac{\mathrm{d}n(p)}{\mathrm{d}E(p)}\right)$$

$$= \frac{\hbar^2}{3}\mathbf{v}\sum_p p^2\left(-\frac{\mathrm{d}n(p)}{\mathrm{d}E(p)}\right) = \rho_N\mathbf{v}, \tag{16}$$

where

$$n(p) = \left[\exp\left(\frac{\hbar\omega(p)}{k_B T}\right) - 1\right]^{-1}, \qquad \omega = sp \tag{17}$$

and s denotes the speed of sound in the superfluid. The proportionality constant ρ_N defines the normal fluid mass density, and the superfluid mass density is then obtained from

$$\rho_s \equiv \rho - \rho_N, \qquad \rho_s(T=0) = \rho. \tag{18}$$

Using Eqs. (5) and (16)–(18) the asymptotic temperature-dependence of the London penetration depth is now readily calculated, yielding Eq. (15) in the form

$$\frac{\lambda(T)}{\lambda(0)} = 1 + \frac{\pi^2(k_B T)^4}{45 s^5 \hbar^3 \rho} = 1 + \frac{16\pi^3(k_B T)^4 e^2 \lambda(0)^2}{45 s^5 \hbar^3 M^2 c^2} = 1 + AT^4. \tag{19}$$

Clearly, this term dominates only at sufficiently low temperatures. At higher temperatures, interaction of the excitations and deviations from the linear dispersion law $(E(\rho) = \hbar\bar{s}p)$ become important. The temperature range in which the long-wavelength description applies can be estimated by introducing in Eq. (16) a cut-off $p_c \approx \pi/a$, where a is the lattice constant. For the long-wavelength regime to be valid it is necessary that the upper limit of the integral in Eq. (16) becomes large. This restricts the temperature range in which Eq. (19) applies to

$$T \ll \frac{\pi\hbar s}{k_B a}. \tag{20}$$

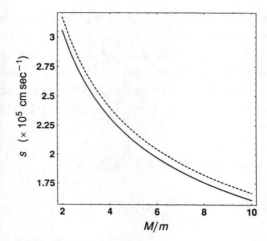

Fig. 7. Estimated superfluid sound velocity s versus pair mass M in units of the electron mass m from $YBa_2Cu_3O_{7-\delta}$ (dashed line) and $Y_{1-x}Pr_xBa_2Cu_3O_{7-\delta}$ (solid line). From [33].

Thus it remains to be shown that condition (20) is satisfied. For this reason we evaluated velocity s from the estimated fit parameter A as a function of the pair mass M. The resulting values are plotted in Fig. 7 in terms of s versus M/m, where m is the mass of the electron. The magnitude of s clearly reveals that condition (20) is satisfied ($\pi\hbar s/(k_B a) \approx 150$ K, $a \approx 4 \times 10^{-8}$ cm, $s \approx 2.5 \times 10^5$ cm s^{-1}). Thus, the experimental data extend to sufficiently low temperatures for the sound wave with frequency $\omega = sp$ to dominate, while excitations corresponding to the binding energy Δ of the pairs do not contribute significantly, because $\Delta > k_B T_c$.

To summarize, there is strong evidence that cuprate superconductors and uncharged superfluids like ^{4}He share the universal 3D x–y critical point properties. We have focused on experiments that probe the x–y nature of the order parameter in terms of critical exponents, critical amplitudes and the universal relations between these amplitudes. For a more complete discussion of the experimental evidence for 3D x–y critical behavior we refer to [14], which includes the magnetic-field-dependence of magnetization and specific heat, the associated universal scaling functions, the fluctuation contribution to the magnetic susceptibility, the decrease in T_c with decreasing film thickness and the fluctuation contribution to the conductivity. Moreover, we have shown that the universal relations between critical amplitudes and T_c, supplemented by the empirical phase diagram (doping-dependence of T_c) predict a doping-dependent mass anisotropy. The most striking. implication, namely, the increasing penetration depth with decreasing T_c in the overdoped regime, is well confirmed by recent kinetic induction measurements on $La_{2-x}Sr_xCuO_4$ films and is consistent with the μSR data. The evidence of uncharged superfluid or 3D x–y behavior has been completed in terms of the asymptotic low-temperature behavior of the penetration depth. Indeed, the sound-wave contribution of the uncharged superfluid accounts remarkably well for the experimental data.

The importance of ascertaining the relevance of fluctuations and the universality class to which the phase transition belongs cannot be emphasized enough. Both microscopic properties and macroscopic behavior depend sensitively on the symmetry of the order parameter and the strength of fluctuations. In this paper we found a remarkably consistent explanation in terms of 3D x–y behavior, revealing that the nature of superconductivity in cuprates is essentially that of superfluidity. The evidence of 3D x–y behavior also implies the following. (i) The fluctuations in the order parameter Ψ, describing the macroscopic wave function of pairs of charge $2e$, are essentially those of an uncharged superfluid and lead to the universal x–y behavior. (ii) The dominant role of these fluctuations invalidates mean-field treatments, and reveals pairing and tightly bound pairs above T_c. (iii) For these reasons, the order parameter $\Psi(r)$ describes long-wavelength and low-energy fluctuations associated with the center-of-mass motion of the pairs. The internal degrees of freedom can be eliminated because their energy scale, set by the binding energy of the pair, exceeds k_BT_c. Because pairs interact and pairing occurs above T_c, the symmetry of the order parameter is no longer simply related to that of the pairing interaction. In a mean-field or BCS treatment, however, the pairs do not interact and there are no pairs above T_c. In this approximation, the symmetry of the order parameter simply follows from the pairing interaction. (iv) Tightly bound and interacting pairs even above T_c, the evidence of 3D x–y universality, the doping-dependence of the transition temperature and of the specific heat singularity all point uniquely to Bose condensation of hard pairs on a lattice as the mechanism that drives the transition from the normal to the superconducting state. Once pairs are formed, the fermionic degrees of freedom adopt the role of spectators, while the pairs become the players.

Acknowledgements

I thank many colleagues for insightful suggestions, questions and comments. Those who contributed include J. G. Bednorz, H. Keller, J.-P. Locquet, M. B. Maple, J. Mannhart, R. Micnas, K. A. Müller, M. H. Pederson, J. J. Rodríguez-Núñez and Ch. Rossel.

References

[1]   J. F. Annett, N. Goldenfeld and S. N. Renn, in *Physical Properties of High Temperature Superconductors II*, Ed. D. M. Ginsberg (World Scientific, Singapore, 1990), p. 571.

[2]   P. Monthoux, A. Balatsky and D. Pines, *Phys. Rev. Lett.* **67**, 3448 (1991).

[3]   D. A. Wollman, D. J. van Harlingen, W. C. Lee, D. M. Ginsberg and A. J. Legget, *Phys. Rev. Lett.* **71**, 2134 (1993).

[4]   P. Chaudhari and S.-Y. Lin, *Phys. Rev. Lett.* **72**, 1084 (1994).

[5]   A. G. Sun, D. A. Gajewski, M. P. Maple and R. C. Dynes, unpublished.

[6]   N. Klein *et al.*, *Phys. Rev. Lett.* **71**, 3355 (1993).

[7]   W. N. Hardy *et al.*, *Phys. Rev. Lett.* **70**, 3999 (1993).

[8]   Z. X. Shen *et al.*, *Phys. Rev. Lett.* **70**, 1553 (1993).

[9]   J. E. Sonier *et al.*, *Phys. Rev. Lett.* **72**, 744 (1994).

[10] A. Kapitulnik, M. R. Beasley, C. Castellani and C. DiCastro, *Phys. Rev.* B **37**, 537 (1988).

[11] D. S. Fisher, M. P. A. Fisher and D. A. Huse, *Phys. Rev.* B **43**, 130 (1991).

[12] T. Schneider and D. Ariosa, *Z. Phys.* B **89**, 267 (1992).

[13] T. Schneider and H. Keller, *Physica* C **207**, 366 (1993).

[14] T. Schneider and H. Keller, *Int. J. Mod. Phys.* B **8**, 487 (1994).

[15] A.Kozlowski, Z. Tarnawski, A. Kolodziejczyk, J. Chmist, T. Scizor and R. Zalecki, *Physica* C **184**, 113 (1991).

[16] A. Buzdin *et al.*, unpublished.

[17] K. Ghiron, M. B. Salamon, M. A. Hubbard and B. W. Veal, *Phys. Rev.* B **48**, 16188 (1993).

[18] V. Privman, P. C. Hohenberg and A. Aharony, in *Phase Transitions and Critical Phenomena* Vol. 14, Ed. C. Domb and J. L. Lebowitzs (Academic, New York, 1991).

[19] J. L. Tallon and N. E. Flower, *Physica* C **204**, 237 (1993).

[20] H. Zhang and H. Sato, *Phys. Rev. Lett.* **70**, 1697 (1993).

[21] R. Morf, T. Schneider and E. Stoll, *Phys. Rev.* B **16**, 462 (1977).

[22] D. R. Harshman and A. P. Mills, Jr., *Phys. Rev.* B **45**, 10684 (1992).

[23] Y. Jaccard *et al.*, private communication.

[24] B. Janossy *et al.*, *Physica* C **181**, 51 (1991).

[25] Ch. Niedermeier *et al.*, *Phys. Rev. Lett.* **7**, 17464 (1993).

[26] Y. J. Uemura *et al.*, *Nature* **364**, 605 (1993).

[27] P. Zimmermann *et al.*, private communication.

[28] L. D. Landau and E. M. Lifshitz, *Statistical Physics Part 2* (Pergamon, Oxford, 1980), especially Ch. 24.

[29] R. P. Feynman, *Statistical Mechanics* (W. A. Benjamin, Reading, 1972) Ch. 11.

[30] P. B. Weichman, *Phys. Rev.* B **38**, 8739 (1988).

[31] A. Barone and G. Paterno, *Physics and Applications of the Josephson Effect* (Wiley, New York, 1982).

[32] T. Schneider and M. Pedersen, *Proc. Int. Conf. on 'Physics and Chemistry of Molecular Oxide Superconductors,' Eugene, OR, July 1993*, in *J. Supercond.* (in press, 1994).

[33] T. Schneider and M. Pedersen, preprint.

18

High-T_c superconductivity with polarons and bipolarons: an approach from the insulating states

SERGE AUBRY

Laboratoire Léon Brillouin (CEA-CNRS), CE Saclay, 91191-Gif-sur-Yvette Cédex, France

Abstract

We review a series of exact, analytical and numerical results obtained on the adiabatic Holstein–Hubbard model, many of which are new and non-trivial. We study next the role of the quantum lattice fluctuations that were initially neglected. The possibility of having high-T_c bipolaronic superconductors is analysed on the basis of these results.

We suggest that models that involve only electron–phonon coupling are very unlikely to produce bipolaronic superconductivity, which is prevented by the spatial ordering of the bipolarons associated with a lattice instability. This is due to a very large effective mass of the bipolarons that can be related to the large Peierls–Nabarro energy barrier required to move these bipolarons through the lattice in the adiabatic limit.

We conjecture that high-T_c superconductivity originates specifically from the exceptionally well-balanced competition between electron–phonon coupling and electron–electron repulsion. In the restricted region in which, within the mean field approach, the energy of a bipolaron is close to those of two polarons, a new type of electron pairing occurs by formation of pairs of polarons in the spin singlet state. Such a polaron pair, called a spin resonant bipolaron (SRB), is not the standard bipolaron (both could exist in the adiabatic case). Its Peierls–Nabarro energy barrier can be sharply depressed, almost to zero. As a result of quantum lattice fluctuations, its tunnelling energy is sharply enhanced. Then, superconductivity could persist for a relatively large electron–phonon coupling, with an unusually large critical temperature as a Bose condensate of SRBs before becoming, at larger coupling and in any case, an insulating spin–Peierls polaronic phase.

1 Introduction

Despite a tremendous amount of experimental observations, the mystery of high-T_c superconductivity persists and stimulates a large number of theories. The main purpose of that paper is not to propose an achieved theory of high-T_c superconductors but, on the basis of solid arguments obtained through deeper understanding of simple models, to indicate new possible directions of research and perhaps to correct some wrong beliefs or assumptions. We demand of our approach that it be intrinsically consistent in itself as much as possible and we reject arbitrary or controversial approximations, which would be motivated by finding appropriate agreement with real experiments. There are no explicit proofs in this paper; instead we focus on the physical ideas that trace out our direction for understanding this problem. Some of these ideas are supported by exact, analytical or numerical results, or have still to be considered as conjectures. The interested reader will have to refer for more details to other papers, some of which are already published, while others are still in preparation.

The early prediction of the existence of polarons goes back to Landau (1933). An electron coupled to a deformable field may gain energy by self-localisation in the potential well generated by its deformation. This intermediate field can be due to lattice deformation due to interaction between electrons and ions (electron–phonon coupling), electric polarisation of the electronic background (ions, molecules ...) coupled with the charge of the electron, magnetic polarisation of a paramagnetic background coupled to the spin of the electron or any other type of excitation. If the direct electronic repulsion is not too large, then two electrons with opposite spins can be trapped self-consistently in the same potential well, and this pair associated with the deformation is called a bipolaron. Polarons and bipolarons are ideally well defined for a single electron (or a pair of electrons) only in the 'adiabatic' limit, where this intermediate field does not fluctuate, that is for a classical field at 0 K. They are in essence localised in real space.

A chemical bond can be viewed as the extreme case of a bipolaron that is locally associated with a given distance between two atoms but in that case the concept is useless because such bipolarons are static objects that cannot tunnel through the lattice and do not hop by thermal fluctuations unless the bond be broken at very high temperature. In practice, the concept of polarons and bipolarons becomes useful when tunnelling and hopping through the lattice becomes possible because of quantum and thermal fluctuations respectively, which result in interesting physical consequences. The concept of polarons and bipolarons remains meaningful on condition that the tunnelling and hopping time scales of these polarons and bipolarons be not too short for electronic self-localisation during a time significantly larger than the time scale of the bare electrons' existence. In other words, the effective mass of the polaron (or bipolaron) has to be significantly larger than the bare mass of one (or two) electron(s) in the band.

Representing superconductivity as a Bose condensation of bipolarons, which in principle are (hard core) bosons, requires that one take into account the quantum fluctuations of the mediating field to form bands of extended bipolarons. In some sense, this problem has a similarity with the problem of describing a band of electrons from localised orbitals (the LCAO approximation). In that simple case, we know that this approximation is rather good for molecular crystals for example and, more generally, any electronic band can be described in that way by using a Wannier function for localised orbitals. However, a serious obstacle to having easily bipolaronic superconductivity, which is not found in that example with non-interacting electrons, comes from the fact that, at non-vanishing density, there is a potential interaction between the bipolarons that could 'kill' their quantum character. They may crystallise as spatially ordered structures (bipolaronic charge density waves) instead of condensing into a superconducting Bose superfluid.

Otherwise, even within the adiabatic approximation, the concept of polarons is clear only for a single electron in a lattice and many questions remain regarding their existence at any density. What happens when they overlap? We studied for many years the many-polaron and bipolaron problems in the adiabatic limit and proved that the concept can be extended unambiguously for adiabatic electron–phonon models in spite of strong overlappings. The conjectured existence at large enough electron–phonon coupling of insulating bipolaronic charge density waves (with many bipolarons) (Aubry 1980) has been observed numerically in one dimension (Le Daëron and Aubry 1983), (Aubry and Quemerais 1989) beyond the transition by breaking of analyticity in several models. Finally, rigorous proofs were obtained (Aubry, Abramovici and Raimbault 1992, Baesens and MacKay 1994) for the adiabatic Holstein model and extensions at any dimension and for any lattice, periodic or not. The ground states of these models were proven to be insulating bipolaronic configurations without magnetic field and mixed polaronic–bipolaronic configurations at large enough magnetic field. Many other exact results concerning their properties were also obtained. Proof of existence of bipolaronic, polaronic and mixed structures has also been recently obtained for the adiabatic Holstein–Hubbard model (Aubry, paper in preparation).

At large electron–phonon coupling, quantum lattice fluctuations appear to be generally extremely small so that superconductivity is practically absent. Discussions concerning this non-superconducting but insulating regime that, however, should be the most probable state in most realistic models with bipolarons and polarons proposed for an (artificial) high-T_c superconductivity, have been discarded in this workshop. Any electronic model, in the strong electron–phonon coupling limit and at any non-vanishing electron density, has to become insulating at zero temperature. In this paper, we focus much attention on the properties of these insulating states that remain, however, highly important for understanding the possible occurrence of superconductivity.

Our approach for understanding bipolaronic superconductivity consists of discussing under which conditions (however, still at moderately large electron coupling), the role of the quantum lattice fluctuations could increase sufficiently to favour a bipolaronic superfluid state instead of these insulating bipolaronic or polaronic structures and reach perhaps an unusually high critical temperature. For that purpose, we propose to study situations in which direct electron–electron interaction competing with electron–phonon coupling can break the standard bipolarons into polarons. In a finite region above the breaking threshold, we proved in our model that strong polaron pairing at short distance is maintained by magnetic interaction in the singlet state. We called the pair of polarons that is obtained a spin resonant bipolaron (SRB). In the adiabatic case, we could then get an insulating spin–Peierls polaronic structure. Then, we remark that the Peierls–Nabarro (PN) barrier of these SRB is sharply depressed with respect to those of standard bipolarons and polarons. We conjecture on the basis of this argument that the quantum tunnelling energy of these SRBs through the quantum lattice is significantly higher than those of standard bipolarons and single polarons, at the same coupling.

If these quantum fluctuations could become large enough, then high-T_c superconductivity could take the place of the spin–Peierls state. If this conjecture turns out to be confirmed, then our theory on high-T_c superconductivity would be specifically the consequence of sharply balanced competition between strong electron–phonon coupling and strong electron–electron repulsion. It should disappear when this balance is broken. Although this kind of superconductivity is obviously far away from the BCS regime, it also does not correspond to the ideal bipolaronic superconductivity as described in the literature. We could expect, not far above the critical temperature, that the normal state could become either a liquid of polarons (broken SRBs) or involve an intermediate state such as a liquid of SRB depending on the ratio of binding and tunnelling energies.

2 The Holstein–Hubbard model

Some of the essential aspects of the propagation of electrons in solids are modelled by the simple Holstein–Hubbard model. It is a one-band model that involves schematically the basic interactions between the electrons and the lattice. The Coulomb interactions of the electrons with the ions of the lattice are represented by linear on-site electron–phonon couplings while the direct Coulomb electron–electron interactions are represented only by their local component as on-site Hubbard repulsions. Direct interactions between the ions of the lattice, including their Coulomb interactions, are represented phenomenologically by an elastic deformation potential.

Let us recall our notations for this Hamiltonian, which is the sum of five terms

$$\mathbf{H} = H_k + H_p + H_{ep} + H_u + H_z, \tag{1}$$

where the periodicity of the underlying lattice is taken into account by considering a single band of electrons described as a tight-binding model. The band Hamiltonian

$$H_k = - \sum_{\langle i,j \rangle, \sigma} T_{i,j} \, c_{i,\sigma}^+ c_{j,\sigma} - \sum_i \mu \, \mathbf{n}_i \tag{2a}$$

is considered on an arbitrary lattice at arbitrary dimensions. $\langle i,j \rangle$ are nearest neighbour sites, $c_{i,\sigma}^+$ and $c_{i,\sigma}$ are the standard creation and annihilation Fermion operators at site i of an electron with spin $\sigma = \uparrow$ or $\downarrow$, the electronic density operator is

$$\mathbf{n}_i = c_{i,\uparrow}^+ c_{i,\uparrow} + c_{i,\downarrow}^+ c_{i,\downarrow} = n_{i,\uparrow} + n_{i,\downarrow} \tag{2b}$$

and μ is the electronic chemical potential or Fermi energy. For a square lattice in d dimensions without magnetic field, $T_{i,j} = T$ is constant and the bandwidth is $B = 2Td$. When a magnetic field is present, $T_{i,j}$ is a complex number

$$T_{i,j} = T \, e^{i\theta_{i,j}} \quad \text{with } T_{i,j} = T_{j,i}^* \text{ or } \theta_{i,j} = -\theta_{j,i} \tag{2c}$$

and the sum

$$\frac{1}{2\pi} \sum_{\langle i,j \rangle \in \mathscr{C}} \theta_{j,i}$$

over the oriented bounds of an arbitrary contour $\mathscr{C}$, is just the magnetic flux through this contour counted in flux units $2\phi_0 = 2\pi\hbar/e$.

The phonon with energy $\hbar\omega_0$ corresponds to uncoupled on-site oscillators with mass m and coupling constant $\frac{1}{2} m \, \omega_0^2$,

$$H_{\mathrm{p}}=\sum_i \hbar\omega_0\left(a_i^+ a_i+\frac{1}{2}\right)=\sum_i \frac{p_i^2}{2m}+\frac{1}{2}\, m\,\omega_0^2 u_i^2. \tag{3}$$

The electron–phonon coupling term with energy constant g is

$$H_{\mathrm{ep}}=g\sum_i \mathbf{n}_i(a_i^+ + a_i). \tag{4}$$

The Hubbard on-site repulsion with constant $\upsilon>0$ is

$$H_{\mathrm{u}}=\sum_i \upsilon\, n_{i,\uparrow}n_{i,\downarrow}=\sum_i \upsilon\, c_{i,\uparrow}^+ c_{i,\uparrow} c_{i,\downarrow}^+ c_{i,\downarrow}. \tag{5a}$$

The Zeeman term in the magnetic field $\mathscr{H}$ is

$$H_{\mathrm{z}}=-\tfrac{1}{2}g_{\mathrm{L}}\mu_{\mathrm{B}}\mathscr{H}(n_{i,\uparrow}-n_{i,\downarrow})\cong -\mu_{\mathrm{B}}\mathscr{H}(n_{i,\uparrow}-n_{i,\downarrow}), \tag{5b}$$

where μ_{B} is the standard Bohr magneton. For studying the large electron–phonon regime, it is convenient to define the dimensionless conjugate operators

$$\mathbf{u}_n=\frac{\hbar\omega_0}{4g}\,(a_n^+ + a_n)\qquad \mathbf{p}_n=\frac{2g}{\hbar\omega_0}\,i(a_n^+ - a_n) \tag{6}$$

with commutator $[\mathbf{u}_n,\mathbf{p}_n]=i$ and to write the Hamiltonian $\mathbf{H}$ as the sum of three terms

$$\hat{H}=\mathbf{H}/E_0=H_{\mathrm{AI}}+tH_{\mathrm{K}}+\beta H_{\mathrm{Q}} \tag{7a}$$

with the (electron–phonon) energy unit

$$E_0=\frac{8g^2}{\hbar\omega_0} \tag{7b}$$

and the dimensionless constants t and β. The main part

$$H_{\mathrm{AI}}=\sum_i \frac{1}{2}\,(\mathbf{u}_i^2+\mathbf{n}_i\mathbf{u}_i)+\mathscr{U}n_{i,\uparrow}n_{i,\downarrow}-\mathfrak{h}\,(n_{i,\uparrow}-n_{i,\downarrow})-\boldsymbol{\mu}(n_{i,\uparrow}+n_{i,\downarrow}) \tag{8a}$$

contains all the potential terms of the initial Hamiltonian with the dimensionless constants

$$\mathscr{U}=\frac{\upsilon\hbar\omega_0}{8g^2},\qquad \mathfrak{h}=\frac{\mu_{\mathrm{B}}\mathscr{H}\hbar\omega_0}{8g^2}\quad\text{and}\quad \boldsymbol{\mu}=\mu\,\frac{\hbar\omega_0}{8g^2}. \tag{8b}$$

This limit has been called anti-integrable because this limit has a strong analogy with those of the anti-integrable limit of a dynamical system and can be studied with similar methods (Aubry *et al.* 1992, Aubry 1994a, b). The electronic kinetic energy term

$$H_k = -\frac{1}{2} \sum_{\langle i,j \rangle, \sigma} e^{i\theta_{i,j}} c_{i,\sigma}^+ c_{j,\sigma} \tag{9a}$$

and the phonon kinetic energy term

$$H_Q = \frac{1}{2} \sum_i \mathbf{p}_i^2 \tag{9b}$$

have dimensionless coefficients

$$t = \frac{T\hbar\omega_0}{4g^2} \quad \text{and} \quad \beta = \frac{1}{4}\left(\frac{\hbar\omega_0}{2g}\right)^4 \tag{10}$$

respectively.

When the electron–phonon coupling is large enough, that is $\hbar\omega_0 \ll g$, we have $\beta \ll 1$, which suggests that we neglect in the first step the phonon kinetic energy term H_Q. This is the adiabatic approximation that can be also viewed as a mean field approximation to the phonon variable. This approximation is improved by the Born–Oppenheimer (BO) approximation, in which the electronic wave functions are supposed to be in equilibrium with the lattice deformation. By elimination of the electronic degree of freedom, an atomic (non-retarded) effective potential is generated. A second-order expansion of this potential yields the phonon modes with renormalised frequencies. We noted (Aubry 1991) that the BO approximation should break down when a gapless (or almost gapless) phonon appears (e.g. a phason mode). To be strictly valid, the adiabatic approximation requires, besides the condition that β be small, that there are no gapless phonons (except of course the acoustic phonons) or equivalently no continuous ground-state degeneracy and vanishing PN energy barrier. As a result, Peierls–Fröhlich charge density waves cannot be treated within this approximation, while it remains consistent for bipolaronic charge density waves provided that phonon softening does not approach too closely the zero frequency.

Superconductivity cannot be obtained in principle within the adiabatic or the BO approximation but is obtained in the opposite 'anti-adiabatic' regime when on the contrary the quantum state of the lattice is supposed to be in adiabatic equilibrium with the electrons. An effective electron–electron (non-retarded) attraction results from elimination of the lattice degree of freedom. In contrast, this approximation becomes valid at small electron–phonon coupling when $g \ll \hbar\omega_0$. the essential problem is to understand the situations that are intermediate between the adiabatic and anti-adiabatic regimes and hence the phase transitions between superconducting states and insulating bipolaronic states that have to take place.

The transfer integral T being usually much larger than the phonon energy $\hbar\omega_0$, the coefficient t is not necessarily small when β is small. The adiabatic Hamiltonian

$$H_{ad} = H_{AI} + tH_k = \sum_i \frac{1}{2}(\mathbf{u}_i^2 + \mathbf{n}_i\mathbf{u}_i) + \mathscr{U}n_{i,\uparrow}n_{i,\downarrow} - \mathfrak{h}(n_{i,\uparrow} - n_{i,\downarrow}) - \boldsymbol{\mu}\mathbf{n}_i$$

$$- \frac{t}{2}\sum_{\langle i,j\rangle,\sigma} e^{i\theta_{i,j}} c_{i,\sigma}^+ c_{j,\sigma} \tag{11}$$

is obtained from (7) by taking $\beta = 0$. Then, $\mathbf{u}_i$ commutes with H_{ad} and can be considered as a scalar variable. For a given set $\{\mathbf{u}_i\}$, H_{ad} has a ground state (which is unknown when $\mathscr{U} \neq 0$) with energy $\Phi_{ad}(\{\mathbf{u}_i\})$. We have proved that, when t is not too large, $\Phi(\{\mathbf{u}_i\})$ have infinitely many local minima (metastable states), corresponding to bipolaronic, polaronic or mixed states.

We often use an equivalent representation for this Hamiltonian (11) for reasons that are partly due to our habitual earlier notation

$$\frac{H_{ad}}{t} = \sum_i \left(\frac{1}{2}\mathbf{v}_i^2 + k\frac{1}{2}\mathbf{n}_i\mathbf{v}_i + Un_{i,\uparrow}n_{i,\downarrow} - h(n_{i,\uparrow} - n_{i,\downarrow}) - \boldsymbol{\mu}\,\mathbf{n}_i\right) - \frac{1}{2}\sum_{\langle i,j\rangle,\sigma} c_{i,\sigma}^+ c_{j,\sigma} \tag{12a}$$

where the new variable is $\mathbf{v}_i = \mathbf{u}_i/\sqrt{t}$ and the dimensionless parameters are

$$k = \frac{1}{\sqrt{t}} = \frac{2g}{(T\hbar\omega_0)^{1/2}}, \qquad U = \frac{\upsilon}{2T}, \qquad h = \frac{\mu_B\mathfrak{h}}{2T} \qquad \text{and} \qquad \mu = \frac{\mu}{2T}. \tag{12b}$$

With these new notations, the energy unit is now $2T$ instead of $E_0 = 2Tk^2$ defined by (7b), and the anti-integrable limit is obtained for $k \to \infty$. When the Hubbard term vanishes ($U = 0$), the electronic state is a Slater determinant. For a given configuration $\{\mathbf{v}_i\}$, the electronic ground-state energy of H_{ad} is

$$\frac{\Phi_{ad}(\{\mathbf{v}_i\})}{t} = \sum_i \frac{1}{2}\mathbf{v}_i^2 + \sum_v \sigma_v(E_v(\{\mathbf{v}_i\}) - 2\mu) - hM, \tag{13a}$$

where $E_v(\{\mathbf{v}_i\})$ are defined as the eigenenergies of

$$-t\sum_{j:\langle i,j\rangle} e^{i\theta_{i,j}}\psi_j^v + \mathbf{v}_i\psi_i^v = E_v(\{\mathbf{v}_i\})\psi_i^v \tag{13b}$$

and σ_v is an occupation number. We have $\sigma_v = 1$ when $E_v < \mu_F - h$ (doubly occupied states), $\sigma_v = 0$ when $E_v > \mu_F + h$ (empty states) and $\sigma_v = \frac{1}{2}$ when $\mu_F - h < E_v < \mu_F + h$ (singly occupied states). The number of these states is M.

Although our exact bounds are more restrictive, the bipolaronic structures are roughly obtained when $4td<1$ (d is the model dimension) while the polaronic and mixed polaronic–bipolaronic structures require t a factor of two smaller. These values physically correspond to moderately large electron–phonon couplings.

3 Single polarons and bipolarons: dimensionality, lattice discreteness and magnetic field

Before discussing the many bipolaron and polaron structures, let us discuss the simpler case of a single electron without magnetic field. In that case, the Hubbard term cannot play any role. A single electron localises in the potential well generated by the lattice deformation when the coupling constant to this lattice is large enough[1]. By definition, this electron associated with its lattice deformation is called a polaron. A pair of electrons in the same eigenstate but with opposite spins (singlet state) generates a deeper potential well and a stronger localisation of the electrons. This pair of electrons associated with the lattice deformation is called a bipolaron. It is physically more stable than two polarons far apart, if the Coulomb electron–electron repulsion is not too large.

The formation of polarons and bipolarons at low electron–phonon coupling depends very much on the dimensionality and discreteness of the lattice. We can illustrate this by considering the above Holstein model without magnetic field where the lattice site $i=r$ is taken as a continuous variable. The model in d dimensions has the energy

$$\Phi(\{\mathbf{u}(r)\}) = \frac{1}{2} \int \mathbf{u}^2(r)\, dr + \frac{1}{2}\, E(\{\mathbf{u}(r)\}), \tag{14a}$$

where $E(r)$ is the lowest eigenenergy of the Schrödinger equation

$$-t\, \Delta\, \Psi(r) + \mathbf{u}(r)\Psi(r) = (E(\{\mathbf{u}(r)\}) + 2t\,d)\, \Psi(r). \tag{14b}$$

Minimisation with respect to $\mathbf{u}$ yields

$$\mathbf{u}(r) = -\tfrac{1}{2}|\Psi(r)|^2, \tag{15a}$$

which implies that $\Psi(r)$ satisfies the non-linear Schrödinger equation

---

[1] We do not consider here the case of large polarons in polarisable insulators (Emin 1995, this proceedings), which originate from long-range Coulomb forces. This concept can hold only for very low densities of polarons. At larger densities, when the polarons overlap, there is a self-screening of the long-range interactions that removes the self-localisation effect. For that reason, we do not expect that the concept of large bipolarons can be useful in high-T_c superconductors in which the carrier density is not very small.

$$-t\Delta\Psi(r)-\tfrac{1}{2}|\Psi(r)|^2\Psi(r)=(E(\{\mathbf{u}(r)\})+2t\,d)\,\Psi(r) \tag{15b}$$

with the normalisation condition $\int|\Psi(r)|^2\,\mathrm{d}r=1$. The one-dimensional model ($d=1$) has the explicit solution

$$\Psi(x)=\left(\frac{\gamma}{2}\right)^{1/2}\frac{1}{\cosh\gamma\,(r-r_0)} \tag{16a}$$

with $\gamma=1/(8t)$ and eigenenergy

$$E=-2t-\frac{1}{64\,t}. \tag{16b}$$

Its total energy is $\Phi=-2t-1/(96\,t)$. There is a positive energy gain $1/(96\,t)$ in forming a polaron instead of having an extended electron with energy $-2t$.

This picture is drastically changed (Emin and Holstein 1976, Emin 1982) when the dimension d is larger than 1, because the non-linear Schrödinger equation has no normalised solution that is easy to prove by dimensionality arguments. Let us assume that the variational energy (14) has some extremum given by the normalised electronic wave function $\Psi(r)$. Let us consider now the family of normalised functions $\Psi_\gamma(r)=\gamma^{d/2}\Psi(\gamma r)$ with the associated lattice distortion $\mathbf{u}_\gamma(r)=-\tfrac{1}{2}|\Psi_\gamma(r)|^2$. The kinetic energy of this wave function

$$K_\gamma=t\langle\Psi_\gamma(r)|\Delta|\Psi_\gamma(r)\rangle=t\int\gamma^d\gamma^2|\nabla\Psi(\gamma r)|^2\,\mathrm{d}r=\gamma^2K>0 \tag{17a}$$

is strictly positive and its potential energy is

$$P_\gamma=\int\left(\frac{1}{2}\,\mathbf{u}_\gamma^2(r)+\frac{1}{2}\,\mathbf{u}_\gamma(r)|\Psi_\gamma(r)|^2\right)\mathrm{d}r=-\frac{1}{8}\int|\Psi_\gamma(r)|^4\,\mathrm{d}r=\gamma^{2-d}P<0 \tag{17b}$$

is strictly negative. K and P are the kinetic energy and potential energy of the initial wave function $\Psi(r)$ (with $\gamma=1$), respectively. Since $\Psi(r)$ corresponds to an extremum of the total energy, the function $\gamma^2K+\gamma^{2-d}P$ should be at its extremum for $\gamma=1$ that implies $2K+(2-d)P=0$.

For $d=1$, this condition implies $2K=-P$, which is indeed fulfilled by the exact solution (16). For $d=2$, this condition implies that $K=0$, which is impossible since K is strictly positive. For $d>2$, we should have $2K=(d-2)P$, which is also impossible because K and P have opposite signs.

Therefore, the continuum approximation is not appropriate for describing the polaron when $d\geq2$. Howevever, the polaron instability does exist in the discrete model provided that t be not too large or equivalently that k be large enough, which can easily be proven. Let us choose as variational state a single-site electronic wave function localised at site 0 and associated with the

Polaron Energy

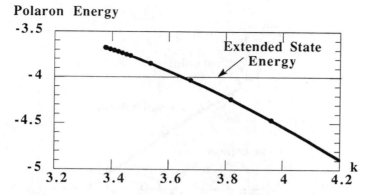

Fig. 1. Polaron energy versus $k = 1/\sqrt{t}$ at zero magnetic field. The energy unit is 2 T. A first-order transition from localised to extended occurs for $k \equiv 3.65$.

distortion $u_0 = -\frac{1}{2}$. The total energy of this state is $-\frac{1}{8}$. Then, the polaronic solution has necessarily less energy than the extended electron energy $-2t \, d/2$ when $t < \frac{1}{8}d$ or $k > 2(2d)^{1/2}$. The numerical calculation (Fig. 1) confirms this result for the adiabatic model in two dimensions. There is a first-order transition as a function of the electron–phonon coupling at $k \cong 3.65$ between the polaronic state (k large or t small) and the extended state.

When the Hubbard term is zero, it is easy to check that the energy of a bipolaron is lower than the energy of two polarons. Using dimensionality arguments, we find for the energy of a bipolaron that $E_{\text{bip}}(k) = 2 \, E_{\text{pol}}(k\sqrt{2})$. Thus, if the bipolaronic instability occurs at $k = k_{\text{b}}$ and the polaronic instability at $k = k_{\text{p}}$, then we have $k_{\text{b}} = k_{\text{p}}/\sqrt{2}$.

There should exist a physically interesting regime close to this adiabatic transition because then the quantum lattice fluctuations have to be taken into account for raising this almost degeneracy. The quantum state of the electron should be a resonant state between localised polarons at each site and extended states. Because the atoms of the lattice are much heavier than the electrons, the quantum life time of the polaron at a given site should be much longer than the life time of a free electron. If one were to assume that a similar resonance can occur in systems with many electrons (which we expect at the Fermi energy for such systems in the intermediate electron–phonon coupling regime), then they should appear experimentally as a mixed valence system. Similar situations are known in systems with transition metals and rare earths (heavy-fermion systems), but as far as we know their standard interpretation discards the role of electron–phonon coupling (at least as being essential).

Although there are no polarons and bipolarons in the continuous model in more than one dimension, this argument breaks down in the presence of a

Serge Aubry

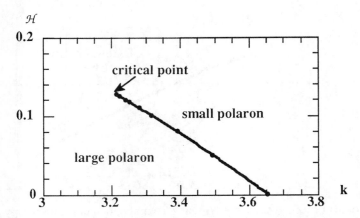

Fig. 2. The first-order transition line of a single polaron ended by a critical point in the plane k, $\mathcal{H}$ (orbital effect). The magnetic field unit corresponds to the number of quantum fluxes per plaquette of the square lattice.

magnetic field. We have proved rigorously for two-dimensional continuous models at 0 K (Aubry and Kuhn, in preparation) that a single electron coupled to a classical deformable medium and submitted to an orthogonal non-zero and uniform magnetic field self-localises as a polaron[2].

This polaronic localisation persists in the 2D discrete Holstein model with a uniform magnetic field. As shown in Fig. 2, the first-order transition between the extended state and the polaronic state becomes a first-order transition line between a large polaron and a small polaron region. When crossing this line by increasing k or $\mathcal{H}$, the large polaron collapses into a small polaron. This first-order transition line in the k, $\mathcal{H}$ diagram ends at a critical point (see Fig. 2) with coordinates $k \cong 3.20$ and $\mathcal{H} \cong 0.13 \times 2\phi_0/s_p$, where s_p is the area of a plaquette of the square lattice. For k and $\mathcal{H}$ small, the energy $\mathcal{E}_p$ of the large polaron is found to be approximately equal to (but strictly smaller than) $\frac{1}{2}\hbar\omega_c[1 - k^2/(8\pi)]^{1/2}$ ($\hbar\omega_c = e\hbar\mathcal{H}/m_e = 2\mu_B\mathcal{H}$ is the cyclotron energy) and its size $\xi \approx \xi_c\sqrt{1-k^2/8\pi}$ remains comparable with the cyclotron radius $\xi_c = [2\hbar/(e\mathcal{H})]^{1/2}$.

When k is moderately large (although not sufficient for generating a polaronic instability by itself without magnetic field), the polaronic localisation energy gain becomes a significant fraction of the cyclotron energy. This

---

[2] A polaronic localisation at small electron phonon coupling, occurs similarly in continuous three dimensional systems as soon as the magnetic field $\mathcal{H}$ is not zero. The localisation length perpendicular to $\mathcal{H}$ is almost the cyclotron radius while the localisation length along the direction of the (weak) magnetic field is much larger and proportional to the inverse square magnetic field. As a result, the localisation energy gain is negligible in practice.

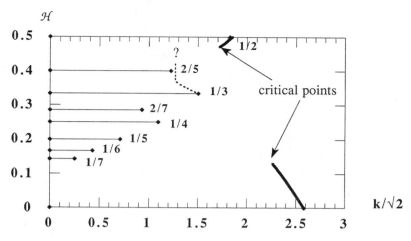

Fig. 3. Some aspects of the phase diagram of a single polaron in a magnetic field and on a two-dimensional discrete lattice versus the electron–phonon coupling k and the magnetic field $\mathcal{H}$ (in quantum flux units as for Fig. 2). On the thin lines, the extended state is the most stable. The end point of this line corresponds to a first-order transition between an extended state and a polaron. First-order transition lines between large and small polarons starting from the end points of the rational magnetic fields have been found for $\mathcal{H} = 0$ (see Fig. 2) for $\mathcal{H} = \frac{1}{2}$ (thick lines) and perhaps for the magnetic field $\mathcal{H} = \frac{1}{3}$ (dotted line).

polaronic localisation effect might become important in two-dimensional systems at low temperature, low electronic density and high magnetic field. At very large magnetic field, the commensurate flux effects due to the discreteness of the lattice become important. When the number of quantum fluxes per plaquette is fixed at a rational number r/s (r and s irreducible), the two-dimensional electronic Hamiltonian becomes periodic with a larger unit cell and, as when the magnetic field is zero, there is a first-order transition at $k_p(r/s)$ between a polaronic state and an extended state. Although Fig. 3 is incompletely calculated, we can see thin straight lines at constant rational field r/s for several rationals, the end points of which are at $k_p(r/s)$.

First-order transition lines ending with a critical point between large and small polarons starting from these end points were observed for $r/s = \frac{1}{2}$ and $\frac{1}{3}$ that are qualitatively similar to the one shown in Fig. 2 at zero magnetic field. There are first-order transition lines ending with critical points at all rational fields. A complete calculation will be presented (Aubry and Kuhn, in preparation). It is also worthwhile to mention that, since the formation of polarons necessarily occurs in low-dimensional systems, it will be favoured especially in quasi-1D systems. In Fig. 4 is shown the variation of the critical value k_p of k versus the anisotropy parameter α, at which the first order transition between

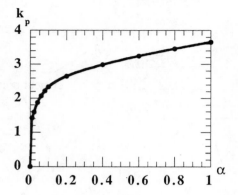

Fig. 4. k_p versus the anisotropy α.

the extended electron and the polaron occurs. This parameter α is defined by replacing the kinetic operator

$$H_k = -\frac{1}{2} \sum_{\langle i,j \rangle, \sigma} c_{i,\sigma}^+ c_{j,\sigma}$$

in the two-dimensional Holstein model (12a) by

$$H_k = -\frac{1}{2} \sum_{\langle i,j \rangle_x, \sigma} c_{i,\sigma}^+ c_{j,\sigma} - \frac{1}{2} \alpha \sum_{\langle i,j \rangle_y, \sigma} c_{i,\sigma}^+ c_{j,\sigma}, \tag{18}$$

where $\langle i,j \rangle_x$ represents the lattice bonds in the x direction and $\langle i,j \rangle_y$ the lattice bonds in the y direction.

Thus $\alpha = 1$ corresponds to the initial isotropic 2D Holstein model and $\alpha = 0$ to the 1D Holstein model. When α goes to zero, k_p also goes to zero. As expected, the polaron shape is very anisotropic and is rather extended along the x chains when the first-order transition to an extended state occurs. This effect should be important for understanding quasi-1D CDWs, which we claim to be necessarily bipolaronic.

When a magnetic field is present, the effect of the anisotropy will be not only to shift the first-order transition line shown in Fig. 2, at smaller values of k but also to flatten the curve at smaller magnetic field. The physical consequence is that, in quasi-one-dimensional systems, this transition line between larger and smaller polarons will be found at much smaller and possibly observable magnetic fields (Aubry and Kuhn, in preparation).

4 Many-polaron and many-bipolaron structures at large electron–phonon coupling

When there are many electrons in the system, it is not clear whether the concept of polarons and bipolarons is meaningful since these objects strongly overlap. One could believe them to be indistinguishable. This is not true because we can prove that the concept of polarons and bipolarons can receive an explicit mathematical definition in that Holstein–Hubbard model with local interactions and then prove their existence at large enough electron–phonon coupling, whatever the electronic density.

It is first useful to recall some well-known features due to the electron–phonon coupling in many-electron systems. One-dimensional systems are special because they are necessarily unstable by a lattice deformation at any non-vanishing electron–phonon coupling, as soon their electronic band is not completely empty or filled. As we saw above, this instability already occurs with a single electron at any electron–phonon coupling. When there are many electrons, the 1D systems develop a periodic (Peierls–Fröhlich) charge density wave (CDW) associated with a periodic lattice distortion (PLD) with wave vector $2k_F$ twice the Fermi wave vector k_F. These theories are usually developed in the limit of small electron–phonon coupling and assume that the CDW–PLD modulation is almost sinusoidal. Similar mean field theories have been developed for predicting the existence of spin density waves (SDW) when electron–electron interactions are taken into account.

In two- and three-dimensional models, the Peierls–Fröhlich instability theory extends almost identically on condition that there is perfect nesting of the Fermi surface by some wave vector corresponding to those of the lattice instability. This situation may happen in very special models, but generally the nesting of the Fermi surface is not perfect even for quasi-1D systems. A lattice instability is found only when the electron–phonon coupling exceeds a certain critical value. Standard theories describe the electronic system in terms of the coexistence of a Fermi liquid due to the pockets of electrons left at the Fermi surface by the imperfect nesting and a CDW–PLD assumed to be almost sinusoidal. We know from our preliminary numerical tests that this standard description is in fact quite different from the real behaviour, although deeper studies are required.

For models in any dimension, the lattice instability becomes obvious at large electron–phonon coupling. This limit is recognised as an 'anti-integrable limit' of the variational form corresponding to the lattice distortion energy. The concept of anti-integrability has many existing and potential applications not only for dynamical systems (Aubry 1994a, b, MacKay and Aubry 1994) but more generally for variational problems, for proving the existence of chaotic solutions. It just consists of proving that the trivial chaotic solutions that can be obtained at this anti-integrable limit can be continued away from this limit, according to the implicit function theorem. The applicability of this theorem requires that one check that the hypothesis of this theorem holds, which is not automatically true in all models and often needs complex proofs (which can be made simple in some paedagogical examples).

We do not present this complete proof for our problem (that has already been described elsewhere), but just describe the anti-integrable limit that makes the result 'physically' intuitive. At the adiabatic limit, the Holstein–Hubbard model becomes a variational problem with respect to the lattice distortion. The large electron–phonon coupling corresponds to the 'anti-integrable' limit $t=0$ of Hamiltonian (11), which yields Hamiltonian H_{AI} that is easily solved because it commutes with the electron density operators $n_{i,\uparrow}$ and $n_{i,\downarrow}$. Then, at $t=0$, the electrons are trivially localised and for a given configuration $\{\mathbf{u}_i\}$, the ground-state energy $\Phi_{ad}(\{\mathbf{u}_i\})$ of H_{AI} is

$$\Phi_{ad}(\{\mathbf{u}_i\}) = \sum_i V(\mathbf{u}_i), \tag{19a}$$

where, by appropriate choice of chemical potential μ, the local potential

$$V(\mathbf{u}_i) = \underset{\sigma_i}{\text{Min}} \left[\tfrac{1}{2}\mathbf{u}_i^2 + \sigma_i \mathbf{u}_i + 2\mathscr{U}\sigma_i(\sigma_i - \tfrac{1}{2}) + 2\,\mathfrak{h}\sigma_i(\sigma_i - 1) + 2\mu\sigma_i\right]$$

$$= \underset{\sigma_i}{\text{Min}} \left[\tfrac{1}{2}(\mathbf{u}_i + \sigma_i)^2 + 2(\mathscr{U} + \mathfrak{h} - \tfrac{1}{4})\sigma_i^2 - \sigma_i(\mathscr{U} + 2\,\mathfrak{h} + 2\mu)\right] \tag{19b}$$

can be either a double or a triple well (instead of a single well) with minima at $-\sigma_i$, where σ_i is an occupation number as shown in Fig. 5. It is defined as follows.

1 $\sigma_i = 1$ if there are two electrons with opposite spins at site i. Then the energy minimum of $V(\mathbf{u}_i)$ is obtained for the lattice distortion $\mathbf{u}_i = -1$. There is a (single site) bipolaron at site i.
2 $\sigma_i = 0$ if there is no electron at site i and the energy minimum is obtained for lattice distortion $\mathbf{u}_i = 0$. There is a hole at site i.
3 $\sigma_i = \tfrac{1}{2}$ if there is a single electron at site i. The energy minimum is then obtained for the lattice distortion $\mathbf{u}_i = -\tfrac{1}{2}$. Then, there is a single polaron at site i.

If potential $V(\mathbf{u})$ indeed has a minimum at $\mathbf{u} = -1$, then the bipolarons are stable at this anti-integrable limit. Identically, if $\mathbf{u} = 0$ is a minimum, holes are stable and if $\mathbf{u} = -\tfrac{1}{2}$ is a minimum, polarons are stable.

When $\mathscr{U} < -\mathscr{H}$, the Min in (19b) can only be obtained for $\sigma_i = 0$ or 1. For

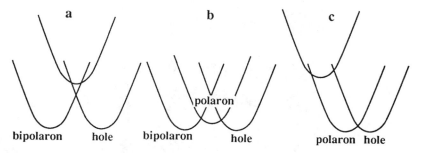

Fig. 5. The scheme of the local adiabatic potential when $\mathcal{U} < -\mathcal{H}$ (a), when $-\mathcal{H} < \mathcal{U} < \frac{1}{2} - \mathcal{H}$ (b) and when $\frac{1}{2} - \mathcal{H} < \mathcal{U}$ (c).

example, $V(\mathbf{u})$ becomes a double well with a symmetry axis at $\mathbf{u} = -\frac{1}{2}$ when $2\mu = \mathcal{U} - \frac{1}{2}$ (see Fig. 5a). Then, only bipolarons and holes are stable.

When $-\mathfrak{h} < \mathcal{U} < \frac{1}{2} - \mathfrak{h}$, the Min in (19b) can be obtained for the three possible values $\sigma_i = 0, \frac{1}{2}$ or 1 and $V(\mathbf{u})$ becomes for example a symmetric triple well when $2\mu = \mathcal{U} - \frac{1}{2}$ (see Fig. 5b). Then, bipolarons, polarons and holes are stable.

When $\frac{1}{2} - \mathfrak{h} < \mathcal{U}$, the Min in (19b) can only be obtained for two values, which can be $\sigma_i = 0$ and $\frac{1}{2}$ or $\sigma_i = 1$ and $\frac{1}{2}$ according to μ (for example $2\mu = -\frac{1}{4} - \mathfrak{h}$ yields $\sigma_i = 0$ and $\frac{1}{2}$).

At the anti-integrable limit ($t=0$), the variational form $\Phi_{\mathrm{ad}}(\{\mathbf{u}_i\})$ has infinitely many extrema that are characterised by their pseudo-spin configurations $\{\sigma_i\}$. Our theorem (Aubry, in preparation), which in part extends early theorems for the adiabatic Holstein model, takes a very simple general form,

Theorem: When the coefficient t of the electronic kinetic energy (with or without magnetic field), is non-vanishing and not too large ($t < t_{\mathrm{m}}$), there exists for each minimum $\{\mathbf{u}_i(0)\} = \{-\sigma_i\}$ of $\Phi_{\mathrm{ad}}(\{\mathbf{u}_i\})$ at $t=0$, at least one uniformly close local minimum $\{\mathbf{u}_i(t)\}$ of $\Phi_{\mathrm{ad}}(\{\mathbf{u}_i\})$, i.e.

$$\lim_{t \to 0} \left(\mathrm{Sup}_i |\mathbf{u}_i(t) - \mathbf{u}_i(0)| \right) = 0.$$

This theorem provides a natural and simple definition of the concept of bipolaronic, polaronic and mixed structures by continuation from their limits at $t=0$, where their existence is obvious. It holds despite the fact that the global electronic wave function becomes a mess and that it is *a priori* impossible to define independent wave functions for the bipolarons and polarons. Thus, one sees that bipolaronic structures exist only for $\mathcal{U} < \frac{1}{2} - \mathfrak{h}$, polaronic structures for $-\mathfrak{h} < \mathcal{U}$ and mixed polaronic structures in the overlap region when $-\mathfrak{h} < \mathcal{U} < \frac{1}{2} - \mathfrak{h}$. This theorem asserts that, when the quantum electronic kinetic energy term ($t \neq 0$) is switched on, an undistorted metallic state is not restored

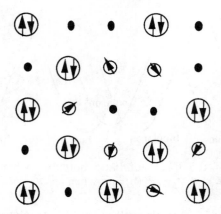

Fig. 6. The scheme of an arbitrary random mixed polaronic-bipolaronic structure at
the anti-integrable limit.

immediately but that *each* bipolaronic, polaronic or mixed structure that exists
at $t = 0$ persists by continuity up to some t_m, whatever the electronic density and
irrespective of whether their distribution is *chaotic* or not (see Fig. 6). In Fig. 7 is
shown qualitatively the region of the t, $\mathcal{U}$ phase diagram for which the theorem
applies. However, this result is incomplete when the Hubbard term is not zero,
because we did not prove the unitariness of the minimum $\{\mathbf{u}_i(t)\}$ for each choice
$\{\sigma_i\}$ and did not find exact physical properties of these solutions. In fact, there
are good reasons for not expecting unitariness of the solution in all cases and
especially for structures without magnetic field involving polarons. The (real)
spin configuration of the polarons is degenerate at $t = 0$, but a magnetic
exchange interaction is expected to appear at t not zero between the localised
(quantum) spins of the polarons. The antiferromagnetic interaction that results
could raise the spin degeneracy. A simple magnetic spin ordering may appear if
the dimension of the model is high enough and if the underlying polaron
structure is ordered, for example periodic. When the underlying polaronic
structure is chaotic, it could be a spin glass with infinitely many magnetic
metastable states associated with it. The resulting structure may also have no
long-range magnetic order in one-dimensional models.

Another possible situation that occurs when $\mathcal{U}$ belongs to some intermediate
range roughly between $\frac{1}{4}$ and $\frac{1}{2}$ is that the polarons form non-magnetic pairs in
their singlet state. In that case, we recognise a spin–Peierls instability for the
magnetic structure. We shall prove in the following that this effect already
occurs with only two polarons provided that the Hubbard coefficient belongs
to some well-determined interval. We consider that this pairing effect is
necessary (although not sufficient) for generating high-T_c superconductors.

At the present stage, our theorem proves the existence of a large number

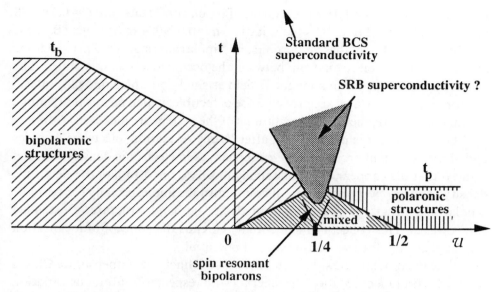

Fig. 7. $(t, \mathcal{U})$ domains of proven existence (qualitative) for the bipolaronic ($\sigma_i = 0$ or 1), polaronic ($\sigma_i = 0$ or $\frac{1}{2}$) or ($\sigma_i = 1$ or $\frac{1}{2}$) and mixed polaronic–bipolaronic ($\sigma_i = 0, \frac{1}{2}$ or 1) structures in the adiabatic Holstein–Hubbard model. There is an overlap region between the domain of existence of bipolaronic and polaronic structures in which mixed polaronic–bipolaronic structures exist. In that region the validity of the adiabatic approximation is expected to break down for rather small values of β. It is conjectured that the quantum lattice fluctuation could produce a high-T_c superconducting phase with a high T_c.

(with non-zero thermodynamic entropy) of bipolaronic, polaronic or mixed metastable states that are local minima of the global variational energy function of the lattice distortions. Discarding the possible magnetic metastable states of the real spins of the polarons, their charge configurations are labelled by the set of pseudo-spins $\{\sigma_i\}$. When t increases beyond t_m, these metastable states $\Phi_{ad}(\{\mathbf{u}_i\})$ are expected to disappear gradually through a complex cascade of bifurcations, which first reduces their total entropy. In two and more dimensions, when t becomes sufficiently large, a metallic state is finally restored through complex transitions and transformations that have not yet been analysed in detail.

5 Properties of bipolaronic and polaronic structures in the adiabatic Holstein model

We have more exact results when the Hubbard term is zero ($\mathcal{U} = 0$). We shall focus now on these results. We have then the proof of unitariness of the continued solution and explicit bounds. Up to now, the best (minimum) bounds that were found for the existence of bipolaronic states are $t_b = 0.181648$ in 1D, $t_b = 0.107982$ in a 2D square lattice, $t_b = 0.079084$ in 3D, etc. (Baesens and McKay 1994). All these polaronic and bipolaronic states are proven to be insulating with an electronic gap between the occupied and empty states. The polaronic states appear in a magnetic field as singly occupied midgap states.

Moreover, these states are proven to be defectible, which means that, unlike in semiconductors, the charge excitations of the system are localised (Aubry *et al.* 1992). Indeed, it is proven that, after lattice relaxation, electrons or holes added to the system produce only local changes in electronic density. The net result is that they appear as exponentially localised distortion of the electronic density. This property holds for all configurations and especially for periodic bipolaronic structures that exhibit an electronic band structure of extended Bloch waves. This concept of defectibility extends the concept of localisation that applies only for electrons in a rigid potential.

Unlike the bipolaronic structures, Peierls–Fröhlich incommensurate CDWs form a continuum of solutions degenerate with respect to continuous variation of the phase. A consequence of this property is the existence of a gapless phonon mode associated with the phase fluctuations, which is called a phason. There is no energy barrier to sliding the phase continuously. Since phase translation is associated with an electric current, such a system is a Fröhlich conductor. Unlike these CDWs, the bipolaronic CDWs and more generally the bipolaronic structures do not form a discrete set of configurations. An explicit physical consequence is that the phonon modes associated with charge fluctuations now exhibit a non-zero gap as well as a non-zero PN energy barrier. This PN barrier is the minimum energy that has to be provided to the system for displacing electric charge and its associated lattice distortion. These features were numerically analysed in detail in the 1D case (Aubry and Quemerais 1989). The existence of this barrier is related at finite temperature to bipolaron hopping and also plays an important role in bipolaron tunnelling when quantum lattice fluctuations are taken into account.

Without magnetic field, the ground state is proven to be a bipolaronic state obtained for a special ordering $\{\sigma_i\}$ of the bipolarons although this is unknown in general. In the special case of the half-filled Holstein model (at any dimensionality), it has recently been proven (Lebovitz and Macris 1994) that the ground state is just the (expected) periodic 'chess board' array of bipolarons. The one-dimensional model is simpler. Numerical analysis suggests that, for irrational band fillings, the ground state is a quasi-periodic array of bipolarons or bipolaronic CDW (Aubry and Quemerais 1989). It becomes periodic with period s when the band filling becomes a rational r/s. This CDW has just the fundamental wave vector $2k_F$ expected from the theory of the

Peierls instability but it does not have all the properties of a Peierls–Fröhlich CDW because the phason spectrum has a non-vanishing gap, and the CDW phase is pinned to the lattice.

It is noteworthy in one-dimensional models that the wave vector of the bipolaronic CDW might not be the Peierls wave vector $2k_F$ in slightly modified Holstein models. For example, when just a first-neighbour coupling $(C/2)(u_{i+1} - u_i)^2$ with C small is added to the elastic energy (3), we prove analytically the existence of phase separations for the ground state between two commensurate bipolaronic phases with different densities and we observe this effect numerically even at relatively small electron–phonon coupling (e.g. $k \approx 1.7$ for $C = 0.1$). On increasing the electron–phonon coupling, there are cascades of phase separations between phases with rational densities following backward the complete sequence of rational approximants of the initial density (Raimbault and Aubry, in preparation). One also expects to observe this infinite cascade of phase separations in reverse order when the temperature is increased from zero. This process should end at some accumulation point at a temperature smaller than that above which the standard Peierls structure is restored. Obviously, in physical situations, long-range Coulomb forces between bipolarons will prevent this phase separation. More complex CDWs with modified wave vectors not those predicted by Peierls appear.

Nevertheless, in the 1D adiabatic Holstein model (and many variations including the model just mentioned), a second-order transition ('transition by breaking of analyticity' or TBA) between an incommensurate bipolaronic CDW (at large coupling) and an incommensurate Peierls–Fröhlich CDW (at small coupling) (with gapless phason mode) occurs on decreasing the electron–phonon coupling, at critical values of k depending on the band filling, but which typically are in the range below but close to 1.58. This transition exhibits the same phenomenology and universality classes as the TBA of the Frenkel–Kontorowa model that was proven to exist on the base of Kolmogorov and Arnol'd Moser theories (Aubry and Quemerais 1989). An effective shape can be defined for the bipolaron at non-vanishing electron density. The size and shape of the bipolarons depend on the electron density. The bipolarons are more extended for incommensurate bipolaronic CDWs. Their size is critical and diverges at the TBA. A consequence is the occurrence of transitions between insulating states and conducting (Peierls–Fröhlich) states when doping from any commensurate (or rational) band filling and when the electron–phonon coupling is not too large.

In contrast, we never numerically observed Peierls–Fröhlich phases in the 2D adiabatic Holstein model at 0 K (even in highly anisotropic quasi-1D cases) at non-zero but arbitrary electronic densities. We are led to conjecture the existence of four regimes as a function of electron–phonon coupling. The first regime is obtained at large electron–phonon coupling, where, according to our theorems, we easily confirm numerically that there are insulating chaotic bipolaronic structures and the ground state has to be such an insulating but ordered bipolaronic structure. When the coupling becomes lower, we found by following continuously the bipolaronic structures obtained at large electron–phonon coupling that there is some evidence for a second regime that involves the coexistence of bipolaronic CDWs, which are possibly chaotic in excited states, and a Fermi surface with extended (or quasi-extended) electronic wave functions. At lower values of electron–phonon coupling, it seems that there is still a third regime, in which CDW modulation of the ground state disappears (no more stable bipolarons) but there is still a local asymmetric double well for the atoms, which was noted by Mott (1993) (see Fig. 1). This local double well disappears in the fourth regime at lower coupling, when a pure metallic state is recovered.

We have not yet explored in detail the phase diagram that deserves complete numerical study in the adiabatic regime. Note that we have implemented new techniques for fast numerical calculation of bipolaronic and polaronic structures (Kuhn and Aubry 1994), which should also be efficient for that purpose in two dimensions. The role of the quantum lattice fluctuations neglected in this adiabatic model remains clearly totally negligible in the first regime at large electron–phonon coupling, but it could play an essential role in the second and third regimes of intermediate electron–phonon coupling because of the possible quantum resonances of the electrons between bipolaronic, polaronic and extended states at the Fermi surface (see Section 4) (there is a local anharmonic double well for the lattice distortion). The fourth regime is of course appropriate for exhibiting standard BCS superconductivity.

Because of the existence of infinitely many metastable states close in energy to the ground state, the low-temperature behaviour of the insulating bipolaronic CDWs (large electron–phonon coupling) has to be described by Ising-like lattice gas models. The bipolaronic interactions at large electron–phonon coupling can be obtained in principle by an expansion with respect to t (the electronic kinetic energy) and yield not only pair interactions but also a series of multispin interactions.

Such expansions were performed at the lowest order by Alexandrov *et al.* (1986) and the next order by Freericks (1993) including also the quantum lattice fluctuations ($\beta \neq 0$ in (10)). These calculations have the flaw that the bipolaron interactions that are taken into account are only those of nearest or next-nearest neighbours, which generates for the Ising part of the Hamiltonian a degenerate ground state when the electron density is low and artificially favours a superconducting state. The role of the small quantum terms of the Hamiltonian could then be drastically enhanced. Moreover, the ground states of these Hamiltonians can be only commensurate structures with short periodicity while the possible incommensurate CDWs are discarded.

We calculated to lowest non-vanishing order in $t = 1/k^2$ the two-body bipolaron interaction J_n at any distance n in the adiabatic Holstein model. The calculation of Aubry and Quemerais (1989) can be extended to any dimension and yields (with energy unit $2T$)

$$J_n \approx 2|n|k^2 \frac{|\phi_n|^2}{|\phi_0|^2}, \tag{20a}$$

where $\{\phi_n\}$ is defined as the bounded state solution of the eigenequation

$$-\Delta\phi_n - k^2\delta_{n,0}\phi_n = E_0\phi_n \tag{20b}$$

that exists only when k is large enough.

In the 1D case[3], we get

$$J_n \approx 2|n|k^2 \left(\frac{1}{k^2}\right)^{2|n|}. \tag{20c}$$

The two-body interaction is repulsive at all distances and decays exponentially with distance. The approximate pseudo-spin Hamiltonian truncated to the two-body interactions is

$$H_{ps} = \frac{1}{2} \sum_{\langle i,j\rangle} J_{i-j}\, \sigma_i\sigma_j - \eta\sigma_i \tag{20d}$$

(bond $\langle i,j\rangle$ is considered as different from $\langle j,i\rangle$, η is a chemical potential). In the 1D case and for $t = 1/k^2$ small, this interaction is convex (i.e. we have $J_{m+1} - 2J_m + J_{m-1} \geq 0$ for $m \geq 2$). Although small, these terms play an essential role in determining the ground state. It is proven (Hubbard 1978, Pokrovsky and Uimin 1978) that the 1D ground states of model (20d) are either incommensurate or commensurate structures (depending on the band filling) with Peierls wave vector $2k_F$. If this interaction is truncated to nearest neighbour interaction ($|n| = 1$), then the pseudo-spin potential interaction $S_i^z S_j^z$ proposed by Alexandrov *et al.* (1986) is recovered but the model has no incommensurate ground state. It exhibits only the dimerised state as commen-

[3] We discovered recently an error in the calculation of J_n (Aubry and Quemerais 1989) because a correction of the electronic energies due to lattice relaxation was lost. This new calculation agrees quite well with the exact numerical calculation (Fig. 8). A similar error was made in calculation of the polaron interaction with a Hubbard term (Aubry 1994c). The correct result is given in (22).

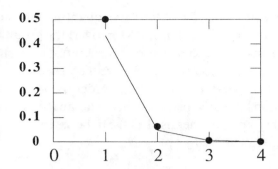

Fig. 8. The interaction J_n between two bipolarons (calculated numerically) in the 1D Holstein model, versus their distance (black dots) compared with the expansion (20b) (thin line) at $k = 2$.

surate ground state in addition to the trivial full and empty band states. The intermediate ground states are obtained as the phase separation between two of these phases. In Fig. 8 our expansion (20c) is compared with the exact interaction calculated numerically between only two bipolarons at distance n in the one-dimensional model. These interactions essentially correspond to the overlapping of the electronic wave function of the bipolarons. They tend to form bipolaronic CDWs with wave vectors close to those predicted by analysis of the nesting of the Fermi surface. In real systems, there should exist two other important interactions between bipolarons that are not taken into account in this calculation, but which should become dominant at large electron–phonon coupling. They are the elastic energy through the phonon dispersion and the electrostatic Coulomb energy. It has been shown (Raimbault and Aubry, in preparation) that these interactions can change the CDW structure drastically.

6 Magnetic field effects on bipolaronic CDWs

Concerning the effect of the magnetic field, orbital effects in two and more dimensions do not change the conclusion of the above theorem. We can say, furthermore, that the upper bounds of validity are enhanced. A magnetic field favours formation of polaronic and bipolaronic structures even when the discreteness of the lattice is not taken into account. We recently proved the following result (Aubry, in preparation) supporting this assertion.

A system of free electrons in two dimensions[4] coupled to a classical continuous deformable elastic medium and submitted to a uniform magnetic

---

[4] As for the single-polaron instability, three-dimensional systems in a magnetic field are in principle unstable with respect to formation of a CDW or SDW because there are many incompletely filled Landau levels. However, this instability is much weaker and its structure is difficult to predict precisely.

field orthogonal to the system is necessarily unstable at 0 K with respect to formation of a 2D polaronic charge density wave (probably a triangular lattice) provided that the last Landau level occupied by the electrons be not completely filled (or empty).

The wave vector of this instability is proportional to the inverse of the electron density on this last occupied Landau level and thus should be rather small in practice. Nevertheless, this effect might become relevant in some physical experiments concerning two-dimensional electronic systems at low density with consequences for the transport properties. Otherwise, and as for the single polaron, this amplitude and the physical consequence of this effect should be strongly enhanced in systems that are highly anisotropic (mostly quasi-one-dimensional but also quasi-two-dimensional ones). This result shows that, in principle, a good nesting condition is not necessary for generating CDWs and SDWs under a magnetic field, but that it certainly helps in magnification of this instability (Gork'ov and Lebed' 1984).

Concerning the Zeeman effect of the magnetic field, we prove that at large electron–phonon coupling and when the magnetic field becomes strong enough, the ground state of the Holstein model becomes a mixed polaronic–bipolaronic structure (Aubry *et al.* 1992) but we have no general prediction concerning the precise structure of this ground state.

We analysed a simple example, that is the one-dimensional half-filled Holstein model (Kuhn and Aubry 1994). In that case, the Peierls prediction would be that, with a magnetic field, two lattice modulations occur, one with wave vector $2k_{F\uparrow}$ corresponding to a gap opening for the electrons with spin $\downarrow$ and equivalently $2k_{F\downarrow}$ for the electron with spin $\downarrow$. Since the electron density per site is just 1, these two wave vectors are equal apart from a reciprocal lattice wave vector. The situation is simple because the two modulations just lock with one another. We calculated numerically the associated mixed polaronic and bipolaronic structures. These structures can be also understood as quasi-periodic arrays of 'neutral solitons' that are localised excitations of half-filled 1D CDWs with no charge but with spin $\frac{1}{2}$ (Su, Schrieffer and Heeger 1979, 1980). These incommensurate structures also exhibit TBAs as a function of the electron–phonon coupling. The magnetisation of the system versus the magnetic field varies as a Devil's Staircase that envelops approximately the Pauli response that would be obtained without electron–phonon coupling. Each plateau of this magnetisation then corresponds to a commensurate array of neutral solitons. This Devil's Staircase is incomplete at low electron–phonon coupling and becomes complete when it becomes large enough ($k > 1.8$). The non-half-filled 1D Holstein model in a magnetic field might have more complex ground states that are not even quasi-periodic.

7 Polaron pairing: spin-resonant bipolarons

We now turn back to the adiabatic Holstein–Hubbard model with $U > 0$. In that case, no methods that are both efficient and exact are known for performing a numerical study of the predicted polaronic and bipolaronic structures. We can, however, easily study a pair of electrons while treating the effect of the Hubbard term exactly. Then, it appears that, in the region corresponding to breaking of a bipolaron into two polarons (within the mean field description), an interesting binding energy between two polarons appears due to the spin fluctuations of the electrons. One gets then what we call a spin-resonant bipolaron (SRB), which is not the same object as the standard bipolaron because it has a different code $\{\sigma_i\}$ at the anti-integrable limit. A bipolaron can be viewed as a pair of polarons located at the same site while a spin-resonant bipolaron is also a pair of polarons but located at different sites. These two objects can coexist in some regions of the phase diagram as metastable states of differing energy.

As shown by the sketch of phase diagram Fig. 7, the bipolarons must break into polarons at large $\mathcal{U}$. In the intermediate region $\mathcal{U} \approx 1/4$ and at the anti-integrable limit $t = 0$ (or $g^2 \approx \frac{1}{2} v \hbar \omega_0$ in the initial notation), the energy of a single-site bipolaron is equal to the energy of two (single-site) polarons that are far apart. When $t = 1/k^2$ is not zero, the energy of a pair of polarons is decreased by raising the spin degeneracy in the singlet state $1/\sqrt{2}\,(|\uparrow\downarrow\rangle + |\downarrow\uparrow\rangle)$. In the limit t small (or k large), the lowest order term of the interaction potential V_n can be calculated for positive U, at any dimension as a function of the lattice distance $|n| = \Sigma_\alpha |n_\alpha|$ between the polarons. We found for $n \geq 1$ (in energy units $2T$)

$$V_n \approx \tfrac{1}{2}\{U - [U^2 + r_n^2 k^2 (k^2 - 2U)]^{1/2}\}, \tag{21a}$$

where $r_n(k)$ is defined as the ratio

$$r_n^2 = \phi_n^2 / \phi_0^2 \tag{21b}$$

for the bound state $\{\phi_m\}$ of the eigenequation

$$-\Delta \phi_m - \frac{k^2}{2}\,\delta_m \phi_m = E \phi_m \tag{21c}$$

that necessarily exists for k large. For $n = 0$, the polaron interaction is just $-k^2/4 + U$ at leading order. This ratio r_n decays exponentially with distance $|n|$. In one dimension, we have $r_n = (2/k^2)^n = (2t)^n$. At long distance n, we have

$$V_n \approx -r_n^2 \frac{k^2}{2}\left(\frac{k^2}{2U} - 1\right). \tag{22}$$

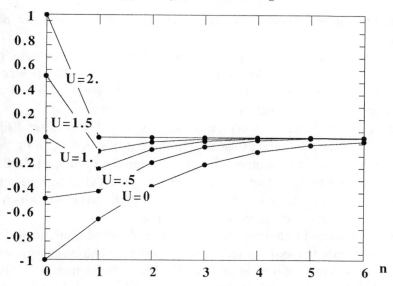

Fig. 9. The interaction potential at $k = 2$ and several values of U between two polarons in the 1D Holstein–Hubbard model (calculated from a large-k expansion), (energy units are 2 T).

This interaction is attractive at long distance when $U < k^2/2$ ($\mathscr{U} < \frac{1}{2}$) and becomes repulsive for $U > k^2/2$. However, at short distance, when $U > k^2/4$ the on-site interaction becomes repulsive and the minimum energy is obtained when the polarons belong to adjacent sites. In Fig. 9 this potential V_n is shown versus the distance obtained from this expansion, for several values of U. Note that the existence of points on this curve for all distances n does not necessarily imply that the corresponding pairs of polarons at all distance n are metastable. (refer to the sketch in Fig. 7 for the region of minimum metastability). In the original units, at large electron–phonon coupling, this region is defined by the condition on the Hubbard energy

$$2\,\frac{g^2}{\hbar\omega_0} \lesssim v \lesssim 4\,\frac{g^2}{\hbar\omega_0}. \tag{23}$$

Exact numerical calculations that are currently being developed for a singlet pair of polarons in the adiabatic Holstein model confirm that the minimum of the interacting energy is indeed obtained for two polarons located at different sites when $k^2/4 \lesssim U \lesssim k^2/2$. When k is not very large, this minimum is not necessarily obtained for nearest neighbour polarons. Moreover, we can determine the exact region of metastability for each polaron pair (which depends on their distance) and those of a standard bipolaron. We thus obtain precisely the contour lines of the phase diagram Fig. 7 for the two polaron configurations.

The interaction between two polarons in the singlet state becomes repulsive at all distances for larger U. When the polarons are not on the same sites, the pairing energy is maximum when $U \approx k^2/4$ and then the polarons are on adjacent sites. The two-dimensional case will also be investigated.

Since we believe that the interaction between two SRBs is repulsive, the ground state of the adiabatic Holstein–Hubbard model will be in that regime and at low enough density there will be an ordered structure of SRBs. This structure appears to be in fact a spin–Peierls magnetic structure because this polaron pairing takes the place of a standard magnetic structure, which would appear within a mean field approach or at larger U.

Another important remark is that is has been observed numerically, at least in the 1D Holstein–Hubbard model, that the Peierls–Nabarro energy barrier of the SRBs is strongly depressed relative to those of the standard bipolaron (Aubry 1993a) in the moderately large coupling regime ($k \approx 2$ in 1D) when the bipolaronic state reaches the SRB regime. This is due to the existence of metastable states of several other SRBs close in energy to those with the minimum energy. There is a rather flat region in energy in the configuration space. Because of that feature the quantum lattice fluctuations could delocalise the SRBs and form an unusual superconducting state. This point motivates our interest in this new type of electron pairing for understanding the high-T_{c} superconductors.

8 Bipolarons with quantum lattice fluctuations

All the above study on polarons and bipolarons was done with a classical lattice deprived of any kind of fluctuations, quantum or thermal, that is by making coefficient β vanish in (7a). We would like now to give some insights about the quantum corrections, which, at the present stage, have been studied with no Hubbard term.

The most standard quantum correction close to the mean field (adiabatic) energy of an adiabatic configuration (e.g. our bipolaronic and polaronic configurations) can be obtained within the Born–Oppenheimer approximation by expanding the electronic energy versus the atomic displacements at second order. The atomic motion treated in a quantum manner in this harmonic effective potential yields renormalised phonon frequencies that are globally softened. We then always obtain a reduction of the zero-point energy of the phonons that thus explicitly depends on the phonon spectrum (in a more general framework, this correction is nothing other than the standard RPA approximation)[5]. However, this correction does not treat the essential question about the quantum tunnelling of bipolarons and polarons.

For that purpose, we note that the adiabatic approximation for the Holstein model with Hamiltonian $\mathbf{H} = H_k + H_p + H_{ep}$ is equivalent to assuming the validity of the mean field approximation $\mathbf{n}_i(a_i^+ + a_i) \approx \langle \mathbf{n}_i \rangle (a_i^+ + a_i) + \mathbf{n}_i \langle a_i^+ + a_i \rangle - \langle \mathbf{n}_i \rangle \langle a_i^+ + a_i \rangle$ in the initial Hamiltonian. Each of the bipolaronic metastable states (with configuration $\{\mathbf{v}_i\}$ and occupied electronic eigenstates φ_i^ν with Hamiltonian (12)) has a mean field wave function denoted

$$|\{\mathbf{v}_i\}\rangle = \exp\left[\sum_i \left(\frac{T}{\hbar\omega_0}\right)^{1/2} \mathbf{v}_i(a_i^+ - a_i)\right] \prod_\nu c_{\nu,\uparrow}^+ c_{\nu,\downarrow}^+ |\varnothing\rangle \qquad (24a)$$

obtained as the result of a product of displacement operators by $\mathbf{v}_i$ and of creation operators of electrons in the occupied electronic states $c_{\nu,\sigma}^+ = \Sigma_i \varphi_i^\nu c_{i,\sigma}^+$ for $\sigma = \uparrow$ or $\downarrow$ to the vacuum state with no electrons and no lattice distortion $|\varnothing\rangle$. The overlap of two arbitrary meanfield wave functions $|\{\mathbf{v}_i\}\rangle$ and $|\{\mathbf{w}_i\}\rangle$ can be explicitly calculated as

$$|\{\mathbf{v}_i\}|\{\mathbf{w}_i\}\rangle = (\det A)^2 \exp\left(\frac{-T}{2\hbar\omega_0} \sum_i (\mathbf{v}_i - \mathbf{w}_i)^2\right), \qquad (24b)$$

where A is the matrix of scalar products $A_{\nu,\nu'} = \Sigma_i \varphi_i^{\nu*} \phi_i^{\nu'}$, where φ_i^ν and ϕ_i^ν are the electronic wave functions of states $|\{\mathbf{v}_i\}\rangle$ and $|\{\mathbf{w}_i\}\rangle$ respectively and

$$\langle\{\mathbf{v}_i\}|\mathbf{H}|\{\mathbf{w}_i\}\rangle = \frac{1}{2}\left(\langle\{\mathbf{v}_i\}|\mathbf{H}|\{\mathbf{v}_i\}\rangle + \langle\{\mathbf{w}_i\}|\mathbf{H}|\{\mathbf{w}_i\}\rangle - T\sum_i (\mathbf{v}_i - \mathbf{w}_i)^2\right) \times \langle\{\mathbf{v}_i\}|\{\mathbf{w}_i\}\rangle. \qquad (24c)$$

If one were to consider only the mean field eigenstates of translated single bipolaronic states $|\{\mathbf{w}_i = \mathbf{v}_{i+n}\}\rangle$ at site n, which are degenerate in energy but are not orthogonal, then one would obtain a matrix D with overlap coefficients $D_{m,n} = \langle\{\mathbf{v}_{i+m}\}|\{\mathbf{v}_{i+m}\}\rangle$ and H with coefficients $H_{mn} = \langle\{\mathbf{v}_{i+m}\}|\mathbf{H}|\{\mathbf{v}_{i+n}\}\rangle$, which yields a band of bipolarons with eigenstates $\Sigma_n \exp(iqn)|\{\mathbf{v}_{i+n}\}\rangle$. Their eigen-energies are $\Phi_b(q) = \Phi_b - \Sigma_n \Gamma_n \cos(qn)$ where Φ_b is the adiabatic energy of the bipolaron without quantum lattice correction (except for the zero-point phonon energy) and Γ_n is the off-diagonal term of $(D^{-1/2}HD^{-1/2})_{i,i+n}$, which is independent of i. This band energy can be introduced as a quantum correction to the pseudo-spin Ising Hamiltonian (20d), where $\sigma_i - \frac{1}{2} = S_i^z$, which is

---

[5] The unitary Wagner transformation (Wagner 1981) yields explicitly the exact correction to the BO Hamiltonian in a somewhat similar way to that in which the Lang–Firsov unitary transformation yields the correction to the anti-adiabatic limit (where the atomic positions stay in adiabatic equilibrium with the electronic state). Then the anti-adiabatic correction to the BO approximation can be estimated and found to give sharp inconsistency in the BO approximation in the case of a gapless phason mode (Aubry 1991).

$$H_Q = -\frac{1}{2} \sum_{\langle i,j \rangle} \Gamma_{i-j} (S_i^x S_j^x + S_i^y S_i^y) \tag{25}$$

(in energy units $2T$). Γ_n decays exponentially with distance $|n|$. In the large-k limit, the main quantum correction is given by the nearest neighbour terms Γ_1 and then one recovers at leading order the quantum Hamiltonian for bipolarons proposed by Alexandrov *et al.* (1986), where

$$\Gamma_1 \approx \frac{2}{k^2} \exp\left[-\left(\frac{2g}{\hbar\omega_0}\right)^2 \right] \tag{26}$$

and $\Gamma_n = 0$ for $|n| > 0$. This term can be associated with the quantum tunnelling energy of a bipolaron through the PN energy barrier generated by the discreteness of the lattice. At large electron–phonon coupling the PN barrier of a bipolaron diverges as $E_{PN} = 2Tk^2/4 = 2g^2/(\hbar\omega_0)$ and this tunnelling energy is approximately $(\hbar\omega_0 T^2/g^2) \exp(2 E_{PN}(\hbar\omega_0)]$ in the original units.

Neglecting this tunnelling term is a good approximation in the bipolaronic regime when $g \gg \hbar\omega_0$, which is usually true in physical systems. Indeed, in more than one dimension, we only have small bipolarons with a large PN barrier, when they appear. It is not only much smaller than the nearest neighbour Ising term J_1 (which is $2/k^2$) but also much smaller than the long-range Ising interactions $J_n \propto 2|n|k^{2-4|n|}$ up to $|n| \approx (g/\hbar\omega_0)^2/\ln k$. As a result, for large electron–phonon coupling, the bipolaronic CDWs remain an insulating bipolaronic structure (unless the bipolaron density becomes extremely low, but then the critical superconducting temperature would also be extremely small).

In two-dimensional and higher models, intermediate mixed structures with both superconductivity and bipolaronic CDWs are expected in the intermediate regimes mentioned in Section 5. There exist real compounds, for example in the series of transition metal dichalcogenides, exhibiting coexistence of CDWs and a metallic state and many other unusual and non-interpreted features reminding one of mixed valence features (e.g. 2H-NbSe$_2$ below 33 K) (Ayache 1994). According to our theory, these bipolaronic CDWs with a Fermi surface must exhibit small bipolarons, unlike the quasi-1D compounds in which the bipolarons could be more extended in the chain directions. (They should exhibit as a consequence a rather weak Kohn anomaly in the phonon spectrum of the CDW phase instead of a gapless phason mode.) These systems also become superconducting (e.g. 2H-NbSe$_2$ below 6 K) but the transition temperature is not exceptionally high.

For high-T_c superconductivity, increasing only the electron–phonon coupling turns out to become inefficient as soon as the electronic pairing occurs in real space and transforms part of the band into localised small bipolarons. Superconductivity totally disappears when the whole band is transformed into bipolarons.

9 BCS and bipolaronic superconductivity: two limits of the same state

A system of electrons can become superconducting when two conditions are both fulfilled, namely the electrons are bound into pairs and the wave functions of these pairs are extended and form a coherent state.

In the standard BCS theory of superconductivity, it is assumed that the electron–phonon coupling reduces to an instantaneous attractive interaction between the electrons. This essential assumption can be proven at leading order, when treating the electron–phonon coupling as a small perturbation. At larger coupling, this interaction becomes more and more retarded, which breaks down the validity of the BCS theory and requires more sophisticated expansions (e.g. the Eliashberg theory), where the phonon frequencies turn out to be renormalised. In the BCS theory, the electrons form pairs (named Cooper pairs) with two electrons with both opposite wave vectors and opposite spins $q\uparrow$ and $-q\downarrow$. The BCS wave function is a combination of pair operators applied to the vacuum state $|\varnothing\rangle$. It has the form $\Pi_q(u_q + v_q c^+_{q\uparrow} c^+_{-q\downarrow})|\varnothing\rangle$, where u_q and v_q are variational coefficients. It is essential to note that, in the BCS theory, pairing of electrons occurs self-consistently in reciprocal space because of the mean field of the other electronic pairs and together with their coherent and thus superconducting ordering. This pairing and the associated gap totally disappear just above the critical temperature. The characteristic size of the Cooper pairs is large relative to the lattice spacing.

In the opposite case, when the electron–phonon coupling is large enough, electron pairing occurs by localisation in real space of two electrons in a potential well created self-consistently by the lattice distortion and small bipolarons form. Unlike the BCS case, this pairing occurs independently for each pair of electrons and does not require any phase coherence between the bipolaronic wave functions. Taking into account quantum lattice effects, the bipolaronic states can be modelled by a hard-core boson model on a lattice described formally by a quantum Hamiltonian involving spin $\frac{1}{2}$ operators $S^z_i = \sigma_i - \frac{1}{2} = +\frac{1}{2}$ when a bipolaron is present around site i and $S^z_i = -\frac{1}{2}$ if no bipolaron is present.

Alexandrov, Ranninger and Robaszkievicz (1986, ARR) proposed an approximation of this bipolaronic Hamiltonian with the form

$$H_{\mathrm{ARR}} = \sum_{\langle i,j \rangle} J S^z_i S^z_j - \Gamma(S^+_i S^-_j + S^-_i S^+_j) - \mu S^z_i, \qquad (27)$$

where $\langle i,j \rangle$ are neighbouring sites, J corresponds to the interaction energy between two nearby bipolarons, Γ corresponds to the tunnelling energy (or coherence energy) of a bipolaron between two nearby sites and μ to the Fermi energy. We have shown that this ARR Hamiltonian can be recovered from our adiabatic approach that also shows that important interactions between bipolarons extending beyond the nearest neighbouring sites were neglected. ARR found that the ground state of this Hamiltonian may have two kinds of ordering as a function of electron–phonon coupling and electron concentration:

1 either the spin $\frac{1}{2}$ orders antiferromagnetically in the z direction and then one gets a dimerised crystal of bipolarons (charge density wave) or
2 the spin $\frac{1}{2}$ orders ferromagnetically in the x–y plane, which yields a bipolaronic superconducting state with a critical temperature energy $k_B T_{sc}$ that is at most Γ.

At large electron–phonon coupling g, we find that the coherence energy Γ given by (26) drops exponentially to zero so that the system reaches the adiabatic limit at which the electronic state follows adiabatically the lattice deformation. (This is the basic assumption of the standard Born–Oppenheimer approximation.) Γ becomes negligible not only relative to the nearest neighbour bipolaron interaction J but also to any interaction at a given distance n between bipolarons (see estimation (20) above). The consequence is that, even without taking into account the extra Coulomb interactions between bipolarons, the ground state has to condense into an insulating bipolaronic structure with spatial ordering (a bipolaronic CDW) for any given non-vanishing bipolaron density and at large enough electron–phonon coupling.

An ideal limit for expecting superconductivity would be obtained in the quite unusual physical situation in which the electron–phonon coupling energy remains much smaller than the phonon energy quanta. This is the anti-adiabatic limit that is the inverse adiabatic limit in which the lattice deformation follows adiabatically the electrons (contrary to the Born–Oppenheimer or adiabatic limit). There is an effective instantaneous attractive interaction between the electrons. Such a situation leads for the Holstein model to a negative-U Hubbard model (Nozières and Schmitt-Rink 1985). Then, in the limit of large electron–electron attraction, $|U|$ is large and one gets localised pairs of electrons, also called bipolarons. The ground state is well described by Bose condensation of these hard-core bosons, namely as a bipolaronic superconductor.

The conditions required for having close to ideal bipolaronic superconductivity in a real system might be exceptionally realised in nature. In contrast, at large electron–phonon coupling, condensation of the electrons into a bipolaronic structure with spatial ordering is the common case and yields an insulating state. Note that, for a similar reason, superfluidity in condensed matter is exceptional because the interaction potential energies involved between most kinds of atoms are usually much larger than their kinetic energies that make the crystal more stable at low temperature. Helium 4 and 3 at low pressure are the only exceptions.

Nevertheless, we believe that the existence of superconductors that can be better understood as bipolaronic rather than BCS superconductors is physically plausible although not frequent. For finding such a situation, there should exist some extra reason, so that the coherence energy Γ (band width) of a bipolaron become very large. If Γ becomes large enough compared to J_1, Hamiltonian (27) should have a superconducting ground state with a pseudo-spin ordering in the x–y plane.

If Γ is smaller than the binding energy Δ of the two electrons forming a bipolaron (possibly a SRB), then this superconductivity has to be considered as a bipolaronic superconductivity. Just above the transition temperature, there is an incoherent state of bipolarons (liquid), which gradually break either into polarons or free electrons at a cross-over temperature $k_B T_{co} \approx \Delta$. If $\Gamma > \Delta$, then pair breaking occurs at the superconducting transition $k_B T_c \approx \Delta$ and the conducting state above T_c is a liquid of either polarons or electrons.

If one makes the electron–phonon coupling g vary from zero, then the pairing energy Δ increases monotonically (and diverges as $k^2/4$ with k when $v = 0$) and the coherence energy Γ decreases monotonically (at zero coupling g the effective mass of an electron pair is twice the mass of the bare electron when $v = 0$) and goes to zero as $(2/k^2)\exp\{[2g/\hbar\omega_0)]^2\}$ when g diverges (and when $v = 0$). At small g, the BCS theory holds because $\Delta \ll \Gamma$. At large g, one must have bipolaronic superconductivity (if it survives into the bipolaronic CDW). Within that picture, the BCS and bipolaronic limits play a role similar to the displacive and order–disorder limits respectively for a model of structural phase transition.

To make this comparison more explicit, let us consider for example the extended 'Ising model' with continuous spins u_i at site i with Hamiltonian

$$H_{st} = \Delta \sum_i (u_i^2 - 1)^2 + \frac{\Gamma}{2} \sum_{\langle i,j \rangle} (u_i - u_j)^2$$

where u_i is a continuous variable at site i, $\langle i,j \rangle$ are neighbouring sites and Δ and Γ are coupling constants. In two and more dimensions, this model has a phase transition at some critical temperature T_c.

The limit $\Delta \gg \Gamma$, is the order–disorder limit. Then, the model is well described by an Ising model, where u_i becomes again a pseudo-spin $\sigma_i = \pm 1$. The critical temperature T_c is essentially determined by Γ. There is an extended critical region exhibiting scaling laws and, above T_c, the local pseudo-spin remains well defined with a rather long life time although its long-range order is broken.

In contrast, the displacive limit is obtained when $\Delta \ll \Gamma$, that is when the Hamiltonian H_{st} is almost harmonic. The transition that is obtained exhibits practically mean field behaviour with only a very narrow critical region close to T_c. In addition, it has a well-defined soft mode, albeit in a narrow region of q space. Unlike the order–disorder case, the fluctuations of local order parameter average very rapidly to zero so that it makes no sense to define a local pseudo-spin above T_c. The transitions of the model can be well understood within mean field approximations, which in spirit are similar to those used by BCS (e.g. the Hartree approximation that consists of replacing the anharmonic terms u_i^4 by $3\langle u_i^2 \rangle u_i^2$).

There is also a continuum of intermediate situations between the order–disorder and the displacive regimes (e.g. see Aubry 1975). The deviation from the displacive regime for a structural phase transition can be qualitatively and easily recognised by the extension of the critical region around the critical temperature in which the mean field behaviour is not valid. By analogy, we argue that there should be no transitions but a continuum of intermediate cases between the bipolaronic and the BCS limits. Unlike the standard BCS superconductors that exhibit mean field behaviour even close to the critical temperature, a characteristic of the tendency of a superconductor to become bipolaronic can be recognised in the existence of non-mean-field critical exponents for the critical quantities (specific heat, penetration depth etc...) observable in a significant region above and below T_c. These critical fluctuations should appear as unusual fluctuations of vortices in some finite region above and below the critical temperature. As shown in this workshop by Schneider (1994), high-T_c superconductors (cuprates) indeed exhibit non-trivial critical exponents. There are also now many other experimental facts suggesting an important role for polarons and (or) bipolarons in high-T_c superconductors (Mott 1993, see also this workshop proceedings).

In fact, there is no conceptual difference between bipolarons (when they form coherent states of delocalised bipolarons) and Cooper pairs. The difference appears in their technical descriptions, which are very different but should finally yield superconducting states that are physically identical.

10 Conjecture: high-T_c superconductivity as a consequence of the competition between strong electron–phonon and Hubbard interactions

For high-T_c superconductivity, both the electronic pairing energy Δ and the coherence energy Γ of the pairs should be strong with magnitude at least comparable to the critical temperature energy. In models involving only electron–phonon coupling, we have seen that these two conditions are rather contradictory. Increasing the electron–phonon coupling leads unavoidably to the system becoming a bipolaronic insulator instead of a bipolaronic superconductor. For that reason, superconductivity can exist in these models only at low electron–phonon coupling and is then close to the BCS type.

For increasing the maximum critical temperature T_c, we suggest that competition between the electron–phonon energy and the Hubbard term must play an essential role in enhancing Γ. We have seen that formation of strongly bound SRBs occurs in a rather limited region of the phase diagram precisely when $2\,g^2/(\hbar\omega_0) \lesssim v \lesssim 4g^2/(\hbar\omega_0)$. Then, the Peierls–Nabarro barrier, which is required to be overcome in order to move the SRBs in the adiabatic lattice, turns out to be strongly depressed (it can become almost zero in the intermediate electron–phonon coupling regime in which bipolarons start to form when v has the appropriate value).

By the effect of the quantum lattice fluctuations, the quantum state of the SRB becomes a highly resonant state with the other metastable states close in energy so that the tunnelling energy Γ should be strongly enhanced. It becomes much larger than it would be without the Hubbard term for the standard bipolaron. If Γ becomes large enough, then it may overcome the potential terms that favour spatial ordering of the SRBs (which should appear as a spin–Peierls structure).

Our conjecture is that, with reasonable values of physical parameters, $T, \hbar\omega_0$, g and v, the coherence energy Γ could become large enough to generate superconductivity at relatively large electron–phonon coupling when usually the system should be a bipolaronic insulator, which could be a bipolaronic CDW (U too small, strongly bound pairs but not enough coherence energy), a polaronic magnetic structure (U too large, no pairing energy or not enough pairing energy) or a spin–Peierls structure (U intermediate, the SRBs form but they have insufficient coherence energy and condense into a structure ordered in real space).

Roughly speaking, the most favourable cases for high-T_c superconductivity should be obtained when the PN barrier of the SRB almost completely vanishes. In that event, this tunnelling (or coherence) energy Γ of the SRB could become comparable to the phonon energy $\hbar\omega_0$. Otherwise, the pairing energy Δ that originates from the magnetic interaction between the neighbouring spins of the polarons could remain comparable to this magnetic interaction provided that the size of the pair is not too large. The characteristic energy $k_B T_{sc}$ of the critical temperature should range about the minimum of the two values Γ and Δ.

In our model, when the electron density is weak enough, this polaron pairing will surely occur if U is in the appropriate interval. This polaronic pairing energy should be highly dependent on the electronic density because it is competing with an antiferromagnetic ordering occuring for polaron close packing. When the size of a SRB becomes comparable to the average distance between polarons, this pairing should become ineffective and we should get a magnetic polaronic structure. In the half-filled case, there is one polaron per site so that the lattice distortion becomes uniform, which yields an apparently standard antiferromagnet. Thus, it is not unreasonable to believe that, by doping, one could get a transition from an antiferromagnet to a superconductor.

The real high-T_c superconductors are cuprates that involve several electronic bands hybridising the oxygen and copper ions. The electron–phonon coupling concerns mostly deformation of the oxygen square while the Hubbard term concerns mostly the copper sites. Although this Holstein–Hubbard model does not involve the detailed band structure of these systems, it mimics one of the most essential characteristics, that is the competition between magnetism and superconductivity. Because there are several bands, doping of the real systems probably also involves charge transfer between oxygen and copper, which in some sense should renormalise the effective electron–phonon coupling and Hubbard terms.

Among several other applications of our ideas, we suggest a plausible scenario in a real system as a function of doping. We assume that, beyond the antiferromagnetic phase, the polaronic binding energy Δ decreases while the coherence energy Γ increases. When they become equal, the maximum critical temperature has been reached. The underdoped regime should correspond to the situation $\Delta > \Gamma$ that corresponds to bipolaronic superconductivity with $k_B T_c \approx \Gamma$. Then, we should have coexistence of polaron pairs (SRBs) in the (poorly) conducting state that breaks gradually into a liquid of magnetic polarons when $k_B T \approx \Delta$. This fact could interpret the decay of the static magnetic susceptibility much above T_c. In the overdoped regime, $k_B T_c \approx \Delta$, the SRBs break into a polaron liquid slightly above the critical temperature. Then the static magnetic susceptibility of the polarons should behave more like a Curie law and thus increase when the temperature decreases with a sharp drop to zero at the superconducting temperature. Finally, when the maximum of the critical temperature is obtained for the intermediate optimised doping, that is when $\Delta \approx \Gamma$, the magnetic susceptibility is roughly constant above T_c.

In summary, although incomplete, our results obtained on the Holstein–Hubbard model as combinations of theorems, analytic expansions or numerical studies reveal at the present stage an incredibly rich and somewhat unexpected variety of behaviours. It turns out that a careful study of the insulating states (which are proved to be necessarily of either polaronic or bipolaronic structure), and of the effect of the quantum lattice fluctuations, could give the key to understanding the high-T_c superconductors.

To paraphrase Sir Nevill Mott, only those who believe in God (the polarons and bipolarons) have some chance to reach heaven (high-T_c superconductivity).

References

Alexandrov A. S., Ranninger J. and Robaszkiewicz S. 1986 *Phys. Rev.* B **33**, 4526.

Aubry S., Abramovici G. and Raimbault J. L. 1992 *J. Statist. Phys.* **67**, 675–780.

Aubry S. and Quemerais P. 1989 in *Low Dimensional Electronic Properties of Molybdenum Bronzes and Oxides* Ed. Claire Schlenker (Kluwer) pp. 295–405.

Aubry S. 1975 Ph.D. dissertation (Université Pierre et Marie Curie, Paris) (see also 1975 *J. Chem. Phys.* **62**, 3217–229 and 1976 *ibid.* **64**, 3392–402).

Aubry S. 1980 *Metal–Insulator Transitions in Id Deformable Lattices* in *Bifurcation Phenomena in* ... Ed. C. Bardos and D. Bessis (Reidel) pp. 163–84.

Aubry S. 1991 in *Microscopic Aspects of Non-Linearity in* ... Ed. A. R. Bishop, V. L. Pokrovsky and V. Tognetti NATO ASI Series, Series B vol. 264 (Plenum), pp. 105–14.

Aubry S. 1993a in *Phase Separation in Cuprate Superconductors* Ed. K. A. Müller and G. Benedek (World Scientific), pp. 304–34.

Aubry S. 1993b *J. Physique* 3 **C2**, 349–55.

Aubry S. 1994a *Physica* D **71**, 196–221.

Aubry S. 1994b in *Chaos, Order and Patterns: Aspects of Nonlinearity*, Como, *September 1993* Ed. G. Casati *Physica* D in press.

Aubry S. 1994c Exact Results and Conjectures in ... in *The Physics and the Mathematical Physics of the Hubbard Model, San Sebastian*, October 1993 NATO ARW in press.

Ayache C. 1994 private communication.

Baesens C. and MacKay R. S. 1994 *Nonlinearity* **7**, 59–94.

Emin D. 1982 *Phys. Today* June 1982 p. 34.

Emin D. 1995 This proceedings.

Emin D. and Holstein T. 1976 *Phys. Rev. Lett.* **36**, 323.

Eliashberg G. M. 1995 This proceedings.

Freericks J. K. 1993 *Phys. Rev.* B **48**, 3881–91.

Gork'ov L. P. and Lebed' A. G. 1984 *J. Physique Lett.* **45**, L433–40.

Hubbard J. 1978 *Phys. Rev.* B **17**, 494.

Kuhn C. and Aubry S. 1994. *J. Phys. Condens. Matter* **6**, 5891–8.

Landau L. 1933 *Phys. Z. Sowjet Union* **3**, 664.

Lebovitz J. and Macris N. 1994 *J. Statist. Phys.* **76**, 91–123.

Le Daëren P. Y. and Aubry S. (1983) *J. Phys.* C **16**, 2497–508.

MacKay R. S. and Aubry S. 1994 Proof of Existence of Breathers ... *Nonlinearity* in press.

Mott N. F. 1993 *J. Phys. Condens. Matter* **5**, 3487–506.

Nozières P. and Schmitt-Rink 1985 *J. Low Temp. Phys.* **59**, 195.

Pokrovsky V. I. and Uimin G. L. 1978 *J. Phys.* C **11**, 3535.

308 *Serge Aubry*

Raimbault J. L. and Aubry S. *Phase Separations in 1d Bipolaronic CDWs* in
 preparation.
Schneider T. 1995 This proceedings.
Su W. P. Schrieffer J. R. and Heeger A. J. 1979 *Phys. Rev. Lett.* **42**, 1968.
Su W. P., Schrieffer J. R. and Heeger A. J. 1980 *Phys. Rev.* B **22**, 2099.
Wagner M. 1981 *Phys. Stat. Sol.* (b) **107**, 617.

19

Coexistence of small-polaron and Anderson localization in high-T_c superconducting materials

JUN TATENO

Japan Atomic Energy Research Institute Tokai-mura, Naka-gun, Ibaraki-ken, 319-11, Japan

Abstract

The electrical features in the normal phase of high-T_c superconducting materials can be explained by the coexistence model of small polarons and Anderson-localized carriers. According to this model, with increasing carrier concentration, the degree of Anderson localization decreases, and then the concentration of coexisting small polarons increases, attains a maximum and decreases with this variation. If T_c is determined by the concentration of bosons (bipolarons) as in Shafroth's formula for Bose condensation, then the shape of the superconducting phase can be explained by this behavior of small polarons. The degree of localization in the oxides without superconductivity is too large for coexistence to be attained.

1 Introduction

In the high-T_c superconducting oxides, the superconducting phase appears just in the composition region in which a metal–insulator transition takes place. This fact leads to the idea that the electronic states proper to this transition are responsible for the origin of superconductivity. This transition accompanying a gradual change in electronic nature in accordance with stoichiometric variation or with carrier doping has the following characteristic features [1].

1 The electrical conductivity can be described by the variable-range hopping (VRH) mechanism at low temperatures in the insulator (semiconductor) region, which means that the carriers in this region are Anderson-localized. The degree of localization decreases with increasing carrier concentration resulting in the occurrence of the metallic phase. So this transition should be classified as an Anderson transition.

2 Plural types of carriers, itinerant and localized ones, coexist in both the metallic and the semiconducting phase. The change in intensity ratio of the Drude component

309

and mid-infrared peak in optical conductivity spectra with carrier doping [2, 3] seems to be evidence for coexistence and mutual transformation between these carriers. One type of carrier is confirmed to be the small polaron by measurement of Raman spectra [4]. Taking account of VRH conduction, the coexisting species are considered to be Anderson-localized carriers and small polarons.

Here we should refer to InO_x as an example to show a close relation between superconductivity and Anderson localization [5, 6]. It was reported that both Anderson localization and superconductivity ($T_c < 3$ K) appear in thin films of InO_x and that the degree of localization as well as the superconducting transition temperature vary with varying degree of disorder k_F, as if the carrier concentration x in the high-T_c superconducting oxides were replaced by the parameter k_F. The fact that the superconductivity can be controlled by changing the degree of disorder as in this case seems to indicate that the common feature of Anderson localization, which appears not only in the high-T_c superconducting oxides but also in other superconducting ones such as InO_x and $BaPb_{1-x}Bi_xO_3$, plays an important role in superconductivity. From this point of view, we analyze the Anderson transition in $La_{2-x}Sr_xCuO_4$ as a typical high-T_c superconductor in the following sections.

2 Anderson transition in $La_{2-x}Sr_xCuO_4$

Using the existing data on electrical resistivity in the normal phase of $La_{2-x}Sr_xCuO_4$, $\log(\rho/T)$ is plotted against T^{-1} in Fig. 1. Good linear relations exist between the plotted quantities in a limited concentration and temperature region. Then we have the nearest neighbor hopping conduction, where the conductivity $\sigma(=\rho^{-1})$ is given by

$$\sigma = (\sigma_0/T)\exp[-E_a/(kT)], \tag{1}$$

where σ_0 is a pre-exponential constant. The activation energy E_a decreases with increasing x.

The plots of $\log\rho$ versus $T^{-1/4}$ are shown in Fig. 2. Good linear relations exist also, which indicates that conduction takes place via VRH. The conductivity is given by [7]

$$\sigma = \sigma_V\exp[-c(T_0/T)^{1/4}], \qquad T_0 = A^3/(kN(E_f)) \qquad \text{and } c \simeq 2, \tag{2}$$

where $N(E_f)$ stands for the density of states at the Fermi level, σ_V for a pre-exponential constant and A^{-1} for the parameter representing the spread of the wavefunction in Anderson localization; it falls off with distance r as $\exp(-Ar)$. When we put d as the distance between nearest neighbor sites, we can represent the degree of localization by the dimensionless parameter Ad.

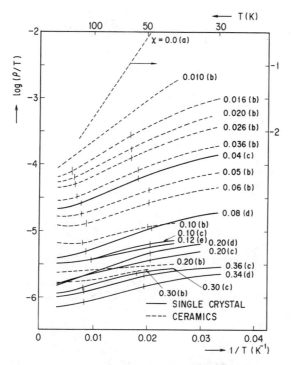

Fig. 1. A plot of $\log(\rho/T)$ against $1/T$ for $La_{2-x}Sr_xCuO_4$. The vertical lines express the limit of the linear relations. The data are adopted as follows: (a) [19], (b) [20], (c) [21], (d) [22] and (e) [23].

The limits of the linear relations are designated by the vertical lines in Fig. 1 and Fig. 2. Using these limiting marks, we can depict the electronic phase diagram classified by the conduction mechanism as shown in Fig. 3 [1]. The boundary designated T_1 in Fig. 3 is depicted using the vertical lines in Fig. 2, because the limit of the linear relation can be seen more clearly here than in Fig. 1.

In the VRH area the characteristic temperature T_0 obtained from Eq. (2) decreases with increasing x and the superconducting phase begins to appear at compositions with $T_0 \simeq 10$ for this material. We can calculate the degree of localization in $La_{2-x}Sr_xCuO_4$ as a function of x, using Eq. (2) and the estimated value of $N(E_f)$ [1]. It is difficult to estimate an accurate value of $N(E_f)$. Nevertheless, we can obtain a rather accuate value of A, because it contains $N(E_f)$ in the form of the third root. The obtained A^{-1} and Ad are shown in Fig. 4. From Fig. 4 Ad is found to vary as $Ad \propto x^{-1}$.

In Fig. 5 the values of Ad in $La_{2-x}Sr_xCuO_4$ are plotted against x. The data for $BaPb_{1-x}Bi_xO_3$, which is a superconducting oxide ($T_c = 13$ K) with VRH

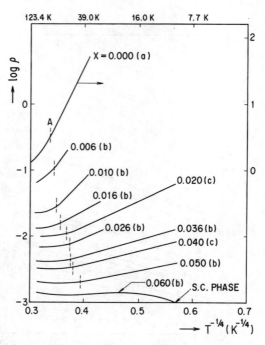

Fig. 2. A plot of $\log \rho$ against $T^{-1/4}$ for $La_{2-x}Sr_xCuO_4$. The vertical lines in Fig. 1 and Fig. 2 express the limit of the linear relations.

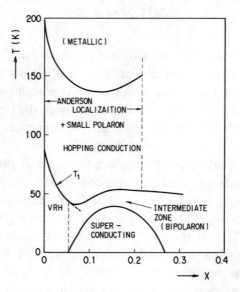

Fig. 3. The electronic phase diagram of $La_{2-x}Sr_xCuO_4$ deduced from the resistivity data. The lower boundary (T_1) is determined by the limiting marks in Fig. 2, and the upper one between metallic and hopping conduction by the marks in Fig. 1.

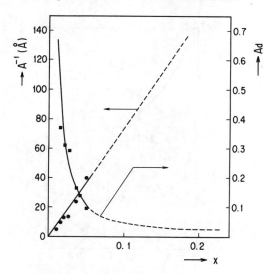

Fig. 4. The spread of the wavefunction of the Anderson-localized carrier in $La_{2-x}Sr_xCuO_4$, A^{-1}, and the dimensionless parameter expressing the degree of localization, Ad, are plotted against x, where d stands for the atomic distance. A is calculated from Eq. (2) and we use $d=4$ Å (the Cu–Cu distance).

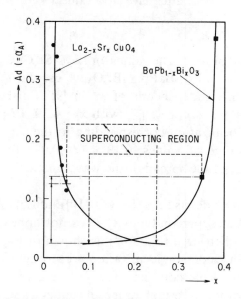

Fig. 5. The degrees of Anderson localization in high-T_c superconducting oxides are plotted against x. The curves are extrapolated using the relations $A \sim x^{-1}$ for $La_{2-x}Sr_xCuO_4$ and $A \sim |x-x_0|^{-1}$ with $x_0=0.39$ for $BaPb_{1-x}Bi_xO_3$, respectively. The composition regions corresponding to superconductivity are also shown.

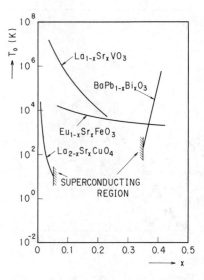

Fig. 6. The values of the characteristic temperature of VRH, T_0 are plotted against x for the superconducting oxides $La_{2-x}Sr_xCuO_4$ and $BaPb_{1-x}Bi_xO_3$ and non-superconducting oxides $La_{1-x}Sr_xVO_3$ and $Eu_{1-x}Sr_xFeO_3$. The characteristic temperatures in the latter have large values.

conductivity, are also plotted. The values for $La_{2-x}Sr_xCuO_4$ are extrapolated as the plot varies with x^{-1}. For $BaPb_{1-x}Bi_xO_3$ we use the T_0 data in [8] and the data of $N(E_f)$ $(=0.1$, independent of $x)$ in [9]. Furthermore we use the extrapolation curve as $Ad \propto |x-x_0|^{-1}$ with $x_0 = 0.39$. From Fig. 5 we realize that the degree of localization corresponding to the superconducting region falls in a certain range of values, $0.13 > Ad > 0.025$ for these oxides. The corresponding range of the spread of the wavefunction is given by $40d > A^{-1} > 7.7d$.

In the case of such oxides as $La_{1-x}Sr_xVO_3$ [10] and $Eu_{1-x}Sr_xFeO_3$ [11], where VRH is observed but superconductivity does not appear, the characteristic temperatures are rather large $(T_0 > 10^3)$ as shown in Fig. 6. In this case the electrons involved in Anderson localization are strongly localized.

The situation that superconductivity appears only when the degree of localization falls within a certain range of values can be explained by the coexistence model described in the following.

3 Coexistence of small-polaron and Anderson localization

It has been pointed out that, in some materials with strong electron–phonon interaction, an itinerant electron in the conduction band (or a hole in the valence band) has a tendency to form a self-trapped state, namely a small polaron. In particular, when the parameter $g = E_L/B$ falls within a certain range of values, itinerant electrons coexist with small polarons, where E_L stands for the lattice relaxation energy due to short-range electron–phonon interaction and B stands for the kinetic energy of the electron corresponding to the band width $2\pi\hbar^2/(2m^*d^2)$, m^* being the effective mass of the electron [12–14]. As a typical case we may mention the coexistence of Anderson localization with small polarons [13] and the coexistence of itinerant and localized states in the Peierls transition [14]. We consider that these treatments are also applicable to the present case.

The energy of such a coexistence system is given as follows [13]. First, we consider the case in the absence of electron–phonon interaction, which means, without deformation of the lattice. We put the degree of localization and the energy (per electron) of the Anderson localization to be α_0 and $E_e(\alpha_0)$, respectively. The dimensionless parameter α_0 is equal to d/r_0 and defined as $1 > \alpha_0 > 0$, where r_0 stands for the spread of the wavefunction of this state. From the variational point of view, the energy of the Anderson localization for an arbitrary degree of localization α can be written as $E_e(\alpha_0) + B(\alpha - \alpha_0)^2 + \ldots$. Next we allow the lattice to deform. The lattice relaxation energy is proportional to α^3 for the short-range electron–phonon interaction, and the total energy is given by

$$E(\alpha_0, \alpha) = E_e(\alpha_0) + B(\alpha - \alpha_0)^2 - E_L\alpha^3, \tag{3}$$

where E_L stands for the lattice relaxation energy at complete localization $\alpha = 1$, corresponding to the small polaron where the electron is localized in the region of one atom. By dropping the term independent of α, Eq. (3) can be rewritten as

$$y = -g\alpha^3 + \alpha^2 - 2\alpha_0\alpha, \tag{4}$$

where y is the α-dependent terms of $E(\alpha_O, \alpha)/B$. The value of y, which is the total energy per electron normalized with respect to B, is shown in Fig. 7 for $g = 0.9$ and for different values of α_0. For large α_0, the only stable state is the state perfectly localized as a small polaron ($\alpha = 1$), though this case is not shown. As the value of α_0 decreases, the curve develops double minima; an Anderson localized state at $\alpha = \alpha_A$ and a small-polaronic state at $\alpha = 1$. The quantity α_A should correspond to Ad described in the preceding section, which is the degree of Anderson localization in real materials. From the fact that Eq. (4) can be rewritten as

$$y = -g\alpha^3 + \alpha^2 - \alpha_A(2 - 3g\alpha_A) \tag{4'}$$

we can realize that the condition of the coexistence is determined by the degree of localization α_A for materials of which g is considered to be constant. The

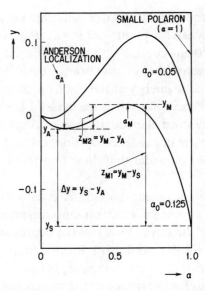

Fig. 7. The values of y in Eq. (4) (the total energy of the coexistence system divided by B) are plotted as a function of the degree of localization, α, for $\alpha_0 = 0.05$ and 0.125, where α_0 stands for the degree of localization without the lattice relaxation due to electron–phonon interaction. $\Delta y = y_S - y_A$ is the energy difference between Anderson localization and a small polaron.

relative energy of the small polaron to that of the Anderson localization, which is given by

$$\Delta y = y_S - y_A = 1 - g - 2\alpha_A + \alpha_A^2 + 3g\alpha_A^2 - 2g\alpha_A^3, \tag{5}$$

decreases with increasing α_A.

In $La_{2-x}Sr_xCuO_4$ the degree of Anderson localization, $Ad\,(=\alpha_A)$, decreases with increasing x as described in the preceding section. The ratio of the number of holes in the Anderson localized state to that in the small-polaronic state is given by the Fermi–Dirac distribution as

$$N_S = N_0\{1 + \exp[(\mu - E_S)/(kT)]\}^{-1},$$
$$N_A = N_0\{1 + \exp[(\mu - E_A)/(kT)]\}^{-1} \qquad \text{with } N_0 = N_S + N_A, \tag{6}$$

where N_S and N_A stand for the number of small polarons (holes) and Anderson-localized holes, respectively, μ for the chemical potential and E_S and E_A for the energy of small polarons and Anderson-localized carriers, respectively. By eliminating μ from Eq. (6), we have

$$N_S/N_A = \exp[-\Delta E/(2kT)], \qquad \Delta E = E_A - E_S = B\,\Delta y. \tag{7}$$

The concentration of the small polarons is given by

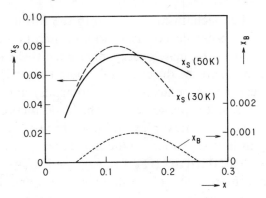

Fig. 8. The concentrations of small polarons, x_S, calculated by Eq. (8) are plotted against x at $T = 30$ and 50 K. The concentration of bipolarons, x_B, estimated from Schafroth's formula, is also shown.

$$x_S = x N_S / N_0. \qquad (8)$$

Using the value of Ad in Fig. 4 and using Eqs. (5), (7) and (8), we can calculate the concentration of small polarons in $La_{2-x}Sr_xCuO_4$ for $T = 30$ K and 50 K as shown in Fig. 8. Here we put the values of parameters B and g as $B = 0.3$ eV and $g = 0.93$.

In Fig. 8 the concentration of bipolarons, x_B, is also depicted as a function of x, which is obtained by postulating that the superconducting transition temperature T_c is the temperature of the Bose condensation $T_c = 3.31 \hbar^2 n^{2/3} / (m^* k)$ given by Schafroth, where n stands for the number of bipolarons per cubic centimeter and m^* for the mass of a bipolaron. Here we use $m^* = 5 m_0$ [1], m_0 being the mass of a free electron. The curves of x_S and x_B have similar appearances with maxima. From this similarity we infer that bipolarons are formed by pairing of some definite fraction of the small polarons that coexist with Anderson-localized carriers and that superconductivity occurs by Bose condensation of these bipolarons. These mechanisms are not clearly known at present. However, as long as T_c is determined as an increasing function of x_B, the curve of x_B has a maximum and the above inference should be correct.

4 Hybridization between itinerant and localized states

It has been reported that the conduction mechanism in Ti_4O_7 is due to bipolarons [15]. In this material conduction of $\exp[c(T_0/T)^{1/4}]$ type was observed [16] but T_0 is so large ($T_0 > 10^5$) that coexistence in the sense of that in Fig. 8 is impossible. No superconductivity is observed in this material. Taking into consideration the mechanism described in the former sections, this fact shows that only when polarons coexist with an itinerant state across a small potential barrier and with small energy difference ΔE can the bipolarons formed from these polarons be the cause of superconductivity. Therefore, the hybrid state between small polarons (or bipolarons) and itinerant carriers (Anderson localization in the present case), which was theoretically introduced by Ranninger *et al.* [17, 18], must play an important role in the occurrence of superconductivity in high-T_c superconducting materials. In particular, when the height of the potential barrier ($z_{M2} = y_M - y_A$ and $z_{M1} = y_M - y_S$ in Fig. 7) is small, both transformation by the tunneling effect and hybridization are considered to take place easily.

5 Conclusion

Weak localization (Anderson localization) in oxides with the electron–phonon interaction produces the conditions for coexistence of itinerant states and small polarons. If bipolarons are formed in this situation, then the bipolarons are hybridized with itinerant carriers. Bose condensation of these bipolarons should be the origin of superconductivity.

Acknowledgements

The author would like to thank Drs K. Sasaki, N. Masaki and A. Iwase for many valuable discussions.

References

[1] J. Tateno, *Physica* C **214**, (1993) 377.
[2] S. Uchida *et al.*, *Phys. Rev.* B **43**, (1991) 7942.
[3] C. H. Ruscher *et al.*, *Physica* C **204**, (1992) 30.
[4] S. Sugai *et al.*, *Solid State Commun.* **76**, (1990) 371.
[5] A. T. Fiory and A. F. Hebard, *Phys. Rev. Lett.* **52**, (1984) 2057.
[6] D. Shahar and Z. Ovadyahu, *Phys. Rev.* B **46**, (1992) 10917.
[7] N. F. Mott and E. A. Davis, *Electronic Processes in Noncrystalline Materials* (Clarendon, Oxford 1979) p. 23.
[8] H. Takagi *et al.*, *Solid State Commun.* **55**, (1985) 1019.
[9] K. Kitazawa, S. Uchida and S. Tanaka, *Physica* B **135**, (1985) 505.
[10] M. Sayer, R. Chen, R. Fletcher and A. Mansingh, *J. Phys.* C **8**, (1975) 2059.
[11] V. Joshi, O. Parkash, G. N. Rao and C. N. R. Rao, *J. Chem. Soc., Faraday Trans.* **75**, (1979) 1199.
[12] Y. Shinozuka and Y. Toyozawa, *J. Phys. Soc. Japan* **46**, (1979) 505.
[13] Y. Shinozuka, *J. Non-Crystalline Solids* **77** & **78**, (1985) 21.
[14] S. Suzuki and Y. Toyozawa, *J. Phys. Soc. Japan* **59**, (1990) 2841.
[15] B. K. Chakraverty and C. Schlenker, *J. Physique* **37**, (1976) C4–353.

[16] C. Schlenker, S. Ahmed, R. Buder and M. Gourmala, *J. Phys.* C **12**, (1979) 3503.

[17] J. Ranninger and S. Robaszkiewicz, *Physica* B **135**, (1985) 468.

[18] S. Robaszkiewicz, R. Micnas and J. Ranninger, *Phys. Rev.* B **36**, (1987) 180.

[19] J. Tateno, N. Masaki and A. Iwase, *Phys. Lett.* A **138**, (1989) 313.

[20] H. Takagi *et al.*, *Phys. Rev.* B **40**, (1989) 2254.

[21] M. Suzuki, *Phys. Rev.* B **37**, (1989) 2321.

[22] T. Ito *et al.*, *Nature* **350**, (1991) 598.

[23] S. Kambe *et al.*, *Physica* C **160**, (1989) 35.

20

Concentration and temperature-dependence of magnetic polaron spectra in the t–J model

N. M. PLAKIDA, V. S. OUDOVENKO and V. YU.
YUSHANKHAI

Joint Institute for Nuclear Research, Dubna 141980, Russia

Abstract

The spectral function and momentum distribution for holes in an antiferro-magnet are calculated on the basis of the t–J model in a slave-fermion representation. The self-consistent Born approximation for a two-time Green function is used to study the dependences on temperature and doping (δ) of the self-energy operator. The numerical calculations show weak dependences on concentration and temperature of the spectral function (quasi-particle hole spectrum) while the momentum distribution function changes dramatically with increasing temperature for $T > T_d$ with $T_d \simeq J\delta$.

1 Introduction

The problem of hole motion in an antiferromagnetic (AF) background has attracted much attention in recent years. That is mainly due to the hope of elucidating the nature of the carriers involved in high-T_c superconductivity in copper oxides. It is believed that the essential features of the problem are described by the t–J model with a Hamiltonian written as

$$H_{t-J} = t \sum_{\langle ij \rangle \sigma} \tilde{c}_{i\sigma}^{+} \tilde{c}_{j\sigma} + J \sum_{\langle ij \rangle} S_i S_j = H_t + H_J. \tag{1}$$

Here $\langle ij \rangle$ indicates nearest-neighbor pairs, $\tilde{c}_{i\sigma}^{+} = c_{i\sigma}^{+} (1 - n_{i, -\sigma})$ are the electron operators with the constraint of no double occupancy. The properties of a single hole doped in the Néel spin background have been analyzed intensively with various numerical and analytical methods. Among them are exact diagonalization of small clusters [1] and variational calculations [2]. A rather transparent description within a 'string' picture has been developed by several authors [3]. A perturbative approach to the problem was proposed by Schmitt–Rink, Varma and Ruckenstein [4] and developed further by Kane, Lee and Read [5] and Martinez and Horsch [6]. In this approach a slave-fermion representation for the *t–J* model was used and hole propagation was treated within the self-consistent Born approximation (SCBA). In this formulation of the *t–J* model, hole motion, which is only possible due to its strong coupling to spin-wave excitations, is regarded as a sort of spin polaron propagation [6, 7].

Now there exists a consensus that a hole doped in a quantum Néel background can propagate coherently with a quasi-particle (QP) band width of the order of the exchange constant *J* and with hole ground-state energy minima at momenta $k = (\pm \pi/2, \pm \pi/2)$. The calculated spectral density function [1, 5, 6] reveals a QP peak of intensity $Z_k \approx J/t$ on the low-energy side of the spectrum and this peak is well separated from the broad incoherent part, which has a width of about (6–7) *t*.

Being so successful in reproducing the single-hole results calculated by the exact diagonalization method, the perturbative approach within the SCBA is expected to provide a reasonable scheme for examining quasi-particle hole properties at finite doping as well. Some progress has been made by Igarashi and Fulde [8]. By developing a standard diagrammatic technique at $T = 0$ they found that, at finite doping, a hole Green function changes in a rather complicated way and new incoherent states below the QP peaks appear. Nevertheless, as was argued in [8] to first order in δ, QP band characteristics change negligibly as compared with the single-hole results [8]. Therefore, the quasi-holes, which are being treated as non-interacting ones, fill up four 'hole pockets' around the degenerate minima at the momenta $k = (\pm \pi/2, \pm \pi/2)$. Some arguments have also been given in [8] that the fraction of the Brillouin zone covered by these pockets is equal to the hole concentration δ.

We present below the results of calculations of a hole's spectral properties based on the same formulation of the *t–J* model as in [5–8]. We assume a quantum Néel state for magnetic background and develop further the self-consistent Born approximation to calculate with more accuracy the spectral density function $A(k, \omega)$ and momentum distribution function $N(k)$ for holes at finite doping and finite temperatures. To this end, the irreducible Green function method will be used to derive the two-time hole Green function.

In Section 2 we describe the effective Hamiltonian in a slave-fermion Schwinger boson representation. Dyson's equation for a hole two-time Green function is derived and the self-energy is obtained in a SCBA. An iteration procedure to solve an integral equation for the self-energy is discussed briefly in Section 3 and numerical results are presented and analyzed. Section 4 contains the concluding remarks.

2 The effective Hamiltonian and Green function for holes

The Hamiltonian (1) of the *t–J* model in the slave-fermion Schwinger-boson representation is written as [5, 6, 8]

$$H_t = \sum_{kq} h_k^+ h_{k-q} [M_1(k,q)\alpha_q + M_2(k,q)\alpha_{-q}^+] - \mu \sum_k h_k^+ h_k, \tag{2}$$

$$H_J = \sum_q \omega_q \alpha_q^+ \alpha_q, \qquad \omega_q = SzJ(1-\delta)^2(1-\gamma_q^2)^{1/2}, \tag{3}$$

where the spinless fermion operator h_k^+ creates a hole in a Bloch state and the boson operator α_q^+ creates an antiferromagnetic magnon with energy ω_q, $\delta = \langle h_i^+ h_i \rangle$ is a hole concentration, $z = 4$ and $\gamma_q = (\cos q_x + \cos q_y)/2$ for a square lattice. The fermion–magnon coupling is given by

$$M_1(k,q) = zt \left(\frac{2S}{N}\right)^{1/2} (u_q \gamma_{k-q} + v_q \gamma_k), \tag{4}$$

$$M_2(k,q) = zt \left(\frac{2S}{N}\right)^{1/2} (u_q \gamma_k + v_q \gamma_{k-q}), \tag{5}$$

where u_q and v_q are the coefficients of the Bogoliubov transformation in the usual linear spin-wave theory. It is worth noting also that an additional term proportional to a chemical potential μ is involved explicitly in (2). A variation of μ, as a function of hole concentration δ and temperature T, should be calculated self-consistently.

Though the form of the Hamiltonian (2), (3) is similar to that of the standard polaron problem, the absence of a free kinetic energy for the spinless fermions should, however, be mentioned. It is the coupling to the spin waves that is the only source for coherent propagation of the QP.

To consider hole propagation we introduce a single-particle two-time retarded Green function

$$\langle\langle h_k(t)|h_k^+(t')\rangle\rangle = -i\theta(t-t')\langle\{h_k(t), h_k^+(t')\}\rangle, \tag{6}$$

where $\theta(t)$ is the step function and $\{,\}$ stands for the anticommutator. The Fourier transform is defined by

$$\langle\langle h_k|h_k^+\rangle\rangle_\omega = \int_{-\infty}^{+\infty} dt\, e^{i\omega(t-t')} \langle\langle h_k(t)|h_k^+(t')\rangle\rangle \equiv G(k,\omega). \tag{7}$$

To obtain an equation of motion for the Green function one may follow [9], differentiating with respect to both times t and t'. Then, after the Fourier transform, we have a set of two equations

$$(\omega + \mu)G(k,\omega) = 1 + \sum_q \langle\langle h_{k-q} B(k,q)|h_k^+\rangle\rangle_\omega, \tag{8}$$

$$(\omega+\mu)\langle\langle h_{k-q}B(k,q)|h_k^+\rangle\rangle_\omega=\sum_{q'}\langle\langle h_{k-q}B(k,q)|h_{k-q'}^+B^+(k,q')\rangle\rangle_\omega, \qquad (9)$$

where

$$B(k,q)=M_1(k,q)\alpha_q+M_2(k,q)\alpha_{-q}^+. \qquad (10)$$

Substituting (9) into (8) and defining the zeroth-order Green function $G_0^{-1}(k,\omega)=\omega+\mu$, we get the equation

$$G(k,\omega)=G_0(k,\omega)+G_0(k,\omega)T(k,\omega)G_0(k,\omega), \qquad (11)$$

where the scattering matrix $T(k,\omega)$ is a higher order Green function

$$T(k,\omega)=\sum_{qq'}\langle\langle h_{k-q}B(k,q)|h_{k-q'}^+B^+(k,q')\rangle\rangle_\omega. \qquad (12)$$

Eq. (11) can be also rewritten as the Dyson equation

$$G^{-1}(k,\omega)=G_0^{-1}(k,\omega)-\Sigma(k,\omega), \qquad (13)$$

where the self-energy $\Sigma(k,\omega)$ is connected with the scattering matrix by the equation $T=\Sigma+\Sigma G_0 T$. Hence, one can see that the self-energy $\Sigma(k,\omega)$ is the irreducible part of $T(k,\omega)$:

$$\Sigma(k,\omega)=T^{(\mathrm{irr})}(k,\omega), \qquad (14)$$

which can be evaluated by a proper decoupling procedure for correlation functions entering in (12) as follows

$$\langle h_{k-q'}^+B^+(k,q')h_{k-q}(t)B(k,q,t)\rangle\simeq\langle h_{k-q}^+h_{k-q}(t)\rangle\langle B^+(k,q)B(k,q,t)\rangle\delta_{qq'}$$
$$\simeq\delta_{qq'}\langle h_{k-q}^+h_{k-q}(t)\rangle\{M_1^2(k,q)\langle\alpha_q^+\alpha_q(t)\rangle+M_2^2(k,q)\langle\alpha_{-q}\alpha_{-q}^+(t)\rangle\}. \qquad (15)$$

Using the spectral representation for Green functions we obtain the following intermediate result for the self-energy:

$$\Sigma(k,\omega)=\sum_q\int_{-\infty}^{+\infty}\frac{d\omega_1}{\pi}\int_{-\infty}^{+\infty}\frac{d\omega_2}{\pi}\frac{e^{\beta(\omega_1+\omega_2)}+1}{(e^{\beta\omega_1}+1)(e^{\beta\omega_2}-1)}\frac{\mathrm{Im}\langle\langle h_{k-q}|h_{k-q}^+\rangle\rangle_{\omega_1+i\eta}}{\omega-(\omega_1+\omega_2)+i\eta}$$

$$\times\{M_1^2(k,q)\,\mathrm{Im}\langle\langle\alpha_q|\alpha_q^+\rangle\rangle_{\omega_2+i\eta}+M_2^2(k,q)\,\mathrm{Im}\langle\langle\alpha_{-q}^+|\alpha_{-q}\rangle\rangle_{\omega_2+i\eta}\}. \qquad (16)$$

Further, we neglect self-energy corrections to the spin-wave Green function that results in the simplest form for the spectral density function

$$-\frac{1}{\pi}\mathrm{Im}\langle\langle\alpha_q|\alpha_q^+\rangle\rangle_\omega=\delta(\omega-\omega_q), \qquad (17)$$

$$-\frac{1}{\pi}\mathrm{Im}\langle\langle\alpha_q^+|\alpha_{-q}\rangle\rangle_\omega=-\delta(\omega-\omega_q).$$

Finally, we obtain

$$\Sigma(k,\omega)=\Sigma_1(k,\omega)+\Sigma_2(k,\omega), \tag{18}$$

$$\Sigma_1(k,\omega)=\sum_q M_1^2(k,q)\int_{-\infty}^{+\infty}d\varepsilon\,\frac{A(k-q,\varepsilon)}{\omega-\varepsilon-\omega_q+i\eta}\,[1-n(\varepsilon)+N(\omega_q)], \tag{19}$$

$$\Sigma_2(k,\omega)=\sum_q M_2^2(k,q)\int_{-\infty}^{+\infty}d\varepsilon\,\frac{A(k-q,\varepsilon)}{\omega-\varepsilon-\omega_q+i\eta}\,[n(\varepsilon)+N(\omega_q)], \tag{20}$$

where $n(\varepsilon)=(e^{\beta\varepsilon}+1)^{-1}$, $N(\omega)=(e^{\beta\omega}-1)^{-1}$ and

$$A(k,\omega)=-\frac{1}{\pi}\,\mathrm{Im}\,G(k,\omega+i\eta) \tag{21}$$

is the spectral density for holes. To make the set of equations (13) and (18)–(21) self-consistent, the following equation for the chemical potential μ should be adopted:

$$\delta=\frac{1}{N}\sum_k\int_{-\infty}^{+\infty}d\varepsilon\,n(\varepsilon)A(k,\varepsilon). \tag{22}$$

The set of equations that we have derived is equivalent at $T=0$ to that obtained by standard diagrammatic techniques within the SCBA. However, in our formulation the temperature-dependence is taken into account explicitly.

3 Numerical solution and discussion

In this section we present the numerical results for the set of self-consistent equations (13), (18)–(22). We consider a two-dimensional square lattice on the 16×16 cluster exploiting all possible symmetries. The frequency space is divided into a mesh of size $0.005t$ to provide sufficient energy resolution. To make the iteration procedure convergent, a small imaginary part $\eta=0.01t$ is added to the frequency in the Green function. The numerical procedure was organized such that, at given δ and T, values for chemical potential μ and spectral density function $A(k,\omega)$ were calculated self-consistently by an iteration procedure. Typically, after 40–50 iterations the chemical potential converged to some fixed value $\mu=\mu(\delta,T)$ and the sum rule on $A(k,\omega)$ was fulfilled with accuracy better then 1%. Below, while presenting $A(k,\omega)$ at given δ and T, we measure the frequency from the chemical potential $\mu(\delta,T)$. By performing calculations with $J=0.4$ and $J=0.2$ (from now on we will refer to all quantities in units of t) we found no qualitative differences. Therefore only the data with $J=0.4$ will be discussed.

3.1 Spectral density A (k, ω)

We started by calculating the spectral density function $A(k,\omega)$ for the single hole at $T=0$ and found substantial agreement with numerical results presented in [6]. Referring further to these results, we discuss how the hole spectral properties will be changed with increasing hole concentration δ and temperature T.

In Fig. 1 we show spectral functions for several k-points calculated at $T=0$ and $\delta=3\%$. Let us first discuss the k-points belonging to the edge of the AF BZ, Fig. 1(a) and (b). There are not any visible changes in the shape of $A(k,\omega)$ as compared with the single-hole case, $\delta \to 0$. More precisely, calculations give a small percentage increase of a QP spectral weight $Z(k)$ (the integrated area under a QP peak). For instance, the value of $Z(\pi/2,\pi/2)$ increases from $Z \simeq 0.33$ at $\delta \to 0$ to $Z \simeq 0.35$ at $\delta=3\%$ and this tendency is kept for higher δ (see Table 1).

For momenta k far from the AF BZ edge, the spectral function $A(k,\omega)$ is modified with doping in another way. Actually, one can see from Fig. 1(c) and (d) that, in comparison with the single-hole case for $k=(0,\pi/2)$ and $(0,0)$ the spectral density is redistributed so that a new incoherent broad structure, as first noted in [10], appears far below the chemical potential. The spectral weight for this new structure is proportional to δ.

The peculiarities in the shape of $A(k,\omega)$ mentioned above become more pronounced with further increase in hole concentration. The spectral density $A(k,\omega)$ calculated at $T=0$ and $\delta=10\%$ is shown in Fig. 2 for the two most representative k-points: $k=k^*=(\pi/2,\pi/2)$ and $k=(0,0)$. Examining the data for k^*, Fig. 2(a) and Table 1, we noticed about 10% increase of the QP spectral weight $Z(k^*)$. For $k=(0,0)$, Fig. 2(b), a strong suppression of the QP peak at $\omega \simeq 0.8$ and formation of a band of incoherent excitations below the chemical potential are seen.

The energy minimum of the QP band was found at $k=k^*$ throughout our computations. In Fig. 3 are shown the doping-dependence of the energy $E(k^*)=\omega_{QP}(k^*)+\mu$ of the lowest QP peak and the chemical potential μ. We think that our finite-cluster computations at extremely low concentration, $\delta < \delta_1$, are not rigorous enough and that this results in the inverse relative positions of μ and $E(k^*)$ i.e. $\mu < E(k^*)$ in this region. Being measured from μ, the position of the lowest QP peak is fitted with good accuracy by the dependence $E(k^*)-\mu=\omega_P(k^*) \approx -1.5J(\delta - \delta_1)$ for $\delta > \delta_1 \approx 1.5\%$.

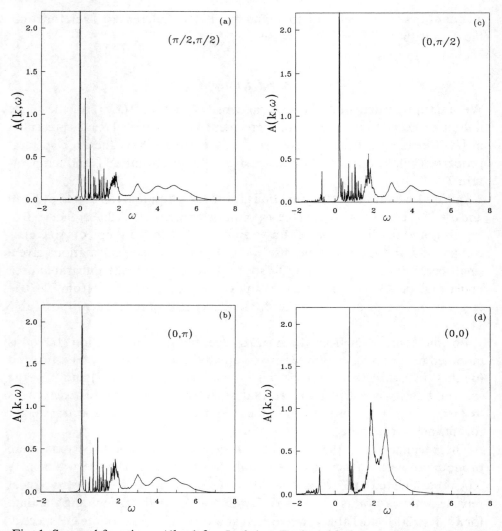

Fig. 1. Spectral functions $A(k,\omega)$ for $J = 0.4$ at $T = 0$ and hole concentration $\delta = 3\%$; $\mathbf{k} = (\pi/2, \pi/2)$ (a); $(0, \pi)$ (b); $(0, \pi/2)$ (c); and $(0,0)$ (d). A frequency ω is measured from the chemical potential μ.

By analogy with the conventional case of a free-fermion gas, we associate the quantity $\mu - E(\mathbf{k}^*)$, taken for $\delta > \delta_1$, with the Fermi energy for a quasi-hole gas. We can also define a degeneracy temperature as $T_d(\delta) = 1.5J\delta$. This means that, at $T > T_d(\delta)$, one should expect a different behaviour for the quasi-hole gas than for the low-temperature limit $T \ll T_d$. We examine this hypothesis in Section 3.2 considering temperature variations for the hole-momentum distribution function $N(k)$.

Table 1. *QP spectral weight Z(**k***) and QP peak position $\omega_{QP}(\mathbf{k}^*)$ at $\mathbf{k}^* = (\pi/2, \pi/2)$ for different hole concentrations and T=0. The concentration $\delta \approx 0.4\%$ corresponds to a single-hole case.*

δ (%)	$Z(\mathbf{k}^*)$	$\omega_{\mathrm{QP}}(\mathbf{k}^*)$
0.4	0.329	0.005
3	0.353	-0.010
10	0.369	-0.055
20	0.373	-0.085

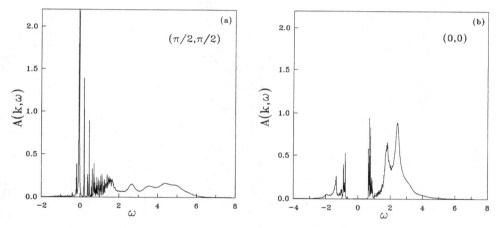

Fig. 2. Spectral functions $A(k,\omega)$ for $T=0$ and $\delta=10\%$ at $\mathbf{k}=(\pi/2,\pi/2)$ (a) and (0,0) (b).

Now we address the problem of the temperature-dependence of quasi-hole spectral properties. On analysing a variety of spectral densities $A(k,\omega)$ calculated over a wide temperature interval and several values of δ, we found no significant changes in the shape of $A(k,\omega)$ up to $T \sim T_{\mathrm{d}}(\delta)$. At a given δ the only temperature effect is an equal shift of $A(k,\omega)$ to higher values of ω for all k. In particular, the lowest QP peak for k^* is close to zero energy, $\omega_{\mathrm{QP}}(\mathbf{k}^*) \simeq 0$, $T \simeq T_{\mathrm{d}}(\delta)$.

Such a behaviour is an effect of temperature variation of the chemical potential $\mu = \mu(\delta, T)$, crossing the bottom of the QP band (i.e. the lowest QP peak energy) when the temperature reaches the critical value $T \simeq T_{\mathrm{d}}(\delta)$. With further increase in T, as can be seen in Fig. 4, the sharp structure of the spectral

N. M. Plakida et al.

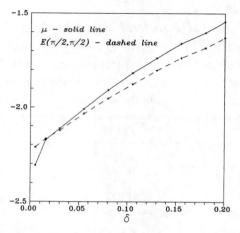

Fig. 3. Chemical potential μ (solid line) and energy of the lowest QP peak $E(\boldsymbol{k}^* = (\pi/2, \pi/2)) = \omega_{\mathrm{QP}}(\boldsymbol{k}^*) + \mu$ versus δ at $T = 0$.

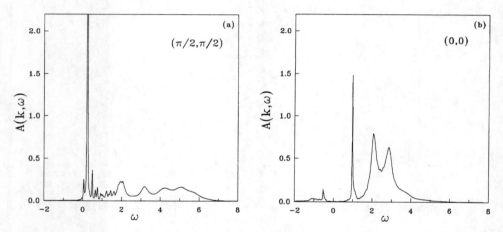

Fig. 4. Spectral functions $A(k, \omega)$ for $T = 0.1$ and $\delta = 3\%$ at $\boldsymbol{k} = (\pi/2, \pi/2)$ (a) and (0,0) (b).

density is smeared out and all QP peaks are shifted and become above the chemical potential ($\omega_{\mathrm{QP}}(k) > 0$).

To get more insight into the problem we compare the data for the spectral density $A(k, \omega)$ and the imaginary part of the self-energy $\Gamma(k, \omega) = -\operatorname{Im} \Sigma(k, \omega)$. In Fig. 5(a) and (b) is shown $\Gamma(\boldsymbol{k}^*, \omega)$ obtained for $\delta = 10\%$ at $T = 0$ and $T = 0.1$, respectively. Calculations at low concentration, e.g. $\delta = 3\%$, give similar results.

In Fig. 5(a) it is shown that $\Gamma(\boldsymbol{k}^*, \omega) \approx 0$ (we found $\Gamma \leq 10^{-14}$) in the frequency interval $|\omega| \leq 0.2$, where the QP peak is located. This clearly indicates that there are no low-energy states to which the quasi-hole can scatter and

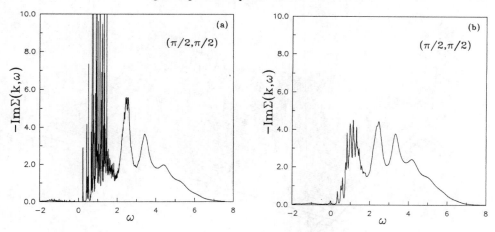

Fig. 5. Imaginary parts of the self-energy operator $\Sigma(k,\omega)$ for $\delta = 10\%$ at $\boldsymbol{k} = (\pi/2, \pi/2)$: $T = 0$ (a) and 0.1 (b).

therefore it has infinite lifetime. For higher frequencies, $0.2 < \omega \leq 7$, $\Gamma(k,\omega)$ grows sharply, giving a wide distribution with a large characteristic amplitude. Since in this region $|\omega - \operatorname{Re}\Sigma(k,\omega)| \ll \Gamma(k,\omega)$, we can write

$$A^{(\text{incoh})}(\boldsymbol{k}^*,\omega) \simeq \frac{1}{\pi\Gamma(\boldsymbol{k}^*,\omega)}. \tag{23}$$

Comparing $\Gamma(\boldsymbol{k}^*,\omega)$ calculated at higher temperature $T = 0.1$, Fig. 5(b), with corresponding $A^{(\text{incoh})}(\boldsymbol{k}^*,\omega)$, we conclude that the same relation (23) holds between them. Though QP width $\Gamma_{\boldsymbol{k}^*} = \Gamma(\boldsymbol{k}^*, \omega = \omega_{\text{QP}})$ increases strongly with temperature and reaches the value $\Gamma \simeq 0.01$ at $T = 0.1$, nevertheless, it still remains negligible. Thus we can infer that quasi-hole states remain well-defined excitations, being stable over a wide range of temperature and hole concentration.

3.2 Momentum distribution $N(k)$

In this subsection we present data for the hole-momentum distribution function $N(k)$ defined as

$$N(k) = \int_{-\infty}^{+\infty} \mathrm{d}\omega\, n(\omega) A(k,\omega). \tag{24}$$

First we calculate $N(k)$ at $T = 0$ for several hole concentrations, Fig. 6 and Fig. 7. One can see four quasi-particle 'pockets' at $\boldsymbol{k} = (\pm\pi/2, \pm\pi/2)$ superim-

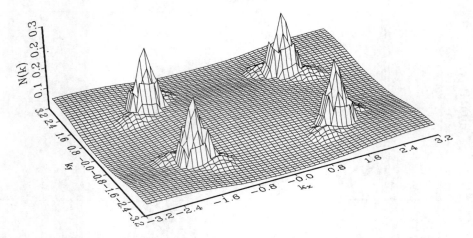

Fig. 6. The hole-momentum distribution function $N(k)$ for $\delta = 3\%$ at $T = 0$. Four quasi-particle hole 'pockets' at $k = (\pm \pi/2, \pm \pi/2)$ are clearly see on a smoothly k-dependent background.

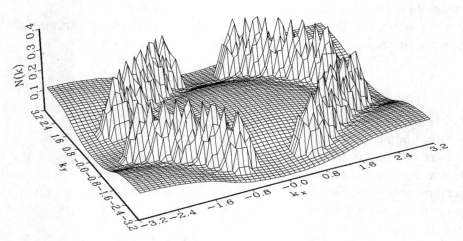

Fig. 7. The hole-momentum distribution function $N(k)$ for $\delta = 10\%$ at $T = 0$.

posed on the smooth, slightly k-dependent background. Let us first discuss an origin for a partial, of order δ, occupancy for k-points far from the edge of the AF BZ. One should take into account a particular shape for the spectral density $A(k, \omega)$ at these points, as seen from Fig. 1(c) and (d) and Fig. 2(b). It is clear that, while calculating $N(k)$ according to (24), the high-energy part of the spectral density is cut off by the Fermi factor and only the low-energy incoherent part $A^{(\text{incoh})}(k, \omega < 0)$ contributes to the background:

$$N(k) = \int\limits_{-\infty}^{+\infty} d\omega \, A^{(\text{incoh})}(k, \omega) = N_B(k). \tag{25}$$

As found numerically, the spectral weight for the low-energy incoherent part $A^{(\text{incoh})}(k, \omega < 0)$ is approximately equal to δ. Hence, it is possible to estimate the background value as $N_B(k) \simeq \delta$.

Now we can consider k-points in the vicinity of the AF BZ edge. As was pointed out above, there is no incoherent part of the spectral density below the chemical potential, i.e. $A^{(\text{incoh})}(k, \omega < 0) \equiv 0$ for this region of BZ. At $T = 0$, QP states for k-points near k^* are located below μ and $\omega_{QP}(k) < 0$. Just these states lead to the appearance of QP 'pockets'. The volume covered by 'pockets' in k-space increases with δ. However, we cannot obtain a sharp Fermi surface at $T = 0$ due to our artificial broadening of $A(k, \omega)$ spectra of order $\eta = 0.01$.

On studying temperature effects at a given concentration δ, we find that the momentum distribution function $N(k)$ changes dramatically when the temperature reaches the value $T \simeq T_d(\delta)$. The 'four-pocket' structure for $N(k)$, which exists at $T \ll T_d$, is almost washed out at $T \geq T_d$. This peculiarity can easily be understood if one takes into account a temperature shift of the QP spectrum $\omega_{QP}(k)$. As discussed in the previous section, for $T > T_d(\delta)$ all the QP peaks occur above the chemical potential and, hence, $\omega_{QP}(k) > 0$. Thus the Fermi-factor leads to a strong reduction of the occupancy weight in comparison with the low-temperature value $Z(k)$, which explains the disappearance of the quasi-hole 'pockets'. We also found that the background value $N_B(k)$ is almost T-independent and $N_B(k) \approx \delta$.

4 Conclusion

Based on the $t–J$ model, we have studied a doped antiferromagnet and calculated the spectrum of hole excitations over a wide range of temperature, $T \leq 0.1$, and hole concentration, $\delta \leq 0.2$. We were mainly interested in the stability of QP hole excitation with respect to hole doping and temperature variation.

For this purpose we have calculated the hole spectral density function $A(k, \omega)$ Eq. (21) and the imaginary part of the hole self-energy $\Gamma(k, \omega)$, whose temperature- and doping-dependences have been studied. For k-points positioned at the AF BZ edge $k = (\pi/2, \pi/2), (0, \pi)$, the spectral density $A(k, \omega)$ clearly shows a quasi-particle peak, with intensity $Z(k) \simeq J$, at $\omega = \omega_{QP}(k)$ (see Fig. 1(a) and (b), Fig. 2(a) and Fig. 4(a)), which is stable over a wide range of T and δ (see Fig. 4 and Fig. 5). A broad band of incoherent states with spectral weight $1 - Z(k)$ at $\omega > \omega_{QP}(k)$ is also seen.

For low temperatures, $T \leq T_d(\delta) \simeq 1.5J\delta$, the QP peak at $\boldsymbol{k} = \boldsymbol{k}^* = (\pi/2, \pi/2)$ is positioned below the chemical potential, $\omega_{QP} = E(\boldsymbol{k}) - \mu < 0$ (Fig. 3), which results in a 'four-pocket' shape for the momentum distribution function $N(\boldsymbol{k})$ (Fig. 6 and Fig. 7). However, at $T > T_d(\delta)$, this picture is completely washed out.

Since the intrinsic widths Γ_k for QP states remain negligibly small at $T \geq T_d(\delta)$, disappearance of the 'pockets' in $N(\boldsymbol{k})$ is not caused by broadening of QP peaks but results from a strong temperature shift of ω_{QP} to positive values, $\omega_{QP} > 0$.

On examining $\boldsymbol{k}$-points far from the AF BZ edge, $\boldsymbol{k} = (0,0), (0, \pi/2)$, we have found a stronger variation of QP peaks with T and δ (Fig. 1(c) and (d) and Fig. 2(b)). In this region of $\boldsymbol{k}$-space the most pronounced effect of doping is the appearance of a band of incoherent states far below μ. These incoherent states manifest themselves as a smoothly $\boldsymbol{k}$-dependent background in the momentum distribution $N(\boldsymbol{k})$.

Therefore the main result of the present calculations is a weak concentration- and temperature-dependence of the spectral function (quasi-particle hole spectrum) while the momentum distribution function is proved to change dramatically with increasing temperature for $T > T_d$ with $T_d \simeq 1.5J\delta$.

In the present investigation we have neglected any renormalization of the spin-wave excitation spectrum. Hence, an extension of the model (2)–(5) [5, 6] to the regime of moderate hole concentration $\delta \geq 10\%$, in which one can expect strong renormalization of the spin-wave excitations [8], seems to be questionable. However, we can argue that, due to the small radius of the AF polaron [7] and finite correlation length of AF fluctuations, even for more highly doped cuprates [11], qualitative description within the framework of the present model is possible.

More reasonable self-consistent calculations allowing for renormalization of the AF spin-fluctuation spectrum will be considered in a separate publication.

Acknowledgements

One of the authors (N. P.) is greatly indebted to P. Fulde, K. Becker and P. Horsh for stimulating discussion.

References

[1] J. Bonča, P. Prelovšek and I. Sega, *Phys. Rev.* B **39**, 7074 (1989); Y. Hasegava and D. Poilblanc, *ibid.* **40**, 9035 (1989); K. J. von Szczepanski, P. Horsch, W. Stephan and M. Ziegler, *ibid.* **41**, 2017 (1990); E. Dagotto, R. Joynt, A. Moreo, S. Bacci and E. Gagliano, *ibid.* **41**, 2585 (1990); *ibid.* **41**, 9049 (1990).

[2] S. A. Trugman, *Phys. Rev.* B **41**, 892 (1990); *Phys. Rev. Lett.* **65**, 500 (1990).

[3] R. Eder and K. W. Becker, *Z. Phys.* B **78**, 219 (1990); R. Eder, K. W. Becker and W. H. Stephan, *ibid.* **81**, 33 (1990); J. L. Richard and V. Yu. Yushankhai, *Phys. Rev.* B **47**, 1103 (1993).

[4] S. Schmitt-Rink, C. M. Varma and A. E. Ruckenstein, *Phys. Rev. Lett.* **60**, 2793 (1988).

[5] C. L. Kane, P. A. Lee and N. Read, *Phys. Rev.* B **39**, 6880 (1989).

[6] G. Martínez and P. Horsch, *Phys. Rev.* B **44**, 317 (1991).

[7] A. Ramšak and P. Horsch, *Phys. Rev.* **48**, 10559 (1993).

[8] J. Igarashi and P. Fulde, *Phys. Rev.* B **45**, 12357 (1992).

[9] N. M. Plakida, *Phys. Lett.* A **43**, 481 (1973); Yu. A. Tserkovnikov, *Theor. Matem. Fiz.* **7**, 250 (1971); *ibid.* **49**, 219 (1981).

[10] M. Ziegler, Thesis, Universität Karlsruhe (1990).

[11] J. Rossat-Mignot *et al.*, *Physica* C **185–189**, 86 (1991).

21

Mass enhancement without band-narrowing in t–t'–J and related models: predictions for Fermi-surface and optical conductivity

S. I. MUKHIN[1,2] AND L. J. DE JONGH[1]

[1]*Kamerling Onnes Laboratory, Universiteit Leiden, P. O. Box 9506, 2300 R. A. Leiden, The Netherlands*
[2]*Moscow Institute of Steel and Alloys, Theoretical Physics Department, Leninskii prospect 4, 117936 Moscow, Russia*

Abstract

An analytically solvable Ansatz set of Migdal-type self-consistent equations is proposed for the coupling between spin and charge degrees of freedom in the strongly correlated electron system described by the t–t'–J (t–J) Hamiltonian. The small parameter validating Migdal's approximation for this problem is found to be $1/\ln(U/t)$ (when $U \gg t$), where U and t are on-site Coulomb repulsion energy and the bare electron hopping integral of the basic Hubbard Hamiltonian ($t' = J = 4t^2/U$). The analytical results, obtained for electron concentrations close to half-filling, demonstrate strong enhancement of the quasi-particle mass, accompanied by a depletion of the *particle* density of states at the Fermi-level (E_F). The spectral density is pushed away from E_F into a broad range of energies (of order t) and possesses a sawtooth form. Theoretical predictions for the optical conductivity are derived, which could provide a qualitative explanation of the mid-infrared anomaly observed experimentally in high-T_c cuprates. The variation in quasi-particle energy over the Brillouin zone is found to be of order J only, in good accord with previous theoretical work and the dispersion observed in recent ARPES experiments in 2212 high-T_c compounds.

1 Introduction and summary of our previous work

It is by now generally accepted that strong Coulomb correlations should play an essential role in the physics of the high-temperature superconductors (HTS). A great number of experimental results obtained since the discovery [1] of the HTSs raise strong doubts about the applicability of 'classical' BCS theory to the description of the superconductivity phenomena in these compounds. Also the spin and charge dynamics in the normal state (i.e. above T_c)

have proved to be highly unusual and have led to renewed interest in the physics of strongly correlated electron systems [2]. For these reasons, studies of transport, specific heat and nuclear magnetic resonance (NMR) properties of high-T_c superconductors in the framework of the local-pair (negative-U) and Hubbard (large positive-U) models are highly valuable.

In its simplest form, the local-pair (negative-U) model can be mapped onto the anisotropic Heisenberg pseudo-spin model in a (temperature-dependent) magnetic field, where the carrier density in the boson problem transforms into the magnetization in the pseudo-spin problem. For a review see [3]. In earlier work (see e.g. [4]) this mapping was exploited to get a qualitative understanding of high-T_c superconductivity. Furthermore, analytical results for e.g. the specific heat [5] and the tunnelling characteristics of a metal superconductor junction [6] were obtained on the basis of the random phase (RPA) approximation to the local-pair Hamiltonian. Later on, Monte Carlo simulations for this model were added [7]. The ordering specific heat anomalies and phase diagrams were calculated for the 3D and quasi-2D local-pair superconductors, and the results compared with the already known RPA results [5, 8], revealing some important differences.

However, the simple model of (hard-core) bosons hopping on a lattice will probably not be sufficient to describe the physical properties of the copper oxide superconductors, in view of the essential role that the antiferromagnetically coupled, (nearly) localized copper spins are thought to play in the normal state and superconducting behaviour. Therefore, we have extended our theoretical investigations to the NMR behaviour of the HTS cuprates on the basis of a more realistic model, where we consider singlet real-space pairs (RSPs) hopping in a quasi-2D background of antiferromagnetically coupled, localized spins [9]. In our opinion the model contains the essential microscopic mechanism of (dynamic) destruction of the antiferromagnetic correlations in CuO_2 layers by moving holes, leading to the strong coupling between spin and charge degrees of freedom that has been found experimentally in all lightly doped high-T_c cuprates. To get analytical results, we have considered the antiferromagnetic Heisenberg Hamiltonian in the Schwinger-boson representation within the mean-field/random phase approximation (MF/RPA). The hopping RSPs (holes coupled into pairs, occupying two adjacent copper sites) were modelled by dynamically broken antiferromagnetic bonds, and treated as a dilute gas of narrow-band particles obeying Boltzman statistics [10, 14]. It was found that the 'gas-parameter' of the RSPs reaches unity at low enough temperature, for any finite concentration of RSPs. This is a consequence of a local coupling of the RSP density fluctuations to the antiferromagnetic fluctuations in the 2D magnetic layers, which slow down in a pure 2D antiferromagnet as the temperature is lowered. Competition between the kinetic energy of the RSPs and the magnetic ordering energy, as the temperature lowers, causes a crossover in the temperature-dependence of the antiferromagnetic correlation length, $\xi(T)$, from the fast exponential growth characteristic for the undoped system, to a weaker (power law) divergence. It is found that this crossover in $\xi(T)$ is reflected as a maximum in the temperature-dependence of the quantity $1/(T_1 T)$, where T_1^{-1} is the nuclear-spin–lattice relaxation rate, and also as a gradual decrease with temperature of the NMR Knight shift K. A simple symmetry argument, valid for the square antiferromagnetic CuO_2 lattice, leads to very different results for the temperature behaviour of $1/(T_1 T)$ for the in-plane oxygen site, in agreement with experiment (in our opinion this provides crucial experimental evidence for the presence of nearly localized Cu spins also in the weakly doped, metallic samples). Possible coupling to the local soft phonon modes, which dynamically breaks this symmetry, was also considered and the resulting contribution to the nuclear-spin–lattice relaxation rate T_1^{-1} at the oxygen site was calculated [12, 13].

Monte Carlo simulations for a quite similar model of RSPs moving in a quasi-2D antiferromagnetic background [11] proved to be consistent with these analytical results, and nicely complement them, since the analytical work so far was restricted to low RSP densities and large ξ. In the simulations it was found that, neglecting Coulomb repulsion between the RSPs, the system of RSPs plus localized spins may either undergo phase separation or remain homogeneous, depending on the RSP density and on the ratio of the antiferromagnetic coupling parameter to the RSP bandwidth. In the first case the RSPs become static, whereas in the second they remain dynamic. Again, excellent qualitative agreement with experimental NMR results (T_1 and ξ as a function of temperature) has been obtained.

The problem of the fluctuation corrections to the thermodynamic quantities of the model studied in [10] was addressed in [12, 13]. It was found that, for the pure 2D antiferromagnet (AFM) at low temperatures, the phase fluctuations of the mean-field (MF) parameter corresponding to the local staggered magnetization lead to infrared divergences of the spin–spin correlation functions. These fluctuations eliminate the long-range order in the 2D AFM at finite temperatures imposed implicitly by the MF/RPA approximation. It was also found that, insofar as doping with mobile RSPs leads to suppression of antiferromagnetic ordering at lowering temperature, the MF approach with respect to the spin degrees of freedom gives more accurate results, although in the doped case the temperature-dependence of ξ differs strongly from that in the pure (undoped) 2D AFM.

Recent photoemission experiments in HTSs [18], showing Fermi-surface-like features in the ARPES (angle-resolved photoemission spectroscopy) data, have encouraged us to investigate the problem of the lightly doped quasi-2D AFM starting with more general assumptions, i.e. RSP formation is not supposed from the start and singly doped fermionic holes are considered. In this way, the t–$t′$–J model, as derived from the large positive-U Hubbard Hamiltonian, was shown to possess a heavy-mass solution for spinless fermionic holes at close to half-filling [15]. The result was obtained by solving a model version of Migdal-type self-consistent equations for the normal state of the strongly interacting spin–hole system.

In the present paper these analytical results are extended and complemented by numerical calculations on the same model. Specifically, the small parameter validating the Migdal approximation for the coupling between the charge and spin degrees of freedom is found to be $1/\ln(U/t)$ for $U \gg t$. We also present numerical predictions for the Fermi surface and the optical conductivity of the lightly doped 2D AFM at various doping concentrations. In the limit of vanishing concentration of doped holes, we find good correspondence of the here-calculated spectral densities and quasi-particle mass enhancement with earlier work on the t–J model [19–21].

In effect, the analytical Ansatz, proposed in [15] and further discussed below, reveals in a straightforward way a number of surprising features. For instance, it shows that the strong-coupling limit (heavy mass) *does not necessarily imply band-narrowing*. Instead, for both the t–J and t–$t′$–J models, we find that strong coupling leads to *depletion* of the particle density of states at the Fermi level (E_F), the spectral density being pushed away from E_F into a broad range of energies (of order t). To the best of our knowledge, this remarkable feature has not been shown analytically before and we suspect that it might constitute an important ingredient of the mechanism leading to superconductivity in high-T_c cuprates.

2 The model and analytical Ansatz

The consequences of the concept of spin–charge separation for the quasi-particle picture in the Hubbard model have been investigated for a long time (see e.g. [16] and references therein). Various decoupling procedures, based on the mean-field approach (MFA), were used to get the new quasi-particle hierarchy. Since, in this approach, fluctuations are neglected, it is strictly applicable only in the multi-component (i.e. classical) limit either for the spin or for the orbital degrees of freedom of electrons [17].

Thus the MFA approach leaves out the basic problem of strong coupling of charge and spin fluctuations in the correlated system. Here the similarity with electron–phonon interaction presents itself. The consequences of strong electron–phonon coupling in conventional metals are well known since the work of Migdal and subsequent authors [24, 25]. These ideas have recently been (qualitatively) reconsidered in relation to the high-T_c cuprates [26] by Eliashberg.

Nevertheless, the nature of the charge–spin coupling in the large positive-U Hubbard model, at energy scales much smaller than U, differs radically from the conventional metal case. Namely, the introduction of the new set of charge and spin operators, as necessitated by the large U (i.e. strong Coulomb correlations), causes an intrinsic incoherence even in the expression for the bare electron kinetic energy. This, in turn, places the system on the edge of the region of applicability of Migdal's procedure (compare with [19–21]). To demonstrate the argument explicitly, we resort to the particular case of the t–t'–J model [27]. The initial Hubbard model is well known:

$$\mathcal{H}^{\text{Hub}} = -t \sum_{\langle i;j \rangle, s} c_{is}^+ c_{js} + U \sum_i \rho_{i\uparrow} \rho_{i\downarrow}. \tag{1}$$

As was shown in [27], the low-energy behaviour of this model may be studied by passing to an effective t–t'–J Hamiltonian, which is obtained by means of second-order perturbation theory, using the ratio t/U as a small parameter:

$$\mathcal{H}^{tt'J} = t \sum_{\langle i;j \rangle} f_i^+ f_j \mathcal{F}_{ij} - \mu \sum_i f_i^+ f_i - \frac{J}{4} \sum_{\langle i;j \rangle} \mathcal{A}_{ij}^+ \mathcal{A}_{ij}(1 - f_j^+ f_j)$$
$$+ t' \sum_{\langle i;jk \rangle} f_k^+ f_i \mathcal{A}_{ij}^+ \mathcal{A}_{kj}(1 - f_j^+ f_j). \tag{2}$$

Here the intra-sublattice transfer integral $t' = J = 4t^2/U \ll t$, and $\langle i;jk \rangle$ denotes next-nearest neighbouring sites. The commuting Schwinger bosons a_i and b_i and slave fermionic holes f_i^+ and f_j together compose 'real' electrons:

$$c_{i\uparrow} \rightarrow a_i f_i^+, \qquad c_{i\downarrow} \rightarrow b_i f_i^+. \tag{3}$$

The $\mathcal{A}$ and $\mathcal{F}$ operators obey the relations $\mathcal{A}_{ij}^+ \mathcal{A}_{ij} = (S_i - S_j)^2 - 1$ and $\mathcal{F}_{ij}^+ \mathcal{F}_{ij} = (S_i + S_j)^2$.

At half-filling the Hamiltonian $\mathcal{H}^{tt'J}$ reduces naturally to the Hamiltonian of the pure quantum Heisenberg antiferromagnet (AFM), with the quantum spin $\frac{1}{2}(S_i)$ on each lattice site. Strong on-site Coulomb repulsion between the charge carriers is modelled in the form of the 'no double-occupancy' constraint: $a_i^+ a_i + b_i^+ b_i + f_i^+ f_i = 1$. The concentration of the doped holes, x, is introduced by the relation: $\langle f_j^+ f_j \rangle = x$, where $\langle \ldots \rangle$ means thermodynamic average.

It follows from (2) that each inter-sublattice hop of the spinless charge with the amplitude t is incoherent, since it is accompanied by backflow of spin-density. The reason is that the t-hop leads to misorientation of the antiferro-magnetically correlated spins in the vicinity of the doped hole. Although the t' term in (2) describes coherent motion of charged spinless holes on the same antiferromagnetic sublattice, it is nevertheless dominated by the main hopping process, since $t \gg t'$. Therefore, the net dispersion of the quasi-particle states is essentially a fluctuation-induced feature, and thus cannot be obtained in the MFA (compare with [19, 21]). This can be readily inferred from the fact (as demonstrated below) that the dispersion function is determined entirely by the spatial variation of the spin–hole coupling matrix elements. For the same reasons the corresponding dimensionless 'electron–phonon coupling strength' λ is of the order of $t/J \gg 1$, and the 'adiabaticity parameter', which arises in the theory of common metals [24] as $(m/M)^{1/2}$ (where m and M are the electron and ion masses respectively), should be replaced here by J/t. Therefore, a rough estimate of the first-order relative spin–hole vertex correction gives the dimensionless parameter $\lambda(m/M)^{1/2} \simeq 1$, instead of $\ll 1$ as for conventional metals. Fortunately, as will be shown in Section 4 below, the actual dimension-less parameter of the present problem proves to be of order of $1/\ln(U/t)$ (when $U \gg t$). This important fact validates the applicability of the Migdal approxi-mation to the present case, and explains (*a posteriori*) the reason for the rather good accord found previously between the numerical self-consistent calcula-tions in the Born approximation [19, 20] and exact diagonalization data for finite-size, doped antiferromagnetic clusters [28].

To demonstrate these arguments, we consider Eq. (2) close to half-filling. The Green function, $D(\Omega)$, of the magnetic subsystem (described by the operators $\mathscr{F}_{ij}$ and $\mathscr{F}_{ij}^{+}$) is modelled by

$$D(\Omega) = \frac{\varDelta}{\Omega^2 + \varDelta^2}. \tag{4}$$

Here the Matsubara representation is used, see e.g. [29]. The parameter $\varDelta$ is regarded as phenomenological, describing a gap in the magnetic excitation spectrum. It follows from Eq. (2) that the energy scale for magnetic excitations is set by $J \simeq t'$, and therefore $\varDelta \lesssim t'$. (Actually, such an Einstein model, with $\varDelta \simeq 300$ K, reflects rather well the magnetic excitation spectrum found in the doped cuprates by neutron scattering experiments [30, 31].) Since the para-meter $1/\ln(U/t)$ is only logarithmically small, the self-consistent equation for the self-energy function of the spinless fermionic holes, $\Sigma(\boldsymbol{q}, \omega)$, should explicitly allow for the $\boldsymbol{q}$-dependence of Σ (usually neglected in the theory of ordinary metals, see [24, 29]). As we are going to investigate the problem qualitatively, we choose the simplest (factorized) form for this $\boldsymbol{q}$-dependence:

$$\Sigma(\boldsymbol{q},\omega)=2t^2(1+\varepsilon(\boldsymbol{q}))T\sum_{\Omega,\boldsymbol{Q}}(1+\varepsilon(\boldsymbol{Q}))D(\boldsymbol{Q},\Omega)G(\boldsymbol{q}-\boldsymbol{Q},\omega-\Omega), \qquad (5)$$

where the prefactor 2 is chosen for the sake of convenience, and where

$$G(\boldsymbol{q},\omega)=\{i\omega-E(\boldsymbol{q})+\mu-\Sigma(\boldsymbol{q},\omega)\}^{-1}. \qquad (6)$$

Summation over $\boldsymbol{Q}$ in Eq. (5) is taken over the whole Brillouin zone, and the frequencies are $\omega=\pi T(2n+1)$; $\Omega=2\pi Tm$; n, $m=0;\pm 1;\pm 2\ldots$. An explicit form of the function $\varepsilon(\boldsymbol{q})$ in Eq. (5), in the limit of one doped hole, may be found from Eq. (2) in the spin-wave approximation (see [19–21]). To match these results we take

$$\varepsilon(\boldsymbol{k})=-\gamma E(\boldsymbol{k})\equiv-\gamma W(\gamma_k^2+\alpha\tilde{\gamma}_k^2), \qquad \gamma_k=\frac{1}{4}\sum_{\boldsymbol{\delta}}\exp{(i\boldsymbol{k}\cdot\boldsymbol{\delta})} \qquad 0<\gamma<(16t')^{-1},$$

$$\qquad (7)$$

where the bare dispersion $E(\boldsymbol{k})$ arises from the intra-sublattice hopping of the hole and $\boldsymbol{\delta}$ runs over the nearest neighbouring sites on the square lattice. The bandwidth $W=16t'$; and $\tilde{\gamma}_k$ are obtained in the same way as γ_k, but taken on one magnetic sublattice only. The coefficient α is $\frac{1}{4}$ in the long-range-ordered, zero-doping limit (see [19]). Actually, at finite doping concentrations α should depend on the remaining amount of short-range antiferromagnetic order. We have found that inclusion of the $\alpha\tilde{\gamma}_k^2$ term is crucial for the appearance of a Van Hove singularity in both the quasi-particle and particle density of states, N_{qp} and N_p. When this singularity crosses the Fermi level, a maximum occurs in the x-dependence of N_{qp} (N_p) (see Fig. 3 and Fig. 4 later), the position of which depends on the value of α. If we take $\alpha=0.05$, then this maximum is at $x_m\approx 0.23$, which is close to the numerical result $x_m=0.262$ found in [22] (experimental data [23] point to $x_m\simeq 0.3$). If we take the zero-doping limit $\alpha=0.25$ (as found in [19] by fitting the quasi-particle dispersion function obtained numerically), then we get the substantially greater value $x_m\simeq 0.4$.

Now we make an analytical continuation of Eq. (5) in the upper half-plane of ω, and arrive at the following integral equation for the retarded function $\Sigma^R(\omega)$, which is an analytical function in the upper half plane of ω (we neglect the coefficient $1-\gamma E(\boldsymbol{k})$ in the integrand):

$$\Sigma^R(\boldsymbol{q},\omega)=\frac{t^2(1-\gamma E(\boldsymbol{q}))}{\pi}\int\frac{d^2k}{(2\pi)^2}\left(\int_0^\infty d\varepsilon\frac{\operatorname{Im}G^R(\boldsymbol{k},\varepsilon)}{\varepsilon+\Delta-\omega-i0^+}+\int_{-\infty}^0 d\varepsilon\frac{\operatorname{Im}G^R(\boldsymbol{k},\varepsilon)}{\varepsilon-\Delta-\omega-i0^+}\right).$$

$$\qquad (8)$$

We look for the solution in the form

$$\operatorname{Re}\Sigma^{R}(\boldsymbol{q},\omega)= -f(\omega)(1-\gamma E(\boldsymbol{q})),\qquad \operatorname{Im}\Sigma^{R}(\boldsymbol{q},\omega)= -\Gamma(\omega)(1-\gamma E(\boldsymbol{q})). \quad (9)$$

By substituting this Ansatz into Eq. (8), we finally obtain two self-consistent equations for the unknown functions $f(\omega)$ and $\Gamma(\omega)$:

$$f(\omega)=\frac{1}{\pi}\left(\mathscr{P}v\int\limits_{-\infty}^{\infty}\mathrm{d}\varepsilon\,\frac{\Gamma(\varepsilon+\varDelta\operatorname{sign}\varepsilon)}{\varepsilon+\varDelta\operatorname{sign}\varepsilon-\omega}\right)$$

$$\Gamma(\varepsilon)\approx\frac{t^{2}}{t'(1+\gamma f(\tilde{\varepsilon}))}\left[\arctan\left(\frac{t'(1+\gamma f(\tilde{\varepsilon}))-(\tilde{\varepsilon}+f(\tilde{\varepsilon})+\mu)}{\Gamma(\tilde{\varepsilon})+0^{+}}\right)\right.$$

$$\left.+\arctan\left(\frac{(\tilde{\varepsilon})+f(\tilde{\varepsilon})+\mu}{\Gamma(\tilde{\varepsilon})+0^{+}}\right)\right]\Theta(\tilde{\varepsilon}\operatorname{sign}\varepsilon), \quad (10)$$

where $\tilde{\varepsilon}=\varepsilon-\varDelta\operatorname{sign}\varepsilon$ and $\Theta(y)=0(1)$, $\operatorname{sign}y=-1(1)$ when $y<0$ (≥0). The symbol $+0^{+}$ in the denominators of the above formula indicates that the zero limit should be approached from the positive side. In the limit $\gamma\to0$, i.e. when the $\boldsymbol{q}$-dependence of Σ is neglected, Eqs. (10) reduce to the well-known equations for the case of strong electron–phonon coupling, see [25]. In the derivation of the equation for Γ we have neglected the $\boldsymbol{q}$-dependent prefactor in the expression for $\operatorname{Im}\Sigma^{R}$, given by Eq. (9). We have also passed to integration over the energy in the $\boldsymbol{k}$-integral in Eq. (8), by introducing a constant density of states $v=1/W$. In a 2D system this is always possible even for small carrier concentration. Finally, from Eqs. (10) onward, we use the notation t' for the bare bandwidth $W=16t'$, since the analytical results presented below will only be valid within an order of magnitude.

3 The self-energy function of the spinless hole

Now, it is convenient to solve Eqs. (10) by starting with $\Gamma(\varepsilon)$. The form of Eqs. (10) suggests a natural division of the ε axis into intervals of width $\varDelta$. Then, due to the presence of the $\Theta(y)$ function, we obtain

$$\Gamma(\varepsilon)\equiv0,\qquad|\varepsilon|<\varDelta. \quad (11)$$

Inserting this result back into Eqs. (10) for $\Gamma(\varepsilon)$, one has

$$\Gamma(\varepsilon)\approx\frac{t^{2}\pi}{t'(1+\gamma f(\tilde{\varepsilon}))}\,\Theta[t'(1+\gamma f(\tilde{\varepsilon}))-(\tilde{\varepsilon}+f(\tilde{\varepsilon})+\mu)]$$

$$\times\Theta(\tilde{\varepsilon}+f(\tilde{\varepsilon})+\mu),\qquad\varDelta<|\varepsilon|<2\varDelta. \quad (12)$$

Inserting again the last result back into the second of Eqs. (10), we can, in

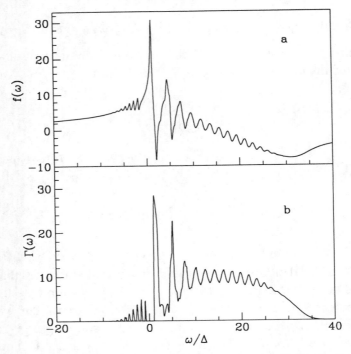

Fig. 1. The dependence on ω of the function $f(\omega)$ (a), and $\Gamma(\omega)$ (b), see Eq. (9), for hole concentration $x = 0.031$. Other parameter values are $t = 10.0$, $W = 5.0$ and $\gamma = 0.1$.

principle, continue this iterative process. Using this idea, we have solved Eqs. (10) numerically for finite concentration of the fermionic holes, Fig. 1(a) and (b) (1600 mesh points within the regarded ω-interval were taken, and a 70-loops iteration process was applied; the convergence achieved was better than 0.1% relative deviation between the ultimate and penultimate iteration runs). Also, it is possible to find an analytical solution of Eqs. (10) for two limiting cases. Namely, when both arguments of the arctan functions in Eqs. (10) are small relative to unity, one has

$$\Gamma(\varepsilon) \approx \frac{t^2}{\Gamma(\tilde{\varepsilon})}, \qquad 2\Delta < |\varepsilon| \ll t, \qquad \Gamma(\varepsilon) \simeq t. \tag{13}$$

Furthermore, when these arguments are very large but finite (and then have opposite signs, as follows from their structure), Eqs. (10) take the form

$$\Gamma(\varepsilon) \approx \frac{t^2 \Gamma(\tilde{\varepsilon})}{[\tilde{\varepsilon} + f(\tilde{\varepsilon}) + \mu - t'(1 + \gamma f(\tilde{\varepsilon}))](\tilde{\varepsilon} + f(\tilde{\varepsilon}) + \mu)} \approx \frac{t^2 \Gamma(\tilde{\varepsilon})}{\varepsilon^2}, \tag{14}$$

when $|\varepsilon| \gg t$, where, as before $\tilde{\varepsilon} = \varepsilon - \Delta \operatorname{sign} \varepsilon$. By converting the last difference equation into a differential one, and solving it, we get

$$\Gamma(\varepsilon) \approx \Gamma_0 \exp\left(-\frac{|\varepsilon|^3}{3t^2 \Delta}\right), \qquad |\varepsilon| \gg t, \tag{15}$$

where $\Gamma_0 \simeq t$. The last result shows that the integration in the first Eq. (10) is actually over the finite interval of order of t, since $\Gamma(\varepsilon) \to 0$ outside this interval. A detailed analysis shows that $\Gamma(\varepsilon)$ never diverges. Therefore, substituting the estimates from Eq. (13) and Eq. (15) into the first of Eqs. (10), we get

$$f(\varepsilon) \simeq -t^2/\varepsilon, \qquad |\varepsilon| \gg t. \tag{16}$$

Here we note that Eq. (14) is valid also in the interval $\Delta < |\varepsilon| \leq t$ close to the points at which $|f(\varepsilon)|$ diverges. This means that, at these points, $\Gamma(\varepsilon) \to 0$. In the opposite case, when $|f(\varepsilon)| \leq \Gamma(\varepsilon)$ (and of course $|\varepsilon| \ll t$), Eq. (13) holds and $\Gamma(\varepsilon)$ rises to its 'saturated' value of about t. Thus, the function $\Gamma(\varepsilon)$ has sawtooth form in the interval $\Delta < |\varepsilon| \leq t$, Fig. 1 (compare with [19, 20]). Having this in mind, we take the function $\Gamma(\varepsilon)$ in a simplified form, which follows from Eqs. (11)–(13) and Eq. (15):

$$\Gamma(\varepsilon) = \begin{cases} 0, & \text{when } |\varepsilon| < \Delta \text{ or } |\varepsilon| > t; \\[2mm] \dfrac{t^2 \pi}{t'(1 + \gamma f(\tilde{\varepsilon}))} \, \Theta(\tilde{\varepsilon} + f(\tilde{\varepsilon}) + \mu)\Theta[t'(1 + \gamma f(\tilde{\varepsilon})) - (\tilde{\varepsilon} + f(\tilde{\varepsilon}) + \mu)], \\[2mm] \quad \text{when } \Delta < |\varepsilon| < 2\Delta; \\[2mm] t\Theta(\tilde{\varepsilon} + \varepsilon(\mu) + n\Delta)\Theta[(\tilde{\varepsilon} + n\Delta)\,\text{sign}\,\varepsilon], \qquad n = 1,2,\ldots, \\[2mm] \quad \text{when } 2\Delta < |\varepsilon| < t, \end{cases} \tag{17}$$

where $\varepsilon(\mu)$ fulfils the condition $\tilde{\varepsilon} + f(\tilde{\varepsilon}) + \mu = 0$ at $\varepsilon = -\varepsilon(\mu) < 0$. It follows, from Eq. (22) below, that $\varepsilon(\mu) \simeq x\Delta \ln(U/t)$, where, as before, x is the density of doped holes. Therefore, the form of Eq. (17) implies that, for negative values of ε, $\Gamma(\varepsilon)$ differs from zero only within intervals of width $\simeq x\Delta \ln(U/t) \ll \Delta$, close to the points $\varepsilon = -n\Delta$ ($n = 1,2,\ldots$). On substituting $\Gamma(\varepsilon)$, as defined in Eq. (17), into the first of Eqs. (10), we find

$$f(\varepsilon) \approx \frac{t^2}{t'(1 + \gamma f(0))} \left(\ln\left|\frac{t}{\Delta - \varepsilon}\right| - \ln\left|\frac{\varepsilon(\mu) + \Delta + \varepsilon}{\Delta + \varepsilon}\right| \right), \qquad |\varepsilon| < 2\Delta. \tag{18}$$

Using Eq. (6), we find the following equation for the poles of the Green function of the spinless fermionic holes (in its retarded form):

$$\omega - E(\boldsymbol{k}) + \mu - \text{Re}\,\Sigma^R(\omega, \boldsymbol{k}) = 0. \tag{19}$$

Solutions of Eq. (19) at $\omega = 0$ give the Fermi surface (FS) of the system, see Fig. 4 later, calculated numerically with the resolution of 700×700 mesh points within the (square lattice) Brillouin zone. On substituting Eq. (17) and Eq. (18)

into the definitions in Eq. (9), and using them in Eq. (19), we get the following equation, which determines the position and dispersion of the quasi-particle sub-band:

$$\varepsilon\left(1+\frac{\partial f(\varepsilon=0)}{\partial \varepsilon}\right)-E(\boldsymbol{k})(1+\gamma f(0))+\mu_1 \approx 0, \qquad \mu_1 \equiv \mu+f(0), \qquad |\varepsilon|<\Delta. \quad (20)$$

Here $\mu_1 \simeq x$ is the shifted chemical potential of the holes. Therefore, $-f(0)$ determines the shift of the band in the limit $x \to 0$. Expanding now Eq. (18) in the interval $|\varepsilon| \ll \Delta$, and using Eq. (20), we find the shift, $-f(0)$, the residue, $z-(1+\partial f(\varepsilon=0)/\partial\varepsilon)^{-1}$, and the dispersion, $\varepsilon(\boldsymbol{k})$, of the renormalized quasi-particle sub-band in the limit $x \ll 1$:

$$-f(0) \approx -t\left(\frac{1}{\gamma t'}\right)^{1/2}\left[\ln\left(\frac{U}{t}\right)\right]^{1/2}, \qquad z^{-1}=1+\frac{\partial f(\varepsilon=0)}{\partial \varepsilon} \approx \frac{U}{t}\left(\frac{1}{\gamma t' \ln(U/t)}\right)^{1/2},$$
$$(21)$$

$$\varepsilon(\boldsymbol{k}) \approx E(\boldsymbol{k})\left[\delta+\gamma\Delta\ln\left(\frac{U}{t}\right)\right], \qquad \varepsilon(\mu)=z\mu_1 \simeq x\Delta\ln\left(\frac{U}{t}\right), \quad (22)$$

where $\delta \simeq \Delta/t \simeq t/U \ll 1$. To derive these results we have used the following relations: $\Delta \leq t' \simeq t^2/U$ and hence $t/\Delta \simeq U/t$.

The results in Eqs. (20)–(22) have remarkable consequences. Firstly, it follows from Eq. (20), in the limit $t' \ll t$ (and $\varepsilon(\boldsymbol{q}) \leq 1$ in the prefactor in Eq. (5), which is just the case at low doping, [19, 21]), that the Fermi velocity, $v_F \simeq (\partial E(\boldsymbol{k}_F)/\partial k)(1+\gamma f(0)) \simeq t[\ln(U/t)]^{1/2}$, is enhanced with respect to the bare value (about t) by a factor $[\ln(U/t)]^{1/2}$. Hence, the *particle* density of states (per unit area) at the Fermi surface, $N_p(0) \simeq v_F^{-1}$, is *depleted* with respect to its bare value by the same factor. This result is in accord with the statements made in [26]. Secondly, it follows from comparison of the two terms in the expression in Eq. (22) that the quasi-particle dispersion $\varepsilon(\boldsymbol{k})$ is predominantly induced by the spin–hole coupling term (t) in the Hamiltonian of Eq. (2) (which contributes to the second term in the brackets in Eq. (22)), and is not sensitive to the contribution due to the bare coherent hopping integral t' (the first term in the same brackets). This means that, in the limit $t' \ll t$, our result should also apply to the t–J model (obtained from Eq. (2) by dropping the t' term, with $t'=J=4t^2/U$). Consequently, the t–J model should have a renormalized quasi-particle band nearly identical to that of the t–t'–J model.

Another important result, which follows from Eq. (21) and Eq. (22) is illustrated in Fig. 2(a) and (b). The coherent (quasi-particle) contribution to the total density of states (shown as the amplitude of the central 'delta-function'-like peak at $\omega=0$) is $z^{-1}(\gg 1)$ times less than it would be for the non-

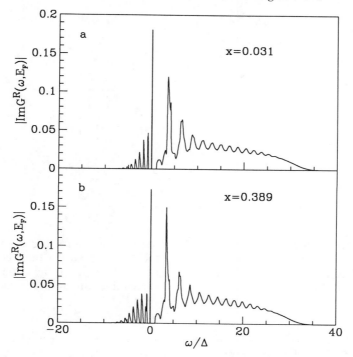

Fig. 2. The calculated density of states $(1/\pi)|\mathrm{Im}\,G^R(\omega, E_F)|$ for hole concentration $x = 0.031$ (a) and 0.389 (b). Other parameters are as for Fig. 1.

interacting system ($z^{-1} = 1$). A substantial part of the spectral density is pushed far outside the Fermi-level region (compare with [26]), and forms a broad incoherent wing. The calculated particle and quasi-particle densities of states at the Fermi level as a function of doping concentration are shown in Fig. 3. The sharp maximum at $x_h \approx 0.23$ arises from a Van Hove singularity in the 2D density of states induced by the next-nearest neighbour term (α) in $E(\mathbf{q})$ in Eq. (7) (see also Fig. 4).

Another peculiar result follows from Eq. (18) and Eq. (19) in the region $0 < \Delta - \varepsilon \ll \Delta$. Having fixed the chemical potential μ by Eq. (20) and Eq. (21), we can rewrite Eq. (19) in the form

$$\varepsilon + \tilde{t}\ln\left(\frac{\Delta}{|\Delta - \varepsilon|}\right) - \xi(\mathbf{k}) \approx 0, \qquad \xi(\mathbf{k}) = \gamma\tilde{t}\ln\left(\frac{U}{t}\right)E(\mathbf{k}) - \mu_1, \qquad (23)$$

where $\tilde{t} \equiv t[\gamma t'\ln(U/t)]^{-1/2}$. From Eq. (23) we find the quasi-particle energy, taken with respect to the chemical potential, close to the top of the quasi-particle sub-band:

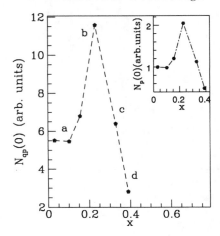

Fig. 3. The calculated *quasi-particle* density of states at the Fermi level for $\alpha = 0.05$, as a function of hole concentration x. Points, which correspond to the Fermi surfaces shown in Fig. 4(a)–(d), are marked with the same indices. Inset: the same, but for the *particle* density of states at the Fermi level. Dashed lines are guides to the eye. Other parameters are as for Fig. 1.

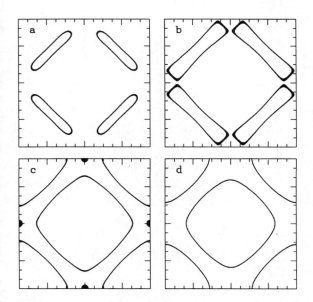

Fig. 4. Calculated Fermi surfaces, in the extended (antiferromagnetic) Brillouin zone scheme, for hole concentration x: 0.10 (a), 0.23 (b); 0.33 (c); and 0.389 (d); $\alpha = 0.05$. Other parameters are as for Fig. 1.

$$\varepsilon(\boldsymbol{k}) \approx \Delta[1 - \exp(-\xi(\boldsymbol{k})/\tilde{t})]. \tag{24}$$

Since $|\xi(\boldsymbol{k})| \leq \tilde{t}$ (as follows from Eq. (23)), Eq. (24) indicates that the quasi-particle energy $\varepsilon(\boldsymbol{k})$ changes by an amount of order $\Delta \leq J$, while the vector $\boldsymbol{k}$ spans over a substantial part of the Brillouin zone. Provided that $\Delta \simeq 300$ K (see [30, 32]), this might explain the experimental data [18], where flat bands were found, that occupy about a fifth of the Brillouin zone and lie within ± 40 meV of E_{F}.

Finally, we expand $f(\varepsilon)$ in Eq. (18) in the vicinity of 2Δ, and then substitute the result into Eq. (19), in order to find the real part of the dispersion function $\varepsilon(\boldsymbol{k})$ of the next, electron-like (damped) sub-band. The imaginary part of the same function is defined by Eq. (9) and Eq. (12). Thus, we have the following results:

$$\mathrm{Re}[\varepsilon(\boldsymbol{k})] \approx -E(\boldsymbol{k})\left[\delta + \gamma \Delta \ln\left(\frac{U}{t}\right)\right] + 2\Delta,$$

$$\mathrm{Im}[\varepsilon(\boldsymbol{k})] \approx \Delta\,\frac{\ln(U/t)}{\ln|t/\{2\Delta - \mathrm{Re}[\varepsilon(\boldsymbol{k})]\}|}, \qquad |2\Delta - \mathrm{Re}[\varepsilon(\boldsymbol{k})]| \ll \Delta. \tag{25}$$

The result in Eq. (25) signifies that the electron-like, damped sub-band is shifted energetically by about 2Δ with respect to the bottom of the previous hole-like sub-band (see Eq. (22)). This may have peculiar consequences for the temperature- and concentration-dependences of the Hall resistivity, due to the different sign of the dispersion function in these two adjacent sub-bands. In addition, the V-shaped form of the damping rate of the quasi-particles as a function of energy in the electron-like sub-band, which follows from Eq. (25) (see Fig. 1(b)), would cause an unusual temperature-dependence of the resistivity. We shall address these issues in detail in a forthcoming paper, and restrict ourselves in this work to the optical conductivity, which we shall discuss in Section 5.

4 The small parameter of the problem

Here we derive a small parameter validating Migdal's approximation implied in Eq. (5) (i.e. use of the bare vertex $\Gamma^0 \simeq t$ instead of the full one). The simplest correction Γ^1 (which involves the full function G) to the vertex Γ^0 may be expressed in a form analogous to the case of electron–phonon interaction [29]:

$$\Gamma^1(\boldsymbol{p},\omega;\boldsymbol{Q},\Omega) \simeq t^3 T \sum_{\omega_1,\boldsymbol{p}_1} G(\boldsymbol{p}_1,\omega_1)G(\boldsymbol{p}_1 - \boldsymbol{Q}, \omega_1 - \Omega)D(\boldsymbol{p} - \boldsymbol{p}_1, \omega - \omega_1), \tag{26}$$

where the functions D and G are determined by Eq. (4) and Eq. (6). As the summation in Eq. (5) goes over the range $|Q| \simeq \pi/a; \Omega \leq t$ (a is the lattice spacing), we evaluate Γ^1 with the arguments $|Q|, \Omega$ in the same range. In the main summation region over ω_1 in Eq. (26), $|\omega - \omega_1| < \Delta$, the function D can be substituted by the constant Δ^{-1}. For a rough estimate we pass from summation over p_1 to integration over energy E in the same way as described after Eqs. (10) above, and replace $E(p_1 - Q)$ in Eq. (26) by $E + E_Q$, where $E_Q \simeq t'$. This allows us to evaluate Γ^1:

$$
|\Gamma^1| \simeq \left| \frac{t^3}{t'} \int_0^{t'} dE \frac{1}{\tilde{f}(\omega) + \mu_1 - E(1 + \gamma f(\omega)) + i\Gamma(\omega)} \right.
$$

$$
\times \left. \frac{1}{\tilde{f}(\omega - \Omega) + \mu_1 - (E + E_Q)[1 + \gamma f(\omega - \Omega)] + i\Gamma(\omega - \Omega)} \right|
$$

$$
\simeq \frac{tv}{\ln(U/t)} \ll t \simeq \Gamma^0, \qquad U \gg t, \tag{27}
$$

where $\tilde{f}(\omega) \equiv \omega + f(\omega) - f(0)$, and Eq. (20) and Eq. (21) have been used to get the final estimate (the numerical factor $v \simeq 1$ incorporates the dimensionless parameter $t'\gamma$, which was generally taken to be about unity). The estimate in Eq. (27) is actually a central result of this paper, as it proves the validity of the Migdal approximation used here and implicitly assumed in previous work [19–21]. It is important to note that the result is obtained here in a self-consistent manner. This follows from Eq. (21) and Eq. (27), which actually show that the small parameter $|\Gamma^1/\Gamma^0| \simeq 1/\ln(U/t)$ arises simultaneously with the depletion of the particle density of states at the Fermi level by a factor $[1/\ln(U/t)]^{1/2}$. The last renormalization arises due to the integration in the first of Eqs. (10) over the energies ε in the 'incoherence interval' of approximate width t, where $\operatorname{Im}\Sigma(\varepsilon) \simeq \Gamma(\varepsilon) \simeq t$ (see Eq. (13)). It was shown recently (Liu and Manousakis 1992) that the lowest order vertex correction Γ^1 is zero, if the Néel ground state is assumed. In this case an estimate gives $|\Gamma^{(2)}/\Gamma^0| \simeq t/\ln^2(U/t)$, where $\Gamma^{(2)}$ is the first non-vanishing vertex correction.

5 Optical conductivity

We start with the tight-binding expression for the electron current (see e.g. [33]):

$$
\hat{j} = \frac{et}{2i} (c_{is}^+ c_{js} - c_{js}^+ c_{is}), \tag{28}
$$

where the notation is the same as in Eq. (1) and e is the electron charge. To allow for the strong spin–charge correlations in the present electronic system, we substitute in Eq. (28) the electron operators c^+, c by their composite spinless fermion–Schwinger boson form, given by Eq. (3). This leads, after a straight-forward derivation using the Feynman diagramatic technique (see e.g. [29]), to the following relations for the real part of the frequency-dependent conductivity, $\sigma'(\omega)$, of the system:

$$\sigma'(\omega) = \frac{\Pi_a(\omega) + \Pi_b(\omega)}{\omega} \tag{29}$$

$$\Pi_i(\omega) = \begin{cases} \dfrac{e^2 t^2}{2\pi^2} \displaystyle\int \frac{d^2k}{(2\pi)^2} \int_{-\infty}^{\infty} d\varepsilon\, \mathrm{Im}\, G^R(\boldsymbol{k},\varepsilon)\,\mathrm{Im}\, G^R(\boldsymbol{k},\varepsilon-\omega) \\ \qquad \times \{\tanh[\varepsilon/2T)] - \tanh[\varepsilon-\omega/(2T)]\}\, M^2(\boldsymbol{k},\omega), \qquad i=b \\[2mm] \dfrac{e^2 t^2}{2\pi^2} \displaystyle\int \frac{d^2k}{(2\pi)^2} \int_{-\infty}^{\infty} d\varepsilon\, \mathrm{Im}\, G^R(\boldsymbol{k},\varepsilon)\,\mathrm{Im}\, \Sigma^R(\boldsymbol{k},\varepsilon-\omega) \\ \qquad \times \{\tanh[\varepsilon/2T)] - \tanh[\varepsilon-\omega/(2T)]\}\, \tilde{M}^2(\boldsymbol{k},\omega), \qquad i=a, \end{cases} \tag{30}$$

where M and $\tilde{M}$ are the current vertices. Here it is important to note that the functions $G^R(\boldsymbol{k},\varepsilon)$ and $\Sigma^R(\boldsymbol{k},\varepsilon)$ are defined in Eq. (5) and Eq. (6), and describe spinless fermionic holes, but not the bare electrons. This fact makes it easier to understand the origin of the anomalous $\Pi_a(\omega)$ term in Eq. (29) and Eq. (30). This term arises entirely from the incoherence of the hopping events of the spinless charge, described by the t term in the Hamiltonian of Eq. (2). The 'traditional' band–particle contribution, $\Pi_b(\omega)$, used in [21], reflects this incoherence only indirectly, by inclusion of all the interactions into the Green functions of the spinless fermionic holes.

Based on the results of the previous sections, and using Eq. (29) and Eq. (30), we obtain the following predictions for the conductivity (here the interval $|\omega| \le 2\Delta$ is considered):

$$\sigma'(\omega) = \sigma'_{1b}(\omega) + \sigma'_{2b}(\omega) + \sigma'_a(\omega), \tag{31}$$

$$\sigma'_i(\omega) \simeq \begin{cases} xe^2 \Delta \delta(\omega), & i=1b; \\[2mm] \dfrac{xe^2 \Delta^3}{\omega[\Delta^2 + (\omega-2\Delta)^2]}, & i=2b, \quad |\omega-2\Delta| \le \Delta; \\[2mm] xe^2[\Delta/(\gamma t'\omega)]\Theta(\omega-\Delta), & i=a, \end{cases} \tag{32}$$

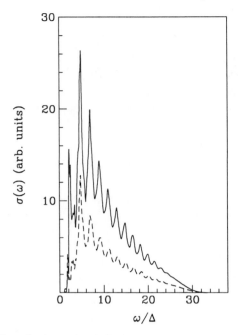

Fig. 5. The calculated optical conductivity $\sigma(\omega)$ (without Drude contribution) for hole concentration $x = 0.102$ (dashes) and 0.389 (solid line). Other parameters are as for Fig. 1.

where $\delta(\omega)$ is the Dirac function, $\varepsilon(\mu)$ is given in Eq. (22) and x is, as before, the concentration of doped holes ($\ll 1$). The term $\sigma'_{1b}(\omega)$ describes the coherent Drude peak due to the lowest hole-like sub-band, given by Eq. (21) and Eq. (22). Another term, $\sigma'_{2b}(\omega)$, arises by convolution of the lowest hole-like sub-band with the broad incoherent part of the spectrum (see Fig. 2(a) and (b)). Finally, the nature of the term $\sigma'_a(\omega)$ is described above in relation to the anomalous term $\Pi_a(\omega)$ in Eq. (29) and Eq. (30). Our numerical results for $\sigma(\omega)$ (without Drude contribution tending to $\delta(\omega)$), obtained directly from Eq. (29) and Eq. (30) for finite doping concentrations, are presented in Fig. 5. We note the proportionality of σ' to $\Delta \simeq 1/m^*$, rather than to $t \simeq 1/m$, which signifies the renormalized (heavy) nature of the quasi-particle mass m^* (see [21]). Finally, in the interval $\Delta \ll |\omega| \leq t$ the sum $\sigma'_{2b}(\omega) + \sigma'_a(\omega)$ provides a broad tail of about $\mathrm{Im}\, G^R(\omega)/\omega \simeq 1/\omega$, as follows from Eq. (17) and Eq. (18) and Eq. (29) and Eq. (30). The tail possesses fine feature with the characteristic energy scale $\delta\omega \simeq \Delta$ resulting from the sawtooth-like behaviour of the function $\Gamma(\varepsilon)$ (see Fig. 1(b)).

6 Conclusions

We have been able to derive a number of surprising new properties characteristic of the strongly correlated electron system, as described by the $t–J$ or $t–t′–J$ models, on the basis of a simplified Ansatz for the set of Migdal-type self-consistent equations. The small parameter validating the Migdal approximation for the present case is found analytically to be $1/\ln(U/t)$. Our approach reveals important qualitative differences between strongly correlated electron systems and conventional metals. The most pronounced feature is the depletion of the particle density of states at the Fermi level (in favour of formation of broad incoherent wings away from E_F), which occurs simultaneously with enhancement of the quasi-particle mass. The fluctuational origin of quasi-particle dispersion has been demonstrated analytically in a straightforward way. Our results are in qualitative accord with recent optical and photoemission measurements for high-T_c cuprates, as well as with the numerical calculations for the low doping limit published previously [19–21].

Acknowledgements

This work is part of the research programme of the Stichting voor Fundamenteel Onderzoek der Materie (FOM), which is financially supported by the Nederlandse Organisatie voor Wetenschappelijk Onderzoek (NWO). Helpful discussions with Professor A. F. Andreev, Professor G. M. Eliashberg and Dr J. Zaanen are gratefully acknowledged. S. I. M. is grateful to David A. van Leeuwen for helpful advice with the computer calculations.

References

[1] J. G. Bednorz and K. A. Müller, *Z. Phys.* B **64**, 188 (1986).
[2] P. W. Anderson, *Science* **235**, 1196 (1987).
[3] R. Micnas, J. Ranninger and S. Robaszkiewicz, *Rev. Mod. Phys.* **62**, 113 (1990).
[4] L. J. de Jongh, *Physica* C **152**, 171 (1988); *Solid State Commun.* **65**, 963 (1988); *Physica* C **161**, 631 (1989); *Eur. J. Solid State Inorg. Chem.* **27**, 221 (1990).
[5] S. I. Mukhin, D. Reefman and L. J. de Jongh, *Physica* C **174**, 455 (1991).
[6] S. I. Mukhin, D. Reefman and L. J. de Jongh, *Physica* C **171**, 42 (1990).
[7] D. Reefman, S. I. Mukhin, and L. J. de Jongh, *Physica* C **199**, 403 (1992).
[8] A. S. Alexandrov, J. Ranninger and S. Robaszkiewicz, *Phys. Rev.* B **33**, 4526 (1986).
[9] S. I. Mukhin, D. Reefman and L. J. de Jongh, *Physica Scripta* T **45**, 47 (1992).
[10] S. I. Mukhin, and L. J. de Jongh, *Physica* C **221**, 77 (1993).
[11] D. Reefman, S. I. Mukhin, and L. J. de Jongh, *Physica* C **221**, 93 (1993).
[12] S. I. Mukhin, LANL Preprint LA-UR-93-1472, Los Alamos, NM (1993).

[13] S. I. Mukhin, *Proceedings of the International Conference 'Physics of Magnetism 93', Poznań (Poland), June 21–24, 1993* (to be published in *Acta Phys. Polonica* 1994).

[14] S. I. Mukhin and L. J. de Jongh, *J. Phys. Chem. Solids* **54**, 407 (1993).

[15] S. I. Mukhin and L. J. de Jongh, *Proceedings of the International Conference 'Superconductivity and Strongly Correlated Electronic Systems', Amalfi (Italy), October 14–16, 1993* (to be published by World Scientific 1994).

[16] G. Baskaran, Z. Zou and P. W. Anderson, *Solid State Commun.* **63**, 973 (1987); L. B. Ioffe and A. I. Larkin, *Phys. Rev. B* **39**, 8988 (1989).

[17] C. Castellani, G. Kotliar, R. Raimondi *et al., Phys. Rev. Lett.* **69**, 2009 (1992).

[18] G. A. Thomas, J. Orenstein, D. H. Rapkine *et al., Phys. Rev. Lett.* **61**, 1313 (1988); D. S. Dessau, Z.-X. Shen, D. M. King *et al., ibid.* **71**, 2781 (1993).

[19] S. Schmitt-Rink, C. M. Varma and A. E. Ruckenstein, *Phys. Rev. Lett.* **60**, 2793 (1988); F. Marsiglio, A. E. Ruckenstein, S. Schmitt-Rink and C. M. Varma, *Phys. Rev. B* **43**, 10882 (1991).

[20] G. Martinez and P. Horsch, *Phys. Rev. B* **44**, 317 (1991).

[21] C. L. Kane, P. A. Lee and N. Read, *Phys. Rev. B* **39**, 6880 (1989).

[22] S. A. Trugman, *Phys. Rev. B* **41**, 892 (1990).

[23] J. B. Torrance *et al., Phys. Rev. B* **40**, 8872 (1989).

[24] A. B. Migdal, *Sov. Phys. JETP* **7**, 996 (1958).

[25] S. Engelsberg and J. R. Schrieffer, *Phys. Rev.* **131**, 993 (1963); D. Dunn, *ibid.* **166**, 822 (1968).

[26] G. M. Eliashberg, preprint Institut für Theoretische Physik C, RWTH/ITP-C 18/93 (1993).

[27] A. Auerbach and B. E. Larson, *Phys. Rev. B* **43**, 7800 (1991).

[28] E. Dagotto, R. Joynt, A. Moreo, S. Bacci and E. Gagliano, *Phys. Rev. B* **41**, 9049 (1990).

[29] A. A. Abrikosov, L. P. Gor'kov and I. E. Dzyaloshinski, *Methods of Quantum Field Theory in Statistical Physics* (Dover Publications, New York, 1975).

[30] J. Rossat-Mignod, L. P. Regnault, C. Vettier, P. Burlet, J. Y. Henry and G. Lapertot, *Physica B* **169**, 58 (1991).

[31] B. Keimer, N. Belk, R. J. Birgeneau, A. Cassanho, C. Y. Chen, M. Greven, M. A. Kastner, A. Aharony, Y. Endoh, R. W. Erwin and G. Shirane, *Phys. Rev. B* **46**, 14034 (1992).

[32] H. Eskes, G. A. Sawatzky and L. F. Feiner, *Physica C* **160**, 424 (1989).

[33] A. A. Abrikosov in: *Fundamentals of the Theory of Metals* (North-Holland, Amsterdam, 1988).

22

Polarons in Peierls–Hubbard models

A. R. BISHOP and M. I. SALKOLA

Theoretical Division, Los Alamos National Laboratory, Los Alamos, NM 87545, USA

Abstract

Motivated by aspects of layered high-temperature superconductors and related quasi-one-dimensional materials, we consider polaron structure, dynamics, and coherence in certain extended Peierls–Hubbard models. We emphasize the qualitative importance of electron–lattice interactions even in the presence of dominant electron–electron correlations, the signatures of polaron structure and dynamics in energy-resolved pair-distribution structure functions, and the effect of disorder on polaron propagation and stability.

1 Introduction

The formation and dynamics of polarons (and bipolarons), despite a half-century of theoretical and experimental study, remain fascinating topics in many-body physics, combining as they do (often competing) aspects of coupled fields with distinct natural time scales (e.g., electron–phonon, spin–phonon, exciton–phonon), electron–electron interactions, lattice discreteness, non-adiabaticity, collective quantum tunneling, thermal fluctuations, competitions between disorder and polaronic localization, etc. Direct observation of polarons through real-space imaging of any of the coupled fields is rare even with the advent of STEM, AFM techniques, etc. Likewise, global measurements such as that of electronic band-structure are insensitive to polaron features. Thus, experiments and theoretical techniques have had to focus on indirect effects on microscopic probes such as transport coefficients, electronic absorption, vibrational spectroscopy, and so forth.

Our purpose here is to briefly review three qualitative effects in polaron physics that have arisen in modeling two components of the lattice structure of the layered cuprate superconductors [1] – namely, extended multi-band Peierls–Hubbard models of (a) the active CuO_2 planes and (b) polarizable

Cu–O clusters in the axial direction (polarizable interplanar medium). The qualitative effects discussed below are (i) the coexistence of lattice- and spin-polaronic distortion upon doping into an antiferromagnetic stoichiometric global background, (ii) the signature of polarons in inelastic neutron scattering and in particular the value of energy-resolved pair-distribution functions, and (iii) the effect of disorder on the dynamics and particle-like collective coherence of polarons. The models below and these effects are, we believe, integral parts of the high-temperature superconducting mechanism. However, the materials control and experimental information remain too incomplete at this point to draw truly definitive conclusions [1]. We also note that *one*-dimensional chain electronic analogs of the Cu–O systems are provided by halogen-bridged transition metal chain complexes. Extensive experimental and many-body modeling studies of these materials have been made, particularly regarding polaron and bipolaron signatures in electronic and optical spectroscopy: the reader is referred to [2] for details.

2　Polarons in the 2D, three-band extended Peierls–Hubbard model

Here, we consider the 2D, three-band extended Peierls–Hubbard model [3],

$$H = \sum_{i \neq j, \sigma} t_{ij}(\{u_k\}) c_{i\sigma}^\dagger c_{j\sigma} + \sum_{i,\sigma} e_i(\{u_k\}) c_{i\sigma}^\dagger c_{i\sigma} + \sum_i U_i c_{i\uparrow}^\dagger c_{i\downarrow}^\dagger c_{i\downarrow} c_{i\uparrow} + \sum_{i \neq j, \sigma, \sigma'} U_{ij} c_{i\sigma}^\dagger c_{j\sigma'}^\dagger c_{j\sigma'} c_{i\sigma}$$

$$+ \sum_l \frac{1}{2M_l} p_l^2 + \sum_{k,l} \frac{1}{2} K_{kl} u_k u_l, \tag{1}$$

where $c_{i\sigma}^+$ creates a hole of spin σ at site i. Sites refer to the Cu(2), O(2), or O(3) in the CuO$_2$ *plane*. Holes in the Cu $d_{x^2-y^2}$, O p_x, and O p_y orbitals interact with each other and couple with the motion of planar O atoms along the Cu–O bonds. We assume a simple spring-constant matrix, $K_{kl} = \delta_{k,l} K$, and couple with the motion of planar O atoms along the Cu–O bonds. We assume that the nearest-neighbor Cu–O hopping is modified by the O atom displacements u_k as $t_{ij} = t_{pd} \pm \alpha u_k$, where the $+(-)$ applies if the bond shrinks (stretches) with positive u_k. We also consider modulation of the Cu site energy by the O atom displacements, $e_i = \varepsilon_d + \beta \sum_k'(\pm u_k)$, where the sum extends over the four surrounding O atoms, and the sign takes $+(-)$ if the bond becomes longer (shorter) with positive u_k. The other matrix elements are O–O hopping $(-t_{pp})$ for t_{ij}, O site energy (ε_p) for e_i with $\Delta = \varepsilon_p - \varepsilon_d$, Cu site (U_d) and O site (U_p) repulsions for U_i, and the nearest-neighbor Cu–O repulsion (U_{pd}) for U_{ij}. We have chosen $t_{pd} = 1$ as a unit of energy and taken $t_{pp} = 0.5$, $U_p = 3$, and changed U_d or U_{pd} with $\Delta = (U_d - U_{pd} - 1)/2$ so as to keep the renormalized site-energy difference between Cu and O nearly a constant value. (When $U_d = 8$

and $U_{pd}=1$, this parameter set is almost in proportion to those derived from constrained LDA calculations for cuprate superconductors [4].) We define dimensionless electron–lattice coupling strengths by $\lambda_\alpha=\alpha^2/(Kt_{pd})$, $\lambda_\beta=\beta^2/(Kt_{pd})$ and vary them. Comparison of our results for local lattice distortion and reduced Cu magnetic moments accompanied by added holes [5] with generalized, inhomogeneous 'LDA + U' calculations for cuprate super-conductors [6] is consistent with values of $U_d=8$, $U_{pd}=1$, $\lambda_\alpha=0.28$, $\lambda_\beta=0.0$, and $K=32t_{pd}$ Å^{-2}.

We use, in this section, a Hartree–Fock (HF) technique for the electronic part, but totally unrestricted in both spin and direct space [3, 5], and a classical treatment for the lattice part. RPA calculations based on linear fluctuations around the generally inhomogeneous HF states then allow us [5] to calculate magnetic, optical and vibronic response functions, including electronic absorption and infrared and Raman scattering. (Large-amplitude (nonlinear) fluctuations around the HF states are necessary to restore translational symmetry, estimate polaron bandwidths or masses, etc.) Self-consistency conditions are imposed for on-site and nearest-neighbor charge and spin densities as well as lattice displacements, without assumptions about the form of these quantities. Calculations were made on systems of 6×6 unit cells with periodic boundary conditions. The undoped system refers to that with one hole per Cu site.

2.1 Rapid crossover from Zhang–Rice to covalent molecular states

Here, we take strong on-site repulsion ($U_d=10$), weak nearest-neighbor repulsion ($U_{pd}=1$), and vary the strength of the intersite electron–lattice coupling λ_α. The effect of finite λ_β is briefly mentioned later.

When electron–lattice couplings are absent ($\lambda_\alpha=\lambda_\beta=0$), each added hole is localized primarily on a single Cu site and four surrounding O sites. The spin density at this Cu site is flipped so that a small ferromagnetic polaron is formed. The spin densities at the four O sites are small and in the opposite direction to the central Cu spin. As λ_α is turned on, the Cu magnetic moment is reduced and the O atoms are displaced toward the central Cu, as shown in Fig. 1(a). Two HF eigenstates per added hole are located deeply inside the charge-transfer gap (Fig. 1(b)), and their associated wave functions are spatially well localized: the higher one corresponds to an oxygen state formed by the four O and has small weight on the Cu. The lower one corresponds mainly to the central Cu state with opposite spin to the higher one. In a Zhang–Rice picture [7] this state mixes quantum mechanically with a similar state in which the spin directions of the central Cu and the four O are reversed.

At the Cu site, where a small polaron resides, reduction of the magnetic

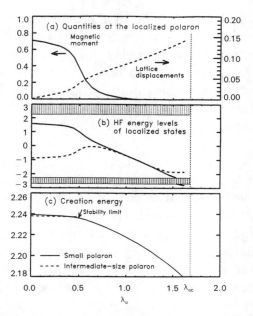

Fig. 1. (a) The magnetic moment on the central Cu site and ratio of lattice displacement of the surrounding O to Cu–O distance (1.89 Å) and (b) gap energy levels, for the small polaron state as a function of λ_α. Parameters are $t_{pd} = 1$, $t_{pp} = 0.5$, $\Delta = 4$, $U_d = 10$, $U_p = 3$, $U_{pd} = 1$, $\lambda_\beta = 0$ and $K = 32t_{pd}$ Å^{-2}. (Shaded bands are extended states.)

moment results in strong local mixing of the four O with the Cu. When λ_α is finite, it gains energy by the O approaching the Cu, locally increasing the covalency between Cu and O. Covalency and O-atom displacements reinforce each other synergistically. Thus, substantially *below* a critical value of λ_α for destruction of the global stoichiometric antiferromagnetic (AF) state (λ_{ac}), both the localized ↑ and ↓ levels become nearly degenerate and the magnetic moment on the central Cu collapses. This quenching with λ_α can be visualized as a rapid crossover from a Heitler–London-like state corresponding to the Zhang–Rice singlet to a highly covalent molecular state. Above λ_{ac}, the *undoped* ground state is replaced by a (non-magnetic) bond-order-wave state.

It should be added that, if $U_d = 10$ and $\lambda_\alpha < 0.4$, then an intermediate-size ferromagnetic polaron may have lower energy than the small polaron above. The former extends to about four Cu sites, and the spin densities at these sites are almost in the same direction and perpendicular to the background AF Cu spins. Tunneling between the small and intermediate size polarons can be anticipated beyond our HF level calculations. (For $U_d = 8$, this intermediate size polaron is unstable in HF for all λ_α, but the above property of the small polaron is not changed.) The parameter sensitivity of polaron type and size

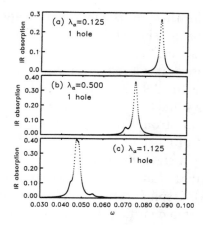

Fig. 2. IR absorption spectra for the one-hole-doped systems with a small polaron: (a) $\lambda_\alpha = 0.125$, (b) $\lambda_\alpha = 0.500$ and (c) $\lambda_\alpha = 1.125$. The other parameters are as in Fig. 1.

may be important in view of the anticipated parameter sensitivity to small structural variations, especially concerning the axial oxygen.

The effect of finite λ_β is very weak up to a certain critical value ($\lambda_{\beta c} \simeq 1.2$) at which the ground state changes to a (non-magnetic) charge-density-wave state. Contrary to the intersite coupling λ_α, which can gain energy by enhancing covalency, the intrasite coupling λ_β can only do so by enhancing double occupancy at Cu sites, directly competing against the strong on-site repulsion U_d. In contrast to Fig. 1, the Cu magnetic moment remains almost constant for $\lambda_\beta < \lambda_{\beta c}$.

Each doping state has distinctive signatures in the magnetic, optical, and vibronic channels [5]. For example, the small hole-polarons are calculated to have specific doping-induced infrared-absorption peaks in addition to the stoichiometric peak – see Fig. 2 – associated with specific local lattice vibrations around the polarons. These features are reminiscent of chemical- and photo-doping measurements on high-temperature superconductors [8].

2.2 Electron–lattice coupling enhanced phase separation

In this section, we first study the effects of U_{pd} (with $\lambda_\alpha = \lambda_\beta = 0$) and then investigate how electron–lattice couplings affect it in the larger U_{pd} regime (varying U_{pd} and λ_α with $\lambda_\beta = 0$ or varying U_{pd} and λ_β with $\lambda_\alpha = 0$). The on-site repulsion is kept strong ($U_d = 8$). We add six holes to the 6×6 systems to focus on phase separation possibilities.

When the electron–lattice couplings are absent ($\lambda_\alpha = \lambda_\beta = 0$), the added holes form isolated small ferromagnetic polarons for $U_{pd} \leq 2.0$; the polarons begin to clump at $U_{pd} \simeq 2.2$ and form three isolated cigar-shaped bipolarons; then the bipolarons clump and form a single rectangular 'domain wall' dividing the two degenerate Cu AF phases for $2.4 \leq U_{pd} < 2.9$; and finally the domain wall and the interior Cu AF phase are replaced by an O-hole-rich phase separated from the Cu-hole-rich phase for $2.9 < U_{pd}$. These various regimes are real-space manifestations of charge fluctuation softening tendencies with U_{pd} found in exact diagonalization at strong-coupling [9] and in weak-coupling homogeneous (k-space) RPA [10].

In the phase-separated state, the Cu and the O spin densities are aligned antiferromagnetically in the Cu- and O-hole rich regions, respectively [5]. The AF Cu spins and the AF O spins are nearly perpendicular to each other, so that both the Cu and the O spin densities on the boundary between the two phases are frustrated and deviate from the respective perfect AF alignments. Intragap HF eigenstates appear, forming a rather wide band of mainly O states in the O-hole-rich regions. The width of the band is due to direct hopping between neighboring O sites (t_{pp}) and small O-site repulsion (U_p).

Both the intersite (λ_α) and the intrasite (λ_β) electron–lattice couplings tend to destabilize the domain wall between the two Cu AF phases. As a result, the electron–lattice couplings lower the critical strength of U_{pd} for the separation into Cu- and O-hole-rich phases ($U_{pd} \simeq 2.9$ for $\lambda_\alpha = \lambda_\beta = 0$; $U_{pd} \simeq 2.7$ for $\lambda_\alpha = 0.5$, $\lambda_\beta = 0$; $U_{pd} \simeq 2.5$ for $\lambda_\beta = 0.8$, $\lambda_\alpha = 0$). In the phase-separated state, the O atoms are displaced on the boundary, enhancing the local covalency between neighboring Cu and O and suppressing the spin densites there [5]. Thus the electron–lattice couplings reduce the frustration of the spin densities on the boundary and isolate the O-hole-rich phase from the Cu-hole-rich phase.

2.3 Spin/charge separation at strong electron–lattice coupling

Here, we take strong intersite electron–lattice coupling ($\lambda_\alpha = 2$) and vary the strength of the on-site repulsion U_d to study their competition. The nearest-neighbor repulsion is kept weak ($U_{pd} = 1$) and $\lambda_\beta = 0$.

The HF undoped states vary [3, 5] between a charge-density-wave state (CDW) accompanied by symmetric lattice distortion with respect to a Cu site for $U_d \leq 4.5$, a bond-order-wave state (BOW) with asymmetric lattice distortion for $5 \leq U_d \leq 11$; a mixed state of spin–Peierls bonds and AF Cu spins (SP + AF) for $12 \leq U_d \leq 13$, a uniform spin–Peierls state (SP) for $U_d \simeq 13$, and the AF state without lattice distortion, relevant to stoichiometric cuprate oxides, for $14 \leq U_d$. In the crossover region (BOW and SP + AF), there are many metastable states with small energy differences, and quantum fluctuation can be expected to destroy some of the mean-field states. Furthermore, long-period (superlattice) phases cannot be resolved in our small system. Similar complexity of crossover phases between charge-density and magnetic regimes in multiband models has been found in one dimension [11] and they are understood in terms of competing nearest- and next-nearest-neighbor interactions (i.e., frustration).

Upon doping, various inhomogeneous mean-field states are found in the respective backgrounds. Doping states in the BOW are especially interesting since they are characterized by *confined separation* of spin and charge [5]. Here, an unpaired spin is located on a Cu site, disfavors and suppresses lattice distortion around it, and redistributes bond charge densities and lattice displacements. Excess charge is in the covalent molecular singlet structure described in Sec. 2.1, which is accompanied by local symmetric lattice distortion with respect to another Cu site *separated* from the unpaired spin – the confinement distance is determined by the mismatch with neighboring rows of atoms. This novel separation of spin and charge is a consequence of both large U_d and large λ_α competing with each other. As doping increases, excess spins, excess charges, and redistributed lattice displacements frustrate each other, affecting delicate energy balances between competing interactions.

Complete details of the calculations described in this section can be found in [3, 5]. As we emphasized earlier, the calculations are at a totally unrestricted HF level (except in one dimension [11]) with its attendant limitations. In the next section we consider a problem in which quantum fluctuations and non-adiabaticity must be treated exactly.

3 Polaron dynamics, coherence, and detection in a three-site cluster

Many generic features of polarons – self-trapping and particle-like band motion due to quantum tunneling – are captured by small-cluster models envisaged to describe electronic correlation effects or electron–lattice interactions, outside the scope of many conventional approaches. Perhaps the simplest model is a two-site cluster that already exemplifies the polaron dynamics arising from the linear coupling of electronic and phonon degrees of freedom [12]. Another example is the local dynamics of an axial-oxygen (O(4)) and chain-copper (Cu(1)) cluster in $YBa_2Cu_3O_7$, for which polaron tunneling has been proposed [13] to be an essential component of the local dynamics. In this approach, the novel features arise from a dynamic double-well structure, inferred from XAFS data [14], and described as polaron tunneling, specifically realizing local polarizability.

We study a Hamiltonian that is naturally divided into three parts:

$$H = H_{el} + H_{ph} + H_{el-ph}. \tag{2}$$

We assume that the dynamics of the O(4)–Cu(1)–O(4) cluster can be effectively separated from the rest of the lattice, because dynamic coupling to the planes is weak due to the long bond lengths, and because some aspects of the coupling along the chain direction can be taken into account as an effective Hamiltonian by integrating out the degrees of freedom associated with the chain oxygens O(1).

First, the electronic part is given by the extended Hubbard model on a three-site cluster:

$$H_{el} = \sum_a \varepsilon_d n_a + \sum_{ab,\sigma} t_{ab}(c_{a\sigma}^\dagger c_{b\sigma} + \text{H.c.}) + U \sum_a n_{a\uparrow} n_{a\downarrow}. \qquad (3)$$

Here, $c_{a\sigma}^\dagger$ creates a hole of spin σ at site a, $n_{a\sigma} = c_{a\sigma}^\dagger c_{a\sigma}$, and $n_a = \sum_\sigma n_{a\sigma}$. Indices $a = 1$ and 3 denote the axial O(4) sites and $a = 2$ is the chain Cu(1) site. The hopping matrix element is $t_{ab} = t$ between the O(4) and Cu(1) sites and $t_{ab} = t'$ for the effective hopping matrix element between the O(4) sites. In the absence of disorder, the site energies can be parameterized by a single parameter ε_0: $\varepsilon_{1,3} = \varepsilon_0$ and $\varepsilon_2 = 0$.

Second, the cluster has two phonon modes corresponding to displacements in the z-direction (the cluster axis) that do not change the center of mass. These bare modes are assumed to be harmonic. If the spring constants between both Cu(1)–O(4) bonds and the two O(4) masses are equal, then parity is a good quantum number. Consequently, one mode is even and the other is odd with respect to inversion, where the symmetric phonon mode is Raman active and the antisymmetric one is infrared active. These are described by boson operators a_R and a_{IR} with bare frequencies ω_R and ω_{IR}, respectively. In terms of these operators, the phonon part of the Hamiltonian is

$$H_{ph} = \hbar\omega_{IR} a_{IR}^\dagger + \hbar\omega_R a_R^\dagger a_R. \qquad (4)$$

It is useful to express the site coordinates in terms of the above normal modes. Let the site coordinates, r_a, be measured relative to their average positions $r_a^{(0)}$: $r_a = r_a^{(0)} + z_a$. Then the normal modes, defined as $u_{IR} = [\hbar/(2M_1\omega_{IR})]^{1/2}(a_{IR} + a_{IR}^\dagger)$ and $u_R = [\hbar/(2M_1\omega_R)]^{1/2}(a_R + a_R^\dagger)$, are given by $u_{IR} = (z_1 - 2z_2 + z_3)/\sqrt{3}$ and $u_R = (z_1 - z_3)/\sqrt{2}$, where we have taken the ratio of the effective copper to oxygen masses $M_2/M_1 = 4$.

Third, we consider linear molecular-crystal-type interaction between electronic and phonon degrees of freedom. In accordance with inversion symmetry, this can be written as

$$H_{el-ph} = \lambda_{IR}(a_{IR} + a_{IR}^\dagger)(n_3 - n_1) + \lambda_R(a_R + a_R^\dagger)(n_1 + n_3 - s_0), \qquad (5)$$

where λ_{IR} and λ_R are the respective coupling constants. The parameter s_0 will be chosen to avoid any artificial shrinkage of the cluster, thus allowing the use of a reduced basis set.

Several structural and spectroscopic consequences of this model have been investigated [13, 17, 18]. Here we focus on just two issues.

3.1 Energy-resolved structure factors

Unlike optical probes, neutron scattering is in principle able to measure dynamic structure factors as a function of momentum transfer. It can be written in the form [15]

$$S(q,\omega) = \sum_{ab} v_{ab} \int_{-\infty}^{\infty} \frac{dt}{2\pi} e^{-i\omega t} \langle e^{-iqr_a(0)} e^{iqr_b(t)} \rangle, \qquad (6)$$

where we have included cross-section coefficients v_{ab}, determined by the scattering lengths of individual nuclei [16]. $S(q,\omega)$ for our cluster is a sum of delta functions in ω at the same frequencies as the infrared or Raman lines; the intensity of each line varies with q (the momentum transfer along the z-axis).

We consider the intermediate strengths of the electron–phonon coupling constants that form the most interesting region, which seems also to be the physically relevant one [13]. The above model is solved by exact diagonalization, and the dynamic structure factors (and their Fourier transforms) are computed exactly numerically. The parameters in H_{el} are taken as $\varepsilon_o = 0.614 \, \text{eV}$, $t = -0.634 \, \text{eV}$, $t' = t/10$, and $U = 4.44 \, \text{eV}$, which are representative values for $YBa_2Cu_3O_7$ [17]. For bare phonon energies, we choose $\hbar\omega_{IR} = 59.3 \, \text{meV}$ and $\hbar\omega_R = 71.5 \, \text{meV}$. We fix $s_o = 1.17$, which is satisfactory for the electron–phonon coupling constants $\lambda_{IR} = 0.155 \, \text{eV}$ and $\lambda_R = 0.260 \, \text{eV}$.

Let us now demonstrate the usefulness of *energy-resolved* correlation functions in probing the physics at the characteristic time scales of our system and detecting dynamic length scales [17, 18]. In Fig. 3, we show three structure factors, $S(r,\omega=0)$, $\int_0^{\omega_T} d\omega \, S(r,\omega)$, and $S(r,t=0)$, as functions of distance r. It is evident from these examples that the usual elastic structure factor contains no clear signature of the bond-length splitting, that the instantaneous structure factor (pair-distribution function) is barely able to resolve the splitting, and that $\int_0^{\omega_T} d\omega \, S(r,\omega)$ shows a pronounced double-peak structure. The difference in resolving power between the pair-distribution function and the partial-frequency sums stems from the fact that the partial-frequency sum emphasizes only the fluctuations at ω_T, whereas $S(r,t=0)$ includes large quantum fluctuations. This difference becomes crucial when the bond-length splitting is small. Note that the ratio of copper to oxygen masses deviates from the optimal value

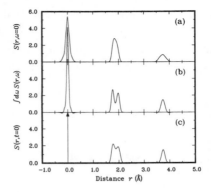

Fig. 3. The dynamic structure factors, (a) $S(r, \omega = 0)$, (b) $\int_0^{\omega_T} d\omega\, S(r, \omega)$, and (c) $S(r, t = 0)$ (in units of Å^{-1}) as a function of distance r. For the chosen parameters, $\delta l = 0.2\ \text{Å}$ and we have used $l_0 = 1.87\ \text{Å}$ for the average copper–oxygen bond length.

in the sense that the relative resolving power of $\int_0^{\omega_T} d\omega\, S(r, \omega)$ is not the best as compared with $S(r, t = 0)$, although it is always better [18].

3.2 Disorder

As we have already emphasized, strongly correlated electron–phonon systems constitute complex many degree-of-freedom systems where nonlinearity and possible non-adiabaticity characterize distinctive, particularly *polaronic*, aspects of collective behavior. In many physically interesting systems, a further complication is caused by *disorder* that combines with nonlinearity, often in a competing fashion [19]. Thus, polarons represent a coherent particle-like motion of correlated electronic and lattice distortions, whereas disorder tends to produce localized wave functions on quite different spatial and temporal scales – either for polaronic or for single-electron states.

Below, we explain specific measures of polaron coherence to establish our main result: the effects of disorder (both diagonal and off-diagonal) on polarons can be understood in terms of two important energy scales, ε_T and w, where ε_T is the polaron tunneling energy (i.e., the polaron bandwidth in our small cluster) and w is the bare electron bandwidth for the lowest energy 'band' (i.e., in the absence of electron–phonon interactions) [20]. The tunneling energy specifies the typical energy scale of disorder at which the coherent motion of polarons is suppressed and they are trapped as composite particles. The bandwidth w, on the other hand, gives the energy scale at which the polaron ceases to be a composite particle of an electron and phonons; i.e., the dressing of the electron with phonons strongly decreases.

To model disorder, we assume that the system has two electrons with opposite spins ($S_z = 0$) on a linear three-site cluster, which is parameterized so that $t_{13} = 0$, $t_{12} = t$, $t_{23} = t + \delta$, $\varepsilon_{1,3} = \varepsilon_0 \pm \Delta/2$, and $\varepsilon_2 = 0$. We also consider the parameter regime in which $0 \leq t \ll \varepsilon_0$, $U - \varepsilon_0$. Large, positive values of the on-site electron–electron interaction, U, and large positive average energies of sites 1 and 3, given by ε_0, inhibit bipolaron formation at low energies and the low-energy physics is characterized by polarons. The symmetry-breaking fields Δ and δ describe the disorder – diagonal and off-diagonal, respectively. These, as well as other parameters, are varied so that the salient disorder-induced features can be studied transparently. Thus, we set λ_R to zero here because the symmetric Raman mode plays only a minor role in the polaron physics below.

While the electronic part of the Hamiltonian, Eq. (3), can be readily diagonalized numerically, it is already too complicated to be solved analytically for two electrons in the general case. However, an effective Hamiltonian can be derived easily when $0 \leq t$, Δ, $\delta \ll \varepsilon_0$, $U - \varepsilon_0$, allowing straightforward physical interpretation. In this parameter regime, the hopping term is a small perturbation that can be eliminated by a unitary transformation: $\tilde{H} = e^{-S} H e^S$. By elimination to second order in t/ε_0 and $t/(U - \varepsilon_0)$, we arrive at the electronic part of the effective Hamiltonian:

$$\tilde{H}_{el} = \sum_{\substack{a=1,3 \\ a=\pm}} \varepsilon_\sigma^{(a)} d_{a\sigma}^\dagger d_{a\sigma} + \sum_{\sigma=\pm} \tau_\sigma (d_{1\sigma}^\dagger d_{3\sigma} + \text{H.c.}), \tag{7}$$

where the effective site energies and hopping matrix elements are defined as

$$\varepsilon_\sigma^{(a)} = \varepsilon_a - \frac{t_{\bar{a}2}^2}{\varepsilon_{\bar{a}}} - 2(1+\sigma)\frac{U t_{\bar{a}2}^2}{U^2 - \varepsilon_a^2}, \tag{8a}$$

$$\tau_\sigma = \frac{1}{2} \sum_{a=1,3} \frac{t_{12}t_{23}}{U - \varepsilon_a}\left(\frac{1+\sigma U}{\varepsilon_a}\right), \tag{8b}$$

and $\bar{a} = 4 - a$. The new fermion operators are $d_{a\pm} = (c_{a\uparrow} \pm c_{a\downarrow})/\sqrt{2}$. The electronic part is thus reduced to a two-site problem with one 'd-particle' in the Fock space of the effective Hamiltonian. Similarly, we can construct the effective electron–phonon interaction. For $\lambda_R = 0$, this becomes

$$\tilde{H}_{el-ph} = \lambda_{IR}(a_{IR} + a_{IR}^\dagger) \sum_{\sigma=\pm} (d_{3\sigma}^\dagger d_{3\sigma} - d_{1\sigma}^\dagger d_{1\sigma}). \tag{9}$$

It is now straightforward to understand how the polaron is affected by weak disorder ($0 \leq \Delta$, $\delta \ll \varepsilon_0$, $U - \varepsilon_0$): (i) for diagonal disorder ($\delta = 0$), the phonon coordinate u_{IR} acquires a negative expectation value, meaning that the polaron begins to localize around site 3; (ii) for off-diagonal disorder ($\Delta = 0$), the on-site interaction has a critical value U_c, which separates the behavior into two distinct regimes – if $U < U_c$, then the polaron tends to localize around site 3; however, if $U > U_c$, then the polaron localizes first around site 1 and then, for sufficiently large values of $\delta (> \delta_c)$, it localizes around site 3. The reason for this behavior for very large values of U and for $\delta \neq 0$ is that it is energetically more favorable for an electron to be at site 1 because of virtual hopping processes between sites 2 and 3, which are not affected by the strong electron–electron interaction. Equation (8a) allows us to evaluate U_c as $U_c/\varepsilon_o = 2 + \sqrt{5}$.

The above analysis is sufficient to describe polaron localization in the presence of weak disorder; it cannot be used to explain how disorder affects the internal stability (coherence) of the polaron. For this, we rely on numerical analysis, namely exact diagonalization of the Hamiltonian, Eqs. (2)–(5). For these calculations, we choose representatively $\varepsilon_o/t = 4$, $\hbar\omega_{IR}/t = 0.08$, and $\lambda_R = 0$, and vary the remainder of the parameters.

For simplicity, we consider here only diagonal disorder, which is conceptually easier to understand; a more complete study of diagonal and off-diagonal disorder is given elsewhere [20]. To monitor the stability and localization of the polaron, we consider two diagnostic quantities.

First, we measure the correlation between the electrons and the phonon coordinate in the ground state. Defining deviations of the operators as $\delta O = O - \langle O \rangle$ and denoting $n_{13} = n_1 - n_3$, we introduce the correlation factor

$$\mathscr{R} = \frac{\langle \delta n_{13} \delta u_{IR} \rangle}{[\langle (\delta n_{13})^2 \rangle \langle (\delta u_{IR})^2 \rangle]^{1/2}}. \tag{10}$$

Although the expectation values may be taken relative to any state, we consider here only the ground-state expectation values that will describe the polaronic state. The closer $\mathscr{R}$ is to unity, the more correlated are the electron and phonon dynamics, implying strong polaronic binding of the electron to its phonon cloud. In contrast, $\mathscr{R} = 0$ signals non-polaronic behavior, where the motions of electrons and phonons are completely decoupled.

Second, we consider the dressing of electrons by phonons in terms of the overlap factors between the ground states of the system in the absence and presence of electron–phonon interaction. For the ground states, the polaronic dressing of electrons is conveniently defined by the overlap factor,

$$Z = |\langle \Psi_o(\lambda_{IR} = 0, \Delta) | \Psi_o(\lambda_{IR}, \Delta) \rangle|^2, \tag{11}$$

where $|\Psi_o(\lambda_{IR}, \Delta)\rangle$ is the ground state of the system for the parameters shown. $Z = 1$ signifies no phonon dressing, and decreasing Z implies increasing dressing.

Our results for diagonal disorder are summarized in Fig. 4, which shows $\mathscr{R}$

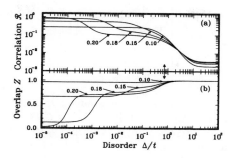

Fig. 4. (a) The correlation factor $\mathscr{R}$ and (b) the ground-state overlap factor Z as functions of the diagonal disorder parameter Δ for $U/t = 16$ and for the electron–phonon coupling constant λ_{IR}/t to 0.10, 0.15, 0.18 and 0.20. In the absence of disorder, the corresponding tunneling energies ε_T/t are 5.3×10^{-2}, 1.5×10^{-2}, 1.2×10^{-3} and 1.1×10^{-4}, respectively. The double-headed arrow marks the energy scale w (see text).

and Z as a function of the disorder strength Δ for $U/t = 16$ and for various values of the electron–phonon coupling constant λ_{IR}. Both $\mathscr{R}$ and Z exhibit one crossover behavior at a larger value of Δ and another one for a smaller value when λ_{IR} is large enough that the system is in the polaron tunneling regime. These results can be understood in terms of two characteristic energy scales: the polaron tunneling energy, ε_T, and w, the energy-splitting of the lowest-energy spin-singlet states (with opposite parity), which plays the role of the lowest-energy bare electron bandwidth in our finite cluster. For $U/t = 16$ and in the absence of disorder, $w/t = 0.72$. As the symmetry-breaking disorder field Δ is increased, the first crossover at $\Delta \simeq \varepsilon_T$ describes the localization of the polaron to essentially one site. As Δ is increased further so that $\Delta \simeq w$, $\mathscr{R}$ decreases sharply to a small value and Z approaches unity. This signals the breaking of the composite character of the polaron, so that the electron is no longer coherently dressed by phonons. Diagonal disorder can be understood as tending to make the electron immobile (by increasing its effective mass) – it causes the electronic and phonon time scales to be so different that polaron wave functions by disorder leads to physically observable implications [18], for example, in dynamic structure factors, such as those measured in neutron scattering [20]. The model also shows an intriguing localization transition, induced by strong electronic correlations, as a function of off-diagonal disorder [20].

Acknowledgements

We have benefited from fruitful discussions on the substance of this research with many colleagues, and particularly from collaborations with J. Mustre de

Leon, J. Lorenzana, M. Spicci, S. Trugman, and K. Yonemitsu. Work at Los Alamos is supported by the USDOE.

References

[1] *Lattice Effects in High-Temperature Superconductors*, Eds. Y. Bar-Yam *et al.* (World-Scientific, Singapore, 1992).
[2] D. Baeriswyl and A. R. Bishop, *Physica Scripta* T **19**, 239 (1987); J. T. Gammel *et al., Phys. Rev.* B **45**, 6408 (1992); X. Z. Huang and A. R. Bishop, *Phys. Rev.* B **48**, 16148 (1993).
[3] K. Yonemitsu, A. R. Bishop, and J. Lorenzana, *Phys. Rev.* B **47**, 8065 (1993).
[4] M. S. Hybertson *et al., Phys. Rev.* B **39**, 9028 (1989).
[5] K. Yonemitsu, A. R. Bishop, and J. Lorenzana, *Phys. Rev.* B **47**, 12059 (1993).
[6] V. I. Anisimov *et al., Phys. Rev. Lett.* **68**, 345 (1992).
[7] F. C. Zhang and T. M. Rice, *Phys. Rev.* B **37**, 3759 (1988).
[8] G. A. Thomas *et al., Phys. Rev. Lett.* **67**, 2906 (1991); G. Yu *et al., Phys. Rev. Lett.* **67**, 2581 (1991).
[9] J. E. Hirsch *et al., Phys. Rev.* B **39**, 243 (1989).
[10] Z. Tesanovic *et al., Solid State Commun.* **68**, 337 (1988); P. B. Littlewood *et al., Phys. Rev.* B **39**, 12371 (1989).
[11] H. Röder, A. R. Bishop, and J. T. Gammel, *Phys. Rev. Lett.* **70**, 3498 (1993).
[12] J. Ranninger and U. Thibblin, *Phys. Rev.* B **45**, 7730 (1992).
[13] J. Mustre de Leon *et al., Phys. Rev. Lett.* **68**, 3236 (1992).
[14] S. Conradson *et al., Science* **248**, 1394 (1990); J. Mustre de Leon *et al., Phys. Rev. Lett.* **65**, 1675 (1990); P. Allen *et al., Phys. Rev.* B **44**, 9480 (1991).
[15] S. Lovesey, *Condensed Matter Physics: Dynamic Correlations* (Benjamin-Cummins, Reading, 1986). We include the cross-sections of individual nuclei in the definition of $S(q,\omega)$. Note that, in the text, by $S(q,\omega)$ we mean $\int_{\omega-\delta}^{\omega+\delta}d\omega' S(q,\omega')$, $\delta=0^+$.
[16] They are normalized so that $\Sigma_a v_{aa}=1$. Because $v_{12}\simeq(v_{11}v_{22})^{1/2}$, where 1 refers to oxygen and 2 to copper, we can approximate $v_{ab}=f_a f_b$, where f_a is proportional to the scattering length of the ath nucleus. This implies that neutron scattering is coherent. For oxygen and copper, $f_1/f_2=0.73$.
[17] M. Salkola, A. Bishop, J. Mustre de Leon, and S. Trugman, *Phys. Rev.* B **49**, 3671 (1994).
[18] M. Salkola, A. Bishop, S. Trugman, and J. Mustre de Leon, LANL preprint (1994).
[19] See, for example, F. Abdullaev, A. Bishop, and S. Pnevmatikos (Eds.), *Nonlinearity with Disorder* (Springer, Berlin, 1992).
[20] M. Spicci, M. Salkola, and A. Bishop, LANL preprint (1994).

23

Exact estimates of inter-polaron coupling constants resulting in bipolaron formation

P. E. KORNILOVITCH

Physics Department, King's College, Strand, London WC2R 2LS, UK

Abstract

Two-polaron states on a square lattice are studied in the presence of on-site repulsion U and inter-site attraction V. In the limit of infinite U the *exact* critical value V^{cr} for bipolaron formation is obtained as a function of the attraction radius R. The results are compared with the continuum limit of the same model. It is shown that if $R \simeq (2-3)$ lattice constants then V^{cr} is of the order of the characteristic phonon frequency in the high-temperature superconductors.

The temperature-dependence of the upper critical field [1, 2], the resistivity and Seebeck coefficient [3], and the universal correlation between the critical temperature and the hole content [4] in the p-type oxide superconductors unambiguously support the validity of the local pair conception for these compounds at low doping $0.06 \leq n \leq 0.12$. Phonons are the most natural candidates for the bosonic field whose interaction with the carriers (polarons) results in the effective interpolaron attraction. However, there are several arguments against the phonon pairing mechanism. One of them is that the phonon-mediated attraction between polarons is much weaker than the short-range Coulomb repulsion, hence creation of *local* pairs is inhibited. The typical estimates for the on-site copper, copper–oxygen, and inter-site copper Coulomb potentials are $U_{dd} \simeq 10$ eV, $U_{pd} \simeq 1$ eV and $U'_{dd} \simeq 0.1$ eV correspondingly [5, 6]. Since the typical phonon frequency ω is of the order of 0.1 eV or less, $\omega \ll U_{dd}, U_{pd}$ and $\omega \simeq U'_{dd}$ and therefore the existence of the local pair in which the polarons are localized on the same lattice site or on the nearest neighbours is impossible. It is clear, however, that at low carrier density, $n \ll 1$, the only consequence of the strong Coulomb correlations is the exclusion of a small part of the phase space, which leads to a relative increase in pair kinetic energy of the order of n. In this case one can produce a bound state by allowing two polarons to attract each other if they are placed on more distant sites.

The aim of this paper is to show that an attraction with a coupling of the order of ω overcomes easily the strong (even infinite) short-range repulsion and creates a bound state for a large enough radius of attraction. Moreover, the bound state should be still regarded as a local pair because its effective radius is smaller than the average distance between pairs. For this purpose I use the single local pair approximation, which is assumed to be valid at $n \ll 1$. The main advantage of this approach is that the two-particle lattice problem can be solved exactly.

Consider for simplicity a two-dimensional square lattice, the lattice constant being the unit of length. The dynamics of the system is determined by the phenomenological two-particle Hamiltonian

$$H = - \sum_{\langle ij \rangle \sigma} a_{i\sigma}^{\dagger} a_{j\sigma} + U \sum_i n_{i\uparrow} n_{i\downarrow} - V \sum_{\langle ij \rangle'} n_i n_j. \tag{1}$$

We regard $a_{i\sigma}^{\dagger}$ ($a_{i\sigma}$) in (1) as the operator that creates (annihilates) a polaron, at site i with spin projection σ. The first term in (1) describes hopping of polarons between the nearest sites (the hopping integral is chosen with unit of energy $t = 1$), the second one is the usual Hubbard term with the potential U and the last term represents the inter-polaron attraction with potential V, which is assumed to result from the strong electron–phonon interaction, that is $V \simeq \omega$.

In what follows we consider the set of Hamiltonians H_i, all of the form of Eq. (1) and differing by the meaning of the notation $\langle ij \rangle'$. Let H_1 be the Hamiltonian with an attraction between polarons placed on the nearest neighbours only, H_2 the one with an attraction between polarons placed on the first *or* the second nearest neighbours, etc. The numbering of the H_i corresponds to inclusion of the attraction on more and more distant lattice sites as shown in Fig. 1.

A system of two particles (with opposite spin projections) is completely described by the two-particle wave function $\psi(r_1, r_2)$ or by its Fourier component $a(k_1, k_2)$. The corresponding Schrödinger equation is

$$[E - \varepsilon(k_1) - \varepsilon(k_2)]a(k_1, k_2) = \sum_i U(l) e^{-ik_1 \cdot l} \sum_q a(q, k_1 + k_2 - q) e^{iq \cdot l}, \tag{2}$$

where $\varepsilon(k) = -4\cos k_x - 4\cos k_y$ is the one-particle spectrum and the function $U(l)$ is given by

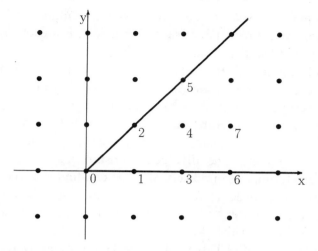

Fig. 1. A fragment of a square lattice. The irreducible part is bounded by the thick lines. The indices 1–7 number the different types of sites as the distance from the origin increases.

$$U(l) = \begin{cases} U, & l=0 \\ -V, & l \in \langle ij \rangle' \\ 0, & \text{otherwise.} \end{cases} \tag{3}$$

Introducing the quantities $C_l(K) = \sum_q a(q, k_1 + k_2 - q) e^{iq \cdot l}$, where $K = k_1 + k_2$ is the total quasi-momentum, one can transform the integral equation (2) into a system of homogeneous linear equations for the $C_l(K)$ quantities:

$$C_l(K) = \sum_{l'} U(l') \left(\sum_q \frac{e^{iq \cdot (l-l')}}{E - \varepsilon(q) - \varepsilon(K-q)} \right) C_{l'}(K). \tag{4}$$

The consistency condition of this system

$$\det \left| U(l') \sum_q \frac{e^{iq \cdot (l-l')}}{E - \varepsilon(q) - \varepsilon(K-q)} - \delta_{ll'} \right| = 0 \tag{5}$$

is the equation for the two-particle spectrum $E(K)$. The corresponding eigenvectors of Eq. (4) determine the two-particle wave function via Eq. (2) and the definition of the $C_l(K)$ terms.

For an arbitrary total quasi-momentum K the dimension d of the system of equations (4) equals the number of vectors l for which $U(l) \neq 0$. Hence, $d = 1, 5, 9, 13, 21, 25, 29$ and 37 for $H_0, H_1, H_2, H_3, H_4, H_5, H_6, H_7$ correspondingly. For the sake of calculational simplicity, we consider in what follows the most symmetric point of the Brillouin zone $K = (0,0)$ only. In this particular case the system (4) becomes a direct sum of several matrices. For each H_i the one that describes the ground state is of dimension $d = i + 1$.

The spectral equation (5) enables one to calculate analytically the condition for occurrence of a bound state (bipolaron). For this purpose one should compare the energy E resulting from Eq. (5) with the ground state energy of two free polarons $E_0 = -8$. The substitution $E = E_0$ leads to a logarithmic divergence of all sums in Eq. (5) near the point $q = (0,0)$. By subtracting from and adding to every integral the divergent constant $z \equiv \Sigma_q [E_0 - 2\varepsilon(q)]^{-1}$ and expanding the determinant in powers of z one can find that the divergence is cancelled in all higher orders but the first one. Since the determinant must be equal to zero, the coefficient of z must also be equal to zero, which yields a condition of pairing. For the case of infinite Hubbard repulsion $U = \infty$ these coefficients are presented in Table 1. For any Hamiltonian H_i the corresponding condition is an algebraic equation of the ith power in V. Its lowest root is the critical value of the attraction parameter V^{cr} i.e. the lowest value of V that results in pairing, even for infinite on-site repulsion. The V_i^{cr} are given in Table 1 and their dependence on the attraction radius R is shown in Fig. 2. It is seen that all V_i^{cr} lie near the curve $V^{cr} \propto 1/R^2$, the latter following from the solution of the continuum Schrödinger equation for a particle in a two-dimensional square attractive potential with a δ-functional repulsion at the origin [7]. The cases of H_4 $(R = \sqrt{5})$ and H_7 $(R = \sqrt{10})$ are the most favourable for pairing, compared with the continuum solution.

In order to estimate the size of a bipolaron one needs to calculate the effective radius of the bound state. The coordinate-dependence of the two-particle wave function is determined by the integral

$$\psi(r_1, r_2) \simeq \sum_q \frac{e^{iq \cdot (r_1 - r_2)}}{E - 2\varepsilon(q)}. \tag{6}$$

In the limit $|r_1 - r_2| \to \infty$ the wave function has exponential asymptotics, which enables one to introduce an effective radius r^*. For the diagonal direction $x_1 - x_2 = y_1 - y_2 = r/\sqrt{2}$ simple calculations yield [8]

$$\psi \simeq e^{-r/r^*}, \qquad r^* = \sqrt{2}\,\mathrm{Arch}^{-1}\left[\frac{1}{2}\left(\frac{E}{4}\right)^2 - 1\right], \tag{7}$$

which means that r^* is completely determined by the bipolaron energy.

Using Eq. (7) and numerical solution of the spectrum equation (5) I obtained the effective radius as a function of V for the $U = \infty$ case of each Hamiltonian H_i. The results are shown in Fig. 3. The interesting quantity is the value of V for

Table 1. *The main characteristics of the two-polaron systems described by the Hamiltonians H_i. R_i is the radius of the attractive potential (in the units of the lattice constant); P_i is the polynomial in V of degree i whose roots correspond to the occurrence of two-polaron bound states; V_i^{cr} is the lowest root of P_i; V_i^R is the attraction parameter value for which the bipolaron radius becomes equal to R_i; V_i^H is the coupling constant of the attractive Hubbard model, which leads to the bipolaron radius R_i.*

H_i	R_i	P_i	V_i^{cr}	V_i^R	V_i^H
H_1	1	$V-2$	2.00000	7.64	7.99
H_2	$\sqrt{2}$	$(32-9\pi)\,V^2+(6\pi-64)\,V+12\pi$	0.90212	3.73	6.36
H_3	2	$(48-15\pi)\,V^3+(168\pi-544)\,V^2+(320-78\pi)\,V-12\pi$	0.57139	2.39	5.32
H_4	$\sqrt{5}$	$(256-81\pi)\,V^4+(1494\pi-4736)\,V^3$ $+(19712-6156\pi)\,V^2+(1776\pi-6656)\,V+96\pi$	0.31280	1.54	5.06
H_5	$\sqrt{8}$	$(417375\pi^2-2646240\pi+4194304)\,V^5$ $+(-7665525\pi^2+48778560\pi-77594624)\,V^4$ $+(31133025\pi^2-200576640\pi+322961408)\,V^3$ $+(-68827950\pi^2+56158080\pi-109051904)\,V^2$ $+(-970200\pi^2+3406080\pi)\,V-25200\pi^2$	0.25075	1.15	4.59
H_6	3	$(5895225\pi^2-37211520\pi+58720256)\,V^6$ $+(-163790550\pi^2+1035242880\pi-1635778560)\,V^5$ $+(1435184100\pi^2-9096049920\pi+14411628544)\,V^4$ $+(-4058384400\pi^2+25929177600\pi-41406169088)\,V^3$ $+(790650000\pi^2-5835356160\pi+10536091648)\,V^2$ $+(5685120\pi^2-182077440\pi)\,V+201600\pi^2$	0.20967	1.01	4.49
H_7	$\sqrt{10}$	$(6974100\pi^2-43938720\pi+69206016)\,V^7$ $+(-246186675\pi^2+1551773760\pi-2445279232)\,V^6$ $+(3061387350\pi^2-19313022720\pi+30459035648)\,V^5$ $+(-15766550100\pi^2+99645730560\pi-157437394944)\,V^4$ $+(28137186000\pi^2-178859873280\pi+284206039040)\,V^3$ $+(-3388467600\pi^2+24163153920\pi-42479910912)\,V^2$ $+(-130233600\pi^2+413767680\pi)\,V-201600\pi^2$	0.15584	0.81	4.41

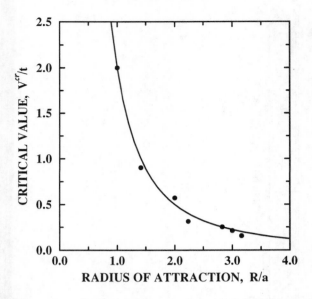

Fig. 2. The critical value of the attraction parameter for bipolaron formation as a function of the radius of the attractive potential. The points correspond to the seven Hamiltonians H_i. The solid line represents the function $V^{cr} = 2/R^2$ (in the continuum the dependence is $V^{cr} \sim 1/R^2$).

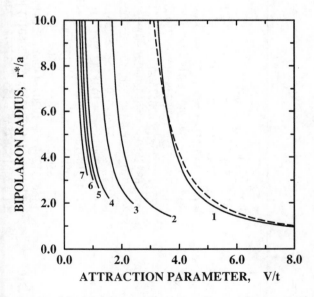

Fig. 3. The dependence of the bipolaron effective radius on the attraction parameter for the seven Hamiltonians H_i. The dashed line corresponds to the attractive Hubbard model.

which the bipolaron radius becomes equal to the radius of the attractive potential, i.e. $r^* = R$. The values V_i^R corresponding to such a localization are presented in Table 1.

It is known that, in the p-type oxide superconductors, the critical temperature increases linearly as n increases in the region $0.06 \leq n \leq 0.12$ [4]. At $n \geq 0.12$ the critical temperature depends weakly on n at the beginning and then decreases sharply [4, 9]. In the local pairs picture, the point $n \approx 0.12$ may be interpreted as a concentration at which the bipolaron radius becomes equal to one half of the average distance between bipolarons. This implies that one bipolaron is distributed over 17 unit cells. Further doping leads to overlapping of the bipolarons, which destroys them and, consequently, the superconductivity.

From this point of view it is interesting to analyse the results of Table 1. First the following estimates of the model parameters are accepted. If the bandwidth is $W = zt \approx 0.7$ eV [10] then the hopping integral, which is used as a unit of energy, is $t \approx 0.1$ eV. The characteristic phonon energy is $\omega \simeq 1000$ K, which is of the order of the hopping integral. In what follows, $\omega = 1$ is taken. The potential of phonon-mediated inter-polaron attraction is of the order of ω and may be written as $V = g^2 \omega$ where g^2 is the dimensionless electron–phonon coupling constant, which should not be much more than unity, because of the exponential increase in polaron effective mass as g^2 increases [11].

One can see from Table 1 that, in the case of short-range attraction, a bipolaron is formed at $g^2 = 2.0$ ($R = 1$) and $g^2 = 0.9$ ($R = \sqrt{2}$). However, localization of the bipolaron within an attractive region needs almost a factor of four increase in g^2 up to the values 7.6 and 3.7 correspondingly. On the other hand, localization with $g^2 \leq 1$, appearing in the cases of H_6 and H_7, leads to a relatively large effective radius and large square occupied by a bipolaron. The case of H_4 seems to be the most interesting one because the localization needs relatively small $g^2 = 1.5$ and the effective radius is $r^* \simeq 2.2$; therefore one bipolaron is distributed over 21 unit cells, which is close to the value mentioned above.

The obtained estimate $V \approx 1500$ K $\simeq 0.15$ eV seems slightly too large for the phonon-induced attraction mechanism. A possible way to reduce V is to consider, instead of a simple square lattice, a two-polaron problem in a CuO_2 plane. In this case the presence of three sites in each unit cell increases by a factor of three the number of attractive centres, which leads to a considerable decrease in V. Unfortunately, exact solution of the two-particle problem in the CuO_2 plane is much more difficult technically because of the multi-band structure of the one-particle spectrum. The critical values of V are known only for two special cases: (i) on-site oxygen attraction, $V^{cr} = 3.77t$ [8] and (ii) attraction between nearest neighbour oxygen sites, $V^{cr} = 1.68t$ [12] (in both cases, on-site Hubbard repulsion on copper sites is infinite).

Finally, one should note that, with appropriate fitting of the coupling, any value of the binding energy and the corresponding effective radius of the bipolaron can be obtained in a simple attractive Hubbard model. The parameters V_i^H, listed in the last column of Table 1, lead in the attractive Hubbard model to the numerical values E and r^* obtained in the models H_i with the parameters V_i^R. V_i^H are several times larger than V_i^R, especially at large R. Thus, the presented calculations provide a justification for using the Hubbard model with a large attractive potential (several times more than ω) representing a local pairing with phonon-induced attraction.

Acknowledgement

The auther is grateful to A. S. Alexandrov and E. G. Klepfish for useful and helpful discussions.

References

[1] A. P. Mackenzie *et al.*, *Phys. Rev. Lett.* **71**, 1238 (1993).
[2] M. S. Osofsky *et al.*, *Phys. Rev. Lett.* **71**, 2315 (1993).
[3] J. B. Goodenough and J.-S. Zhou, *J. Chem. Phys. Rev.* B **47**, 5275 (1993).
[4] H. Zhang and H. Sato, *Phys. Rev. Lett.* **70**, 1697 (1993).
[5] F. Mila, *Phys. Rev.* B **38**, 11358 (1988).
[6] A. K. McMahan, J. F. Annett and R. M. Martin, *Phys. Rev.* B **42**, 6268 (1990).
[7] The exact solution of the continuum problem yields $V^{cr} = 13.7/R^2$ instead of the function $V^{cr} = 2.0/R^2$ drawn in Fig. 2.
[8] A. S. Alexandrov and P. E. Kornilovitch, *Z. Phys.* B **91**, 47 (1993).
[9] Y. J. Uemura *et al.*, *Phys. Rev. Lett.* **62**, 2317 (1989).
[10] M. Sato, R. Horiba and K. Nagasaka, *Phys. Rev. Lett.* **70**, 1175 (1993).
[11] A. S. Alexandrov and A. B. Krebs, *Sov. Phys. Usp.* **35**, 345 (1992).
[12] P. E. Kornilovitch, unpublished.

24

Coulomb interaction and the criteria for bipolaron formation

D. KHOMSKII

Laboratory of Solid State Physics, Universiteit Groningen, Nijenborgh 4, 9747 AG Groningen, The Netherlands and P. N. Lebedev Physical Institute, Moscow, Russia

Abstract

The possibility of bipolaron formation is studied taking into account both electron–phonon interaction and direct Coulomb repulsion. Starting from the Bethe–Salpeter equation with a rather general interaction of the form $V(\mathbf{k}, \omega) = 4\pi e^2/\varepsilon(\mathbf{k}, \omega)$, and using spectral representations for both the bipolaron wave function and the interaction, the effective Schrödinger-type equation is obtained with the new effective potential, which is non-local and which parametrically depends on the binding energy. It is shown that, if the static response function $1/\varepsilon(\mathbf{k},0)$ is non-negative, then there is no bound state, i.e. in this case bipolarons do not form (electron–lattice interaction is not sufficient to overcome direct Coulomb repulsion). Possible ways out are discussed, among them the possibility of a negative static dielectric function or more general form of the effective electron–electron interaction.

1 Introduction

The problem of the state of electrons in crystals with strong electron–phonon interaction is now attracting considerable attention. It is well known that one of the possibilities in this case is the formation of polarons [1, 2]. The conditions for their existence and their properties have been studied in numerous publications.

Much less studied (and more controversial) is the next possible step – formation of bipolarons in certain cases. The possibility of bipolaron formation was probably first pointed out by Vinetskii and Giterman [3]; later, bipolrons were suggested as possible candidates to explain some properties of Ti_4O_7 [4]. The problem of the existence of bipolarons has recently acquired special significance in view of the suggestions that they may exist in cuprates

375

and may possibly explain the phenomenon of high-temperature superconductivity in them (see especially [5–7]).

For the high-T_c materials, there exist arguments both for and against the idea of local pair superconductivity there. Also, the exact nature of the forces that could in principle provide the mechanism of binding is unclear at present: it may be interaction with the lattice (standard bipolarons) [3, 7], or, e.g. interaction with the magnetic subsystem (magnetic polarons or bipolarons [5], spin bags [8] etc.)

Very often, in studying formation of bipolarons, only electron–lattice interaction is taken into account, and direct Coulomb repulsion between electrons is ignored. In other studies they are both treated, but in an approximate way (most often variationally). As the region of existence of bipolarons and their binding energies in these calculations are rather small [9], it is highly desirable to get some exact statements, even in the simplest cases. That is one of the aims of the present treatment.

Competition between phonon-induced attraction of electrons and direct Coulomb repulsion is well known in the theory of superconductivity. It is shown there (see e.g. [10]) that net attraction (overscreening) is possible only in the presence of large Fermi energy ε_F (or large Fermi velocity v_F) due to the fact that the attractive and repulsive parts of the total interaction have drastically different time- (or frequency-) dependence: electron–phonon interaction is retarded (the characteristic timescale is $\tau_{e\text{-}ph} \simeq \hbar/\omega_D$) whereas Coulomb repulsion is practically instantaneous ($\tau_{Coul} \simeq \hbar/(\varepsilon_F, \omega_{pl})$, where ω_{pl} is the plasmon frequency). This is reflected in the well-known expression for the critical temperature:

$$T_c = \omega_D \exp[-1/(\lambda - \mu^*)], \tag{1}$$

$$\mu^* = \frac{\mu}{1 + \mu \ln(\varepsilon_F/\omega_D)}, \tag{2}$$

where λ is the dimensionless constant of phonon-induced electron–electron attraction, and μ is the corresponding Coulomb constant. In the simplest models like the jellium model (see e.g. [11] and below) $\lambda < \mu$, and it is only the strong reduction of the repulsive interaction (transition to a 'Coulomb pseudo-potential' $\mu^* \ll \mu$) that makes superconductivity possible. However, for such a reduction to take place, the existence of a Fermi surface of Fermi energy $\varepsilon_F \gg \omega_D$ is crucial, see Eq. (2).

From these arguments one may already see that the situation typically encountered in the problem of bipolaron formation is drastically different. In this case we are actually interested in the possibility of the binding of two electrons in otherwise empty conduction bands (or two holes in a valence band). In this case there is no Fermi surface, which in ordinary metals restricts possible values of the momenta of the electrons and guarantees that the characteristic electron velocities are much larger than the velocity of sound (this is crucial to reduce Coulomb repulsion). Consequently, one may suspect (and it turns out to be true) that this repulsion acts here at full force, and that it would be very difficult, if not impossible, to overcome it. This is essentially the conclusion of the present investigation: we will show that, in the conventional model that is widely used for description of bipolarons, the Coulomb repulsion always dominates and there is no bipolaronic bound state. One evidently has to go beyond the standard treatment to be able to get bipolarons. In the last part of the paper, possible factors that can in principle change this conclusion and help to form a bound state of two electrons are shortly discussed.

2 The model

We consider the situation with two electrons in an otherwise empty conduction band, interacting with the lattice and also directly between themselves by Coulomb interaction. We do not specify the form of the electron–lattice interaction; we will need only some general properties of the response functions such as their analyticity and causality.

A convenient way to describe the resulting electron–electron interaction is via introduction of the dielectric function $\varepsilon(\boldsymbol{k}, \omega)$ [12, 11]. A rather general (but not the most general, see below) form of the electron–electron interaction can be written as

$$V(\boldsymbol{k}, \omega) = \frac{4\pi e^2}{k^2 \varepsilon(\boldsymbol{k}, \omega)}. \tag{3}$$

This interaction definitely takes care of the Coulomb interaction between electrons. However, by properly choosing $\varepsilon(\boldsymbol{k}, \omega)$ we can incorporate in this expression also the interaction via phonons, excitons etc. Thus, in a jellium model $1/\varepsilon(\boldsymbol{k}, \omega)$ for $\omega < k v_F$ is given by the expression

$$\frac{1}{\varepsilon(\boldsymbol{k}, \omega)} = \frac{k^2}{k^2 + \mathit{\ae}^2} \left(1 + \frac{\omega_k^2}{\omega^2 - \omega_k^2} \right), \tag{4}$$

where ω_k is the frequency of the longitudinal oscillations

$$\omega_k^2 = \frac{\omega_i^2 k^2}{k^2 + \mathit{\ae}^2},$$

ω_i is the plasma frequency of ions and $æ^{-1}$ is the Debye screening length.

One can also use a more general form of $1/\varepsilon(\boldsymbol{k},\omega)$. The necessary requirement is only that $1/\varepsilon$ satisfies the dispersion (Kramers–Kronig) relation

$$\frac{1}{\varepsilon(\boldsymbol{k},\omega)}=1-\int_0^{\infty}\frac{\rho(\boldsymbol{k},x)\,\mathrm{d}x^2}{x^2-\omega^2-\mathrm{i}\delta},\qquad \rho(\boldsymbol{k},x)\geq 0. \tag{5}$$

It is clear from Eq. (5) and from the example (4) that the interaction (3) is in general strongly energy-dependent; in particular there are regions of ω in which $V(\boldsymbol{k},\omega)$ is negative (regions of anomalous dispersion). It is just these attractive parts of the interaction that can provide pairing of electrons. However, for large enough ω, $1/\varepsilon(\boldsymbol{k},\omega)$ and $V(\boldsymbol{k},\omega)$ are always positive and tend to a bare unscreened Coulomb repulsion, see (3) and (5). (In principle one may also include in these expressions high-frequency screening ε_∞.) In ordinary super-conductors we really can make use of these regions of anomalous dispersion to form Cooper pairs whereas the repulsive part of the interaction is suppressed, see Eq. (2). The question is whether this is also possible for bipolarons, in the absence of the Fermi surface. In studying this problem, we follow the method developed in [12], adapting it to our present problem. For the effective interaction we use the shortened notation

$$V(\lambda,\omega)=V(\boldsymbol{k})/\varepsilon(\boldsymbol{k},\omega), \tag{6}$$

which slightly generalizes (3) in that we can include in (6) e.g. also $\varepsilon_\infty \neq 1$.

3 The Bethe–Salpeter equation and its transformation

The equation that describes the possible two-particle bound state in the case of frequency-dependent (retarded) interaction is the well-known Bethe–Salpeter equation [13], which for an interaction of type (3) has the form

$$\left(\frac{1}{2}E+\frac{\boldsymbol{p}^2}{2}+\varepsilon-\mathrm{i}\delta\right)\left(\frac{1}{2}E+\frac{\boldsymbol{p}^2}{2}-\varepsilon-\mathrm{i}\delta\right)\Psi(\boldsymbol{p},\varepsilon)$$
$$=-\frac{1}{2\pi\mathrm{i}}\int\mathrm{d}\boldsymbol{k}\,\mathrm{d}\omega\,V(\boldsymbol{p}-\boldsymbol{k},\varepsilon-\omega)\Psi(\boldsymbol{k},\omega). \tag{7}$$

Here $\Psi(\boldsymbol{p},\varepsilon)$ is the generalized 'wave function' of the bound state, which depends not only on relative coordinates $\boldsymbol{r}-\boldsymbol{r}'$ or, in momentum representation, on relative momentum $\boldsymbol{p}$, but also on respective times $t-t'$, or, in the Fourier transform, on ω. E is the binding energy, which is taken positive. We put $\hbar$ and electron mass m equal 1.

For an instantaneous interaction $V(k)$, which is independent of ω, one can introduce the function

$$\varphi(p) = \int_{-\infty}^{\infty} \Psi(p,\varepsilon)\,d\varepsilon \tag{8}$$

(the wave function at coinciding times) and, putting this expression into Eq. (7), one obtains for $\varphi(p)$ the equation

$$(E+p^2)\varphi(p) = -\int V(p-k)\varphi(k)\,dk, \tag{9}$$

which is nothing other than the ordinary Schrödinger equation for a potential $V(k)$.

The difficulty in analyzing equation (7) lies in the fact that the kernel of the equation $V(p-k,\varepsilon-\omega)$ can in principle have different signs, so that it is not easy to tell whether the bound state does exist. There is a convenient method to deal with this problem, suggested in [12], which permits us to get rid of the negative-frequency parts of the interaction and gives the equation with a kernel of simpler properties. The method consists in using the spectral representations both for the interaction and for the wave function itself. We will use the spectral representation (5) for $1/\varepsilon$, and we write down a similar representation for $\Psi(p,\varepsilon)$:

$$\Psi(p,\varepsilon) = \frac{1}{2\pi i}\int_0^{\infty} \frac{A(p,x)\,dx^2}{x^2-\varepsilon^2-i\delta}. \tag{10}$$

The 'instantaneous' wave function φ (8) is expressed through spectral function A by the relation

$$\varphi(p) = \int_0^{\infty} A(p,x)\,dx. \tag{11}$$

We will not reproduce here all the technical details of the treatment that permits us to transform Eq. (6) into the equation for the spectral densities. After rather lengthy but straightforward calculation, we get finally

$$(E+p^2)\varphi(p) = -\int dk \int_0^{\infty} dz\, V(p-k)\left(1-\int_0^{\infty} \frac{dx^2\rho(p-k,x)}{x(x+z+E/2+p^2/2)}\right)A(k,z). \tag{12}$$

Eq. (12) is still not a closed equation for a function of one variable. Again, in the case of an instantaneous interaction, $\rho - 0$, and keeping in mind relation (11), we immediately recover from (12) the Schrödinger equation (9). One may also obtain a generalization of this equation, if one puts all the variables on the 'mass shell', see below. We shall discuss this approximation later, but now we want to show that already from the form (12) one can draw important conclusions. One can show that, similar to $\rho(k, x)$ (5), the spectral density $A(k, x)$ corresponding to the ground state wave function is also non-negative (it follows from the connection of Ψ with a two-particle Green function). It is also clear that $\varphi(p)$ (11) is also ≥ 0.

Eq. (12) is non-singular, and the existence of a solution depends on the sign of a kernel of this equation: if the quantity in parentheses in Eq. (12) is positive, then there is no solution for positive $V(k)$ (bare repulsion). (Logarithmic singularity of the corresponding equation due to the presence of the Fermi surface in the case of superconductors may change the situation, see [12].)

Using the spectral representation of $1/\varepsilon(k, \omega)$ (5), we immediately see that the expression entering the kernel of Eq. (12) satisfies the inequality

$$1 - \int_0^\infty \frac{dx^2 \rho(p-k, x)}{x(x + z + E/2 + p^2/2)} \geq 1 - \int_0^\infty \frac{dx^2 \rho(p-k, x)}{x^2} = \frac{1}{\varepsilon(p-k, 0)}, \qquad (13)$$

Thus, if the static response function $1/\varepsilon(k, 0)$ is positive, the kernel of Eq. (12) will be positive too, and there will be no solution, i.e. no bound state.

The condition

$$\frac{1}{\varepsilon(k, 0)} \geq 0 \qquad (14)$$

is usually taken as a criterion of the stability of a system [14, 9]: the standard argument is that the energy of static electric fields is given by the expression

$$\mathscr{E} \simeq \int \frac{|D_k|^2}{\varepsilon(k, 0)} \, dk, \qquad (15)$$

where D is the electric induction, and, if $1/\varepsilon(k, 0)$ would be negative for certain k, then the system would be unstable with respect to spontaneous generation of the electric field with the corresponding period.

This statement was recently disputed (see e.g. [15] and discussion below); at the moment we only want to stress once again that, with the asusmption that is very often made (interaction of form (3) or (6), positive static dielectric constant), one can quite generally and rigorously prove the absence of a bound state of two electrons (bipolaron), irrespective of the detailed form of the electron–lattice interaction, structure of electron and phonon spectra etc.; one cannot overcome in this case the direct Coulomb repulsion.

One can obtain a somewhat simpler equation by taking all the variables 'on the mass shell', i.e. by putting in Eq. (12)

$$A(k,x) = \delta[x^2 - (E/2 + k/2)^2]f(k). \qquad (16)$$

Putting (16) into Eq. (12) and taking into account (11), we get finally

$$(E + p^2)\varphi(p) = -\int dk \, \mathscr{V}(p, k|E)\varphi(k), \qquad (17)$$

where the effective 'potential' (the kernel of Eq. (17)) is

$$\mathscr{V}(p, k|E) = V(p - k)\left(1 - 2\int_0^\infty \frac{dx\,\rho(p - k, x)}{x + E + p^2/2 + k^2/2}\right). \qquad (18)$$

We see that, in this approximation, we obtain the analog of the Schrödinger equation but with the non-local effective potential $\mathscr{V}$ (it depends not on $p - k$ but on p and k separately), which, moreover, parametrically depends on the eigenvalue E.

Probably this equation may be used e.g. to analyze the influence of electron–phonon interaction (and all other effects that may be incorporated in $\varepsilon(k, \omega)$) on the electron–hole bound states in crystals – excitons, but for our purpose, the treatment of possible bipolaron formation, the conclusion is again the same: it is clear from comparison of (18) and (5) that if only $1/\varepsilon(k,0) \geq 0$, then the effective potential $\mathscr{V}(p, k|E)$ has the same sign as $V(p - k)$ – i.e. in our case (two electrons) it is repulsive, and there is no bound state.

One can illustrate the corresponding results by writing down the specific form of $\varepsilon(k, \omega)$ that is often used in studies of polarons and bipolarons:

$$\varepsilon(\omega) = \varepsilon_\infty - \frac{\Omega^2(\varepsilon_0 - \varepsilon_\infty)}{\omega^2 - \Omega^2} \qquad (19)$$

or

$$\frac{1}{\varepsilon(\omega)} = \frac{1}{\varepsilon_\infty}\left(1 + \frac{b}{\omega^2 - \tilde{\omega}^2}\right) \qquad (20)$$

where

$$\tilde{\omega}^2 = \Omega^2 \frac{\varepsilon_0}{\varepsilon_\infty}, \qquad b = \left(\frac{\varepsilon_0}{\varepsilon_\infty} - 1\right)\Omega^2 = \left(1 - \frac{\varepsilon_\infty}{\varepsilon_0}\right)\tilde{\omega}^2. \qquad (21)$$

We would take here $V(k) = 4\pi e^2/k^2$. In this case the effective potential takes the form

$$\mathscr{V}(p, k|E) = \frac{4\pi e^2}{(p-k)^2 \varepsilon_\infty} \left(1 - \frac{\tilde{\omega}(1 - \varepsilon_\infty/\varepsilon_0)}{\tilde{\omega} + E + p^2/2 + k^2/2} \right). \tag{22}$$

One sees that the effective potential is indeed positive, and there is no bound state for two electrons. This conclusion would not change if one were to take a more general form of $\varepsilon(k, \omega)$, including the contribution of oscillators with arbitrary frequencies and oscillator strengths; independent of the specific details, general requirements of analyticity and causality (the Kramers–Kronig relation) put several restrictions on the possibility of formation of bound states.

4 Conclusion: possible ways out

As we quite generally have shown above, if the interaction between electrons has the form (3) or (6) and if the static response function $1/\varepsilon(k, 0)$ is positive then there is no bound state of two electrons (bipolarons) irrespective of the strength of electron–phonon interaction. However, our results do not exclude the possibility of bipolarons (or, more generally, bi-electrons: the binding may – and possibly should – be not only due to electron–phonon interaction, but may be caused also by other mechanisms, e.g. magnetic ones). There are several points at which our conclusions may be modified.

First of all there is the question of whether the condition $1/\varepsilon(k, 0) \geq 0$ is really a necessary condition for stability of the system. There are arguments, most extensively presented in [15], that it may not be the case. In particular, in real crystals the response function is not just a function of one momentum k, but is actually a matrix in *umklapp* wave vectors. In this case calculation of an effective electron–electron interaction is a highly non-trivial problem, involving inversion of large matrices. This, as claimed in [15], can in principle lead to inversion of the sign of an effective interaction. Physically these effects are known as local field corrections. Their possible role is very difficult to determine in general, but in principle they can result in a more complicated behavior of certain particular systems.

Probably more important is another limitation of our treatment. We used as an effective interaction (the kernel of the Bethe–Salpeter equation) an expression of the form (9), which depends only on transferred 4-momentum $q = (k, \omega)$. In a diagrammatic language it corresponds to keeping in an irreducible vertex part, which enters the Bethe–Salpeter equation, of diagrams of the type

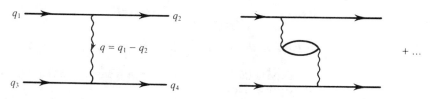

However, in principle a general irreducible vertex depends not only on transferred momentum $q = q_1 - q_2$, but also on total momentum $Q = q_1 + q_3$. For instance, such are the contributions of the diagram

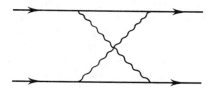

or certain vertex corrections.

What could be the effect of such terms is an open problem. One cannot draw any general conclusion about their possible role. It is quite probable that the effects described by such terms may indeed provide a mechanism for formation of bipolarons. However, the nature of the corresponding forces would in this case be more complicated than ordinary polarization of the lattice, it would probably involve strong exchange effects and other factors (which, for instance, in chemical language are responsible for formation of valence pairs and, maybe, for some other specific phenomena). In any case, the message of the treatment presented above is that formation of bipolarons is a highly non-trivial problem, so one should be very cautious in using conventional models and in relying on results of approximate treatments. Such approaches, which are very often used, are definitely inadequate; one should look rather deeply into specific features of particular substances to establish the possibility of formation of bipolarons in them. However, this may be just what makes this problem even more interesting and challenging.

Acknowledgement

I am grateful to Professor G. Sawatzky and Dr M. Czyzyk for useful discussions. This investigation was supported by the Netherlands Foundation for Fundamental Research on Matter (FOM).

References

[1] S. J. Pekar, *Untersuchungen über die Electronentheorie der Kristalle*, Akademie-Verlag, Berlin 1954.

[2] N. F. Mott and E. A. Davis, *Electronic Processes in Non-Crystalline Materials*, Oxford University Press, New York 1979.

[3] V. Vinetskii and M. Giterman *Zh. Eksp. Theor. Fiz.* **40**, 1459 (1961) (*Sov. Phys. – JETP* **13**, 1023 (1961)).

[4] S. Lakkis *et al.*, *Phys. Rev.* B **14**, 191 (1993).

[5] N. F. Mott, *Physica* C **205**, 191 (1993).

[6] L. J. de Jongh, *Physica* C **161**, 631 (1989).

[7] A. S. Alexandrov, *J. Low Temp. Phys.* **87**, 721 (1992).

[8] J. R. Schrieffer, X. G. Wen and S. C. Zhang, *Phys. Rev.* B **39**, 11 663 (1989).

[9] J. T. Devreese, G. Verbist and F. M. Peeters, preprint.

[10] J. R. Schrieffer, *Theory of Superconductivity*, Benjamin, New York 1964.

[11] *High-Temperature Superconductivity*, Eds. V. L. Ginzburg and D. A. Kirzhnits, Consultant Bureau, New York 1982.

[12] D. A. Kirzhnits, E. G. Maksimov and D. I. Khomskii, *J. Low Temp. Phys.* **10**, 79 (1973).

[13] E. E. Salpeter and H. A. Bethe, *Phys. Rev.* **84**, 1232 (1951).

[14] D. Pines and P. Nozières, *The Theory of Quantum Liquids*, Benjamin, New York 1966.

[15] O. V. Dolgov, D. A. Kirzhnits and E. G. Maksimov, *Rev. Mod. Phys.* **53**, 81 (1981).

25

Large bipolarons and high-T_c materials

J. T. DEVREESE*, G. VERBIST and F. M. PEETERS

Departement Natuurkunde, Universiteit Antwerpen, Universiteitsplein 1, B-2610 Antwerpen, Belgium

Abstract

A Feynman path-integral type of treatment is developed to determine under which conditions the energy of a bipolaron is lower than the energy of two polarons. A detailed analytical and numerical study of the Fröhlich bipolaron is presented, resulting in a phase diagram for the stability of the bipolaron in terms of the electron–phonon coupling strength α and the strength U of the Coulomb repulsion. The stability region for two- and three-dimensional bipolarons is examined for several materials.

It is shown that the bipolaron binds more easily in 2D than in 3D and that its radius is only a few ångström units. Alexandrov, Bratkovsky and Mott have recently stressed the importance of this confinement, as derived by the present authors, for high-T_c superconductivity. We analyze as an example the occurrence of bipolarons in La_2CuO_4. First results on optical absorption of bipolarons are also presented.

Bednorz and Müller's discovery of the high-temperature superconductors stimulated both experimental and theoretical efforts to determine the mechanism responsible for superconductivity in these new materials. Bipolarons (dielectric and spin) have been invoked as possible 'Cooper pairs' at the basis of high-T_c superconductivity (Alexandrov, Bratkovsky and Mott [1]).

Bipolarons (large and small) had been studied before [2, 5–10] also in the context of superconductivity [3]. Emin proposed Bose–Einstein condensation of *large* two-dimensional bipolarons as a possible mechanism responsible for superconductivity in these materials.

---

* Also at the Technische Universiteit Eindhoven, The Netherlands.

In the present paper a path-integral study of *large bipolarons* is presented in two and three dimensions. Conditions will be discussed under which bipolarons can exist in the copper oxides. Some experimental difficulties in determining material parameters, e.g. the band mass, are also discussed.

Bipolarons are closely related to (single) polarons. A polaron is an electron together with its polarization field, due to its interaction with the lattice vibrations of the surrounding solid. Two main different cases can be distinguished:

1. the *small* polaron, where the polaron is localized;
2. the *large* polaron, which behaves much like a free electron, but with a larger effective mass and with possible internal degrees of freedom.

The bipolaron is a state of two polarons bound to each other through the mediating polarization interaction, despite Coulomb repulsion. The distinction between small and large bipolarons is analogous to that for single polarons. The present paper concentrates on large bipolarons, in which the electrons interact with the longitudinal optical (LO) phonons, and the interelectronic potential is taken to be Coulombic.

In the prsent contribution we discuss the results of Verbist, Peeters and Devreese [4] and extend them to optical absorption (in collaboration with P. Vansant).

Earlier studies of this system used the Rayleigh–Ritz variational principle and the adiabatic approximation for the polarization cloud. The estimates for the bipolaron ground-state energy were compared with (twice) the polaron ground-state energy, calculated within the same formalism. Since the latter quantity compares poorly to known results, e.g. from the path-integral formalism [5, 6], very favorable conditions were obtained for the existence of bipolarons. In [7] this difficulty was resolved by comparing with path-integral results for the single-polaron energy. This approach to the bipolaron consists of numerical minimization of the ground-state energy using a singlet wave function after performing a unitary transformation. A narrow stability region was found for the bipolaron in three dimensions. The authors of [8] have recently generalized this approach and they also reported results in two dimensions, where a larger stability region was found than in three dimensions.

Calculations within the path-integral formalism allow for exact elimination of the phonon degrees of freedom, just like for single polarons [5]. Because the free energy, rather than the ground-state energy, is calculated, generalization to non-zero temperatures is straightforward. In [9] the electron–phonon interaction with both longitudinal optical and longitudinal acoustic phonons was included. They only considered the three-dimensional case and emphasized the role of acoustic phonons for self-localization of the bipolaron. In a supercon-ducting state, however, the bipolarons should be mobile and therefore self-localization should be prevented. Since we want here to study mobile bipolar-ons, we concentrate on the LO-phonon interaction. Analytical strong coupling results have already been reported in three dimensions for the two-polaron problem [10], but here we consider the whole coupling range in both two and three dimensions.

The system studied is described by the following Hamiltonian:

$$H = \sum_{j=1,2} \frac{p_j^2}{2m_b} + \sum_i \hbar\omega_{LO} a_k^\dagger a_k + \frac{U}{|r_1 - r_2|} + \sum_{j=1,2} \sum_k (V_k a_k e^{ik\cdot r_j} + \text{H.c.}),$$

where m_b denotes the band mass of the electrons and ω_{LO} the frequency of the LO phonons. In what follows, a system of units is used in which $\hbar$, m_b and ω_{LO} have the numerical value of unity. The problem is characterized by two parameters:

1. the strength U of the Coulomb repulsion; and
2. the electron–phonon coupling constant α hidden in the Fröhlich [11] interaction coefficients V_k, given by

$$|V_k|_{3D}^2 = \frac{2\sqrt{2\pi\alpha}}{k^2}, \qquad |V_k|_{2D}^2 = \frac{\sqrt{2\pi\alpha}}{k}.$$

Both these parameters are related to the static (ε_0) and high-frequency (ε_∞) dielectric constants of the crystal:

$$\alpha = \frac{e^2}{\sqrt{2}} \left(\frac{1}{\varepsilon_\infty} - \frac{1}{\varepsilon_0} \right), \qquad U = \frac{e^2}{\varepsilon_\infty}.$$

Since the dielectric constants are positive quantities, only the region $U \geq \sqrt{2}\alpha$ of the (α, U)-plane has physical significance.

For the path-integral treatment of this system, a similar procedure was followed to that introduced by Feynman for the single polaron. After elimination of the phonons, the Jensen–Feynman inequality was used to determine a variational upper bound to the free energy. As a trial system, two harmonic oscillators were introduced to represent the phonon cloud around each electron. All interactions were then replaced by harmonic interactions, which results in a quadratic trial action, which is a generalization of the one in [9]. This trial action is translationally invariant (i.e. total linear momentum is a conserved quantity), as is necessary to have *mobile* bipolarons. Details of the calculation are given in [4, 12]. In this paper we only summarize the main numerical results for the stability region of bipolarons at zero temperature.

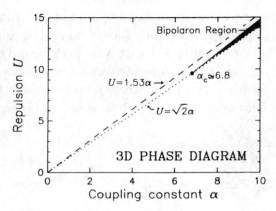

Fig. 1. The stability region for bipolaron formation in 3D. The dotted line $U = \sqrt{2\alpha}$ separates the physical region $U > \sqrt{2\alpha}$ from the non-physical one. The shaded area is the stability region in physical space. The dashed line is determined by $U = \sqrt{2\alpha}(1 - \varepsilon_\infty \varepsilon_0)$ for La_2CuO_4. The critical point $\alpha_c = 6.8$ is represented by a full dot [4].

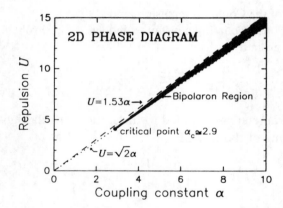

Fig. 2. The same as Fig. 1, but for 2D. The critical point is $\alpha_c = 2.9$ [4].

The phase diagram for bipolaron formation is shown in three (Fig. 1) and two (Fig. 2) dimensions. In both cases the non-physical region $U \leq \sqrt{2\alpha}$ is entirely bipolaronic. Below a *critical value* α_c the bipolaron–polaron transition is continuous and occurs at $U_c = \sqrt{2\alpha}$. For $\alpha \geq \alpha_c$ the transition is discontinuous, and bipolarons do exist for physically acceptable values of α and U. The stability region for bipolarons is indicated by the shaded area. The dotted line ($U = \sqrt{2\alpha}$) separates the physical region from the non-physical one. Numerically it was found that $\alpha_c \approx 2.9$ in two dimensions, and $\alpha_c \approx 6.8$ in three dimensions, indicated by the full dots. Note that the stability region is larger in two than in three dimensions, and, more importantly, that bipolarons in two dimensions exist at smaller and more realistic values of the coupling constant α.

In order to discuss the relevance of bipolaron formation in the copper oxides, realistic estimates of α and U have to be obtained. Denoting by λ the ratio of the *effective* Rydberg to the LO-phonon energy

$$\lambda = \frac{e^4 m_b / (2\hbar^2)}{\hbar \omega_{LO}}$$

α and U are given by

$$\alpha = \sqrt{\lambda} \left(\frac{1}{\varepsilon_\infty} - \frac{1}{\varepsilon_0} \right), \qquad U = \frac{(2\lambda)^{1/2}}{\varepsilon_\infty}$$

but, as in single-polaron physics, accurate values of the band mass m_b are hard to obtain, since electron–phonon interaction increases the band mass to the *effective polaron mass* m^*, which is the one measured experimentally (e.g. from cyclotron resonance). However, the band-mass-dependence can be eliminated if λ is eliminated between α and U, which gives the relation

$$U = \sqrt{2\alpha} \left(1 - \frac{\varepsilon_\infty}{\varepsilon_0} \right),$$

which is shown as a dashed line in Fig. 1 and Fig. 2, where the ratio $\varepsilon_\infty / \varepsilon_0 = 0.08$ was taken, as is relevant for La$_2$CuO$_4$ at zero temperature [14]. With these parameters, three-dimensional bipolaron formation is not expected in La$_2$CuO$_4$; but two-dimensional bipolaron formation seems quite possible, since at low temperatures ε_0 is found to increase with increasing temperature [14] and to be highly anisotropic (with a value of 50 in the (a, b)-plane, and 23 in the c-direction). In order to have bipolarons at $\alpha = 4(5)$, the ratio $\varepsilon_\infty / \varepsilon_0$ should be around 0.04(0.055). Assuming an LO-phonon energy of 70 meV and $\varepsilon_\infty = 4$, the band mass in units of the free-electron mass should equal 1.4(2.3). It is hard to speculate whether such conditions are realized in other high-temperature superconductors. Anisotropy effects are averaged out in polycrystalline measurements [15], e.g. on YBa$_2$Cu$_3$O$_{6+x}$, but, at least in the copper oxides, we have given evidence that bipolaron formation is not unrealistic. Furthermore, it is expected that inclusion of short-range phonons (e.g. optical [3] or acoustic [9] deformation) will enlarge the stability region for bipolaron formation.

Recently (in collaboration with P. Vansant) the present authors have also determined the optical absorption spectrum of bipolarons. Here we present the result for $\alpha = 7$ (Fig. 3). The discussion will be presented in a separate paper.

In summary, we have analyzed the stability of large bipolarons and show that they are more easily formed in 2D than in 3D; also their radius is only a few

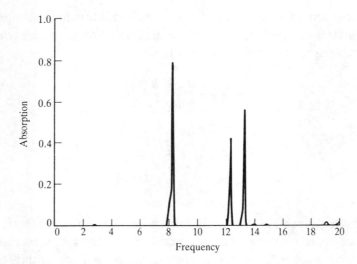

Fig. 3. The optical absorption spectrum of a bipdaron for $\alpha = 7$.

ångström units. Alexandrov, Bratkovsky and Mott have stressed the importance of this fact. Also, optical absorption of bipolarons is reported.

References

[1] A. S. Alexandrov, A. M. Bratkovsky and N. F. Mott, *Phys. Rev. Lett.* **72**, 1734 (1994).
[2] V. L. Vinetskii, *Zh. Eksp. Teor. Fiz.* **40**, 1459 (1961) [*Sov. Phys. – JETP* **13**, 1023 (1961)]; S. G. Suprun and B. Ya. Moizhes, *Fiz. Tverd. Tela* **24**, 1571 (1982) [*Sov. Phys. – Semicond.* **16**, 700 (1982)]; F. M. Peeters and J. T. Devreese, *Periodic solutions of the classical polaron and bipolaron system, J. Math. Phys.* **21**, 2302 (1980); H. Hiramoto and Y. Toyozawa, *J. Phys. Soc. Japan.* **54**, 245 (1985).
[3] M. R. Schafroth, *Phys. Rev.* **100**, 463 (1955); B. K. Chakraverty, D. Feinberg, Z. Hang and M. Avignon, *Solid State Commun.* **64**, 1147 (1987); A. S. Alexandrov and J. Ranniger, *Phys. Rev.* B **23**, 1796 (1981); *ibid.* B **24**, 1164 (1981); A. S. Alexandrov, J. Ranniger and S. Robaszkiewicz, *Phys. Rev.* B **33**, 4526 (1986); J. de Jongh, *Physica* C **161** (1989); D. Emin, *Phys. Rev. Lett.* **62**, 1544 (1989); D. Emin and M. S. Hillery, *Phys. Rev.* B **39**, 6575 (1989).
[4] G. Verbist, F. M. Peeters and J. T. Devreese, *Solid State Commun.* **76**, 1005–7 (1990).
[5] R. P. Feynman, *Phys. Rev.* **97**, 660 (1955).
[6] F. M. Peeters and J. T. Devreese, *Phys. Rev.* B **25**, 7281 (1982).
[7] J. Adamowski, *Acta Phys. Polonica* A **73**, 345 (1988); *Phys. Rev.* B **39**, 3649 (1989).
[8] F. Bassani, M. Geddo, G. Jadonisi and D. Ninno, *Phys. Rev* B **43**, 5296 (1991).
[9] H. Hiramoto and Y. Toyozawa, *J. Phys. Soc. Japan.* **54**, 245 (1985).

[10] E. A. Kochetov, S. P. Kuleshov, V. A. Mateev and M. A. Smondyrev, *Teor. Mater. Fiz.* **30**, 183 (1977) [*Theor. Mater. Phys.* 1978)].

[11] H. Fröhlich, *Adv. Phys.* **3**, 325 (1964).

[12] G. Verbist, F. M. Peeters and J. T. Devreese, *Phys. Rev. B* **43**, 2712 (1991).

[13] J. W. Hodby, G. P. Russell, F. M. Peeters, J. T. Devreese and D. M. Larsen, *Phys. Rev. Lett.* **58**, 1471 (1987).

[14] D. Reagor, E. Ahrens, S.-W. Cheong, A. Migliori and Z. Fisk, *Phys. Rev. Lett.* **62**, 2048 (1989).

[15] G. A. Samara, W. F. Hammeter and E. L. Venturini, *Phys. Rev. B* **41**, 8974 (1990).

26

Collective excitations in the ground state of a two-dimensional attractive Fermi gas

S. V. TRAVEN[1]

Department of Physics, University of Warwick, Coventry CV4 7AL, UK

Abstract

The spectrum of collective (pair) excitations in the ground state of a dilute two-dimensional (2D) attractive Fermi-gas is studied within the functional integral formalism. The linearized equations for the fluctuations about the non-trivial saddle point are analyzed for all coupling regimes, which are characterized by the ratio $\varepsilon_0/\varepsilon_F$, where ε_0 is the two-fermion binding energy and ε_F is the Fermi energy. The approximation takes into account propagation of the fluctuations and their interaction with the condensate. In the strong-coupling, or 'Bose-gas', regime ($\varepsilon_0/\varepsilon_F \gg 1$) the spectrum is continuous and has the Bogolubov form, but in the weak-coupling limit ($\varepsilon_0/\varepsilon_F \ll 1$) there are two types of excitations (different from the two-fermion scattering states): (i) long-wavelength sound-like excitations with the cut-off at momentum $q_c \simeq 1/\xi_0$ (where ξ_0 is the Cooper pair size), and (ii) pair excitations with $q \simeq (8m\mu)^{1/2}$, where μ is the chemical potential and m is the fermion mass. The crossover between weak- and strong-coupling behavior of the excitation spectrum is found to occur at the value of the coupling parameter $\varepsilon_0/\varepsilon_F \simeq \frac{1}{4}$.

1 Introduction

Since the discovery of high-T_c superconductivity there has been growing interest in studying 2D Fermi gases with attractive interaction, especially in the crossover regime, when the pair size ξ_0 is of the order of the interparticle distance. The importance of such a model, which can be regarded as a semi-phenomenological model of a 2D superconductor, is highlighted by experimental evidence that the high-T_c superconductors, most of which have a

[1] Permanent address: Moscow Engineering Physics Institute, 31 Kashirskoe shosse, Moscow, 115409, Russia.

layered structure, have a short coherence length, so that $k_F \xi_0 \simeq 1$, where k_F is the Fermi momentum. Although the model has been intensively investigated by several authors in recent years [1–4], the analysis is still far from complete. As was shown by Schmitt-Rink *et al.* [1], the main difficulty arising, for example, in the random-phase approximation (RPA) to the thermodynamics of the system at low temperatures (close to the BCS T_c), comes from the divergent contribution of the long-wavelength Cooper-pair fluctuations. This occurs because the RPA does not take into account interaction between these fluctuations. Extension of the RPA results so as to take into account repulsive (because of the Pauli exclusion principle) interaction between the fluctuations leads to $T_c = 0$ for arbitrary coupling strength [4], which is probably an artifact of the approximation. Indeed, superfluidity in the system at finite T should exist at least in the strong-coupling limit, where the problem is reduced to that of the superfluidity of 2D composite bosons interacting via hard-core repulsive interaction. In this latter case, evaluation of T_c requires calculation of the superfluid density, which could then be used in the Berezinskii–Kosterlitz–Thouless renormalization [5]. Such an analogy suggests an approach, which has not yet been investigated in detail, namely to start from $T < T_c$ and to calculate the superfluid density of the 2D attractive Fermi gas. This scheme, however, requires one to have a perturbation theory for the pair fluctuations. To provide a basis for such a perturbation theory, it would be natural to investigate first the quadratic ('Gaussian') fluctuations of the pairing amplitude about its condensate value Δ in the ground state. This approach, which describes propagation of the pair fluctuations and their interaction with the condensate, should take into account the main features of the excitation spectrum of these fluctuations. The problem is investigated with the help of the functional integral formalism, which was applied recently to the study of the crossover from the weak-coupling to the strong-coupling limit of a 3D attractive Fermi gas [6, 7]. The analysis is given for arbitrary coupling strength, which is characterized by the parameter $\varepsilon_0/\varepsilon_F$. The repulsive interaction between pair fluctuations, arising from the Pauli exclusion principle, is shown to have a range of the order of the pair size $\xi_0 \simeq (m\varepsilon_0)^{-1/2}$. It is shown that, for all values of $\varepsilon_0/\varepsilon_F$, the long-wavelength part of the excitation spectrum is a sound-like collective mode. In the strong-coupling regime the spectrum is continuous over the entire range of the pair momentum q, but in the weak-coupling limit the spectrum consists of two different types of excitations. The long-wavelength part has a cut-off at the momentum $q_c \simeq 1/\xi_0$, where the pair excitation energy reaches the value 2Δ, the threshold for decay into two quasi-particles. For $q \gtrsim (8m\mu)^{1/2}$ there exist pair excitations with energy $\omega \geq 2\Delta$. General analysis of the properties of the pair excitation spectrum near the decay threshold reveals that, in contrast to 3D systems, in two dimensions decay of an excitation into two quasi-particles with parallel non-zero momenta k_1 and k_2 cannot occur at $T=0$, except for the case in which the quasi-particle spectrum has minima at k_1 and k_2.

2 The general formalism

The initial Hamiltonian of the attractive Fermi gas is taken in the standard form

$$\hat{H}=\int dx \left[\bar{\Psi}_\sigma \left(-\frac{\nabla^2}{2m} - \mu \right) \Psi_\sigma + u_0 \bar{\Psi}_\uparrow \bar{\Psi}_\downarrow \Psi_\downarrow \Psi_\uparrow \right]$$

with the short-range attractive potential u_0 (its range R must satisfy the diluteness condition $k_F R \ll 1$). Potential u_0 is eliminated in favor of the t-matrix of the two-fermion scattering in the usual way, so that the actual parameter that characterizes the strength of the attraction is the absolute value of the binding energy of two fermions in vacuum ε_0 related to u_0 via

$$\frac{1}{u_0} = - \int \frac{d^2k}{(2\pi)^2} \frac{1}{2\varepsilon_k^0 + \varepsilon_0},$$

where $\varepsilon_k^0 = k^2/(2m)$. The divergence of this integral at the upper limit is fictitious and is cancelled with that of the two-quasi-particle Green function. For the magnitude of the binding energy one then obtains $\varepsilon_0 - 2\Lambda \exp[4\pi/(mu_0)]$, where $\Lambda \simeq 1/(mR^2)$ is the ultraviolet cut-off (see, for example, [2]). The partition function is given by the imaginary time functional integral [8]: $Z = \int\int \exp\{S\} D\bar{\Psi} D\Psi$ with the action $S = \int dr[\bar{\Psi}_\sigma \partial_\tau \Psi_\sigma - \hat{H}(\tau)]$. After the Hubbard–Stratonovitch transform with the auxiliary field $\Delta(x)$, where $x = (\boldsymbol{x}, \tau)$, the fermion variables are integrated out, so that $Z = \int\int \exp\{S_{\text{eff}}\} D\Delta^* D\Delta$ with

$$S_{\text{eff}} = \frac{1}{u_0} \int dx |\Delta(x)|^2 + \text{Tr} \ln \tilde{G}^{-1}[\Delta(x)].$$

Here $\tilde{G}^{-1}$ denotes the inverse Nambu propagator for the fermions:

$$\tilde{G}^{-1}(x,x') = \begin{bmatrix} \partial_\tau + \dfrac{\nabla^2}{2m} + \mu & -\Delta^*(x) \\[2ex] -\Delta(x) & \partial_\tau - \dfrac{\nabla^2}{2m} - \mu \end{bmatrix} \delta(x-x').$$

At $T=0$ there is a condensate in the system, and the effective action is expanded in terms of the fluctuation of the field $\Delta(x)$ about the condensate value:

$\Delta(x) \to \Delta + \tilde{\Delta}(x)$, so that the partition function will have the form $Z = Z_0 \cdot \iint \exp\{\tilde{S}_{\text{eff}}[\tilde{\Delta}^*, \tilde{\Delta}]\} D\tilde{\Delta}^* D\tilde{\Delta}$. Here Z_0 is the contribution from the fermions in the state described by the non-trivial saddle point solution for Δ, the condition for which is $(\delta \tilde{S}_{\text{eff}}/\delta \tilde{\Delta})|_{\tilde{\Delta}=0}=0$. This equation, together with the evaluation of the particle density from the fermion Green function in the broken-symmetry state, leads to the expressions for the chemical potential and the pairing amplitude:

$$\mu = \varepsilon_F - \varepsilon_0/2, \qquad \Delta = (2\varepsilon_0\varepsilon_F)^{1/2}, \tag{1}$$

where the Fermi energy is related to the fermion density as $\varepsilon_F = \pi n/m$. This result was previously obtained by the diagram method [2]. Gaussian fluctuations about the saddle point are described by the quadratic part of the effective action

$$\tilde{S}_{\text{eff}}^{(2)} = \sum_{q\omega_n} \left\{ A|\tilde{\Delta}(\boldsymbol{q}, i\omega_n)|^2 - \frac{1}{2} [B\tilde{\Delta}(\boldsymbol{q}, i\omega_n)\tilde{\Delta}(-\boldsymbol{q}, -i\omega_n) + \text{H.c.}] \right\}, \tag{2}$$

where at $T=0$ the sum over the Matsubara boson frequencies ω_n should be transformed into the integral. After analytic continuation $i\omega_n \to \omega + i0$ functions A and B can be written down in the form

$$A(\boldsymbol{q}, \omega) = \frac{1}{4} \int \frac{d^2k}{(2\pi)^2} \left[\left(1 + \frac{\xi_+}{\varepsilon_+}\right)\left(1 + \frac{\xi_-}{\varepsilon_-}\right) \frac{1}{\varepsilon_+ + \varepsilon_- - \omega - i0} \right.$$
$$\left. + \left(1 - \frac{\xi_+}{\varepsilon_+}\right)\left(1 - \frac{\xi_-}{\varepsilon_-}\right) \frac{1}{\varepsilon_+ + \varepsilon_- + \omega + i0} \right] - \int \frac{d^2k}{(2\pi)^2} \frac{1}{2\varepsilon_k^0 + \varepsilon_0}, \tag{3}$$

$$B(\boldsymbol{q}, \omega) = \frac{1}{4} \int \frac{d^2k}{(2\pi)^2} \frac{(\Delta^*)^2}{\varepsilon_+ \varepsilon_-} \left(\frac{1}{\varepsilon_+ + \varepsilon_- - \omega - i0} + \frac{1}{\varepsilon_+ + \varepsilon_- - \omega + i0} \right), \tag{4}$$

where $\xi_\pm = (\varepsilon_{q/2\pm k}^0 - \mu)$ and $\varepsilon_\pm = (\Delta^2 + \xi_\pm^2)^{1/2}$. Eq. (3) and Eq. (4) can of course be obtained with the diagram technique as well, by summation of the infinite sequences of graphs of two-quasi-particle scattering in the ladder approximation.

The effective action, Eq. (2), describes the propagation of the pair fluctuations and their interaction with the condensate. Using the standard definition [8], one can now introduce the 'normal' and 'anomalous' Green functions of the pair fluctuations $G(\boldsymbol{q}\omega)$ and $G_1(\boldsymbol{q}\omega)$, respectively. In the first approximation given by the quadratic part of the effective action, Eq. (2), they are

$$G = \frac{A(-\boldsymbol{q}, -\omega)}{A(\boldsymbol{q}, \omega)A(-\boldsymbol{q}, -\omega) - |B(\boldsymbol{q}, \omega)|^2}, \quad G_1 = \frac{B(\boldsymbol{q}, \omega)}{A(\boldsymbol{q}, \omega)A(-\boldsymbol{q}, -\omega) - |B(\boldsymbol{q}, \omega)|^2}, \tag{5}$$

and the excitation spectrum can then be determined from the poles of G and G_1.

3 The pair excitation spectrum

Let us discuss now the excitation spectrum of the pair fluctuations, which is different from the continuum of two-quasi-particle scattering states. The long-wavelength part of the spectrum can be calculated analytically for *all values* of the coupling parameter $\varepsilon_0/\varepsilon_F$. Expanding $A(q, \omega)$ and $B(q, \omega)$, given by Eq. (3) and Eq. (4), to second order in q^2 and ω^2, one arrives, after some algebra, at the result

$$\omega^2(q) = \left(\frac{\varepsilon_F}{m}\right) q^2 + \left[\frac{3\varepsilon_0^2 - 2\varepsilon_0\varepsilon_F - 4\varepsilon_F^2}{3\varepsilon_0(\varepsilon_0 + 2\varepsilon_F)}\right]\left(\frac{q^2}{4m}\right)^2 + o(q^5). \tag{6}$$

As can be seen from Eq. (6), the low-energy part of the spectrum is linear in q, which is closely related to the presence of the condensed pairs. In the weak-coupling (BCS) limit (when $\varepsilon_0 \ll \varepsilon_F$) the sound velocity evaluated from Eq. (6) equals $v_F/\sqrt{2}$, where v_F is the Fermi velocity. This result differs from that in 3D [9] only by a numerical factor ($\sqrt{2}$ appears instead of $\sqrt{3}$). Existence of such undamped low-energy Bose excitations in the weakly interacting Fermi gas is possible only for attractive interaction between the fermions, when there is a gap in the single-particle excitation spectrum, and the sound excitation cannot decay into two quasi-particles. The second term in Eq. (6) is negative in the same limit, making the spectrum unstable at the momentum $q_c \simeq 1/\xi_0$, so that the wavelength of the sound-like excitations cannot be smaller than the Cooper-pair size. In the opposite, strong-coupling (or 'Bose gas') limit, the coefficient in front of the second term equals unity, and the spectrum has essentially the Bogolubov form for a Bose gas of particles of mass $2m$. The boson chemical potential $\mu_B = 2\mu$ consists of the 'ideal Bose gas' value ($-\varepsilon_0$) and the repulsive contribution from the interaction with the condensed bosons $\delta\mu_B = 2\varepsilon_F$, which can be rewritten as $\delta\mu_B = N_B U(0)$, where $N_B = n/2$ is the 'boson' density, and $U(0) = 4\pi/m$ is the Fourier transform of the corresponding repulsive interaction, which has range of order ξ_0. For the limit $k_F\xi_0 \ll 1$ such a 'kinematic' repulsion can be obtained by analyzing the non-boson commutators for the local-pair operators [10]. This repulsive part then appears in the square of the sound velocity in the Bogolubov spectrum, Eq. (6). For large q, where the integration in Eq. (3) and Eq. (4) can also be done analytically, one finds that the Green functions, Eq. (5), always have real poles corresponding to the two-fermion bound states. Indeed, at large q the effect of the Pauli exclusion principle is removed, and the corresponding vertex part of the two-particle Green function is equal to the t-matrix of two-particle scattering problem in vacuum, where there always exists a bound state of two fermions in 2D, however weak the attraction is (in contrast to the 3D case, in which the existence of the bound state is of threshold character). For the normal state of the attractive 2D Fermi gas the existence of two-particle bound states with large momenta was discussed in [1]. Therefore, for large q in the weak-coupling limit the pair excitation energy is just equal to that of two fermions in a vacuum $\omega_b^{(0)}(q) = q^2/(4m) - 2\mu - \varepsilon_0$. In the strong-coupling regime for pair momenta $k_F \ll q \ll 1/\xi_0$ there is also a correction ($\simeq 2\delta\mu_B$, which then vanishes when $q \gg 1/\xi_0$) arising from the hard-core repulsive interaction with bosons in the condensate. The corresponding Hartree–Fock term in the weak-coupling limit is of order μ_0, and can be neglected in the present approximation.

To illustrate the behavior of the excitation spectra in between long-wavelength and large-q limits for the various coupling regimes, the corresponding integrals in Eq. (3) and Eq. (4) have been evaluated numerically for a set of coupling strength values. For the imaginary parts $\text{Im}\{A\}$ and $\text{Im}\{B\}$ integration over the angle in Eq. (3) and Eq. (4) can be done by hand. After calculation of the remaining integral over the energy, one obtains the corresponding real parts from the Kramers–Kronig dispersion relations. As an example of the excitation spectrum in the strong-coupling limit, the result for $\omega(q)$, which was then obtained from the poles of the Green functions, Eqs. (5), is plotted in Fig. 1(a). The coupling parameter equals $\varepsilon_0/\varepsilon_F = 8$. All energies shown are measured in units of ε_F, and the values of the momentum are given in units of $k_F/\sqrt{2}$. In the weak-coupling limit the spectrum has a completely different behavior. In Fig. 1(b) is shown an example of such a spectrum for $\varepsilon_0/\varepsilon_F = \frac{1}{8}$. Its long-wavelength part terminates at q_c, as the excitation energy reaches the value 2Δ, when decay of the pair excitation into two quasi-particles becomes possible. For momenta $q \geq q_0$, where $q_0 = (8m\mu)^{1/2}$, there exist two-quasi-particle bound states, for which the spectrum approaches that of two-particle bound states in vacuum at large q. The crossover between two different coupling regimes (i.e. when the excitation spectrum changes its character from the strong-coupling form to that in the weak-coupling limit) occurs, as revealed by the numerical analysis, at $\varepsilon_0/\varepsilon_F \simeq \frac{1}{4}$. For this critical value of coupling strength the low-energy part of the spectrum only touches the two-quasi-particle continuum at $q_c = q_0$. This is illustrated by Fig. 1(c). For both $q \geq q_0$ and $q \leq q_0$ the behavior of the excitation spectrum near q_0 obeys, as follows from the analysis given below, $\omega - 2\Delta \simeq (q - q_0)^2$.

4 Properties of the spectrum near the decay threshold

The properties of the excitation spectrum near the threshold points, q_c and q_0, can be analyzed within the approach given by Pitaevskii in 3D for the general problem of the decay of an excitation into two quasi-particles [11]. For our

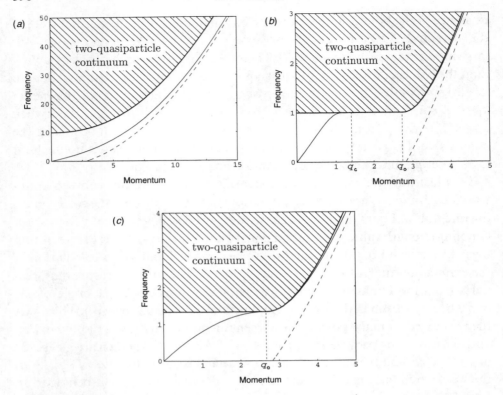

Fig. 1. Pair excitation spectra $\omega(q)$ for the various coupling regimes. Results of the numerical evaluation of $\omega(q)$ are shown by solid lines. Dashed lines correspond to the two-fermion energies in vacuum $\omega_b^{(0)}(q) = q^2/4m - 2\mu - \varepsilon_0$. The energy gap to the two-quasi-particle continuum is equal to 2Δ for $q \leq (8m\mu)^{1/2}$ and $\{4\Delta^2 + [(q^2/4m) - 2\mu)^2\}^{1/2}$ otherwise. (a) For $\varepsilon_0/\varepsilon_F = 8$ (the strong-coupling, or 'Bose-gas', regime) the excitation spectrum has the Bogolubov form. (b) For $\varepsilon_0/\varepsilon_F = \frac{1}{8}$ (the weak-coupling limit) the long-wavelength part of the spectrum has a cut-off at $q_c = 1/\zeta_0$. There also exist pair excitations with $q > (8m\mu)^{1/2}$. (c) For the critical value of the coupling parameter, $\varepsilon_0/\varepsilon_F = \frac{1}{4}$, the excitation spectrum only touches the continuum at $q = (8m\mu)^{1/2}$.

quasi-particle excitation spectrum $\varepsilon = (\Delta^2 + \xi^2)^{1/2}$ there may be two different regimes of such decay: (i) both new quasi-particles have zero velocity, so that decay occurs into two quasi-particles with momenta $|\boldsymbol{k}_1| = |\boldsymbol{k}_2| = (2m\mu)^{1/2}$, which have relative momentum perpendicular to $\boldsymbol{q}$ (in this case the total momentum $q \leq q_0$); (ii) new excitations have the same non-zero velocities and zero relative momentum, so that the quasi-particles 'coalesce' with each other (this requires $q > q_0$). The cut-off point at q_c shown in Fig. 1(b) can easily be identified with case (i). The singularities of $A(\boldsymbol{q}, \omega)$ and $B(\boldsymbol{q}, \omega)$, associated with this process, are contained in the corresponding denominators, which appear in Eq. (3) and Eq. (4). Integration near $q = q_c$ and $\omega = 2\Delta$ gives for the principal non-regular contributions

$$\delta A \simeq \delta B \sim \frac{1}{(q_0 - q_c)^{1/2}} \ln \left(\frac{\Delta}{2\Delta - \omega} \right).$$

The remaining parts of A and B are regular (analytic) functions of q and ω near the decay threshold, and can be expanded in powers of $(q - q_c)$ and $(2\Delta - \omega)$. The excitation spectrum, obtained from the poles of the Green functions, Eq. (5), has then the following form near the threshold point:

$$\omega(q) = 2\Delta - \gamma \exp \left(-\alpha \frac{(q_0 - q_c)^{1/2}}{q_c - q} \right), \tag{7}$$

with some constants α and γ, so that the spectrum approaches this point with a horizontal tangent of infinite order. The spectrum therefore cannot be continued beyond q_c and $\omega(q_c) = 2\Delta$, and the point q_c is thus a cut-off point, where the long-wavelength part of the excitation spectrum terminates (see Fig. 1(b)). Although, for this type of the threshold, case (i), the behavior of the spectrum, given by Eq. (7), is the same as it would be in 3D, a completely different situation occurs in case (ii). Suppose that our pair excitation spectrum (which definitely exists for $q \gg q_0$) has a decay threshold at some $q_c > q_0$ with $\omega(q_c) = \omega_c$ and velocity v_c. Again, one can analyze a possible singular behavior of A and B near this point by evaluation of the corresponding integrals:

$$\delta A \simeq \delta B \simeq \int \frac{d^2 k}{(2\pi)^2} \left[\delta + 2 \frac{v_c}{q_c} k^2 + \alpha (q_c k)^2 \cos^2 \phi \right]^{-1},$$

where $\delta = \omega_c \omega + v_c (q - q_c)$, and ϕ is the angle between $\boldsymbol{k}$ and $\boldsymbol{q}$. Integration over the angle gives

$$\delta A \simeq \delta B \simeq \int \frac{k \, dk}{[\delta(\delta + \beta k^2)]^{1/2}}.$$

This integral contains no singularity at the lower limit for small δ, which is in contrast to the 3D case, where additional power of k in the above integral leads to the square-root singularity, so that there δA and δB would be about $\sqrt{\delta}$ [11]. In two dimensions, as we now see, both $A(\boldsymbol{q}\omega)$ and $B(\boldsymbol{q}\omega)$ are regular functions near such a type of the threshold and can be expanded in powers of $(q - q_c)$ and $(\omega_c - \omega)$. The corresponding equation for the spectrum $\omega(q)$ near the threshold would then have the general form

$$(\omega_c - \omega) + v_c(q - q_c) + \alpha_1 (q - q_c)^2 + \beta_1 (\omega_c - \omega)^2 = 0.$$

The solution of this equation reads

$$\omega \simeq \omega_c + v_c(q - q_c) + (\alpha_1 + v_c^2 \beta_1)(q - q_c)^2, \tag{8}$$

where, in order to satisfy the condition that, for $q > q_c$ the spectrum lies below the continuum of two-quasi-particle states, one has to require that α_1 and β_1 are such that

$$\alpha_1 + v_c^2 \beta_1 < \frac{1}{4} \left(\frac{\partial^2 \varepsilon}{\partial k^2} \right)_{|k = q/2}.$$

However, for momenta $q < q_c$ the same solution, Eq. (8), would then give the energy to be also less than that of the two-quasi-particle continuum. One may now conclude that, if such a point existed, then the spectrum could be continued beyond q_c, and the excitation energy would remain real (this differs from the case 3D, in which the root singularity does not allow continuation [11]). Therefore, the only point at which the decay may occur, and the spectrum may terminate, is that of type (i). For our model this means that the large-q part of the pair excitation spectrum starts at $q = (8m\mu)^{1/2}$ and $\omega(q_0) = 2\Delta$ (see Fig. 1(b)). It is important to emphasize that this conclusion about the impossibility of the decay of an excitation into two quasi-particles with parallel momenta k_1 and k_2 and non-zero velocities (unless the quasi-particle spectrum has minima at k_1 and k_2) is general for 2D. Indeed, as was shown by Pitaevskii [11], singularities associated with the corresponding denominators in Eq. (3) and Eq. (4) are the only ones to appear in general equations for the Green function near such a decay threshold. This result can be applied, for example, to the two-dimensional repulsive Fermi gas, where existing (for $q < 2k_F$) two-hole bound states merge with the continuum at $q = (8m\mu)^{1/2}$ [12]. One should mention, however, that there may exist another instability in the excitation spectrum, associated with the possibility of a decay into two excitations, one of which is a phonon with zero momentum [11]. In the present model this could be the case of a decay of a pair excitation into two other pair excitations, one of which is to be a long-wavelength 'phonon'. Such a process, which is not present in our Gaussian form of the action, will be discussed elsewhere [13].

5 Summary

In conclusion, collective (pair) excitations in the ground state of a dilute $(k_F R \ll 1)$ two-dimensional attractive Fermi gas have been studied by analyzing the linearized equations for the pair fluctuations about the non-trivial saddle point. The latter takes into account the non-perturbative effects of pair formation and condensation, and includes, to first order, the effect of the repulsive interaction between pairs in the condensate. The range of the effective repulsive interaction between pairs is of the order of ξ_0. The long-wavelength excitations are sound-like for all values of $\varepsilon_0/\varepsilon_F$. In the strong-coupling regime the excitation spectrum is continuous for all q, and has the Bogolubov form. In the weak-coupling limit the low-energy part of the spectrum has a cut-off at $q_c \simeq 1/\xi_0$, with $\omega(q_c) = 2\Delta$. For $q \geq (8m\mu)^{1/2}$ there exist pair excitations of energy $\omega \geq 2\Delta$, approaching that of a two-particle bound state in vacuum at large q. It is demonstrated that, in 2D, in contrast with the 3D case, there is no decay, at $T = 0$, of a pair excitation into two quasi-particles, which 'coalesce' with each other, except for the special case in which the momentum of the initial excitation equals $(8m\mu)^{1/2}$. Further work is needed to construct a corresponding perturbation theory for the pair fluctuations investigated here in the Gaussian approximation.

Acknowledgements

The author is grateful to A. S. Alexandrov, M. A. Baranov, B. L. Gyorffy, and J. B. Staunton for useful and helpful discussions, and acknowledges financial support from the Royal Society.

References

[1] S. Schmitt-Rink, C. M. Varma, and A. E. Ruckenstein, *Phys. Rev. Lett.* **63**, 445 (1989).

[2] K. Miyake, *Prog. Theor. Phys.* **69**, 1794 (1983); M. Randeria, J. Duan, and L. Shieh, *Phys. Rev. Lett.* **62**, 981 (1989); *Phys. Rev. B* **41**, 327 (1990).

[3] H. Fukuyama, Y. Hasegawa, and O. Narikiyo, *J. Phys. Soc. Japan* **60**, 2013 (1991).

[4] A. Tokumitu, K. Miyake, and K. Yamada, *J. Phys. Soc. Japan* **60**, 380 (1991); *Phys. Rev. B* **47**, 11 988 (1993).

[5] V. N. Popov, *Theor. Mater. Phys.* **11**, 565 (1972); D. S. Fisher and P. C. Hohenberg, *Phys. Rev. B* **37**, 4936 (1988). For zero temperature the model was investigated by M. Schick, *Phys. Rev. A* **3**, 1067 (1971).

[6] A. S. Alexandrov and S. G. Rubin, *Phys. Rev. B* **47**, 5141 (1993).

[7] C. A. R. Sá de Melo, M. Randeria, and J. R. Engelbrecht, *Phys. Rev. Lett.* **71**, 3202 (1993).

[8] V. N. Popov, *Functional Integrals in Quantum Field Theory and Statistical Physics* (Reidel, Dordrecht, 1983).

[9] V. M. Galitskii, *Sov. Phys. JETP* **7**, 698 (1958).

[10] A. S. Alexandrov, D. A. Samartchenko, and S. V. Traven, *Sov. Phys. JETP* **66**, 567 (1987).

[11] L. P. Pitaevskii, *Sov. Phys. JETP* **9**, 830 (1959).

[12] J. R. Engelbrecht and M. Randeria, *Phys. Rev. B* **45**, 12 419 (1992).

[13] S. V. Traven, to be published.

27

Strong two-band electron self-trapping, state hybridization effects and related pressure-induced phenomena in semiconductors

M. I. KLINGER and S. N. TARASKIN

Department of Chemistry, University of Cambridge, Lensfield Road Cambridge CB2 1EW, UK

Abstract

A review is presented of a recent theory of strong two-band electron self-trapping in a semiconductor, for which hybridization of the related electron state with the band states is essential and gives rise to new features of both electron and atomic dynamics. Pressure-induced phenomena in such materials predicted in the theory are discussed.

1 Introduction

As demonstrated in many works (see, e.g. [1–9]), the type of self-trapping of quasi-particles, e.g., electrons or holes in semiconductors, depends on properties of the materials. Whatever the origin of self-trapping, in most papers [1–4, 9] generally important contributions come from states of a single energy band, e.g. of the conduction band for electrons. Hence, single-band self-trapping has largely been considered in most works, which holds true, insofar as the characteristic self-trapping energy W_{ST} (<0) is substantially less in magnitude than the interband, or mobility, gap width $E_g^{(0)}$, $|W_{ST}| \ll E_g^{(0)}$. However, there are realistic semiconducting materials in which two-band self-trapping occurs, in the sense that contributions from states of both conduction and valence bands are important and hybridization of the states in the gap gives rise to new effects [5–8]. For instance, a single-band self-trapping energy for a single electron in a harmonic atomic lattice $W_{ST} = W_1 \simeq -W$, $W \partial Q_d^2/(2ka_0^2)$ may be comparable in magnitude to $E_g^{(0)}/2$, as self-trapping occurs at a soft 'defect' exhibiting a small effective atomic spring constant $k \ll k_0$, for typical values $Q_d \approx 3$–5 eV, $E_g^{(0)} \approx 1$–3 eV, and $k_0 \approx 30$–50 eV Å^2 (hole self-trapping can be treated in a similar way, with trivial substitutions of conduction band states for valence band ones and vice versa). In those cases the theory includes as a basic parameter $E_g^{(0)}$, unlike the theory of single-band self-trapping. A realistic example is two-band self-trapping in a glassy semiconductor, which, as much as any glass, is expected to contain a high atomic concentration $c_d \leq 0.1$ of soft 'defects', local soft atomic configurations with $k \leq k_0 \eta^*$, at $\eta^* \approx 0.1$–0.2. The latter can be responsible for atomic low-energy dynamics and low-temperature properties of glasses (see, e.g. [5]). Semiconductors of this type are mainly implied in what follows. Another example might be two-band self-trapping at a deep level soft defect in a narrow-gap crystalline semiconductor like PbTe.

In such materials self-trapping of a singlet electron pair turns out to be energetically most favourable, as the pair self-trapping energy W_2 is large enough in magnitude the pair correlation energy appears to be negative and comparable to the interband gap width,

$$U = W_2 - 2W_1 + U_d < 0, \tag{1}$$

for $|W_2| - 2|W_1| \simeq E_g^{(0)}/2$ and typical values $U_d \leq 0.3$ eV $\ll E_g^{(0)}$ of the Hubbard repulsion at the 'defect' in question (see, e.g. [10, 15, 5]).

For two-band self-trapping, inter-band interactions and related states' hybridization in the gap are important, and in some respects decisive. In fact, the matrix elements V_{id} of hybridization between the bare 'defect' electron state ψ_d, of energy ε_d, under self-trapping and largely extended band states ψ_i, of energy ε_i, substantially contribute to the true, renormalized state

$$\Psi_{d\sigma} = C_{dd}^{(\sigma)} \psi_d + \sum_i C_{id}^{(\sigma)} \psi_i, \tag{2}$$

with $|C_{dd}^{(\sigma)}|^2 \partial \gamma_{d\sigma}^2 = 1 - \Sigma_i |C_{id}^{(\sigma)}|^2$. This is the case as the energy level $E_{d\sigma}$ of the state $\Psi_{d\sigma}$ is close to the (mobility) edge of the alternative band of extended states, e.g. of the valence band for electrons.

The main purpose of the present paper is to discuss a recent theory of strong two-band electron self-trapping in semiconductors [6–8]. Another purpose is to describe briefly some implications of the theory, concerning pressure-induced phenomena related to two-band self-trapping and predicted even for not very high hydrostatic pressures $p \approx 10^4$–10^5 bar [12]. In what follows, the discussion is confined to low temperatures T for which temperature-dependent effects are negligible.

2 The model: basic relations and approximations

The system under discussion, electrons interacting with a local 'defect' atomic configuration, is modelled by electrons at the 'defect', which interact with a single representative, most important, atomic motion mode x. For a glassy semiconductor the latter is the mode (soft mode) characteristic of a soft configuration [5]. The model Hamiltonian is as follows [6, 7]:

$$\hat{H} = \hat{V}_{at}(x) + \hat{H}_{el}(x) \equiv \hat{V}_{at}(x) + \hat{H}_{el}^{(0)} + \hat{H}_{el-at}(x), \tag{3}$$

where, for actual, not very large displacements $|x| \equiv |u|/a_0 \leq 1$

$$V_{at} \simeq A(\eta x^2 + \xi x^3 + \gamma x^4), \tag{4}$$

with $\gamma = 1$, $A = k_0 a_0^2/2 \approx 15\text{--}25$ eV and $\{|\eta|, \xi^2\} \leq 1$;

$$\hat{H}_{el}^0 \equiv \hat{H}_{HA} = \sum_{i,\sigma} \varepsilon_i \hat{n}_{i\sigma} + \sum_{\sigma} \varepsilon_d^{(0)} \hat{n}_{d\sigma} + \frac{1}{2} U_d \sum_{\sigma \neq \sigma} \hat{n}_{d\sigma} \hat{n}_{d\sigma'} + \sum_{i,\sigma} \{V_{id}^{(0)} \hat{a}_{i\sigma}^+ \hat{d}_\sigma + \text{H.c.}\}, \tag{5}$$

with $\hat{n}_{j\sigma} \equiv \hat{a}_{j\sigma}^+ \hat{a}_{j\sigma}$ and $\{\hat{a}_{j\sigma}, \hat{a}_{j\sigma}^+\}$ the electron operators, $j \equiv (i, d)$, $\hat{a}_{d\sigma} \equiv \hat{d}_\sigma$; and

$$\hat{H}_{el-at} = \Delta\varepsilon_d(x) \sum_\sigma \hat{n}_{d\sigma} + \sum_{i,\sigma} \{\Delta V_{id}(x) \hat{a}_{i\sigma}^+ \hat{d}_\sigma + \text{H.c.}\}, \tag{6}$$

with $\quad \Delta\varepsilon_d(x) \equiv \varepsilon_d - \varepsilon_d^{(0)} \equiv \varepsilon_d(x) - \varepsilon_d(0) \simeq -Q_d x \quad$ and $\quad \Delta V_{id}(x) \equiv V_{id} - V_{id}^{(0)}$ $\equiv V_{id}(x) - V_{id}(0) \simeq Q_{id} x$. The substrate atomic potential energy $V_{at}(x)$ is that of a local 'defect' configuration (effectively neutral or singly positively charged) with respect to displacements in the mode x and, generally speaking, is anharmonic. The basic parameters $\{\eta, \xi\}$ generally can vary in the range $\{|\eta|, \xi^2\} \leq 1$. For soft configurations in a glassy semiconductor η and ξ are random parameters of either sign, varying in the range $\{|\eta|, \xi^2\} \leq \eta^* \approx 0.1\text{--}0.2$ and obeying a distribution density $F(\eta, \xi)$, which exhibits a main maximum for most atoms at $\eta = \bar{\eta} \approx 1$ and $\xi = \bar{\xi} \approx 0$ with nearly harmonic $V_{at} \simeq V_{at}^{(0)}(x)$ $= A\eta x^2 \equiv kx^2 a_0^2/2$ and $k \approx k_0$. The operator $\hat{H}_{HA}^{(0)}$ is the Haldane–Anderson Hamiltonian [13]. It takes into account the hybridization ($V_{id}^{(0)}$) of a bare defect state ($\psi_d^{(0)}, \varepsilon_d^{(0)}$) with the states of both bands (ψ_i, ε_i) in an atomic configuration with fixed positions at $x = 0$. In what follows the bare state ψ_d is assumed to be orbitally non-degenerate, so it can be occupied by n electrons at $n = 0$, 1 or 2. This assumption not only makes the theory simpler but also appears to be relevant for amorphous semiconductors like glassy semiconductors. It is worth noting that Coulomb interactions of electrons in the band states ψ_i, as well as of 'defect' electrons (ψ_d) with electrons in the band states, are not taken into account explicitly. These interactions, and the associated polarization of the 'defect' environment in the host 'lattice' when the 'defect' is charged, appear to reduce the bare Hubbard energy to U_d in Eq. (5) with typical relatively small values $U_d < E_g^{(0)}$ in semiconductors [10, 11].

Finally, $\hat{H}_{el-at}(x)$ describes the electron–mode interaction, both diagonal ($\Delta\varepsilon_d(x)$) and non-diagonal ($\Delta V_{id}(x)$), which as usual are approximated linearly in x. The higher order expansion terms (up to x^4) might be accounted for in renormalization by, e.g., $|\delta\eta| \approx O(Q_d/A) \ll 1$, $|\delta\xi| \approx O(Q_d/A)$ and $|\delta\gamma| \simeq Q_d/A$ of the basic parameters $\{\eta, \xi, \gamma = 1\}$, not changing the order-of-magnitude estimations.

The bare electron state ψ_d is considered, as in [13], as a site 'defect' state of which the energy level $\varepsilon_d^{(0)}$ in the reference 'defect' configuration at $x = 0$ belongs either to the gap or to the conduction band, not too high above its (mobility) edge E_c [7],

$$E_v \equiv 0 < \varepsilon_d^{(0)} < E_c \equiv E_g^{(0)} \qquad \text{or} \qquad E_c < \varepsilon_d^{(0)} < E_c + W,$$

with the single-band self-trapping energy magnitude $W \ll E_g^{(0)}$.

Generally speaking, the bare energy $\varepsilon_d \equiv \varepsilon_d(x) \simeq \varepsilon_d^{(0)} - Q_d x$ and hybridization matrix elements $V_{id} \equiv V_{id}(x) \simeq V_{id}^{(0)} + Q_{id} x$ are random quantities depending on x, obeying appropriate probability distribution densities $P_0(\varepsilon_d^{(0)}, Q_d)$ and $P_1(V_{id}^{(0)}, Q_{id})$, which reduce to δ-function distribution densities like $\delta(\varepsilon_d^{(0)} - \bar{\varepsilon}_d^{(0)})\delta(Q_d - \bar{Q}_d)$ and $\delta(V_{id}^{(0)} - \bar{V}_{id}^{(0)})\delta(Q_{id} - \bar{Q}_{id})$ for a defect in a crystalline lattice. For disordered systems, like glassy semiconductors, it is assumed in what follows that

$$P_0(\varepsilon_d^{(0)}, Q_d) \simeq P_0'(\varepsilon_d^{(0)})P_0''(Q_d) \qquad \text{and} \qquad P_1(V_{id}^{(0)}, Q_{id}) \simeq P_1'(V_{id}^{(0)})P_1''(Q_{id}),$$

where $P_0''(Q_d)$ and $P_1''(Q_{id})$ are narrow distributions around some $\bar{Q}_d$ and $\bar{Q}_{id}$, whereas $P_0'(\varepsilon_d^{(0)})$ varies weakly for $\varepsilon_d^{(0)}$ around E_c and drops rapidly in the mobility gap for $E_c - \varepsilon_d^{(0)} > w_t$, the band tail width, and $P_1'(V_{id}^{(0)})$ is largest around a characteristic $\bar{V}_{id}^{(0)}$. It is implied that

$$|Q_d| \approx |\bar{Q}_d| < D_{c,v}, \qquad \text{e.g., } |Q_d| \lesssim 3\text{--}5 \text{ eV},$$
$$V_{id}^{(0)} \neq 0 \qquad \text{or} \qquad V_{id}^{(0)} = 0, \qquad \text{while } |Q_{id}| < |Q_d|,$$

where $D_{c,v}$ stands for the conduction/valence band width, which actually exceeds $E_g^{(0)}$ for the semiconductors in question, e.g. glassy semiconductors.

The basic problem is to calculate the correlation energy U of a self-trapped singlet electron-pair state in a 'defect' atomic configuration and to find out whether and under what conditions U is negative for the system under discussion, for which states hybridization is essential or even decisive. The correlation energy can be expressed in terms of adiabatic potentials $\Phi_n(x)$ of the configuration distorted at self-trapping in the bare 'defect' state for different nominal electron occupation $n = 0$, 1 or 2 [7, 14]:

$$U = \Phi_2(\bar{x}_2) + \Phi_0(\bar{x}_0) - 2\Phi_1(\bar{x}_1) \equiv U(\eta, \xi). \tag{7}$$

A related problem is to calculate the self-trapping energies

$$W_n = \Phi_n(\bar{x}_n) - \Phi_n(0) \equiv W_n(\eta, \xi), \tag{8}$$

which are negative as self-trapping occurs. The potential minima x_n, including the equilibrium self-trapping displacements $\bar{x}_n$ along the mode x, are found as usual from the equation

$$\frac{\mathrm{d}\Phi_n(x)}{\mathrm{d}x} = 0 \qquad \text{at } x = x_n \equiv x_n(\eta, \xi). \tag{9}$$

The Green function method is applied and the total electron energy $E_{\mathrm{el}}^{(n)}(x)$ of the system (3) is calculated, in the Hartree–Fock mean-field approximation taking into account the Hubbard interactions. Then the total electron Hamiltonian in (3) is approximated in a standard way as [7, 13]

$$\hat{H}_{\mathrm{el}}(x) \equiv \hat{H}_{\mathrm{el}}[\varepsilon_{d\sigma}(x), V_{id}(x)] \simeq \hat{H}_{\mathrm{HF}}(x) \equiv \hat{\tilde{H}}_{\mathrm{el}}[\varepsilon_{d\sigma}^{\mathrm{eff}}(x), V_{id}(x)] - \delta E, \tag{10}$$

where

$$\hat{\tilde{H}}_{\mathrm{el}}[\varepsilon_d, V_{id}] \equiv \hat{H}_{\mathrm{el}}[\varepsilon_d, V_{id}] - \frac{1}{2} U_d \sum_{\sigma \neq \sigma'} \hat{n}_{d\sigma}\hat{n}_{d\sigma'}, \qquad \delta E \equiv \frac{1}{2} U_d \sum_{\sigma \neq \sigma'} v_{d\sigma}v_{d\sigma'}.$$

Here

$$\varepsilon_{d\sigma}^{\mathrm{eff}} = \varepsilon_{d\sigma}^{\mathrm{eff}}(x) = \varepsilon_d(x) + U_d v_{d,} - \sigma(x) \tag{11}$$

is the effective single-particle energy level substituted for the bare energy level $\varepsilon_{d\sigma}(x)$ in $\hat{H}_{\mathrm{el}}(x)$, so that $\hat{H}_{\mathrm{HF}}(x)$ is bilinear in the electron operators, whereas $v_{d\sigma}(x)$ is the true occupation of the bare 'defect' state (ψ_d). This approximation is assumed to be relevant, since in the semiconductors in question the Hubbard energy $U_d \ll E_g^{(0)}$ [10, 11, 5] while the resulting pair correlation energy U is negative and much larger in magnitude, $|U| \simeq E_g^{(0)}$, for most of the pairs in the soft 'defect' configurations in question. In the Green function approach the electron-energy contribution $E_{\mathrm{el}}^{(n)}(x)$ to the adiabatic potential $\Phi_n(x)$ and the true occupations $v_{d\sigma}$ are described by the following relations [7, 13]:

$$\Delta\Phi_n(x) = E_{\mathrm{el}}^{(n)}(x) \equiv \Phi_n(x) - V_{\mathrm{at}}(x) = \sum_\sigma \int\limits_{(R_n)} \mathrm{d}\omega\, \omega\pi^{-1} \operatorname{Im} \operatorname{Tr} \hat{G}^{(\sigma)}(\omega) - \delta E$$

$$= \sum_\sigma \int\limits_{(R_n)} \mathrm{d}\omega\, \omega\pi^{-1} \operatorname{Im} \hat{G}_{dd}^{(\sigma)}(\omega) \left(1 - \frac{\mathrm{d}\Sigma_d'(\omega)}{\mathrm{d}\omega}\right) - \delta E, \tag{12}$$

$$v_{d\sigma}(x) \equiv \langle n_{d\sigma}(x)\rangle = \int\limits_{(R_n)} \mathrm{d}\omega\, \pi^{-1} \operatorname{Im} \hat{G}_{dd}^{(\sigma)}(\omega). \tag{13}$$

Here (R_n) stands for the range of energies ω of the occupied states only, and the following formula is applied [13]:

$$\mathrm{Tr}\, \hat{G}^{(\sigma)}(\omega) = \hat{G}_{dd}^{(\sigma)}(\omega)\left(1 - \frac{d\Sigma_d'}{d\omega}\right).$$

In Eq. (12) and Eq. (13),

$$\hat{G}_{dd}^{(\sigma)}(\omega) = \langle\psi_d|\hat{G}^{(\sigma)}(\omega)|\psi_d\rangle = (\omega - \varepsilon_{d\sigma}^{\mathrm{eff}} - \Sigma_d)^{-1} \tag{14}$$

is the respective diagonal matrix element of the Green function operator for the Hamiltonian (10), whereas the self-energy part

$$\Sigma_d \equiv \Sigma_d(\omega, x) = \sum_i |V_{id}(x)|^2 (\omega - \varepsilon_i)^{-1} = \int d\varepsilon\, \Delta(\varepsilon, \omega)(\omega - \varepsilon)^{-1}, \tag{15}$$

with

$$\Sigma_d' \equiv \mathrm{Re}\,\Sigma_d, \qquad \Sigma_d'' \equiv \mathrm{Im}\,\Sigma_d,$$

is determined by the effective hybridization energy

$$\Delta(\varepsilon, \omega) = g_0(\varepsilon)|V(\varepsilon, \omega; x)|^2 \equiv \Delta(\varepsilon, \omega; x), \tag{16}$$

with $V_{id}(x) \equiv V(\varepsilon, \omega; x)$ and $g_0(\varepsilon)$ the band electron density of states (DOS). Moreover, an important characteristic of the two-band self-trapping is the degree of hybridization of the bare 'defect' state with band states (see Eq. (2))

$$0 \le \gamma_{d\sigma}^2(x) \equiv |C_{dd}^{(\sigma)}|^2 = \int_{E_v}^{E_c} d\omega\, \pi^{-1}\,\mathrm{Im}\,\hat{G}_{dd}^{(\sigma)}(\omega) \simeq \left(1 - \frac{d\Sigma_d'}{d\omega}\right)^{-1}\bigg|_{\omega = E_{d\sigma}(x)}. \tag{17}$$

The true, hybridization renormalized single-electron energy level $E_{d\sigma} \equiv E_{d\sigma}(x)$ is the pole of $\hat{G}_{dd}^{(\sigma)}(\omega; x)$ in the (mobility) gap and can essentially differ from the effective energy level $\varepsilon_{d\sigma}^{\mathrm{eff}}(x)$ for large displacements $|x| \simeq 1$:

$$E_{d\sigma} - \varepsilon_{d\sigma}^{\mathrm{eff}}(x) - \Sigma_d'(E_{d\sigma}) = 0, \tag{18}$$

where $\Sigma_d'(\omega) \equiv \mathrm{Im}\,\Sigma_d(\omega) = 0$ for $E_v \equiv 0 < \omega = E_{d\sigma} < E_c \equiv E_g^{(0)}$. The last approximate relation on the right-hand side of (17) just follows from Eq. (14) and Eq. (18).

Some remarks are worth adding. Eqs. (7)–(9) take into account adiabaticity of the self-trapping electron motion with respect to atomic motion in the mode x, as the related parameter

$$\frac{\varepsilon_{\mathrm{exc}}}{|E_{d\sigma}(x) - \varepsilon_d^{(0)}|} \approx \frac{\varepsilon_{\mathrm{exc}}}{E_g^{(0)}} \ll 1, \tag{19}$$

for actual large ST displacements $|x| \approx |\bar{x}_n| \simeq 1$ and low energies $\varepsilon_{\mathrm{exc}}(< \hbar\omega_D)$ of the motion-mode excitations [5, 7]. Since V_{id} and ε_d depend on x, v_d and Σ_d are also functions of x, so the adiabaticity of the atomic motion (x) with respect to variations of the state occupation that is often supposed for self-trapping electrons (see e.g., [14]) is violated in this sense. The degree of hybridization $\gamma_{d\sigma}^2(x)$ is the probability $|C_{dd}^{(\sigma)}|^2$ of finding the bare 'defect'-state electron in the cloud of the true state $\Psi_{d\sigma}$ (Eq. (2)) with energy $E_{d\sigma}$ in the gap. In particular, $\gamma_{d\sigma}^2(x) \simeq 0$ corresponds to nearly complete hybridization at large displacements $|x| \simeq 1$ while $\gamma_{d\sigma}^2(x) \simeq 1$ corresponds to negligible hybridization at small x. Finally, unlike the effective energy level $\varepsilon_{d\sigma}^{\text{eff}}(x)$, the true level is not related to the defect state occupation so it is sensitive neither to the nominal occupation nor to the true one.

Solving the problem of the strong two-band self-trapping characteristics and related hybridization effects is reduced to calculations of $E_{\text{el}}^{(n)}(x)$, $E_{d\sigma}(x)$ and $v_d(x)$, $\gamma_{d\sigma}^2(x)$, as well as of $\bar{x}_n(\eta, \xi)$. The basic parameters of the theory, besides $\{\varepsilon_d^{(0)}/E_g^{(0)}, U_d/E_g^{(0)}, \Delta(\varepsilon, \omega; x=0)/E_g^{(0)}\}$ [13], are also [7, 16]

$$\{\Delta(\varepsilon, \omega; x)/E_g^{(0)}, Q_d/E_g^{(0)}, A/E_g^{(0)}\}. \tag{20}$$

The calculations are performed by applying the following approximation for $\Delta(\varepsilon, \omega; x)$ [7, 13]:

$$\Delta(\varepsilon, \omega; x) \simeq \Delta_v(x)\theta(\varepsilon - E_v')\theta(E_v - \varepsilon) + \Delta_c(x)\theta(E_c' - \varepsilon)\theta(\varepsilon - E_c), \tag{21}$$

where $\theta(z) \equiv \{1, \text{ at } z>0; \ 0 \text{ at } z<0\}$, while $\Delta_{v,c}(x)$ are, generally speaking, x-dependent characteristics of extended states of the valence and conduction bands, with $E_v' \equiv E_v - D_v$, $E_c' \equiv E_c + D_c$ and $D_{c,v}$ the band widths. It is assumed in fact in Eq. (21), in accordance with [13], that details of the dependence of $\Delta(\varepsilon, \omega; x)$ on the extended state energy ε do not affect the basic features and estimations obtained. Moreover, hybridization of the band tail states with the 'defect' state of energy ω in the gap are argued [8] to be relatively unimportant for the features of two-band self-trapping, so the ω-dependence of Δ can approximately be neglected in Eq. (21). Typical values of $\Delta_{c,v}(x)$ can be estimated as [7]

$$\Delta_{c,v}(x) \approx N|V_{id}(x)|^2 g_0(E_{c,v}) \approx |V_1|^2/D_{c,v} \approx 0.01\text{--}0.1 \text{ eV}, \tag{22}$$

with $V_{id} \equiv V_1 N^{-1/2}$, $|V_1| \approx 0.3\text{--}1.0$ eV and $g_0(E_{c,v}) \simeq 1/D_{c,v}$, $D_{c,v} \approx 5\text{--}10$ eV (N stands for the total number of atoms in the system in question).

Straightforward calculations of $E_{\text{el}}^{(n)}(x)$, $E_{d\sigma}(x)$ and $v_{d\sigma}$, $\gamma_{d\sigma}^2(x)$, as well as a transformation of Eq. (9) give rise to the following relations [7, 16]:

$$\Phi_n(x) \simeq V_{\text{at}}(x) + E_{\text{VB}}^{(0)} - \frac{1}{2} U_d \sum_\sigma v_{d\sigma} v_{d-\sigma} + \Delta E_{\text{VB}}(x) + \sum_{\sigma(\text{occupied})} E_{d\sigma}(x), \tag{23}$$

$$v_d(x) = \sum_\sigma v_{d\sigma}(x) \simeq \frac{1}{\pi} \sum_\sigma \int_{E'_{\mathrm{v}}}^{E_{\mathrm{v}} \equiv 0} \mathrm{d}\omega \left(-\frac{\mathrm{d}\phi_\sigma(\omega;x)}{\mathrm{d}\varepsilon_{d\sigma}^{\mathrm{eff}}(x)} \right) + \sum_{\sigma(\text{occupied})} \gamma_{d\sigma}^2(x), \qquad (24)$$

$$\gamma_{d\sigma}^2(x) \simeq \left\{ 1 + \frac{\Delta_{\mathrm{v}} D_{\mathrm{v}}}{(\omega - E_{\mathrm{v}})(\omega - E'_{\mathrm{v}})} + \frac{\Delta_{\mathrm{c}} D_{\mathrm{c}}}{(E_{\mathrm{c}} - \omega)(E'_{\mathrm{c}} - \omega)} \right\}_{E_{\mathrm{v}} < \omega = E_{d\sigma}(x) < E_{\mathrm{c}}}, \qquad (25)$$

$$\frac{\mathrm{d}V_{\mathrm{at}}(x)}{\mathrm{d}x} + v_d(x) \frac{\mathrm{d}\varepsilon_d(x)}{\mathrm{d}x} = 0 \qquad \text{at } x = x_n(\eta, \xi), \qquad (26)$$

where

$$\Delta E_{\mathrm{VB}}(x) \simeq -\frac{1}{\pi} \sum_\sigma \int_{E'_{\mathrm{v}}}^{E_{\mathrm{v}} \equiv 0} \mathrm{d}\omega\, \phi_\sigma(\omega;x) \simeq -\sum_\sigma (E_{d\sigma}(x) - \varepsilon_{d\sigma}^{\mathrm{eff}}(x)),$$

$$\phi_\sigma(\omega;x) = \frac{\pi}{2} + \arctan\left(\frac{\omega - E_{d\sigma}^{\mathrm{eff}}(x) - \Sigma'_d(\omega)}{\pi \Delta_{\mathrm{v}}} \right), \qquad (27)$$

$$\Sigma_d(\omega) - \Delta_{\mathrm{v}} \ln\left| \frac{\omega - E'_{\mathrm{v}}}{E_{\mathrm{v}} - \omega} \right| + \Delta_{\mathrm{c}} \ln\left| \frac{E_{\mathrm{c}} - \omega}{\omega - E'_{\mathrm{c}}} \right| + i\pi\Delta_{\mathrm{v}}\theta(E_{\mathrm{v}} - \omega)\theta(\omega - E'_{\mathrm{v}}) \qquad (28)$$

and $E_{\mathrm{VB}}^{(0)}$ is the total energy of the valence-band states in the absence of hybridization. The last term in Eq. (23) stands for the total energy of occupied true 'defect' energy levels calculated from Eq. (18) and Eq. (28), whereas $\Delta E_{\mathrm{VB}}(x)$ accounts for the gain in total energy of the valence-band states due to hybridization with the 'defect' state (the last expression on the right-hand side of Eq. (27) results when neglecting contributions of the conduction-band states to $\Sigma_d(\omega)$ in Eq. (28)). In fact, as $E_{d\sigma}(x)$ approaches E_{v} at large displacements $|x| \approx x_{\mathrm{g}} = E_{\mathrm{g}}^{(0)}/|Q_d|$, the conduction-band states hardly contribute to $\Phi_n(x), v_d(x), \gamma_{d\sigma}^2(x)$, i.e. to $\Sigma_d(\omega = E_{d\sigma})$. For this case,

$$\gamma_{d\sigma}^2(x) \simeq \left(1 + \frac{\Delta}{|E_{d\sigma}(x) - E_{\mathrm{v}}|} \right)^{-1}, \qquad (29)$$

corresponding to practically complete hybridization with the valence-band states at $0 < E_{d\sigma}(x) - E_{\mathrm{v}} < \Delta$ while to negligible hybridization at $E_{d\sigma}(x) - E_{\mathrm{v}} \simeq \varepsilon_{d\sigma}^{\mathrm{eff}}(x) - E_{\mathrm{v}} \gg \Delta \equiv \Delta_{\mathrm{v}}$ (see Eq. (19)).

3 Results of calculations and discussion

The equations (18) and (26) have been numerically solved and the basic characteristics (7), (8) and (23)–(25) have been calculated for self-trapped states

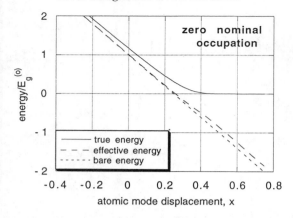

Fig. 1. Dependences of bare $(\varepsilon_d(x))$, effective $(\varepsilon_{d\sigma}^{\mathrm{eff}}(x))$ and true $(E_{d\sigma}(x))$ energies on x at $\varepsilon_d^{(0)}/E_g^{(0)}=1$, $Q_d/E_g^{(0)}=4$, $Q_{id}=0$, $V_{id}^{(0)}/E_g^{(0)}=0.7$, $U_d/E_g^{(0)}=0.15$ and $D_v/E_g^{(0)}=5$.

at $x = \bar{x}_n(\eta, \xi)$ and different nominal occupations $n = 0$, 1 or 2 of the 'defect' state [7]. In what follows the main results are briefly presented and discussed, mainly for a nominally free $(n = 0)$ 'defect' state as an example most characteristic of the peculiarities of two-band self-trapping and related hybridization effects. The latter are most pronounced for substrate soft configurations (4) for which the anharmonic features are essential, limiting the equilibrium self-trapping displacements to realistic $|\bar{x}_n| \leq 1$. On the other hand, analytical approximations are derived only for harmonic substrate configurations, and their estimations are found to be close in scale to those for anharmonic configurations. In this connection, some formulae for the harmonic substrate configurations are applied to interpret the self-trapping in question and for some estimations.

Three most important effects can be revealed, which are not characteristic of single-band self-trapping and are due to inter-band interactions and hybridization of states [6–8, 16].

First, as the electron level $\varepsilon_{d\sigma}^{\mathrm{eff}}(x) = \varepsilon_{d\sigma}(x) + U_d v_{d,-\sigma}(x)$ approaches under self-trapping the (mobility) edge E_v, increasing hybridization of states gives rise to the true level $E_{d\sigma}(x)$ being repelled away from E_v. As can be seen from Fig. 1, hybridization is negligible as $\varepsilon_{d\sigma}^{\mathrm{eff}}(x)$ is far from E_v,

$$E_{d\sigma}(x) - E_v \simeq \varepsilon_{d\sigma}^{\mathrm{eff}}(x) - E_v \gg \varDelta, \qquad v_{d\sigma}(x) = 1 - \gamma_{d\sigma}^2(x) \ll 1, \qquad (30)$$

whereas it becomes decisive as $\varepsilon_{d\sigma}^{\mathrm{eff}}(x)$ approaches E_v and penetrates the band, $\varepsilon_{d\sigma}^{\mathrm{eff}}(x) \leq E_v$. In the latter case, $E_{d\sigma}(x)$ is stopped near E_v,

$$E_v < E_{d\sigma}(x) < E_v + \varDelta, \qquad v_{d\sigma}(x) \simeq 1. \qquad (31)$$

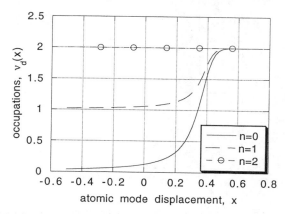

Fig. 2. The true occupation of the 'defect' state at different nominal occupations as a function of x (for the same parameter values as in Fig. 1).

In fact, the effective electron-mode coupling energy $Q_{d\sigma}^{(\mathrm{eff})}(x)$ can be defined, at least for large displacements and $E_{d\sigma}(x)$ close to E_v, as

$$Q_{d\sigma}^{(\mathrm{eff})}(x) \equiv \mathrm{d}E_{d\sigma}(x)/\mathrm{d}x \simeq \gamma_{d\sigma}^2 Q_d, \tag{32}$$

as follows from Eqs. (11), (18) and (31). It can be seen that $Q_{d\sigma}^{(\mathrm{eff})}(x) \ll Q_d$ and $\gamma_{d\sigma}^2(x) = 1 - v_{d\sigma}(x) \ll 1$ for large $|x|$, in accordance with Eq. (24). The essential decrease in $Q_{d\sigma}^{(\mathrm{eff})}(x)$ can be considered as the ultimate cause of the true energy level being repelled away from E_v. It follows from Eq. (18) and Eq. (28) that $E_{d\sigma}(x)$ approximately approaches the (mobility) edge E_v in an exponential way [6], e.g.

$$E_{d\sigma}(x) - E_v \approx D_v \exp\left(-\frac{E_v - \varepsilon_{d\sigma}^{\mathrm{eff}}(x)}{\varDelta}\right) \tag{33}$$

for $|\varepsilon_{d\sigma}^{\mathrm{eff}} - E_v| = -(\varepsilon_{d\sigma}^{\mathrm{eff}} - E_v) \gg \varDelta$.

Second, hybridization of states is also responsible for an increase in 'defect' state occupation from the nominal $n = 0$ to the true occupation $v_{d\sigma} > 0$ (Fig. 2). This corresponds to a change in equilibrium self-trapping displacement from $x_0 \simeq 0$ to a finite $x_0 \neq 0$, as follows from Eq. (26) and also discussed further. Roots of Eq. (26) are illustrated in Fig. 3 for different more or less realistic values of the substrate softness parameter η in Eq. (4), and for $\xi = \bar{\xi} \approx 0$ corresponding to the main maximum of the distribution density $F(\eta, \xi)$. A single root $x_0 = x_{01} \simeq 0$ only occurs for large enough $\eta > \eta_0^* \equiv \eta_{n=0}^*$, at weak hybridization, with $v_{d\sigma}(x_{01}) = 1 - \gamma_{d\sigma}^2(x) \approx \varDelta/E_g^{(0)} \ll 1$. However, an additional root $x_{02} \neq 0$ appears for smaller $\eta < \eta_0^*$, for a large displacement $|x_{02}| \simeq x_g$

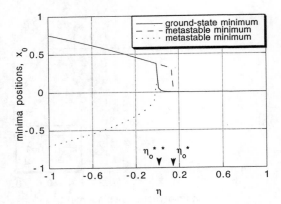

Fig. 3. Minima positions of the adiabatic potential $\Phi_0(x)$ depending on the softness parameter η at $\varepsilon_d^{(0)}/E_g^{(0)}=1$, $Q_d/E_g^{(0)}=4$, $Q_{id}=0$, $V_{id}^{(0)}/E_g^{(0)}=0.2$, $U_d/E_g^{(0)}=0.15$ and $D_v/E_g^{(0)}=5$, $A/E_g^{(0)}=40$ and with $\zeta=0$ and $\gamma=1$ (the same set of parameters is used in the next figures).

$=E_g^{(0)}/Q_d \lesssim 1$, which corresponds at first to a metastable minimum and, for still smaller η, to an equilibrium minimum, as discussed further. Thereby, the effective level $\varepsilon_{d\sigma}^{\text{eff}}(x)$ reaches E_v and penetrates the band, the true level $E_{d\sigma}(x)$ is repelled from E_v and nearly complete hybridization occurs, $\gamma_{d\sigma}^2(x) \ll 1$, so x_{02} should be close to the equilibrium self-trapping displacement x_2 for a nominally doubly occupied state ($n=2$) in the hybridization-free case:

$$x_{02} \simeq x_2, \qquad |x_{02}| \le 1, \tag{34}$$

for sufficiently soft configurations.

The characteristic value η_0^*, corresponding to appearance of an inflection point for $\Phi_0(x;\eta)$, is found to be $\eta_0^* \simeq 0.08$ for some representative values of parameters (20) used in Fig. 3. A value $\eta_{0h}^* \approx 0.10$ close in scale to η_0^* is obtained from an approximate expression

$$\eta_{0h}^* \approx Q_d/A \approx 0.1 \tag{35}$$

for a harmonic substrate configuration.

It is worthy of note that self-trapping of the nominally free 'defect' state ($n=0$) is entirely determined by hybridization. The latter originally is weak ($v_{d\sigma}(x) \simeq \Delta/E_g^{(0)} \ll 1$ for $|x| \ll 1$) but finite, and then increases with growing $|x|$ up to $x \approx x_{02}$, due to increasing contributions from extended valence-band states. The resulting self-trapped pair state is still localized, but its effective size ρ increases as compared to the bare state size $\rho_0 \approx (1-2)a_1$, a_1 being the average nearest neighbour separation ($a_1 \approx 2.0$ Å). A rough estimation of ρ/ρ_0 can be obtained from Eq. (29) and the approximate relation

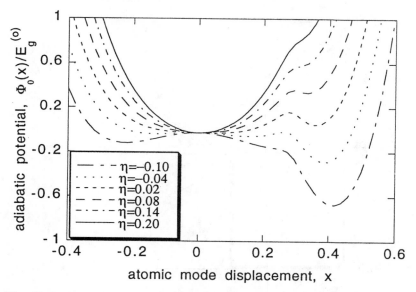

Fig. 4. The dependence of the adiabatic potential $\Phi_0(x)$ on x for different values of the softness parameter η.

$$\gamma_{d\sigma}^2(x_{02}) \approx \int_0^{\rho_0} dr\, 4\pi r^2 |\Psi_{d\sigma}(r|x=x_{02})|^2 \leq \frac{1}{6}\left(\frac{2\rho_0}{\rho}\right)^3, \tag{36}$$

with $\Psi_{d\sigma}(r|x_{02}) \equiv C_{d\sigma}\exp(-r/\rho)$ and $\psi_d = C_0\exp(-r/\rho_0)$, so that

$$1 < \rho/\rho_0 \approx [1 + \Delta/(E_{d\sigma}(x_{02}) - E_v)]^{1/3} \leq 5$$

for $1 < \Delta/(E_{d\sigma}(x_{02}) - E_v) \leq 10^2$.

Third, competition between increasing substrate atomic potential energy and electron energy gain due to growing hybridization, with increasing $|x|$, can generate additional anharmonic features of the adiabatic potential for sufficiently soft substrate configurations (small $|\eta| \ll 1$):

$$\Phi_0(x) = V_{at}(x) + E_{VB}^{(0)} - U_d(v_{d\sigma}(x))^2 - 2(E_{d\sigma}(x) - \varepsilon_{d\sigma}^{eff}(x)), \tag{37}$$

and thus of the related local atomic dynamics. A signature of the anharmonicity is softness and/or appearance of an extra potential minimum as a metastable one for an excited self-trapped state at $\eta < \eta_0^*$, and then as a lowest-energy minimum for a ground state at smaller $\eta < \eta_0^{**} < \eta_0^* \ll 1$ (Fig. 4). The value $\eta_0^{**} \simeq 0.01$ is found for the representative values of parameters (20) mentioned in Fig. 4. An approximate expression of η_{0h}^{**} for a harmonic substrate follows from [7]:

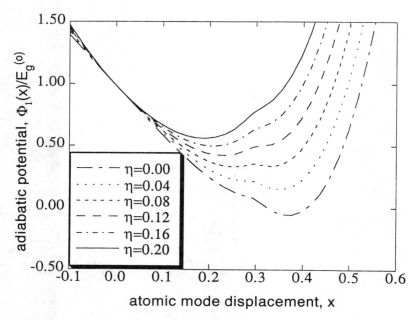

Fig. 5. The dependence of the adiabatic potential $\Phi_1(x)$ on x for different values of the softness parameter η.

$$\Phi_0(x_{01} \simeq 0) \simeq 2\Delta \ln (D_v/E_g^{(0)}) \simeq \Phi_0(x_{02} \simeq x_2) \simeq 2E_g^{(0)} - Q_d^2/A\eta_{0h}^{**} + U_d,$$
$$\eta_{0h}^{**} \simeq Q_d^2/[A(2E_g^{(0)} + U_d)], \tag{38}$$

so that again $\eta_{0h}^{**} \simeq 0.05$ is not far in scale from η_0^{**}.

One can conclude that the lowest-energy minimum position of the adiabatic potential depends on the value of η/η_0^{**}, unlike the universal single minimum at $x_0 = 0$ for single-band self-trapping. For $\eta < \eta_0^{**}$, an electron ground state related to a lowest-energy potential minimum at $x = x_{02}$ is the self-trapped state for a singlet electron pair $(v_d(x_{02}) \simeq 2)$, so self-trapping rather requires overcoming of the inter-well barrier. The anharmonic features introduced by hybridization of states include both an extra potential minimum and an additional softness for some intermediate values of η and x_{02} (Fig. 4). It is seen that actually η_0^{**} is very small, so the related anharmonic features appear only for very soft 'defect' configurations occurring in a glassy semiconductor.

The situation shown in Fig. 5 for a nominally singly occupied 'defect' state $(n = 1)$ is mainly similar to that for $n = 0$. The main difference is that the respective η_1^* and η_1^{**} are larger and characteristic of more numerous and less soft 'defect' configurations: $\eta_1^* > \eta_0^*$ and $\eta_1^{**} > \eta_0^{**}$, at $\eta_1^{**} < \eta_1^*$, e.g. $\eta_1^* \simeq 0.12 > \eta_0^* \simeq 0.08$ and $\eta_1^{**} \simeq 0.05 > \eta_0^{**} \simeq 0.01$. Of course, in this case self-trapping occurs even without hybridization while it is essentially changed by the latter for sufficiently soft configurations.

The situation is essentially different for a nominally doubly occupied 'defect' state ($n=2$). Such a state is weakly sensitive to hybridization, as the state is filled without hybridization. Therefore,

$$\Phi_2(x) \simeq V_{at}(x) + E_{VB}^{(0)} + U_d + 2\varepsilon_{d\sigma}(x), \tag{39}$$

$$\Phi_2(x_2) \simeq E_{VB}^{(0)} + 2\varepsilon_d^{(0)} + U_d + W_2,$$

as if the influence of the valence-band states were to vanish. Then, the adiabatic potential has the same structure as the substrate one $V_{at}(x)$. For instance, it would be harmonic, with

$$W_{2h} \equiv W_{2h}(x_2) \simeq -Q_d^2/(A\eta a_0^2) \tag{40}$$

at $x_2 \simeq Q_d/(A\eta)$ if $V_{at}(x) \simeq V_{at}^{(0)}(x) = A\eta x^2$.

The hybridization-induced changes in $\Phi_0(x)$ and $\Phi_1(x)$, as well as in the x_n, for the lowest-energy minima, give rise to significant deviations of the total electron energies $E_{el}^{(n)}(x)$, self-trapping energies $W_n(n=0,1)$ and pair correlation energy U in Eqs. (7)–(9) from the standard formulae [1–3, 15]

$$\begin{aligned} W_0 &= 0, \qquad W_1 W_2/4 = -W = -Q_d^2/(2ka_0^2), \\ E_{el}^{(n)} &= n(\varepsilon_d^{(0)} - Q_d x) + \delta_{n2} U_d, \qquad U = -2W + U_d \end{aligned} \tag{41}$$

for self-trapping in a harmonic substrate configuration $V_{at}(x) \simeq V_{at}^{(0)}(x) = A\eta x^2 = kx^2 a_0^2/2$. In Eqs. (41) U decreases linearly with W and is negative for $k < Q_d^2/(2U_d a_0^2) \leq k_0$. The deviations of W_0, W_1 and $E_{el}^{(n)}(x)$ ($n=0,1$), U from Eqs. (41) are shown schematically in Fig. 6 and Fig. 7. Those become essential for small enough $\eta < \eta_1^{**}$ as the electron ground state, nominally singly occupied ($n=1$), is related to the extra minimum at $x_1 = x_{12} \simeq x_2$ for $\Phi_1(x)$ and is practically doubly occupied ($v_d(x_{12} \simeq 2)$). This behaviour might be illustrated for self-trapping in harmonic substrate configuration with W_{2h} from Eq. (40),

$$W_1 = E_g^{(0)} - 4W + U_d = W_2(x_2) + E_g^{(0)} + U_d, \quad U = 4W - U_d - 2E_g^{(0)} < 0 \tag{42}$$

for $\eta_{1h}^{**} > \eta > \eta_1^{**}$ and typical $\varepsilon_d^{(0)} \simeq E_c \equiv E_g^{(0)}$ (with accuracy to within small corrections $O(\Delta/E_g^{(0)})$). In Eq. (42), W_1 decreases linearly with $W \simeq \eta^{-1}$, whereas the negative U passes through the lowest value U_m at $\eta = \eta_1^{**}$, increasing in the range $\eta_{1h}^{**} < \eta < \eta_1^{**}$. Further decreases of η ($\leq \eta_1^{**}$) changes the lowest-energy minimum position for $\Phi_0(x)$ from $x_0 = x_{01} \simeq 0$ to $x_0 \simeq x_{02} \simeq x_{12} \simeq x_2$, so that (with accuracy to small corrections $O(\Delta/E_g^{(0)})$) $\Phi_2(x_2) \simeq \Phi_1(x_{12}) \simeq \Phi_0(x_{02})$ and

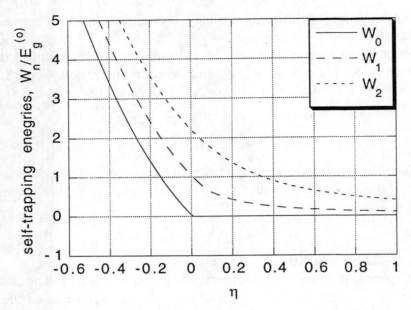

Fig. 6. Self-trapping energies: dependences on the softness parameter η.

$$U_{\rm h} \geq -\varDelta, \qquad \text{i.e. } |U_{\rm h}| \leqq \varDelta \ll E_{\rm g}^{(0)}. \tag{43}$$

In fact, U asymptotically tends to zero, not becoming positive, at $\eta < \eta_{\rm 1h}^{**}$. Then the negative correlation energy U exhibits a non-monotonic dependence on W and a lowest value

$$U_{\rm mh} = U_{\rm h}(\eta_1^{**}) \simeq -\frac{2}{3}\, E_{\rm g}^{(0)} \left(1 + \frac{U_d}{2E_{\rm g}^{(0)}}\right), \tag{44}$$

practically close to $(-E_{\rm g}^{(0)}/2)$ for $E_{\rm g}^{(0)}/2 \gg U_d$.

As noted above, the basic features of $U(\eta)$ shown in Fig. 6 (solid line) for self-trapping in anharmonic substrate configurations (4) are in fact similar to those of the non-monotonic dependence $U_{\rm h}(\eta)$ from formulae (42)–(44). On the other hand, even in the absence of hybridization, the behaviour of $U(\eta)$ shown in Fig. 6 (dashed line) for self-trapping in anharmonic configurations with double-well potentials is also similar to that described by the solid line in Fig. 6 and by formulae (42)–(44). This similarity can be interpreted as that hybridization effects give rise to anharmonic features of the same type as the original anharmonicity of Eq. (4), which limits the self-trapping equilibrium displacements to realistic $|x| \leq 1$. The difference is that hybridization prevents occurrence of positive correlation energies for anharmonic configurations ($\eta < \eta_{\rm h}^{**}$),

(a)

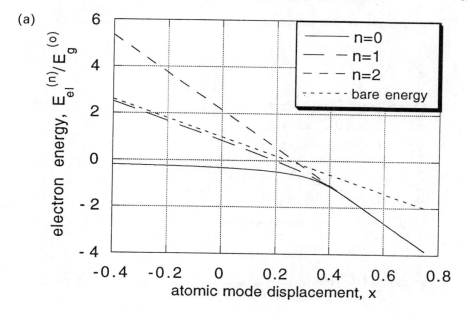

(b)

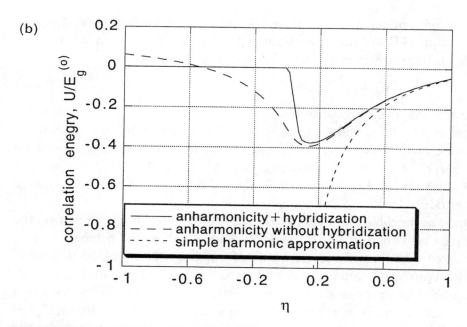

Fig. 7.(a) The total electron energy gain $E_{el}^{(n)}$: dependences on x for different nominal occupations. (b) The dependence of the pair correlation energy on the softness parameter η for different cases.

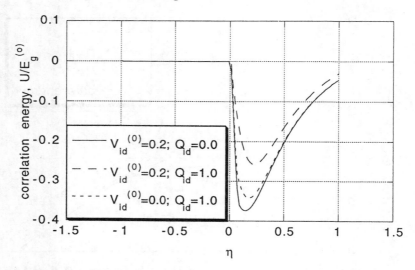

Fig. 8. Variations of the pair correlation energy behaviour for different dependences of the hybridization energy on x.

whereas anharmonicity itself does not. A quantitative difference exists between U_{mh} in Eq. (44) and the lowest value U_m for self-trapping in anharmonic configurations, which is, however, relatively small, so that

$$0 < U_m - U_{mh} \ll |U_{mh}|, \qquad \text{i.e. } U_m \simeq - E_g^{(0)}/2, \tag{45}$$

as can be seen from Fig. 6 as well.

It is worth adding that the behaviour of $U(\eta)$ and its hybridization-induced features, such as the scale of the magnitude, do not essentially depend on whether $V_{id}^{(0)} = 0$ or $V_{id}^{(0)}$ is finite (e.g. $|V_{id}^{(0)}|N^{1/2} = |V_1| \leq 1$ eV) in $V_{id}(x) \simeq V_{id}^{(0)} + Q_{id}x$ as shown in Fig. 8. Presumably this holds because of the relatively large magnitude $|x_2| \simeq 1$.

One can conclude that hybridization in two-band self-trapping changes the relations between $E_{d\sigma}(x_n)$, W_2 and W_1, W_0, as well as the behaviour of U, compared with those for single-band self-trapping. Moreover, occurrence of stable negative-U centres, i.e. doubly occupied self-trapped states, becomes favourable for energies $E = E_2/2$ (per particle) around the Fermi level in a characteristic energy stripe of width $\delta\varepsilon \simeq w_t \simeq 0.1E_g^{(0)}$. The position of ζ, as usual, follows from the electro-neutrality equation, taking into account the true DOS $g(E)$ in the (mobility) gap. The latter has been shown [5, 17] (in an approximation accounting only for the true electron level being repelled away from the edge E_v) to exhibit its largest values around the gap middle

$$\bar{E} = E_v + E_g^{(0)}/2 \equiv E_g^{(0)}/2, \qquad g(E) \leq g(\bar{E}), \tag{46}$$

for $E_v \equiv 0 < E < E_c \equiv E_g^{(0)}$. Moreover, the Fermi level is also found to be close to $\bar{E}$, $\zeta - E_v \equiv \zeta \simeq E_g^{(0)}/2$, in accordance with a symmetry of the gap spectrum with respect to substituting electrons (e) for holes (h) or vice versa (with accuracy to within corrections $O(U_d/E_g^{(0)}, \Delta/E_g^{(0)}) \ll 1$), as $U_{mh}^{(e)} \simeq E_g^{(0)}/2 \simeq U_{mh}^{(h)}$ in Eq. (44). The Fermi level is pinned near the gap middle, at the high DOS $g(\bar{E}) \approx g(\zeta)$ $\lesssim 3 \times 10^3$ eV^{-1}, with respect to increasing temperature (T) and impurity concentration (c_i), so no temperature and impurity effects are essential for [5, 17]

$$T \ll L_L^{(0)} \equiv T_L(E_g^{(0)}) \equiv \Lambda^{-1} E_g^{(0)} \quad \text{and} \quad c_i \ll c_{iL}^{(0)} = \Lambda^{-1}, \qquad (47)$$

with $\Lambda \approx 10 \ln(g_0(E_c)E_g^{(0)}/c_2^{(0)}) \gg 1$.

The concentration $c_2^{(0)}$ of the negative-U centres, for both electron and hole pairs, can be described as follows

$$c_2^{(0)} = \int\int\int \mathrm{d}\varepsilon_d^{(0)} \, \mathrm{d}\eta \, \mathrm{d}\xi \, g_0(\varepsilon_d^{(0)}) \phi_2(\varepsilon_d^{(0)}; \eta, \xi) F(\eta, \xi). \qquad (48)$$

Here the Gibbs pair occupation factor is

$$\phi_2(\varepsilon_d^{(0)}; \eta, \xi) = Z^{-1} \exp\left(\frac{2\zeta - E_2(\eta, \xi)}{T}\right),$$

$$Z = 1 + 2\exp\left(\frac{\zeta - E_1(\eta, \xi)}{T}\right) + \exp\left(\frac{2\zeta - E_2(\eta, \xi)}{T}\right)$$

and $E_n(\eta, \xi) = \Phi_n[x_n(\eta, \xi)] - \Phi_0[x_0(\eta, \xi)]$, at $n = 0, 1$ or 2, is the ground-state energy of the two-band system in question, nominally occupied by n electrons self-trapped in the substrate atomic configuration (4). Moreover $F(\eta, \xi)$ can plausibly be approximated as

$$F(\eta, \xi) \approx F_0 \exp\left[-\left(\frac{\bar{\eta} - \eta}{(\Delta\eta)_0}\right)^2\right] \equiv \exp\left[-\frac{1}{2\eta_c^0}\left(1 - \frac{\eta}{\bar{\eta}}\right)^2 - \frac{\xi^2}{(\xi_c^0)^2}\right], \qquad (49)$$

where, for typical atomic concentrations $c_d \lesssim 0.1$ of soft configurations, or 'defects', the variation scales are

$$\eta_c^0 = ((\Delta\eta)_0/\bar{\eta})^2 \approx 0.1 \approx (\xi_c^0)^2, \qquad (50)$$

for $|\eta| \ll 1$ at least. For the low $T \ll T_L^0$ in question, the effective ground-state energy $E_2^{\text{eff}}(\eta, \xi)$ of most of the negative-U centres can be obtained from the relation

$$2\zeta - E_2^{\text{eff}}(\eta, \xi) = 0 = 2\zeta - 2\varepsilon_d^{(0)} - W_{\text{eff}}^0 - U_d, \qquad (51)$$

which corresponds to the main contribution of $\phi_2(\varepsilon_d^{(0)};\eta,\xi)F(\eta,\xi)$ to the integral in Eq. (48) for actual $\varepsilon_d^{(0)} - E_v \equiv \varepsilon_d^{(0)} \simeq E_c - E_v = E_g^{(0)}$. With $\xi = \bar{\xi} \approx 0$ for the main maximum of $F(\eta,\xi)$ in Eq. (49), the effective value η_{eff}^0 of η for most of the negative-U centres is found approximately from Eq. (40) for $W_2(\eta, \xi = 0)$ and Eq. (5) within the frame of harmonic approximation:

$$\eta_{\text{eff}}^0 \simeq \bar{\eta}_{\text{eff}}^{(0)}\theta(\bar{\eta} - \bar{\eta}_{\text{eff}}^{(0)}) + \bar{\eta}\theta(\bar{\eta}_{\text{eff}}^{(0)} - \bar{\eta}), \tag{52}$$

$$\bar{\eta}_{\text{eff}}^{(0)} = \frac{Q_d^2}{A(E_g^{(0)} + U_d)}, \qquad \bar{\eta} \approx 1$$

and $W_{\text{eff}}^0 = W_2(\eta_{\text{eff}}^0, \xi = 0) \simeq -Q_d^2/(A\eta_{\text{eff}}^{(0)})$.

Basic gap-related electronic processes in glassy semiconductors, such as photoluminescence and in part the fundamental optical absorption, in glassy and similar semiconductors, are related to local centres, which are electronic excitations of the negative-U centres. The processes are characterized by intensities I_q^0 largely proportional to the concentration of centres $c_q^0 \propto c_2^0$ and by energies, such as E_{PL}^0 and $E^{*(0)}_{\text{opt}}$, determined by the effective value U_{eff}^0 of the pair correlation energy U for most of the pair states near the Fermi level [5]:

$$E^{*(0)}_{\text{opt}} \simeq 2|U_{\text{eff}}^0| + U_d \simeq E_g^{(0)} \simeq E_{\text{PL}}^0 + |U_{\text{eff}}^0| + U_d, \tag{53}$$

$$U_{\text{eff}}^0 = U(\eta_{\text{eff}}^0, \xi = 0), \qquad \text{e.g. } U_{\text{eff}}^0 = U(\bar{\eta}_{\text{eff}}^{(0)}, 0) \simeq -E_g^{(0)}/2 + U_d.$$

For c_2^0 and the related concentration c_{ST}^0 of the pair states the following relations can be obtained:

$$c_2^0 \simeq c_{\text{ST}}^0, \qquad c_2^0 \leq 10^{-3} < c_{\text{ST}}^0 \approx g(\zeta)E_g^{(0)}/2 < 10^{-2}, \tag{54}$$

so in Eq. (49) $T_L^0 \simeq 10^3$ K and $c_{iL}^0 \simeq 0.1$. Then, the overlap $I_2(R_{\text{ST}}^0) \approx \exp(-2R_{\text{ST}}^0/\rho)$ of the self-trapped pair states, on average the separation $R_{\text{ST}}^0 = a_1(c_{\text{ST}}^0)^{-1/3}$, is weak, $I_2(R_{\text{ST}}^0) \leq 0.1$, for a typical $2R_{\text{ST}}^0 \geq 10a_1 > \rho$ and a typical state size $\rho \leq 5\rho \leq 10a_1$ in such materials (see Eq. (36)) [7]. In this respect the mentioned characteristic energy stripe of width $\delta\varepsilon(\simeq w_t)$ around the Fermi level, containing the most essential self-trapped pair states and the related negative-U centres responsible for basic gap-related electron properties of glassy semiconductors at least, is similar to a partly filled 'impurity band' in the gap, of which the states are localized.

4 Pressure-induced phenomena

The problem under discussion in what follows is whether hydrostatic pressure p gives rise to significant changes in basic characteristics of the two-band self-trapping and related hybridization effects for the pair states and negative-U

centres, and to new effects, even though the gap width $E_g = E_g(p)$ is still finite ($E_g \neq 0$) and no global pressure-induced semiconductor–metal transition occurs [18]. In fact, it is experimentally established at least for glassy semiconductors [19, 20] that the gap width E_g monotonically decreases with p up to $E_g = 0$ at a characteristic pressure $p_g = q_1 \times 10^5$ bar, $q_1 \approx 0.7$–2.1, mainly as

$$E_g \equiv E_g(p) \simeq E_g^{(0)}(1 - p/p_g) \tag{55}$$

for $p \leq p_g^* \equiv p_g(1 - \Delta_1/p_g) \simeq 0.99 p_g$, with $E_g^{(0)} \equiv E_g(0)$. Another important conclusion [5, 21] extracted from experimental data [19] is that the distribution density $F(\eta, \xi)$ in Eq. (51) narrows with increasing p, its width $\Delta\eta$ and variation scale η_c in Eq. (49) monotonically decreasing as

$$\eta_c = \eta_c(p) \simeq \eta_c^0(1 - p/p_\eta) + \eta_c^* \tag{56}$$

for $p \leq p_\eta^* = p_\eta(1 - \Delta_2 p/p_\eta) \approx 0.99 p_\eta$, while $\eta_c \simeq \eta_c^* \leq 0.1 \eta_c^0$ for $p > p^*$, with the characteristic pressure $p_\eta = q_2 \times 10^5$ bar, $q_2 \approx 1$.

On the other hand, analysis of the pressure-dependence of the other parameters Q_d, A, U_d and $\Delta(\varepsilon, \omega)$ of the two-band self-trapping under discussion, by taking into account experimental data [5, 10, 21] and theoretical estimations [5, 18], gives rise to a conclusion that the variation scales p of these parameters with varying p substantially exceed both p_g and p_η so their p-dependence can be neglected for $p < \{p_g; p_\eta\}$. With this conclusion in mind and Eq. (55) and Eq. (56) the problem of pressure-induced effects can be answered by calculating the effective correlation energy $U_{\text{eff}} \equiv U_{\text{eff}}(p) \equiv U(\eta_{\text{eff}}, \xi = 0)$, which is characteristic of most of the pair states and negative-U centres in the energy stripe around the Fermi level, and the related concentrations $c_{\text{ST}} \equiv c_{\text{ST}}(p)$ and $c_2 \equiv c_2(p)$, in the (mobility) gap [18]. The calculations take into account the formulae (34), (48)–(50), (52) and (53), substituting E_g and η from Eq. (55) and Eq. (56) for $E_g^{(0)}$ and η_c^0, and are performed both by computer simulations (solid lines in the related Fig. 9 and Fig. 10) and approximate analytical derivations (dashed lines) for the typical values of the parameters. Since the Fermi level as before is pinned near the gap middle, with E_g decreasing as p grows (Eq. (57)), temperature and impurity effects are neglected for sufficiently low $c_i \ll c_i^0 \approx 0.1$ and $T \ll T_L(p) \equiv T_L^0(E_g) \ll 0.1 E_g$. The results of calculations for $U_{\text{eff}}(p)$ and $c_2(p)/c_2(0)$ are presented in Fig. 9 and Fig. 10.

As can be seen from Fig. 9, $U_{\text{eff}}^0 \equiv U_{\text{eff}}(0) \simeq -0.43 E_g^{(0)}$, monotonically growing with p up to $U_{\text{eff}}(p) < 0$ and small magnitudes $|U_{\text{eff}}(p)| \leq \Delta \leq 0.1 E_g^{(0)}$ for $p_g > p \geq p_U \simeq 0.9 p_g$. The result, that U_{eff} does not reach positive values even for $p > p_U$, is related, as noted in Sec. 3, to contributions of the hybridization of states. An estimation of p_U, which agrees with the above noted results of numerical calculations, can be derived from Eq. (29):

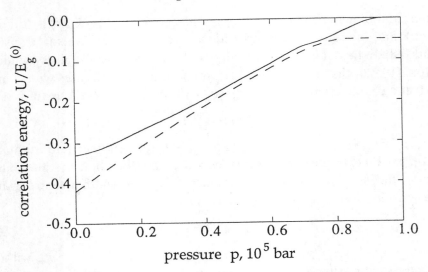

Fig. 9. The dependence of the effective correlation energy on pressure (solid line, anharmonic substrate; dashed line, harmonic substrate).

$$p_U \simeq p_g(1-\kappa), \tag{57}$$

where $\kappa \equiv \max\{U_d/E_g^{(0)}; 2\Delta/E_g^{(0)}\}$ and typically $\kappa \simeq 0.1$, e.g., with $U_d \simeq 0.15 E_g^{(0)}$ and $\Delta \leq 0.1 E_g^{(0)}$, i.e. with $U_d \geq 2\Delta$. Indeed, Eq. (29) means that the hybridization of states is nearly complete ($\gamma_{d\sigma}^2 \simeq 0$) as $E_g(p)$ is reduced to (or more narrow than) the two energy stripes of total width $2\Delta \ll E_g^{(0)}$. In the latter case, for $p_\Delta = p_g(1 - 2\Delta/E_g^{(0)}) \leq p < p_g$, the 'defect' states in the gap become negative-U centres with small $|U| \leq \Delta$ and the related local 'defect' atomic dynamics becomes essentially anharmonic, as are the adiabatic potentials.

As shown in Fig. 10, $c_2(p)/c_2(0)$ exhibits a non-monotonic behaviour, with a minimum at $p = p_{\min}(\approx 0.4 p_g)$ and with $c_2(p)/c_2(0) \geq$ at $p \geq p_0(\simeq 0.6 p_g)$. This behaviour can be interpreted as follows. Let us distinguish at $p = 0$ 'strong' and 'weak' negative-U centres. The former correspond to strong self-trapping in soft configurations ($k \leq 0.1 - 0.2 k_0$), for which $U_{\text{eff}}^0 < 0$ and $|U_{\text{eff}}^0| \simeq E_g^{(0)}/2(\gg U_d)$ and the concentration $c_2^*(p) \simeq c_a(p) \simeq \exp(-\eta_c^{-1}(p))$, as well as the concentration $c_d(p)$ of soft configurations, decreases monotonically with narrowing distribution density $F(\eta, \xi)$ for growing p [5, 21]. Then the 'strong' centres and related effects should practically vanish for $p > p_\eta \eta_c^0 \approx 0.1 p_\eta \approx 10^4$ bar. On the other hand, the 'weak' centres are associated with pair self-trapping in more rigid configurations ($k_0 > k > 0.1 k_0$) for which $U(\eta) < 0$ still, but $|U(\eta)| \ll |U_{\text{eff}}^0|$. Their concentration $c_2^{(w)}(p)$ rapidly increases with p, due to increasing $\eta_{\text{eff}}(p)$ up to $\eta_{\text{eff}} \leq \bar{\eta} \approx 1$ and then to increasing distribution density $F(\eta \simeq \eta_{\text{eff}}, \xi = 0)$. Competition between the two effects just gives rise to the

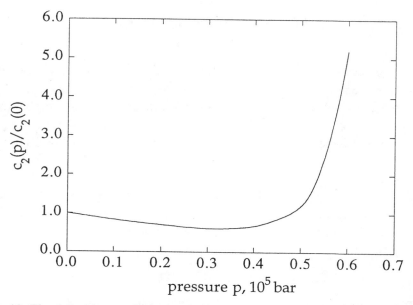

Fig. 10. The dependence of the concentration of negative-U centres on pressure.

non-monotonic behaviour of $c_2(p)/c_2(0)$. This interpretation can be illustrated by approximating

$$\frac{c_2(p)}{c_2(0)} \approx \frac{F(\eta_{\text{eff}}(p), \xi \simeq \bar{\xi} \approx 0)}{F(\eta_{\text{eff}}^0, 0)}, \tag{58}$$

as follows from Eqs. (48)–(54) and Eq. (55) and Eq. (56). The related estimations for $p_{\min}$ and p_0,

$$p_{\min}/p_{\text{g}} \simeq (1 - \lambda)/(1 + \lambda), \qquad p_0/p_{\text{g}} \simeq 1 + \beta - \lambda, \tag{59}$$

e.g. $p_{\min}/p_{\text{g}} \simeq 0.4$ and $p_0/p_{\text{g}} \simeq 0.7$ for typical values $\beta \equiv U_d/E_{\text{g}}^{(0)} \simeq 0.1$ and $\lambda = 2Q_d^2/(A\bar{\eta}E_{\text{g}}^{(0)}) \simeq 0.4$, agree reasonably well with the values obtained from numerical calculations $p_{\min}/p_{\text{g}} \simeq 0.4$ and $p_0/p_{\text{g}} \simeq 0.7$. Note that $c_2(p)\partial c_2^{(w)}(p)$ rapidly increases while the average centre separation $R_2(p) = a_1 c_2^{-1/3}(p)$ rapidly decreases with growing $p > p_{\min}$. It follows from Eq. (53) and Eq. (55) and Fig. 9 that $E_{\text{PL}}(p)$ and $E_{\text{opt}}^*(p)$ decrease with p largely as $1 - p_U/p_{\text{g}}$, up to about zero for $p > p_U$, while $I_{\text{PL}} \propto c_2(p)/c_2(0)$ exhibits the non-monotonic p-dependence described above. It is worth adding that the p-dependence of $c_{\text{ST}}(p)$ and $R_{\text{ST}}(p)$ is expected to be similar to that of $c_2(p)$ and $R_2(p)$, so $R_{\text{ST}}(p) = a_1 c_{\text{ST}}^{-1/3}(p)$ rapidly decreases and the overlap of the self-trapped pair states $I_2(R_{\text{ST}}) \approx \exp(-2R_{\text{ST}}/\rho)$ rapidly increases with p and $p > p_0$, so that overlap becomes significant,

$$I_2(R_{ST}) \geq 0.3 \qquad \text{as } 2R_{ST}(p) \leq \rho,$$
$$\text{for } p \geq p_I, \qquad 2R_{ST}(p_I) = \pi. \tag{60}$$

Then, delocalization of the self-trapped pair states in the characteristic energy stripe around the Fermi level can be expected at high enough p, $p_g > p \geq p_c \geq p_I$, which is similar to the Anderson–Mott 'insulator–metal' transition in an 'impurity band', so long as the characteristic pressure $p_c < p_g$. The latter can roughly be estimated from the related criterion (see [18, 21])

$$2zJ_2(p_c)/\delta\varepsilon(p_c) = \lambda_{cr} \approx 1.5\text{–}2.0, \tag{61}$$

where $J_2(p) \approx J_2^* \exp(-2R_{ST}(p)/\rho)$, $J_2^* \approx (D_c^2/E_g^{(0)})\exp(-\phi)$, $\phi(p) \leq E_g(p)/(2\hbar\omega_D)$, and $\delta\varepsilon(p) \approx (2e)^2/(\kappa_0 R_{ST}(p))$, with κ_0 the static dielectric susceptibility and ω_D the Debye frequency. As found from numerical calculations, it follows from Eq. (61) that p_c is close to both p_I and p_U,

$$p_g > p_c \geq \max\{p_I; p_U\} \qquad \text{e.g. } p_c \approx 0.9 p_g \tag{62}$$

for typical values of related parameters, including $\kappa_0 = 10$ and $\phi = 10$, at $E_g^{(0)} = 2$ eV.

Insofar as the relations (62) hold true, a quasi-metallic conductivity $\sigma_m(p)$ of the delocalized local pairs might dominate over the standard activation-type conductivity $\sigma_a = \sigma_0 \exp(W_\sigma(p)/T)$ with $W_\sigma(p) = |U_{eff}(p)| \leq \Delta$, at high p, $p_c \leq p < p_g$, and low temperatures $T < T_\sigma(p) \approx W_\sigma(p)/\ln(\sigma_0/\sigma_m) < W_\sigma(p) \leq \Delta$. The quasi-metallic conductivity might be changed by a kind of superconductivity due to the delocalized local pairs at low enough $T < T_c(p)$, if $T_c(p) < T_\sigma(p)$, or even be suppressed by such a superconductivity, if $T_c(p) > T_\sigma(p)$. Although local pairs of small size ρ are of importance here, it is not easy to relate such a hypothetical superconductivity to one of the mechanisms available, since the very motion of the local pairs is associated with the substantial overlap, which might give rise to their instability [23]. If, nevertheless, the bipolaron Bose-condensation model is applied [24], then

$$T_c(p) \approx \hbar^2/(M(p)R_2^2(p)) \approx 1\text{–}10 \text{ K}, \tag{63}$$

at plausible local-pair mass $M(p) \approx m_e \exp(\phi(p)) \approx (10^3\text{–}10^2)m_e$ for $p \approx p_c$. Moreover, $dT_c/dp > 0$ as $dR_2/dp < 0$ and $dM(p)/dp < 0$ in the range $p_c \leq p < p_g$ implied.

It is worthy of note that recent data on pressure-induced superconductivity in some glassy semiconductors (g-Ge$_{33}$As$_{12}$Se$_{55}$, and g-As$_2$Te$_3$) for $p_g > p > p_c^{(exp)} \approx 10^5$ bar and low $T < T_c^{(exp)}(p)$ show that $T_c(p) \approx 1\text{–}8$ K and $dT_c/dp > 0$ up to $T_{c,\max}$ for p approaching p_g. If it is supposed that $T_{c,\max} = T_c(p_g)$ is reached at the global semiconductor–metal transition, with $E_g(p) \simeq 0$ at $p = p_g$, then

$dT_c/dp > 0$ for $p > p_g$ may perhaps indicate BSC-type superconductivity. Of course, the remarks concerning pressure-induced superconductivity, are still nothing more than possible indications.

5 Concluding remarks

To conclude, the electron (hole) self-trapping under discussion is of essentially two-band origin, being related to hybridization of states and giving rise to largely self-trapped local singlet pairs, which determine new features of both ground state and excitations of the electrons (holes) and of the related local atomic dynamics. Strong pressure-induced effects can be predicted in the materials in question, including a non-monotonic behaviour of intensities of electronic processes and a possible 'insulator–metal' transition, which might give rise to superconductivity of overlapping local pairs, even before the global 'semiconductor–metal' transition occurs.

Acknowledgements

We are grateful to the Department of Chemistry, University of Cambridge, and particularly to Dr S. R. Elliott for his warm hospitality. One of us (M. I. K) highly appreciates the kind hospitality and support of Trinity College, University of Cambridge. Another author (S. N. T.) is grateful to the Royal Society, London for support.

References

[1] C. G. Kuper and G. D. Whitefield, eds., *Polarons and Excitons*, Oliver & Boyd, Edinburgh, 1963.
[2] D. Emin, *Adv. Phys.,* **24** (1975) 305.
[3] M. I. Klinger, *Problems of Polaron Transport Theory*, Pergamon Press, Oxford, 1979.
[4] J. T. Devreese and F. M. Peeters, eds., *Polarons and Excitons in Polar Semiconductors*, Plenum Press, New York, 1984.
[5] M. I. Klinger, *Phys. Rep.,* **94** (1983) 183; *ibid.,* **165** (1988) 275.
[6] M. I. Klinger and S. N. Taraskin, *Solid State Commun.,* **79** (1991) 231; *Proc. Int. Conf. on Glass Physics*, Zinante, Riga (1991) p. 25.
[7] M. I. Klinger and S. N. Taraskin, *JETP Lett,* **56** (1992) 209; *ibid.,* **57** (1993) 210; *Phys. Rev.* B **47** (1993) 10235.
[8] M. I. Klinger and S. N. Taraskin, *Phys. Lett.,* A **170** (1992) 225; M. I. Klinger, V. N. Chukov and S. N. Taraskin, *Phys. Lett.,* A **176** (1993) 259.
[9] A. S. Alexandrov and N. F. Mott, *J. Supercond.* **7** (1994) 599.
[10] A. G. Milnes, *Deep Impurities in Semiconductors*, Wiley, New York, 1963.
[11] W. Fowler and R. J. Elliott, *Phys. Rev.* B **34** (1986) 5525.
[12] M. I. Klinger and S. N. Taraskin, *J. Non-Cryst. Solids,* **164–166** (1993) 391.

[13] F. D. M. Haldane and P. W. Anderson, *Phys. Rev.,* B **13** (1976) 2553.
[14] P. W. Anderson, *Phys. Rev. Lett.,* **34** (1975) 953.
[15] W. A. Harrison, *Electronic Structure and the Properties of Solids,* Freeman, San Francisco, 1980.
[16] M. I. Klinger and S. N. Taraskin, *J. Phys. Condens. Matter,* to be published.
[17] M. I. Klinger, L. I. Shpinar and I. I. Yaskovets, *Fiz. Tverd. Tela,* **28** (1986) 470.
[18] M. I. Klinger and S. N. Taraskin, *J. Non-Cryst. Solids,* to be published.
[19] K. Tanaka, *Phys. Rev.,* B **30** (1984) 4549.
[20] I. V. Berman and N. V. Brandt, *Fiz. Niz. Temperatur,* **16** (1990) 1227 (in Russian).
[21] M. I. Klinger, *Solid State Commun.,* **65** (1988) 449; *ibid.,* **70** (1989) 939.
[22] N. F. Mott and M. Kaveh, *Adv. Phys.,* **34** (1985) 329.
[23] D. Emin, *Physics and Materials Science of High Temperature Superconductors –* *II,* R. Kossowsky, N. Ravean and S. Patapis, eds., Kluwer, Dordrecht, 1992.
[24] N. F. Mott, *Adv. Phys.,* **39** (1990) 55; A. S. Alexandrov and N. F. Mott, *Supercond. Sci. Technol.,* **6** (1983) 215.

28

Bismuth disproportionation in super- and semi-conducting barium bismuthates

N. C. PYPER[1] and P. P. EDWARDS[2]

[1]*University Chemical Laboratory, Lensfield Road, Cambridge CB2 1 EW, UK*
[2]*The School of Chemistry, The University of Birmingham, Edgbaston, Birmingham B15 2TT, UK*

Abstract

It is explained why, despite the strong exothermicity of the reaction generating two Bi^{4+} ions from one Bi^{3+} and one Bi^{5+} ion in the gas phase, solid barium bismuthate ($Ba_2Bi^VBi^{III}O_6$) contains bismuth in two different oxidation states and has a structure distorted from that of a perfect cubic perovskite by displacements of the oxide ions towards the Bi^V species. The predicted difference between the energy of this and that of an undistorted undisproportionated cubic perovskite ($BaBiO_3$) containing only Bi^{IV} is in good agreement with the activation energy measured for conductivity by thermal hopping. It is also shown why, for $x \gtrsim 0.3$, all materials of stoichiometry $K_xBa_{1-x}BiO_3$ contain only one type of bismuth (Bi^{IV}) in a perovskite structure without the distorting oxide displacements of the undoped material. The distorted disproportionated structure is predicted to be degenerate with the undistorted undisproportionated structure when $x = 0.34$. A possible connection between this degeneracy and high-temperature superconductivity is discussed.

1 Observations for explanation

The primary object of this paper is to explain two related experimental observations concerning the structure of the semi-conductor barium bismuthate and the materials that result from its doping with K^+ ions. We then comment on possible connections with the high-temperature superconductivity (HTS) exhibited by the latter materials at sufficiently high levels of doping.

The first observation requiring an explanation is that there are two different types of bismuth site in the undoped material of stoichiometric formula $BaBiO_3$[1–7]. This compound thus has the formula $Ba_2Bi^VBi^{III}O_6$[1–7], where the superscript denotes formal oxidation state. The occurrence of two bismuth

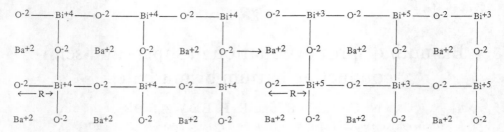

Fig. 1. Plan view of the hypothetical disproportionation reaction $2Ba^{2+}Bi^{4+}O_3 \rightarrow Ba_2^{2+}Bi^{3+}Bi^{5+}O_6$ without oxide ion displacement. Ba^{2+} is located at distance R above the plane of the paper.

sites rather than just one requires explanation because the internal energy change (ΔE) for the disproportionation reaction

$$2Bi^{4+} \rightarrow Bi^{3+} + Bi^{5+}, \tag{1}$$

where the superscripts denote true ionic charges, is strongly endothermic in the gas phase. Standard atomic data [8] show that the gas phase value (ΔE_{gas}) for this energy change is 10.7 eV. This shows that it must be the in-crystal environment that stabilizes barium bismuthate in the disproportionated state $Ba_2Bi^VBi^{III}O_6$, corresponding to a negative ΔE, through a mechanism that is to be elucidated.

The second observation to be explained is that materials of stoichiometry $K_xBa_{1-x}BiO_3$, prepared by replacing some Ba^{2+} ions by K^+ ions, contain only one type of bismuth (Bi^{IV}) if the doping level x is greater than approximately 0.3 [5–7, 9]. Such materials are cubic, having the perovskite structure, and show HTS [5, 6, 9–14]. The occurrence of only one type of bismuth site implies in the fully ionic description, where the material is written $K_x^+Ba_{1-x}^{2+}Bi^{4+}O_3^{2-}$, that ΔE must be positive (as in the gas phase) for reasons that remain to be elucidated.

2 The structure of undoped barium bismuthate

The disproportionation in undoped barium bismuthate can be understood by considering a hypothetical fully ionic perovskite of formula $Ba^{2+}Bi^{4+}O_3^{2-}$ composed of Ba^{2+}, Bi^{4+} and O^{2-} ions. Such a material represents the structure that barium bismuthate almost certainly would adopt if it did not disproportionate. In the plan view of such a perovskite shown on the left in Fig. 1 with the closest Bi^{4+}–O^{2-} distance denoted R, all the ions lie in the plane of the paper except for the Ba^{2+} ions, which are located at a height R equal to half a unit cell side above the page. Occurrence of reaction (1) would convert the $Ba^{2+}Bi^{4+}O_3^{2-}$ perovskite into the cubic structure shown on the right in Fig. 1, where alternate Bi^{4+} ions have been reduced to Bi^{3+} by gaining an electron from a Bi^{4+} neighbour to produce a structure containing alternate Bi^{3+} and Bi^{5+} ions. Relative to the charges in the undisproportionated material, each Bi^{5+} ion has a charge of $+1$ and each Bi^{3+} ion a charge of -1. The charge differences between the undisproportionated and disproportionated structures therefore constitute a rock-salt (NaCl) lattice having charges of unit magnitude and a lattice spacing of $2R$. If only the leading contributions to ΔE are considered, namely ΔE_{gas} and the purely electrostatic inter-ionic interactions treated on the basis of non-overlapping charge distributions, it then follows with R in a.u. that ΔE is given in electron-volts by

$$\Delta E = \Delta E_{gas} - 27.211 \times 1.74756456/(2R) \tag{2}$$

provided that the disproportionation does not change the ionic positions. Although the inter-ionic interaction term in (2) reduces ΔE from its gas phase value, reaction (1) is still predicted to be highly endothermic since (2) yields a value of 4.98 eV for ΔE when R is taken to be 4.157 a.u., which is the average of the $Bi^{5+}-O^{2-}$ and $Bi^{3+}-O^{2-}$ distances measured by both neutron diffraction [6] and EXAFS [7] in the disproportionated material.

Reaction (1) converts one half of the Bi^{4+} ions with their (core $6s^1$) electronic configurations into Bi^{5+} ions having no 6s electrons whilst changing the remaining bismuth ions into Bi^{3+} species with (core $6s^2$) electronic configurations. The absence of a 6s electron in Bi^{5+} would be expected to allow each O^{2-} ion to move a distance Δ towards each Bi^{5+} without encountering an excessive increase in the short-range repulsions generated by overlapping of the electronic charge clouds. Although each bismuth ion is still octahedrally coordinated by six O^{2-} ions in the resulting lattice, this differs slightly from the experimentally determined structure of $Ba_2Bi^VBi^{III}O_6$ depicted on the left in Fig. 2 because the bismuth octahedra are slightly tilted in the latter. However, this small difference need not be considered because the essence of the problem is revealed by studying the reaction (illustrated in Fig. 3) in which the disproportionated material is generated from the $Ba^{2+}Bi^{4+}O_3^{2-}$ perovskite by displacements of the O^{2-} ions parallel to the $Bi^{4+}-O^{2-}-Bi^{4+}$ axes. EXAFS measurements have shown that the displacement (Δ) of each O^{2-} ion towards its closest Bi^{5+} neighbour is 0.17 a.u. over the entire range of temperatures (13–295 K) examined [7]. If Δ is zero in the disproportionated material, then the bismuth–O^{2-} interactions contribute zero to ΔE because, compared with the $Bi^{4+}-O^{2-}$ attractions in $Ba^{2+}Bi^{4+}O_3^{2-}$, the increase in $Bi^{5+}-O^{2-}$ attractions is exactly balanced by the decrease in $Bi^{3+}-O^{2-}$ interactions. However, if Δ is positive then the enhancement of $Bi^{5+}-O^{2-}$ interactions more than outweighs the decrease in $Bi^{3+}-O^{2-}$ attractions, thereby causing the bismuth–O^{2-}

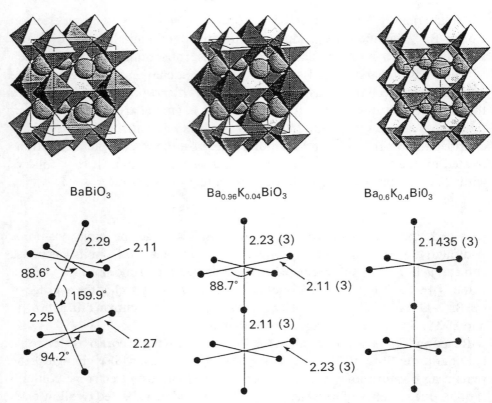

Fig. 2. Experimentally determined structures of the materials with stoichiometries $BaBiO_3$, $K_{0.04}Ba_{0.96}BiO_3$ and $K_{0.4}Ba_{0.6}BiO_3$. All taken from [5] excepting the Bi–O distances in $BaBiO_3$ that were taken from [7] (all distances in ångström units).

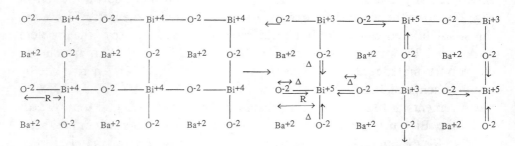

Fig. 3. A plan view of the hypothetical disproportionation reaction $2Ba^{2+}Bi^{4+}O_3 \rightarrow Ba_2^{2+}Bi^{3+}Bi^{5+}O_6$ in which the oxide ions become displaced by a distance Δ in the product. Ba^{2+} is located at distance R above the plane of the paper.

interactions to contribute negatively to ΔE. This contribution is found to be -7.48 eV for the experimentally determined values of R ($=4.157$ a.u.) and Δ ($=0.17$ a.u.) when only the closest bismuth–O^{2-} interactions are considered. The large negative value of this contribution to ΔE reveals that such ionic displacements provide a mechanism for driving the disproportionation. However, since these displacements change many inter-ionic interactions besides the closest bismuth–O^{2-} attractions, it is necessary to consider all these changes to predict a value of ΔE sufficiently accurate for understanding the fundamentals of the problem.

The total electrostatic contribution to ΔE arising from the changed inter-ionic interactions caused by a finite Δ has to be evaluated as the sum of all such changes. This total sum is composed of changes in

1. all the bismuth–O^{2-} attractions,
2. all the O^{2-}–O^{2-} repulsions and
3. all the Ba^{2+}–O^{2-} attractions.

Although all the interactions in each of these three classes must be summed, the sign of each of these three is determined by that of the closest interactions. The contribution to ΔE from each of these three classes was evaluated computationally by summing the terms generated by displacing from the undistorted structure ions in successively more distant shells until encountering those making only negligible contributions. This procedure yielded the following contributions to ΔE.

1. The bismuth–O^{2-} attractions are enhanced to contribute -5.980 eV to ΔE, the sign being determined by the sum of the closest Bi^{5+}–O^{2-} and Bi^{3+}–O^{2-} attractions in $Ba_2^{2+}Bi^{5+}Bi^{3+}O_6^{2-}$ being 7.48 eV greater than the closest Bi^{4+}–O^{2-} attractions in $2Ba^{2+}Bi^{4+}O_3^{2-}$. This is slightly less than the approximate result -6.0 eV obtained [15] for $\Delta = 0.15$ a.u.
2. The O^{2-}–O^{2-} repulsions are increased to contribute $+0.619$ eV to ΔE with the sign being determined by the repulsion between those O^{2-} ions displaced towards the same Bi^{5+} neighbour being 0.744 eV greater than the decreased repulsion between those O^{2-} ions in the same unit cell that are displaced towards different Bi^{5+} ions.
3. The Ba^{2+}–O^{2-} attractions are reduced, thus contributing $+0.143$ eV to ΔE, the sign being determined by the reduction in attraction between each Ba^{2+} ion and the O^{2-} ions located in the same unit cell.

Addition to the relation (2) of these numerical results, which are all for the case $R = 4.157$ a.u., $\Delta = 0.17$ a.u., yields

$$\Delta E = \Delta E_{gas} = 27.211 \times 1.74756456/(2 \times 4.157) - 5.980 + 0.619 + 0.143$$
$$= -0.24 \text{ eV,} \tag{3}$$

where the last three terms arise from displacements of the O^{2-} ions. The absence from (3) of short-range repulsions originating from ion wavefunction overlap does not imply that one has made the clearly false assumption that these are negligible but only that they have been assumed to be the same in both the disproportionated and undisproportionated materials, thus contributing zero to ΔE. We are currently testing this assumption by computing these short-range forces with the Relativistic Integrals Programme [16] using methods that have been sucessfully used to study other polar solids [17–19].

The prediction (3) of a negative value (-0.24 eV) for ΔE implies that the disproportionation reaction (1) is exothermic, thereby explaining why there are two different bismuth sites in barium bismuthate. Relation (3) places the symmetrical undisproportionated perovskite structure containing only Bi^{4+} ions 0.24 eV in energy above that of the $Ba_2Bi^VBi^{III}O_6$ ground state. This not only explains why this material is a semi-conductor but also agrees well with the activation energy of about 0.2 eV measured [3] for the thermally activated hopping mechanism of electrical conduction.

3 Doped barium bismuthate

The replacement in $Ba_2Bi^VBi^{III}O_6$ of some Ba^{2+} ions by K^+ ions to produce a material of stoichiometry $K_xBa_{1-x}BiO_3$ inevitably involves some oxidation of either bismuth or oxide ions. There is evidence [7] from XANES spectroscopy that it is the O^{2-} ions which are oxidized to produce some singly charged O^- ions whilst Seebeck measurements [20] again indicate that it is the O^{2-} ions which are oxidized when $Ba_2Bi^VBi^{III}O_6$ is doped with lead ions to produce a material of stoichiometry $BaPb_{1-x}Bi_xO_3$. All the oxidation in the generation of $K_xBa_{1-x}BiO_3$ will therefore be taken to occur on O^{2-} ions. Overall electrical neutrality considerations show that the average charge q carried by an oxide ion is related to the variable x defining the composition through

$$q = 2 - (x/3) \qquad (4)$$

so that $q=2$ in the undoped material but is $\frac{5}{3}$ in the hypothetical fully doped $KBi^{IV}O_3$. Reaction (1) is only exothermic in undoped $BaBiO_3$ on account of the increased attraction arising when a doubly charged O^{2-} ion is displaced a distance Δ ($=0.17$ a.u.) towards a Bi^{5+} ion. This energy gain will be diminished if the magnitude of the oxide ion charge is decreased, which raises the possibility that ΔE might not be negative in a doped ($x>0$) material. For fixed R and Δ, the reduction of the average oxide charge from 2 to q would reduce by a factor of $q/2$ the magnitude of the bismuth–O^{2-} contribution to ΔE whilst reducing the contribution of the oxide–oxide repulsion by a factor of

$q^2/4$. Since the average charge on the cation sites occupied by K^+ or Ba^{2+} ions is $2-x=3q-4$, the magnitude of the contributions to ΔE from the interactions between these ions and the oxide ions is reduced by a factor of $q(3q-4)/4$. For doped materials having $R=4.157$ a.u. and $\Delta=0.17$ a.u., these observations show, when taken in conjunction with result (3), that ΔE, which is now a function of q ($\Delta E(q)$), is given in electron-volts by

$$\Delta E(q) = \Delta E_{gas} - 27.211 \times 1.74656456/(2 \times 4.157)$$
$$- 5.980(q/2) + 0.619(q^2/4) + 0.143[q(3q-4)]/4. \tag{5}$$

The term arising from the changes in bismuch–bismuth interactions, which is that appearing last in the first line of (5), is independent of the doping level (x). Furthermore although the repulsions between the cations on the sites occupied by K^+ or Ba^{2+} in the doped material are less than those between the Ba^{2+} ions in the undoped solid, these cation–cation interactions do not contribute to $\Delta E(q)$ because for fixed q these repulsions are the same in both undispropor-tionated and disproportionated compounds. Similarly, the contribution to $\Delta E(q)$ from interactions of these cations with bismuth ions is zero. The use of average charges in derivation of (5) is merely a convenient device. The same result (5) is obtained if the K^+ and Ba^{2+} ions are treated as carrying their actual charges and fractions $x/3$ and $1-(x/3)$ of the oxide ions are taken as being respectively O^- and O^{2-}, provided that there is no special site ordering of the oxidized species. There is X-ray crystallographic evidence for absence of ordering of K^+ and Ba^{2+} ions [5].

Result (5) yields a positive value of 0.49 eV for $\Delta E(q)$ in the fully doped material $KBiO_3$ having $x=1$ ($q=\frac{5}{3}$), thus generating the prediction that $KBiO_3$ will not disproportionate but will retain the undistorted perovskite structure shown on the left in Fig. 1 and Fig. 3. Relation (5) also predicts that $\Delta E(q)$ changes from positive at the lowest value ($\frac{5}{3}$) of q to negative ($= -0.24$ eV.) at the highest (2) q value, as shown in Fig. 4 with $\Delta E(q)$ being zero at $q=1.887$ corresponding to $x=0.34$. This explains the experimental observations (see Fig. 2) that materials of stoichiometry $K_xBa_{1-x}BiO_3$ are disproportionated, having two types of bismuth site at low and moderate doping levels, where x is less than about 0.3 but that such materials are not disproportionated and have only one type of bismuth site at all higher doping levels [5–7, 9].

4 Possible links with superconductivity

It is observed on increasing the doping level x from zero that the materials $K_xBa_{1-x}BiO_3$ remain disproportionated and semi-conducting but that, once a critical doping level of around $x=0.3$ is reached, the materials become

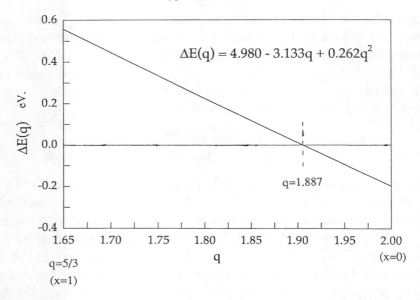

Fig. 4. Prediction from equation (5) of the dependence on average oxide ion charge q of the energy $\Delta E(q)$ of the disproportion reaction $2Ba^{2+}Bi^{4+}O_3 \rightarrow Ba_2^{2+}Bi^{3+}Bi^{5+}O_6$ for $R = 4.157$ a.u. with oxide ions displaced through $\Delta = 0.17$ a.u. in the product. The rearrangement of (5) appears in the figure.

superconducting and non-disproportionated [5–7, 9–14]. Although the different experiments yield slightly different x values for the onset of HTS, they all agree that the conductivity as a function of x changes discontinuously from semi- to superconducting at a critical x value for which the superconducting transition temperature T_c is a maximum. Furthermore, the experiments agree that, above this critical value, T_c decreases monotonically with increasing x although the structure remains non-disproportionated, containing only Bi^{IV}. There is evidence that the critical x value for the onset of HTS is 0.37 [6, 12] and that slightly lower values reported in other publications can be attributed to the samples containing more than one phase.

In the last section it was shown that relation (5) predicts that, for the composition $x = 0.34$, a static crystal having the composition $K_{0.68}^{+}Ba_{1.32}^{2+}Bi^{5+}Bi^{3+}O_6^{2-}$ with $\Delta = 0.17$ a.u. is degenerate with a static crystal $K_{0.68}^{+}Ba_{1.32}^{2+}Bi_2^{4+}O_6^{2-}$ having the slightly different geometry $\Delta = 0$. This suggests that the true wavefunction of the entire crystal is not describable within the usual Born–Oppenheimer approximation and that this function must therefore be written

$$\Psi(r_e, r_n) = c_1 \Psi_{el}^{dis}(r_e)\chi_{nuc}^{dis}(r_n) + c_2 \Psi_{el}^{undis}(r_e)\chi_{nuc}^{undis}(r_n). \tag{6}$$

Here r_e and r_n represent the electronic and nuclear coordinates respectively and

$\Psi_{el}^g(r_e)$ and $\chi_{nuc}^g(r_n)$ are respectively the electronic and nuclear wavefunctions for both the disproportionated (g = dis) and undisproportionated (g = undis) materials. The mixing of the two wavefunctions in (6) will be maximal ($c_1 = c_2 = 1/\sqrt{2}$) when they are degenerate, which is predicted to occur in the vicinity of $x = 0.34$ since this is the value for which we predicted that the two purely electronic functions are degenerate. Increase of x beyond the value generating maximal mixing in (6) would augment c_2 at the expense of c_1. If HTS were to be associated with the mixing between these two states, then such a mechanism would be most efficient at the maximal mixing (around $x = 0.34$) and would become less effective as x increased, thereby reducing the mixing. This would explain the observed behaviour of T_c as a function of x. Furthermore, on account of the vibrational wavefunctions in (6) this idea also predicts that the superconductivity would be influenced by the nuclear masses, although this dependence would be expected to be less than that in a purely phonon mechanism of the BCS type. This naturally explains the experimental observations that the superconductivity appears to be affected by isotopic substitution although the changes are less than those predicted from the standard BCS phonon mechanism [21].

5 Conclusion

It has been shown that the major structural features of both barium bismuthate and the materials produced by potassium doping can be explained by considering only the leading energetic contributions, namely the gas phase energies of bismuth ions and the purely electrostatic inter-ionic interactions evaluated as non-overlapping spherical charge distributions. The disproportionation in the undoped material was shown to be driven by the energy gained through displacement of O^{2-} ions towards Bi^{5+} ions. These energy gains are decreased if the average charge carried by an oxide ion is reduced, thus explaining the experimental observations that doped materials $K_xBa_{1-x}BiO_3$ having x greater than about 0.3 remain undisproportionated with the oxide ions undisplaced. The distorted disproportionated and undistorted undisproportionated structures are predicted to be electronically degenerate in the material having stoichiometry $K_{0.34}Ba_{0.66}BiO_3$. The possible significance of this degeneracy for HTS was discussed.

Acknowledgement

We thank Dr C. H. Rüscher for drawing our attention to the measurement of the activation energy for conduction in undoped barium bismuthate.

References

[1] D. E. Cox and A. W. Sleight 1976 *Solid State Commun.* **19** 969.

[2] D. E. Cox and A. W. Sleight 1979 *Act. Cryst.* B **35** 1.

[3] G. Thornton and A. J. Jacobson 1978 *Acta Cryst.* B **34** 351.

[4] A. Chaillout, J. P. Santoro, J. P. Remeika, A. S. Cooper, G. P. Espinosa and M. Marezio 1988 *Solid State Commun.* **65** 1363.

[5] L. F. Schneemeyer, J. K. Thomas, T. Siegrist, B. Batlogg, L. W. Rupp, R. L. Opita, R. J. Cava and D. W. Murphy 1988 *Nature* **335** 421.

[6] S. Pei, J. D. Jorgensen, B. Dabrowski, D. G. Hinks, D. R. Richards, A. W. Mitchell, J. M. Newsam, S. K. Sinha, D. Vaknin and A. J. Jacobson 1990 *Phys. Rev.* B **41** 4126.

[7] S. Salem-Sigui Jr., E. E. Alp, S. M. Mini, M. Ramanathan, J. C. Campuzano, G. Jennings, M. Faiz, S. Pei, B. Dabrowski, Y. Zheng, D. R. Richards and D. G. Hinks, 1991 *Phys. Rev.* B **43** 5511.

[8] C. E. Moore, Ionization potentials and ionization limits derived from the analysis of optical spectra 1971 NSRDS-NBS **34** (USA Government Printing Office).

[9] D. G. Hinks, B. Dabrowski, J. D. Jorgensen, A. W. Mitchell, D. R. Richards, S. Pei and D. Shi 1988 *Nature* **333** 836.

[10] L. F. Mattheiss, E. M. Gyorgy and D. W. Johnson Jr. 1988 *Phys. Rev.* B **37** 3745.

[11] R. Cava, B. Batlogg, J. J. Krajewski, R. C. Farrow, L. W. Rupp, A. E. White, K. T. Short, W. F. Peck and T. Y. Kometani 1988 *Nature* **332** 814.

[12] H. Sato, S. Tajima, H. Takayi and S. Uchida 1989 *Nature* **338** 241.

[13] H. Sato, T. Ido, S. Tajima, M. Yosida, K. Tanabe and S. Uchida 1991 *Physica* C **185–189** 1343.

[14] C. H. Ruscher, A. Heinrich and W. Urland 1994 *Physica* C **219** 471.

[15] J. O. Sofo, A. A. Aligia and M. D. Nunez-Reguerio 1989 *Phys. Rev.* B **40** 6955.

[16] C. P. Wood and N. C. Pyper 1986 *Phil. Trans. Roy. Soc.* A **320** 71.

[17] N. C. Pyper 1986 *Phil. Trans. Roy. Soc.* A **320** 107.

[18] N. C. Pyper 1994 *Chem. Phys. Lett.* **220** 70.

[19] N. C. Pyper 1994 *Phil. Trans. Roy. Soc.* in press.

[20] T. Tani, T. Itoh and S. Tanaka 1980 *J. Phys. Soc. Japan. Suppl.* **49** 309.

[21] For a recent review see R. J. Cava in *Chemistry of Superconducting Materials* ed. T. A. Vanderah, Noyes Publication, New Jersey 1991, p. 380.

29

Magnetic polarons in concentrated and diluted magnetic semiconductors

S. VON MOLNÁR, I. TERRY†, and T. PENNEY‡

Center for Materials Research and Technology, Florida State University, 406 Keen Building, Tallahassee, FL 32306–4000, USA
†Department of Physics, University of Durham, Science Laboratories, South Road, Durham DH1 3LE, UK
‡IBM Research Division, T. J. Watson Research Center, P. O. Box 218, Yorktown Heights, New York, 10598, USA

Abstract

We describe the influence of local magnetization on electron localization in concentrated and diluted magnetic semiconductors. This includes a review of transport and optical evidence such as the magnetic-field-induced insulator–metal transition in concentrated systems, i.e. Eu chalcogenides and $Gd_{3-x}v_xS_4$ (where v = vacancy). It also includes a brief review of salient experimental evidence for polaron formation in the diluted magnetic semiconductor $Cd_{1-x}Mn_xTe$:In. In addition, static and dynamic photo-induced magnetic measurements in $ZnTe/Cd_{1-x}Mn_xSe$ heterostructures are presented and their relevance to high-T_c materials discussed.

1 Introduction

This paper deals with the effects of local exchange due to interaction of carrier spins with the ionic spins in magnetic semiconductors. We are specifically concerned with the 4f shell of Gd^{3+} and Eu^{2+} ($S=\frac{7}{2}$) and the 3d shell of Mn^{2+} ($S=\frac{5}{2}$), which contribute to the magnetic character of the concentrated and diluted systems respectively. Towards this end, the body of the paper begins with a brief review of the relevant exchange mechanisms leading to polaron formation. In the concentrated systems, this leads to giant magneto-resistive effects in the Eu chalcogenides [1] and very distinctive luminescence [2]. It also leads to the spectacular magnetic-field-induced metal–insulator phase transition in the $Gd_{3-x}v_xS_4$ compounds [3–5]. In the dilute system, polarons account for novel effects in spin-flip Raman scattering [6, 7] and low-temperature transport properties [8, 9], and for the best examples to date of photo-induced magnetism. The latter properties have been studied both by magneto-optical [10] and by very sensitive magnetic techniques [11, 12]. Furthermore, new femtosecond dynamic studies have given information about the formation and decay of the polarons themselves [10, 12]. One of the striking results of these last experiments has been the observation that photo-induced magnetic polaron formation leaves a magnetic imprint in the material long after the photo-excited charges have recombined [13].

Finally, we will attempt to make a link with the high-T_c materials, in which magnetic polarons and bipolarons may be important [14, 15], and we speculate about an experiment that might make polarons observable.

2 Magnetic exchange

The extraordinary interactions that make magnetic systems interesting are those involving exchange coupling J between Gd^{3+} or Mn^{2+} moments S, and the spin of the electrons s. For extended states, these lead to large red shifts in the optical absorption, first discovered by Busch and Wachter [16] in the Eu chalcogenides and studied in detail in the large-band-gap diluted magnetic semiconductors $Cd_{1-x}Mn_xTe$ by Gaj *et al.* [17]. These effects are due to the large spin splitting of the conduction and valence bands. The magnetic contribution to an s-band electron is given by

$$E_s = g^* \mu_B s H + 2Js\langle S\rangle_\alpha, \tag{1}$$

where g^* is the spectroscopic splitting factor for the carriers, μ_B the Bohr magneton and H the applied magnetic field. $\langle S\rangle_\alpha$ is the average value of the local moments, with the average being taken only over the region occupied by the s-band electrons. It is clear that the second term of Eq. (1) can produce enormous splittings, amounting to fractions of an electronvolt in Eu com-pounds below the magnetic ordering temperature and as large as 1 meV kOe^{-1} in (paramagnetic) diluted magnetic semiconductors. However, when the electronic states are not extended, the exchange J can lead to additional localization beyond the normal Coulomb binding and correlations observed in non-magnetic semiconductors. In an antiferromagnet (a paramagnet is much the same) this can lead to formation of 'ferromagnetic' clusters, so-called magnetic polarons [18, 6, 7].

The concept of spin-dependent electron transfer was first invoked in order to explain the conductivity-induced ferromagnetism in $LaMnO_3$ by the double-exchange mechanism [19]. Spin polrons were first introduced to understand the physics of these materials by de Gennes [20], but we will present here arguments for the stability of a spin polaron first given by Kasuya [21]. These involve the 'Gedanken' experiment in which a single electron or hole is introduced into an otherwise antiferromagnetic or paramagnetic lattice. If there are no defects or impurities present, then the differential free energy, ΔF_f, of a polarized cluster in this background may be expressed as

$$\Delta F_f = \frac{E_0}{\gamma^2} - \frac{J^x}{2\gamma^3},\tag{2}$$

where $E_0 = \pi^2 h^2/(2ma^2)$, $\gamma = R/a$ and a is defined by the equation $4\pi a^3/3 = 1/N_{Eu}$. Minimizing this free energy with respect to the radius results in a solution that is only stable at infinity or R approaching zero. A more realistic model introduces the additional energy term found in Eq. (3) due to the Coulomb attraction of an electron to an oppositely charged center:

$$\Delta F = \Delta F_f - \frac{e^2}{\varepsilon R}.\tag{3}$$

Of course, this is the situation most often encountered, since the magnetic semiconductors are insulators, unless doped, and the dopant acts as an attractive center, just as it would in any ordinary semiconductor. Under these conditions, minimization with respect to R yields a stable configuration, and Kasuya referred to this as the bound magnetic polaron or magnetic impurity state. It is easy to see that, below the Néel temperature of the antiferromagnet, the polarons will not easily move in the presence of an electric field, since they have to drag their polarization cloud along, which leads to a large effective mass (diffusive motion) or localization. It also becomes clear how an applied magnetic field can produce an insulator–metal transition. The magnetic contributions to the localization come about from the difference between the antiferromagnetic order of the host and the ferromagnetic order of the polaron. In a magnetic field, the antiferromagnetic sublattice develops an ever increasing magnetization, which, when saturated, has the same magnetic structure as the polaron cluster and the carrier becomes unbound. If there are any defects present, then this polaron, which ordinarily might be describable as a very heavy but band-like particle moving diffusively [22] will typically be bound to defects such as the Coulomb potential described earlier. Thus, transport under these conditions can only occur via 'hopping', which is also magnetic-field-dependent and will be described later in this paper.

Clearly, local magnetic order lowers the state energy of the bound carrier, Eq. (2), with respect to some conduction or valence band. This effect is most clearly observed in optical experiments (e.g. [7]). Finally, the time evolution of this magnetic polarization, when a carrier is introduced into the magnetic background, can be observed in exquisite detail by time-dependent spectroscopy, which will form the last part of our discussion concerning magnetic polarons in magnetic semiconductors [13].

3 Experimental evidence for magnetic polarons

Magnetic polarons were first invoked in ferromagnets by von Molnár and Methfessel in their study of the giant negative magneto-resistance in ferromagnetic $Eu_{1-x}Gd_xSe$ [1]. As already mentioned, Kasuya *et al.* [18], and, independently, Nagaev [23], gave theoretical treatments for this magnetic impurity state. An analysis by von Molnár and Kasuya of critical scattering near the ferromagnetic transiton temperature in the Eu compounds [24] gave further credence to this model. Other than the transport properties already mentioned, experimental evidence for magnetic polarons in concentrated magnetic semiconductors came from optical studies, reported first by Busch and his collaborators in 1970 [2], in which they studied the emission and excitation spectrum of the photoluminescence of EuTe, an antiferromagnet. These authors found that an observed increase in intensity with decreasing temperature could be accounted for by formation of magnetic polarons. The physical idea was that the polaron would shrink the wave function, and thereby increase the probability of recombination. These authors also discovered that applying a large magnetic field reversed this process. The intensity became much smaller, presumably because the magnetic polaron was no longer bound, thereby reducing the probability of recombination. von Molnár *et al.* [3–5] studied the transport properties of $Gd_{3-x}v_xS_4$. Their results are shown in Fig. 1, where they have plotted the logarithm of the resistivity at various fields and temperature. The striking result (see sample 2) is that, whereas the resistivity rises without bound with decreasing temperature when no magnetic field is applied, in the presence of a field of approximately 3.2 T the material appears to be metallic with a resistivity independent of temperature below 10 K. Also, it was their study, with Penney [26], of the magnetic properties of these compounds with various doping concentrations that provided the first magnetic proof that magnetic polarons in this otherwise antiferromagnetic material did exist. These results, given in Fig. 2, show the magnetization of several samples as a function of applied magnetic field at 4.2 K. The most significant result provided by these data is that the pure material without any carriers shows a magnetization rising linearly from zero with field, as would be expected for a canting antiferromagnet, whereas with increasing dopant concentration the extrapolated slope of magnetization versus field to zero field has a positive intercept on the magnetization axis. This remnant ferromagnetic component represents the polaron, which has been created by the strong exchange interaction between the dopant carriers and the surrounding Gd^{3+} spins.

In some ways the most spectacular consequences of these ideas were the studies on the same compound that demonstrated the magnetic-field-driven

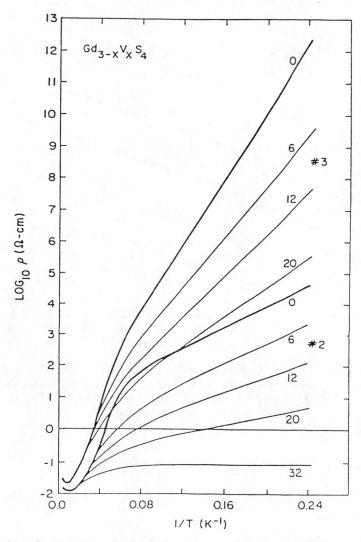

Fig. 1. The log of resistivity versus reciprocal temperature for samples 2 and 3 in applied fields $H = 0$, 6, 12, 20 and 32 kOe (after [26]).

insulator–metal transition. This remarkable result is shown in Fig. 3. The plot is that of σ, the conductivity, which was taken to be the reciprocal of the experimental resistivity measured, as a function of applied magnetic field. A linear dependence of σ on H over a large range of applied magnetic fields is observed. The extrapolation to $\sigma = 0$ represents a critical field below which the material is insulating. These studies, which were carried out at 6 mK [5], demonstrate the physical ideas mentioned earlier in the text and show that the

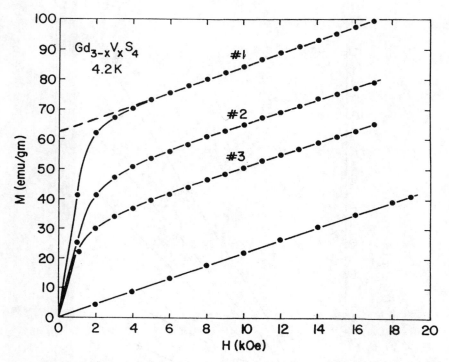

Fig. 2. The magnetization, M, versus applied field, H, for samples 1, 2 and 3 together with the pure antiferromagnetic insulator. The dashed line indicates the extrapolation to M_0 (after [26]). #1, $n = (2.5 \pm 0.2) \times 10^{20}$ cm^{-3}, #2, $n = (1.6 \pm 0.5) \times 10^{20}$ cm^{-3} and #3, $n = (8.7 \pm (0.8) \times 10^{19}$ cm^{-3}.

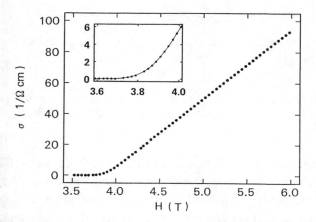

Fig. 3. The conductivity of sample 2 as a function of magnetic field at $T = 6$ mK. The critical field defined by a linear extrapolation of the data to $\sigma = 0$ is $H_c = 39$ kOe. The insert is an expanded view of the transition region showing the rounding of the transition (after [5]).

electron, which is bound principally by the magnetic interaction, can become unbound if its magnetic state becomes similar to the magnetic state of the sublattice in which it finds itself.

Thus far we have reviewed experimental evidence for magnetic polarons in concentrated magnetic semiconductors, that is, those in which each ionic site contains a magnetic ion. Another class, diluted magnetic semiconductors such as the Mn-doped II–VI compounds (of which $Cd_{1-x}Mn_xTe$ is the prototype), also exhibit evidence of polaron formation. Although there exist magneto-transport studies, much like the ones already discussed, in Mn-doped II–VI semiconductors, which are interpretable in terms of magnetic polaron formation [8, 9], the most direct evidence has come from optical studies [6, 7, 10, 12, 13]. However, before delving into a summary of both static and dynamic studies using optical techniques, we would like to review a transport experiment [27–29] and its concomitant magnetic study [30], which depend on the persistent photoconducting properties of these materials. $Cd_{1-x}Mn_xTe$ doped with In is a persistent photoconductor. It is possible to produce long-lived mobile carriers in the material by illuminating with sub-bandgap radiation at low temperatures. This permits continuous tuning of the carrier concentration by photo-doping. For concentrations 4×10^{16} cm$^{-3} < n < 2 \times 10^{17}$ cm^{-3}) in which carrier transport occurs by 'hopping' at low temperatures, it was found that the temperature-dependence of the logarithm of resistivity between 0.3 and 30 K changes from a $1/T^{1/2}$ law, characteristic of variable-range hopping with Coulomb interactions, to simple activation, $1/T$, with decreasing temperature. This type of transport has been observed only in very few system [31] and is a very unusual result, quite contrary to normal expectations. Terry and co-workers [27, 29] have been able to explain this effect in terms of magnetic polaron formation. In Fig. 4 is shown the important experimental result. In zero applied magnetic field, the plot of the log of the resistivity as a function of $1/T^{1/2}$ strongly deviates from a straight line as indicated. When a modest field of 8 T is applied, however, this behavior is completely eradicated, and the curve once again follows a straight $1/T^{1/2}$ law over the entire temperature range. It is clear from this picture that the magnetic field has changed the mode of transport.

It will become evident from discussion of Fig. 5 [32] how such an experimental observation may be understood. In Fig. 5(a) is shown a schematic illustration of the electron 'hopping' process near the Fermi energy, E_F, at low temperatures. The 'hopping' trajectory will be as indicated (Fig. 5(a)). In this process, the charged particle hops from a state in which the magnetic environment is more or less aligned to another state in which the magnetic environment is presumably disordered or antiferromagnetic. Only after a

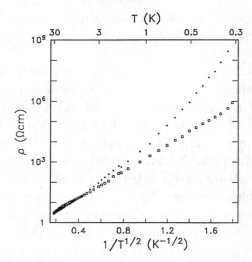

Fig. 4. A plot of the log of the resistivity as a function of inverse temperature to the power $\frac{1}{2}$, for a carrier concentration $n = 4.3 \times 10^{17}$ cm^{-3}: ($\bigcirc$) $H = 0$ and ($\bullet$) $H = 80$ kOe (after [32]).

certain relaxation time will the total energy of the particle plus its magnetic environment go from that of the dashed line to that of the solid curve. It is not difficult to convince oneself that there is a spectrum of energies, given by the difference between unrelaxed and magnetically relaxed states, which are not available to the 'hopping' electron. Thus, the relaxation effect produces an energy gap in the excitation spectrum at lowest temperatures as indicated in Fig. 5(c), which is a modification of the single particle density of states in the presence of Coulomb interactions sketched out in Fig. 5(b). The picture clearly explains why the mode of transport changes from variable range hopping to $1/T$ type with decreasing temperature. The physics leading to recovery of the $1/T^{1/2}$ law upon application of large magnetic fields is equally obvious from the schematic diagram (Fig. 5(a)). When all the manganese spins are aligned, there no longer is a mechanism for relaxation and 'hopping' will occur without the gap in the density of states. Although we believe that this is quite convincing evidence for magnetic polarons, supporting direct magnetic evidence has been obtained in an analogous diluted magnetic photoconductor by Wojtowicz *et al.* [30], who demonstrate that they can produce magnetization steps at low temperatures when the material is photo-doped by successively turning a sub-bandgap light on and off.

The first optical evidence for magnetic polaron formation in Mn-doped II–VI semiconductors came from work by Deitl *et al.* [6] and Heimann *et al.* [7], independently. Both groups developed theories involving magnetic polarons to

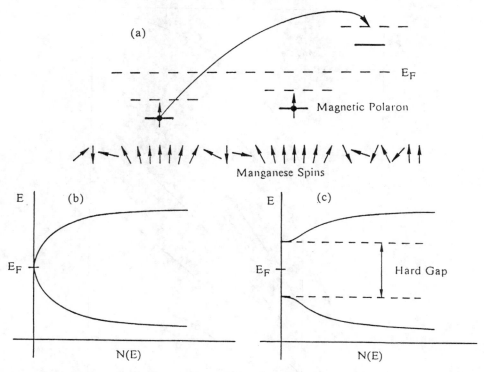

Fig. 5. (a) A schematic illustration of electron hopping in the presence of Coulombic interactions and magnetic polarons. A solid circle represents the electron, E_F is the Fermi energy and the arrows represent the spins of the Mn^{2+} ions. (b) The single-particle density of states in the presence of Coulombic interactions only indicating the soft gap where $N(E) = 0$. (c) The density of states for a system with Coulombic interactions and polaronic relaxation giving rise to the hard gap where $N(E) = 0$ (after [32]).

explain spin-flip Raman data, an example of which is shown in Fig. 6. It is to be noted, in particular, that the spin-flip energy is finite at zero field. This result proves the existence of local magnetization fluctuations and bound magnetic polarons, since spin-flip processes of the carriers require finite energy only in the presence of a local magnetic field.

Arguably, the most exciting new developments in the study of magnetic polarons have come with the advent of time-resolved techniques. Dynamical optical processes in transmission and luminescence have been studied at picosecond time scales and have been related to polaron formation by monitoring the time evolution of the polaron-binding energy [33]. The most spectacular experimental work, however, comes from direct observations of spin dynamics responsible for polarons, which have been developed principally

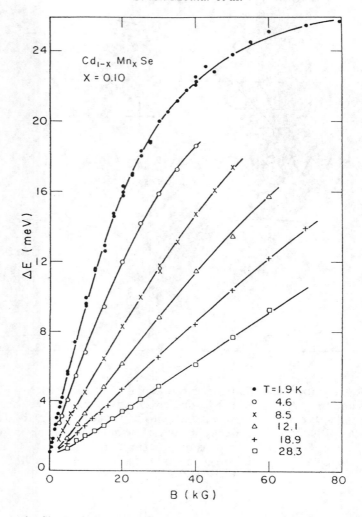

Fig. 6. The spin-flip Stokes energy ΔE versus applied magnetic field B from Raman scattering in $Cd_{0.9}Mn_{0.1}Se$ (after [7]).

by Awschalom and his co-workers [34]. Early experiments addressed the time evolution in bulk materials using both time-resolved Faraday rotation [10] and an elegant magnetic spectroscopy, which employs a planar DC SQUID as a detector to measure the static and dynamic induced magnetization as a function of impinging radiant energy. It is this latter technique applied to new results in MBE-grown magnetic semiconductor superlattices [13] that will now be discussed. The samples consist of multiple quantum wells of $Cd_{1-x}Mn_xSe$ confined by ZnTe barriers. Under the MBE growth conditions, the $Cd_{1-x}Mn_xSe$ stabilizes in the zinc-blende phase, the cubic phase of ZnTe. It has

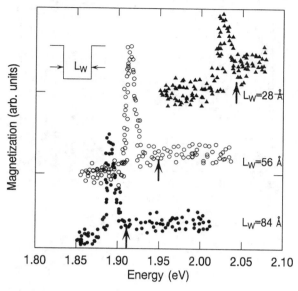

Fig. 7. Magnetic excitation spectra of three multiple quantum well heterostructures having well widths of 2.8, 5.6 and 8.4 nm, and ZnTe barrier thicknesses of 16 nm. The resonance-like peaks are the signatures of magnetic polaron formation, while the energy gaps as determined by reflectance spectroscopy are indicated by the arrows. The y axis marks indicate the relative zeros of magnetization. $T = 1.5$ K, the power is 100 μW, number of periods 30, 15 and 10, respectively (after [13]).

been shown that these magnetic heterostructures are of type II, that is, optically excited electrons are confined to the $Cd_{1-x}Mn_xSe$ layer and holes to the ZnTe layer. Samples constituting a series of well widths (2.8, 5.6, and 8.4 nm) confined by fixed width (16 nm) barriers were studied for two different magnetic concentrations ($x = 0.13, 0.23$). The experiment then consists of optically exciting electrons from the valence to the conduction bands of the heterostructure and monitoring the magnetization as a function of excitation energy, both statically and dynamically. The static results are shown in Fig. 7. Here the time-averaged magnetization induced by circularly polarized light is shown for the $x = 0.23$ series of structures as a function of excitation energy in the vicinity of the quantum well ground state level. The magnetic peak, which occurs for energies somewhat lower than the $n = 1$ quantum level (indicated by arrows for the three wells shown), is due to magnetic polaron formation. Physically, this means that, for energies lower than the polaron energy, absorption is minimal, magnetization rises with the onset of absorption from the radiation field into the impurity level at which polarons form, and the subsequent decrease for energies higher than the peak indicates that the initial carrier spin orientation is lost to a larger degree through spin scattering as the

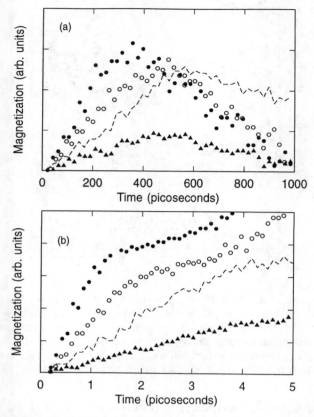

Fig. 8. The time-dependence of the optically induced magnetization as determined by time-resolved magnetic spectroscopy for above-bandgap excitation energies. (a) Spin–lattice relaxation of the magnetization; the solid circles, open circles and solid triangles, respectively, represent the 8.4, 5.6 and 2.8 nm well samples of Fig. 7, while the dashed line is a measurement of an 8.4 nm, 13% Mn well structure. (b) The initial rise in magnetization in response to an excitation pulse of 150 fs duration (after [13]).

energies relax, and the polarons thus formed point in all directions, producing a much smaller net magnetization in a direction perpendicular to the pick-up loop of the SQUID. One final observation is important: the polaron peak exists despite the fact that the bulk polaron radius is estimated to be approximately 9 nm, whereas the thinnest quantum well is only 2.8 nm wide. Clearly, the polaron can no longer be thought of as being spherical, but must by necessity take on a pancake-like shape. We dwell on this point because it seems to us that the case for high-T_c superconductors, in which the copper oxide layers are thought to be two-dimensional, is not far from the physically tractable model of the thinnest of these quantum wells.

The time-dependent magnetization for the three wells (Fig. 8) shows yet another surprising feature of the magnetic polaron. These data were taken with pulsed excitation at the peak of the polaron energy shown in Fig. 7. Without discussing the details of these curves, two important observations deserve comment. The first is that the observed magnetization grows and decays at relatively long time scales, on the order of 400 ps. These results are shown in Fig. 8(a) and reflect the fact that the polaron is created and decays at time scales characteristic of spin–lattice relaxation times. A more detailed look at the rise time of the magnetization is shown in Fig. 8(b) and shows the additional time constant at the smallest time scales, which is interpreted to be due to electron–ion spin-flip scattering. It is clear from this picture (and the knowledge that recombination times of photoexcited carriers are on femtosecond time scales) that this process leaves a magnetic imprint, which lasts far longer than the lifetime of the photoexcited electron that caused it in the first place. The exquisite detail afforded by optically induced magnetization studies in these essentially two-dimensional systems should, in principle, also be applicable to the high-T_c superconductors. In particular, one might expect to see the results of photo-induced magnetism in the copper–oxygen planes. The effects might be much smaller than in the magnetic semiconductor systems, however, because of the small magnetic moment of the copper and the unknown coupling of the hole spin to the Cu^{2+} spin.

Theoretical speculations of polaron formation in $La_{2-x}Sr_xCuO_4$ were originally advanced by Aharony *et al.* [35] and, independently, by Gor'kov and Sokol [36]. Although there have been many additional discussions concerning magnetic polarons and their influence in describing the superconductivity in high-T_c materials, much of which is being reviewed at this conference by the inventors of these schemes, we would like to present as our last illustration (Fig. 9) an example of a calculation that depicts the form that such a magnetic polaron might take based on the so-called t–J model [37]. In principle, of course, the holes could reside either on the copper or oxygen sites, but hybridization is required in any event such that the t–J model is an effective model, which relies on the idea that the Cu^{2+} and O_2 holes form a singlet state. It might be pointed out that, within this model, motion of the holes through the spin degrees of freedom is similar to that which has been described throughout this paper to account for the physical properties of the magnetic semiconductors. There, the spins of the conduction or valence band electrons couple to the rare earth or transition metal ion spins, which are viewed as being localized on cation sites. The picture represented in Fig. 9 is a more detailed study and elaboration of earlier work [38] and shows the distortion in the plane of the spins residing on the copper ions for both sublattices of the antiferromagnet. It also implies (and the calculations bear this out) that a ferromagnetic moment

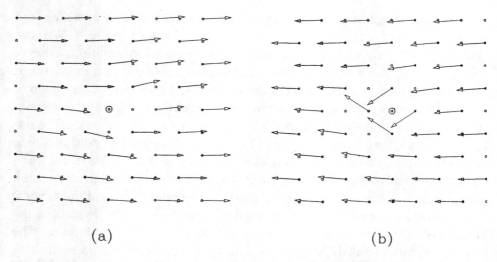

$$(a) \qquad\qquad\qquad\qquad (b)$$

Fig. 9. Distortion of the antiferromagnetic alignment for the 16×16 lattice, at $k = (\pi/2, \pi/2)$ and $t/J = 1$. (a) Refers to the even sublattice, where the hole is residing, and (b) refers to the odd sublattice. The arrows represent the expectation values of the spin at the lattice sites (after [37]).

perpendicular to the Cu–O planes is created near the hole. It appears, therefore, that a sensitive magnetic probe that can measure the out-of-plane component of the magnetization should be able to see these magnetic fluctuation effects.

This then brings us back to the original point of this paper. We have tried to show various examples of experimental observation of magnetic polarons in three-dimensional magnetic semiconductors. We have also shown convincing data of magnetic polaron formation in essentially two-dimensional quantum wells, where the dimension of the quantum well is smaller than the expected polaron radius. Finally, we showed one example of a calculation based on the *t–J* model, which also shows that, in the presence of a hole, a two-dimensional antiferromagnet lattice will develop a small magnetic component perpendicular to the plane in which the magnetic ions sit.

Acknowledgements

We thank the organizers of this conference for focusing on these magnetic correlations and their application to high-T_c, and are grateful to E. Manousakis for his considerable help in clarifying various aspects of theory. Finally, we are grateful to Professor Sir Nevill Mott, who in many ways influenced much of the research carried out by us and reviewed here.

References

[1] S. von Molnár and S. Methfessel, *J. Appl. Phys.* **38**, 959 (1967).

[2] G. Busch, P. Streit and P. Wachter, *Solid State Commun.* **8**, 1759 (1970).

[3] S. von Molnár, F. Holtzberg, T. R. McGuire and T. J. A. Popma, *AIP. Conf. Proc.* **5**, 869 (1972), S. von Molnár and F. Holtzberg, *ibid.*, **10**, 1259 (1973).

[4] S. von Molnár, A. Briggs, J. Flouquet and G. Remenyi, *Phys. Rev. Lett.* **51**, 706 (1983).

[5] S. Washburn, R. A. Webb, S. von Molnár, F. Holtzberg, J. Flouquet, and G. Remenyi, *Phys. Rev.* B **30**, 6224 (1984).

[6] T. Dietl and J. Spalek, *Phys. Rev. Lett.* **48**, 355 (1982).

[7] D. Heimann, P. A. Wolff and J. Warnock, *Phys. Rev.* B **27**, 4848 (1983).

[8] M. Sawicki, T. Dietl, J. Kossut, J. Igalson, T. Wojtowicz and W. Plesiewicz, *Phys. Rev. Lett.* **56**, 508 (1986).

[9] T. Wojtowicz, T. Dietl, M. Sawicki, W. Plesiewicz and J. Jaroszynski, *Phys. Rev. Lett.* **56**, 2419 (1986).

[10] D. D. Awschalom, J. M. Halbout, S. von Molnár, T. Siegrist and F. Holtzberg, *Phys. Rev. Lett.* **55**, 1128 (1985).

[11] H. Krenn, W. Zawadzki and G. Bauer, *Phys. Rev. Lett.* **55**, 1510 (1985).

[12] D. D. Awschalom, J. Warnock and S. von Molnár, *Phys. Rev. Lett.* **58**, 812 (1987).

[13] e.g. D. D. Awschalom, M. R. Freeman, N. Samarth, H. Luo and J. K. Furdyna, *Phys. Rev. Lett.* **66**, 1212 (1991).

[14] N. F. Mott, *Contemp. Phys.* **31**, 373 (1990) and references therein.

[15] N. F. Mott, *Supercond. Sci. Technol.* **7**, 48 (1994).

[16] G. Busch and P. Wachter, *J. Phys. Condens. Matter* **5**, 232 (1966); *Z. Angew, Phys.*, **26**, 1 (1968).

[17] J. A. Gaj, R. Planel and G. Fishman, *Solid State Commun.* **29**, 435 (1979).

[18] T. Kasuya and A. Yanase, *Rev. Mod. Phys.* **40**, 684 (1968).

[19] C. Zener, *Phys. Rev.* **82**, 403 (1951).

[20] P. G. de Gennes, *Phys. Rev.* **118**, 141 (1960).

[21] T. Kasuya in *Proc. 10th Int. Conf. on the Physics of Semiconductors, Cambridge, Mass., 1970*, edited by S. P. Keller, J. C. Heusel and F. Stern, CONF-700801 (USA AEC Division of Technological Information, Springfield, VA 1970), p. 243; S. von Molnár and T. Kasuya, *ibid.*, p. 233.

[22] T. Kasuya, *Solid State Commun.* **8**, 1635 (1970); T. Wolfram and J. Callaway, *Phys. Rev.* **127**, 1605 (1962).

[23] E. L. Nagaev, *Phys. Status Solidi* (b) **65**, 11 (1974); *J. Magn. Magn. Mater.* **110**, 39 (1992) and references therein.

[24] S. von Molnár and T. Kasuya, *Phys. Rev. Lett.* **21**, 1757, (1968).

[25] Y. Shapira, S. Foner, N. F. Oliveira Jr., and T. B. Reed, *Phys. Rev.* B **5**, 2647 (1972).

[26] T. Penney, F. Holtzberg, L. J. Tao and S. von Molnár, *AIP Conf. Proc.* **18**, 908 (1974).

[27] I. Terry, T. Penney, S. von Molnár and P. Becla, *Phys. Rev. Lett.* **69**, 1800 (1992).

[28] I. Terry, S. von Molnár, A. Torressen and P. Becla, *Phil. Mag.* B **65**, 1245 (1992).

[29] I. Terry, T. Penney, S. von Molnár and P. Becla, in *Hopping and Related*

Phenomena 5, edited by C. J. Adkins, A. R. Long and J. A. McInnes, World Scientific, Singapore, 1994, p. 248.

[30] T. Wojtowicz, S. Kolesnik, I. Miotkowski and J. K. Furdyna, *Phys. Rev. Lett.* **70**, 2317 (1993).

[31] e. g. P. Dai, T. Zhang, and M. P. Sarachik, *Phys. Rev. Lett.* **69**, 1804 (1992); A. N. Aleshin, A. N. Ionov, R. V. Parfen'ev, S. Shlimak, A. Heinrich, J. Schumann and D. Elefant, *Fiz. Tverd. Tela* **30**, 696 (1988) [*Sov. Phys. Solid State* **30**, 398 (1988)].

[32] S. von Molnár, T. Penney, I. Terry and P. Becla, *Mater. Res. Soc. Symp. Proc.* **290**, 335 (1993).

[33] J. H. Harris and A. V. Nurmikko, *Phys. Rev. Lett.* **51**, 1472 (1983).

[34] e. g. D. D. Awschalom, J. F. Smyth, N. Samarth, H. Luo and J. K. Furdyna, *J. Lumin.* **52**, 165 (1992).

[35] A. Aharony, R. J. Birgeneau, A. Coniglio, M. A. Kastner and H. E. Stanley, *Phys. Rev. Lett.* **60**, 1330 (1988).

[36] L. P. Gor'kov and A. V. Sokol, *J. Physique* **50**, 2823 (1989).

[37] M. Boninsegni and E. Manousakis, *Phys. Rev.* B **43**, 10353 (1991).

[38] B. I. Shraiman and E. D. Siggia, *Phys. Rev.* B **42**, 2485 (1990).

30

Energy scales of exotic superconductors

Y. J. UEMURA

Department of Physics, Columbia University, New York, New York 10027, USA

Abstract

An energy scale of the superconducting condensate, which we call the effective Fermi temperature T_F, can be derived from the magnetic field penetration depth λ determined by muon spin relaxation (μSR) measurements. We classify various superconductors in the crossover from Bose–Einstein (BE) to BCS condensation, based on the ratio T_c/T_F. The phase diagram of high-T_c cuprate superconductors, as a function of carrier doping, can be understood in the context of this BE–BCS crossover, if we identify the 'pseudo-gap' as the signature of pair formation in the normal state. In particular, the universal linear relation between T_c and n_s/m^* (superconducting carrier density/effective mass), found in the underdoped region, comes from a general feature expected in the BE condensation of pre-formed pairs. The optimal T_c occurs around the doping concentration at which the condensate energy scale T_F becomes comparable to the energy scale $\hbar\omega_B$ of the pair-mediating interaction. A surprising decrease of n_s/m^* with increasing carrier doping was found in the overdoped Tl 2201 system. This behavior suggests that evolution to the BCS region does not occur in a simple way, but rather is associated with a possible microscopic separation between superconducting and residual normal metallic phases.

During the past several years, we have performed measurements of the magnetic field penetration depth λ of high-T_c cuprate and other superconducting systems using the muon spin relaxation (μSR) technique [1–5]. The μSR relaxation rate σ, observed in a transverse field H_{ext} ($H_{c1} \ll H_{ext} \ll H_{c2}$) is given by

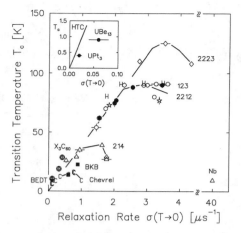

Fig. 1. A plot of T_c versus the muon spin relaxation rate $\sigma(T \to 0) \propto n_s/m^*$ of high-T_c cuprates (214, 123, 2212 and 2223), (Ba, K)BiO$_3$ (BKB; closed square), Chevrel phase systems (C), (BEDT-TTE)$_2$Cu(NCS)$_2$ (closed star), doped fullerenes (buckie ball symbols); heavy-fermion systems (inset figure, derived from bulk measurements) and Nb ([1–4]; and references therein). The straight line in the inset corresponds to the linear relation found in the underdoped cuprates.

$$\sigma \propto \frac{1}{\lambda^2} = \frac{4\pi n_s e^2}{m^* c^2} \frac{1}{1 + \xi/l}, \tag{1}$$

where n_s is the superconducting carrier density, and m^* is the effective mass, while the correction term due to the ratio of the coherence length ξ and the mean free path l becomes negligible in the clean limit $\xi/l \ll 1$. Since most of the 'exotic' superconductors discussed in the present paper lie within the clean limit, the relaxation rate σ directly represents the spectral weight of the superconducting condensate (or supercurrent density) n_s/m^*. In Fig. 1 is shown a plot of T_c versus $\sigma(T \to 0)$. We found a linear relation between T_c and $\sigma \propto n_s/m^*$ common to various different cuprate systems in the underdoped region, followed by saturation and suppression of T_c as the doping progresses [1]. It is also demonstrated in Fig. 1 that points from BKBO, doped-C$_{60}$, organic BEDT, Chevrel phase, and heavy-fermion systems lie close to the linear trend in cuprates, suggesting that the high ratio $T_c/(n_s/m^*)$ is a common feature of these 'exotic superconductors' [2–4].

The spectral weight n_s/m^* represents an energy scale of the superconducting condensate, as can be easily understood if we remember that the Fermi temperature of a 2D electron gas is proportional to n/m^*, where n denotes the normal state carrier density. Then, we can derive an 'effective Fermi temperature' T_F by substituting n_s/m^* for n/m^* in 2D systems. For 3D systems, only a

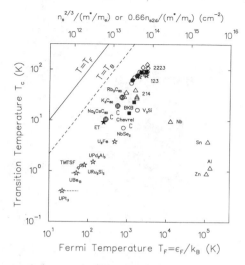

Fig. 2. A log–log plot of T_c versus effective Fermi temperature T_F derived from the results of σ in Fig. 1 (combined with the interplanar distance for 2D and with the Sommerfeld constant γ for 3D systems)[2–4]. The values of T_F for Al, Sn and Zn are based on knowledge from other estimates. The broken line represents the BE condensation temperature T_B of an ideal boson gas with corresponding $n_s^{2/3}/m^*$, the boson mass $2m^*$ and density $n_s/2$.

minor correction is required; $T_F \propto n_s^{2/3}/m^*$ can be given as $T_F \propto \sigma^{3/4}\gamma^{-1/4}$, where $\gamma \propto n^{1/3}m^*$ is the Sommerfeld constant. These procedures allow us to convert the horizontal axis of Fig. 1 into an energy scale T_F. In Fig. 2 is shown a plot of T_c versus T_F thus derived from n_s/m^* [2–4]. The Bose–Einstein condensation temperature T_B, calculated for non-interacting point-like bosons of mass $2m^*$ and density $n_s/2$, is shown by the broken line in Fig. 2. Points from 'exotic superconductors' indicate high ratios of T_c/T_F in these systems, as high as $T_c/T_B \simeq \frac{1}{4}$. This reduction of T_c from T_B can be expected for doped holes in cuprates, which considerably overlap in real space (there are several pairs per coherence area ξ^2 of the CuO_2 planes in the cuprates). From Fig. 2 we can classify superconductors in terms of a crossover between the limits of BE condensation ($T_c \to T_B \propto n_s^{2/3}/m^*$, non-retarded interaction, pairing in real space) and BCS condensation ($T_c/T_F \ll 1$, retarded interaction, pairing in momentum space). On this plot, many of the 'exotic superconductors' lie close to the BE limit. The linear relation $T_c \propto n_s/m^*$ observed in the underdoped region of the cuprates can be ascribed to the relation expected in BE condensation.

The energy scale T_F calculated in this way does not necessarily correspond to the energy difference between the bottom of the conduction band and the

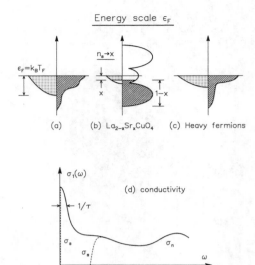

Fig. 3. A schematic illustration of the density of states $N(\varepsilon)$ in (a) a complicated electronic system; (b) high-T_c cuprates; and (c) heavy-fermion systems. The real $N(\varepsilon)$ is drawn on the right-hand side, while the left-hand side shows how these systems are represented in the calculation to obtain the effective Fermi temperature T_F described in the text. The same phenomena can be seen in optical conductivity (d), where σ_n denotes the normal-state conductivity and σ_s the conductivity in the superconducting state.

Fermi surface, especially in strongly correlated electron systems with complicated densities of states. As illustrated in Fig. 3, the super-current density is determined by the low-energy electronic structure near the Fermi level [4]. The width of this renormalized low-energy density of states corresponds to what we call T_F here. In frequency space, many strongly correlated metals exhibit a complicated optical conductivity $\sigma(\omega)$, typically consisting of a sharp Drude part at low energies coexisting with a strong response at higher energies, such as the mid-infrared reflection (MIR) observed in cuprates. The energy scale T_F shown in Fig. 2 selectively reflects the spectral weight of the sharp Drude part, which condenses into a δ function at $\omega = 0$ below T_c. Moreover, one can derive T_F from n_s/m^* even for a purely bosonic system without a Fermi surface. Thus, the horizontal axis of Fig. 2 should be regarded as the characteristic energy scale of the superfluid condensate, a concept valid both in bosonic and in fermionic systems.

Let us now consider general aspects expected in the evolution from BE to BCS condensation. Usually, this is done by fixing the particle density n and varying the energy scale of the attractive interaction [6]. Here, we would like to fix the energy scale $\hbar\omega_B$ of the pairing interaction, and vary the particle density n. The expected evolution is illustrated in Fig. 4. When there is an effective

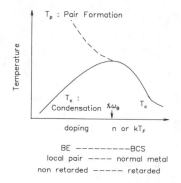

Fig. 4. A schematic illustration of the crossover from Bose–Einstein to BCS condensation expected for a fixed energy scale $\hbar\omega_B$ of the interaction mediating pairing while the particle density n is varied. The solid line shows the superconducting condensation temperature T_c, while the broken line shows the pair-formation temperature T_p. The BE-like region and BCS-like region are separated around the particle density with which the effective energy scale of particles $k_B T_F$ becomes comparable to $\hbar\omega_B$. The interaction is non-retarded in the BE region but retarded in the BCS region.

attractive interaction with the energy scale E_a in the dilute limit $n \to 0$, local pairs will be formed in the normal state below the 'pair formation temperature' $T_p \simeq E_a/k_B$. This pair binding is an activated process, leading to a spin susceptibility proportional to $\exp[-E_a/(k_B T)]$ if the pairing is of spin singlet type. The pairing interaction is non-retarded as long as the energy scale of the particles T_F is smaller than $\hbar\omega_B$. These pre-formed pairs in the normal state condense into superfluid at $T_c \propto T_B$ via BE condensation in this region. Thus we can expect a linear behavior $T_c \propto n_s/m^*$ in this 'underdoped' region. The pair formation temperature T_p will decrease with increasing doping, since pair formation and condensation should occur at the same temperature in the BCS region where the interaction is retarded ($k_B T_F > \hbar\omega_B$). Details of the evolution of T_c in this 'overdoped' side are unclear. However, if the effective attractive interaction E_a keeps decreasing with increasing n or $k_B T_F$, then we would expect the situation illustrated in Fig. 4.

The phase diagram shown in Fig. 4 can explain various observed results in high-T_c cuprates. Pair formation above T_c is manifested as the 'pseudo-gap' observed in NMR [7], neutron [8], and specific heat [9] measurements, as well as the optical conductivity $\sigma_c(\omega)$ measured along the c-axis [10]. The decrease in spin response below the 'pseudo-gap temperature' T^*, observed in the underdoped region, results from singlet pair formation. T^* decreases rapidly with increasing doping, and merges with the T_c line around the optimal T_c region, as indicated by the Korringa relation $1/(T_1 T) = $ constant observed by NMR in the optimal T_c and overdoped regions [11], where pair formation and condensation

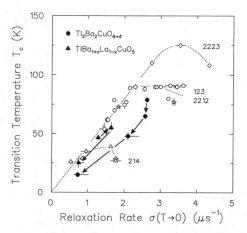

Fig. 5 A plot of T_c versus muon spin relaxation rate $\sigma(T \to 0)$ in $Tl_2Ba_2CuO_{6+\delta}$ (Tl 2201) and $TlBa_{1+x}La_{1-x}O_5$ (Tl 1201) compared with previous results in 214, 123, 2212 and 2223 systems. The arrow indicates order of increasing doping (from [5]).

are expected to occur at around the same temperature. In the underdoped region, $\sigma_c(\omega)$ decreases with decreasing temperature below T^*: this can be understood if we assume that tunneling between adjacent CuO_2 planes for unpaired carriers occurs with much higher probability than for 2e pairs. The underdoped side is then characterized by predominantly bosonic features. This explains why T_c is proportional to n_s/m^*, with the universal ratio common to various different series of cuprates: the Bose condensation temperature is determined only by the density and mass of the condensing particles, regardless of detailed differences in the pairing interactions. This may also explain why the universal relation in cuprates can be extended to many other exotic superdconductors.

If we assume that the optimal T_c region exists around $k_B T_F \simeq \hbar\omega_B$, then the energy scale of $\hbar\omega_B$ of the pair-mediating bosons can be expected to be about 1000–2000 K from the plots in Fig. 1 and Fig. 2. The mid-infrared reflection (MIR) of cuprate systems peaks around the frequency of 1000 K below T_c after the clean Drude part condenses into the $\delta(\omega = 0)$ function. Therefore, the MIR reflection may directly represent the pairing interaction.

μSR measurements in the overdoped Tl 2201 systems have recently been carried out by our group [5] and by Niedermayer *et al.* [12]. The results shown in Fig. 5 indicate, surprisingly, that the condensate spectral weight n_s/m^* decreases with increasing doping in the overdoped region. We have also observed the same behavior in Hg 1201 systems [13]. Since the Tl 2201 systems are known to have a very long mean free path $l \simeq 600$ Å in the normal state above T_c, it is not possible to ascribe this behavior to the term ξ/l in Eq. (1). We consider that this is due to a possible phase separation, where only a part of the normal state carriers n condenses as n_s. Then, we would expect increasingly 'gapless' features in the overdoped region with a residual normal metal response below T_c. Such features were observed in recent optical experiments in some 123 specimens, which presumably lie close to the overdoped side [14]. These results demonstrate that the cuprate systems do not evolve smoothly to simple BCS behavior in the overdoped region. In the usual retarded interaction and BCS condensation, we cannot expect n_s to be different from n. Therefore, the overdoped region of the cuprates seems to be more complicated than illustrated in Fig. 4.

Even in the overdoped region, we see a rather monotonic relation between T_c and n_s/m^* in Fig. 5, systems with lower T_c having lower n_s/m^*. This might suggest that a feature reminiscent of Bose condensation still remains in the overdoped region. Reduction of T_c with increasing doping in the overdoped region may then possibly be ascribed to 'overcrowding' of carriers, which suppresses formation of local pairs, and thus reduces the condensate spectral weight. It requires further experimental and theoretical work to fully explain the situation in the overdoped region. Similarly, it would be interesting to perform detailed studies of organic and doped-C_{60} systems, where superconductivity disappears when the system approaches a simpler metal (under the application of pressure), a behavior analogous to the case in overdoped cuprate systems.

Acknowledgements

The author acknowledges S. Uchida, G. M. Luke, and many other scientists in [1–5] for collaboration and discussion, and the Packard, Mitsubishi, and Yamada Foundations for financial support. The author is also particularly grateful to Sir Nevill Mott for his interest and encouragement of our work.

References

[1] Y. J. Uemura *et al.*, *Phys. Rev. Lett.* **62**, 2317 (1989); *Phys. Rev. B* **38**, 909 (1988).
[2] Y. J. Uemura *et al.*, *Phys. Rev. Lett.* **66**, 2665 (1991).
[3] Y. J. Uemura *et al.*, *Nature* **352**, 605 (1991).
[4] Y. J. Uemura, L. P. Le and G. M. Luke, *Synthetic Metals* **55–57**, 2845 (1993); Y. J. Uemura and G. M. Luke, *Physica B* **186–188**, 223 (1993).
[5] Y. J. Uemura *et al.*, *Nature* **364**, 605 (1993).
[6] Ph. Nozières and S. Schmitt-Rink, *J. Low Temp. Phys.* **58**, 195 (1985).
[7] H. Yasuoka, T. Imai and T. Shimizu, *Spin Correlation and Superconductivity*, ed. H. Fukuyama *et al.* (Springer, 1989), p. 254; M. Takigawa *et al.*, *Physica C* **162–164**, 853 (1989); *Phys. Rev. B* **43**, 247 (1991).

[8] J. Rossat-Mignod *et al.*, *Physica* C **185–189**, 86 (1991); *Phys. Scripta* T **45**, 74
     (1992).
[9] J. W. Loram *et al.*, *Phys. Rev. Lett.* **71**, 1740 (1993).
[10] C. C. Homes *et al.*, *Phys. Rev. Lett.* **71**, 1645 (1993).
[11] Y. Kitaoka *et al.*, *Physica* C **179**, 107 (1991).
[12] Ch. Niedermayer *et al.*, *Phys. Rev. Lett.* **71**, 1764 (1993).
[13] Y. J. Uemura *et al.*, unpublished work.
[14] S. Tajima *et al.*, presented at the APS March Meeting, 1994.

Index